biology

SIXTH EDITION

bju press®

Greenville, South Carolina

NOTE: The fact that materials produced by other publishers may be referred to in this volume does not constitute an endorsement of the content or theological position of materials produced by such publishers. Any references and ancillary materials are listed as an aid to the student or the teacher and in an attempt to maintain the accepted academic standards of the publishing industry.

BIOLOGY Student Edition
Sixth Edition

Writers
David M. Quigley, MEd
Christopher D. Coyle
Kelly Driskell

Writer Consultants
Jeff S. Foster, MS
Jessica Minor, PhD
Georgia Purdom, PhD
Renton Rathbun, PhD
Caitlin Ueckert

Biblical Worldview
Chris Collins, MDiv
Bryan Smith, PhD
Tyler Trometer, MDiv

Academic Integrity
Jeff Heath, EdD

Instructional Design
Rachel Santopietro, MEd
Danny S. Wright, DMin

Editor
Rick Vasso, MDiv

Designers
Terra Baughn
Rachel Nichols

Cover and Lead Designer
Sarah Lompe

Illustrators
Giulia Borsi c/o Lemonade
    Illustration Agency
John Cunningham
KJA artists
David Lompe
Elizabeth Matías
Rachel Nichols
Troy Renaux
Van Saiyan/agoodson.com

Production Designer
Maribeth Hayes

DesignOps Coordinators
Kaitlyn Koch
Lesley Ramesh

Permissions
Sharon Belknap
Sarah Gundlach
Stacy Stone

Project Coordinators
Heather Chisholm
Chris Daniels

Postproduction Liaison
Peggy Hargis

Indexer
Golden Rule Indexing

Photo credits appear on pages 622–27.

The text for this book is set in Adobe Minion Pro, Adobe Myriad Pro, Bungee by David Jonathan Ross, FF Uberhand by Jens Kutilek, Free3of9 by Matthew Welch, LiebeRuth by Ulrike Rausch, Playfair by Claus Eggers Sørensen, Raleway by The League of Moveable Type, and Sketchnote by Mike Rohde.

The cover photo is a close-up of the feathers of a green-winged macaw.

© 2024 BJU Press
Greenville, South Carolina 29609
Fifth Edition © 2017 BJU Press
First Edition © 1980 BJU Press

ISBN 978-1-64626-111-6

15 14 13 12 11 10 9 8 7 6 5 4 3 2 1

CONTENTS

CONTENTS

(continued)

CONTENTS

J. Michael
Fay

Megabiology

J. Michael Fay does big biology. As an Explorer-in-Residence for the National Geographic Society, he works as an ecologist and conservationist. In 1999, he hiked over 3000 miles across the drainage basin of the Congo River in Africa, cataloging every species of plant and wildlife that he encountered and observing the human impact on the rainforest. The documenting of the biodiversity and ecological health over a large tract of land is called a *megatransect*. "To quantify a stroll through the woods" is how Fay describes his efforts. After 455 days of hiking with a group of Pygmies, Fay was inspired to push through governmental red tape to establish the first thirteen national parks in Gabon to protect the area he had hiked through and had come to love.

In 2004 Fay surveyed Africa in a different way—from the air. He conducted his *MegaFlyover*, logging 60,000 miles flying at low altitude. He took a high-resolution picture every 20 seconds, collecting over 100,000 aerial images. Many of Fay's photos have been used in Google Earth. Fay compiled the results of his MegaTransect of the Congo Basin and his MegaFlyover of the African continent to give information to conservationists and government officials.

In 2007 and 2008 Fay was at it again, this time doing a megatransect of the redwood forests of California and Oregon, forests containing the tallest trees in the world. He hiked 1800 miles, painstakingly taking notes on the health of this ecosystem and networking with other professionals to collect more data.

But why did Fay do all this? He is motivated to help all of us wisely use resources so that we can protect wildlife and beautiful, wild places for future generations. Some natural resources can be sustainably used, but some strategic natural resources must be protected for the health of society and for the benefit of future generations.

Is this a good thing? It is if you believe the Bible. Genesis 1 tells us that God placed Earth in our hands to care for and wisely use. The earth and all living things in it are God's creation, but He has entrusted them to us. Fay doesn't claim to be a Christian, but he is trying to do something that the Bible calls worthy. He is spending his life trying to help people wisely use the earth. Let's learn how biology can help us keep the big picture in mind as we take care of the planet God has lovingly placed us on.

FEATURES OF THIS BOOK

This book is just for you!

We've designed it to help you learn. Flip through the following pages to see the features that we've designed into this textbook to help you succeed in biology. In the back of the book you'll see appendixes, a glossary, and an index.

Opener—a short article that highlights issues and developments in biology that need to be examined from a biblical worldview

Extreme Life

We live on a planet that throbs with life to the extreme. Sometimes life takes hold in the most unexpected places—spots on Earth with no sunlight or air, some with oppressively hot or bitterly cold temperatures. Microbes have been found in the deepest ocean trenches, in acid lakes, and even buried deep under Antarctic ice.

Evolutionists look at these forms of life, called *extremophiles*, and wonder whether life like this exists outside of Earth. In fact, in 2024 NASA plans to send a probe to Europa, a moon of Jupiter with an icy surface. They suspect that this moon has liquid water beneath its icy surface and a tectonic structure like Earth's. But what will they find? Will it change the way we think about life on our planet?

3.1 OUR LIVING PLANET
The Biosphere

The search for life originating outside our living planet on places like the moon, Mars, and *astrobiology.* gy. It is a hot, force in NASA's space program. The problem for astro-biologists is that they haven't found life that comes from anywhere other than Earth despite spending millions of dollars developing the finest technology. None at all! So why look for life in space when there is so much around us?

Scientists are looking to extend their beliefs about life on Earth to space. If life and everything in the universe is a product of chance and a big bang, why wouldn't we see life elsewhere in the universe? In their view of biology, evolution is life's designer. The late Carl Sagan, former professor and astronomer at Cornell University, once said, "The universe is a pretty big place. If it's just us, seems like an awful waste of space." Life should be easy to find wherever we look, but that's not what we observe.

God designed life, and He made Earth for life. We see this in the wording and unfolding of the Creation account. The heavens, the earth, the waters, and the stars, sun, and moon are all mentioned in direct connection to the living things that God created. The stars are mentioned almost as an afterthought because life on this planet is the centerpiece of God's creation, though all of creation declares His glory (Ps. 19:1; Rom. 11:36). Earth is the shelter, the haven, the home for God's precious, living creation in the deathly wasteland of space. When we look around our living planet, we see the fingerprint of a God who cares and who provides for the needs of His creation. The realm of life on Earth, called the **biosphere,** extends from a few kilometers into the atmosphere to a few kilometers into Earth's crust. This thin shell around us is the only place we know of where life can occur.

?

What makes Earth a good place for life?

Questions

How do ecologists categorize the living and nonliving parts of Earth?

How are living things affected by their environment?

Terms
biosphere
ecology
biodiversity
biome
ecosystem
habitat
niche
abiotic factor
biotic factor
population

Essential Question—the "big question" that you will learn about in a section

Key Questions—the smaller questions that you can ask along the way through a section to help you answer the essential question

Italicized Terms—other important science terms

Bold-Faced Terms—vocabulary terms that you need to know

Ecology **51**

GENETICALLY MODIFIED FOODS ETHICS

Researchers who wear personal protection equipment are usually required by federal regulations to do so when working with certain chemicals that are used with both GM and non-GM crops, including some organic crops.

ISSUE

Genetically modified foods, otherwise known as "GM foods," have been the center of ethical debate for quite some time. Those who support modifying the genes of crops see value in altering foods to fit the needs of people and support a growing population. Others question the long-term effects on people and animals who consume GM foods. Many GM foods have been created with the intent of increasing a healthy food supply by reducing insects, pests, and diseases that are harmful to crops. They also have the benefits of adding nutritional value to foods and making produce more attractive.

When the first GM tomato became available in 1994, the market for GM foods began to widen. Research companies began expanding their testing and new GM crops began popping up. Since that

time, teams of researchers have studied the health effects of GM foods. Even with research suggesting that GM foods are safe to consume, people often wonder whether there has been enough testing and research done to truly understand the long-term effects. With all we know and don't know about GM foods, should we eat them?

Work through this issue using the guiding questions from the biblical ethics triad.

1. What information can I get about this issue?
2. What does the Bible say about this issue?
3. What are the acceptable and unacceptable options of consuming GM foods?
4. What are the motivations of the acceptable options?
5. What action should I take?

Use the ethics box to answer Question 33.

33. Using the biblical ethics triad questions above, formulate an essay on the Christian position of consuming genetically modified foods. Be sure to address each leg of the triad: biblical principles, biblical outcomes, and biblical motivations.

Ethics Features—opportunities to apply a biblical worldview to ethical issues related to biology

MINI LAB

Starch and Fat Test

We have just learned about the properties and structures of organic compounds, but can we identify them in the foods we eat? Identifying whether foods have fats and starches can actually be quite an easy and fun task. Let's take a look at different foods and try to see whether we can correctly identify them as fats or starches.

How can we spot fats and starches in the food we eat?

Materials
4–6 fat food samples • 4–6 starch food samples • water • large brown paper bag • small cups • permanent marker • iodine

PROCEDURE

Fat Test

A. Cut the brown paper bag so that you can spread it out and lay it on a flat, covered surface.

B. Using your permanent marker, divide the bag into sections according to the number of food items that you will be testing. Label each section with the name of the food sample being tested in that section. Be sure to include water as your control sample.

C. Place each fat food sample into its designated section on the paper. Some of the items will need to be spread onto the paper, while others can simply be placed into their sections. The samples will need to sit for about 15 minutes to allow them to dry. Use this time to make predictions as to whether these items will contain fat.

D. After the time is up, remove each food sample. Check whether any oily residue is left behind. An oily residue indicates that the food sample contains fat. Compare your predictions to the results from each trial.

1. Were your predictions correct? What results surprised you?

Starch Test

E. Place the small cups on a flat, covered surface according to the number of food samples that you will be using. Label each cup with the name of the sample being tested and place the samples in their respective cups. Be sure to include water in one cup as your control sample.

F. Make predictions as to whether the samples will contain starch.

G. Put three drops of iodine on each sample. If the sample contains starch, the iodine will turn blue-black. Be sure to record the results.

2. Were your predictions correct? What results surprised you?

GOING FURTHER

3. The substance amylose, present in starch, causes the color of iodine to change from yellow-brown to blue-black. What do you think causes this?

Mini Labs—short hands-on exercises to get you thinking and working like a scientist

FEATURES OF THIS BOOK (continued)

Worldview Investigations—inquiry-based investigations that help you think through controversial areas of biology through the lens of Scripture

worldview investigation

CREATURES AND CLIMATE CHANGE

Introduction

Several years ago it was reported that malnourished baby sea lions off the coast of California were coming onshore in droves. Many wondered whether this was related to a change in climate. What could have caused this odd behavior?

Task

Choose a migrating animal, such as a specific bird, butterfly, or sea creature, and try to determine whether there's a connection between the animal's behavior and climate change. Then decide whether humans should be part of the solution for helping it get back on its feet.

Procedure

1. Once you've chosen an animal, research to find out the following information about it:
   - typical migration patterns, eating habits, and habitat preferences
   - new or changing migration patterns, eating habits, and habitat preferences
   - possible causes for the changing behaviors
   - ways that people are helping it recover

   Pay attention to words such as *predicted*, *projected*, *forecasted*, and *potential*. Many statistics that you'll find about the effect of climate change on animals may be projections into the future and not observed data.

2. Create a map that compares the animal's migration pattern in the past (ten to twenty years ago) and its migration pattern now. Include on your map a summary of the research that you did.

3. Suggest a possible way that your animal can be helped. For example, there are now several sea lion rescue groups in California that nurse baby sea lions back to health and then release them. There are also new fishing restrictions to help increase the fish populations that sea lions like to eat.

4. Include a few sentences about how worldview might affect a person's interpretation of data. Base your sentences on the two questions below.
   - Why do many scientists believe that changing migration patterns are caused by humans?
   - What is the biblical role of people when it comes to fulfilling the Creation Mandate?

Conclusion

It makes sense that as Earth's climate changes, animal populations change too. We should try to help struggling animals as much as we can because they are resources that are under our care. We should also keep in mind that "the earth is the Lord's" (Ps. 24:1) and that He is in control of every change.

Case Studies—opportunities to investigate specific areas in biology to apply what you have learned in a chapter

case study

SAILOR BUG

Crustaceans are the arthropods of the sea, and insects are the arthropods of the land, right? Don't tell that to the ocean strider, a member of the true bug family and one of the very few marine insects. Though it lives on the ocean, it doesn't swim—It walks on water! Ocean striders eat plankton off the surface of the ocean. They have little air-trapping hairs on their legs that act as life jackets to keep them afloat. They had better look out though! Seabirds in the air and crustaceans below the surface are constantly trying to scoop them up.

Ocean striders are getting an unexpected lift from the Great Pacific Garbage Patch, an area of tiny artificial particles suspended in the water column trapped in a circular current in the North Pacific Ocean. Some experts estimate the patch

to be the size of Texas. Much of the garbage has been broken down into very small pieces after being continually bashed by wave action. Even though the particles are creating a problem for aquatic animals, they are providing a handy place for ocean striders to lay their eggs on the open sea. Thanks to the garbage patch, ocean striders are thriving.

Ecologists are concerned that the biodiversity of the ocean could be affected by all this artificial material in the ocean. And in another twist, scientists are finding that the ocean strider acts as a storage vehicle for heavy metals present in the ocean, such as cadmium. This could give scientists some clues about how pollution and toxic metals spread in the oceans and what we can do to keep our waters and insects pollutant-free.

17.3 SECTION REVIEW

1. Give three ways that you can identify the walking stick shown below as an insect.

2. How does an insect use its wings to maintain homeostasis?

3. How does an insect breathe?

4. How does the shape of an insect's mouthparts relate to its function?

5. How does an insect find a mate?

6. Does an insect reproduce sexually or asexually? Explain.

7. How does an insect change over its life?

8. What are some ways that we can control insects without using pesticides that could be harmful to the environment?

Use the case study above to answer Questions 9–11.

9. How are ocean striders able to survive on the ocean?

10. How does floating debris on the water affect marine life?

11. Explain how studying ocean striders may help scientists find new ways to keep our water clean.

394 Chapter 17

Review Questions—Questions at the end of each section and chapter will give you practice in applying what you've learned in a section or a chapter. Problem-solving and extra-thought questions are marked with a purple box—you may need to think a little harder or do some research to answer these questions.

---

17 CHAPTER REVIEW
Chapter Summary

17.1 ARTHROPOD INTRODUCTION AND CHELICERATES

- Arthropods have open circulatory systems, compound or simple eyes, segmented bodies, nervous systems, jointed appendages, and exoskeletons.
- Chelicerates are arthropods with clawlike appendages called chelicerae. Most have one or two body sections and four pairs of walking legs.
- Chelicerates use pedipalps to eat, mate, and sense their environment.
- Aquatic chelicerates reproduce sexually through external fertilization; terrestrial chelicerates reproduce through internal fertilization.
- Some chelicerates may be parasites or predators. Some chelicerates serve as food for other animals, while some are used for the treatment of diseases and infections.

Terms
arthropod • thorax • abdomen • cephalothorax • compound eye • molting • antenna • chelicera • pedipalp • book lung • spinneret

17.2 CRUSTACEANS

- Most crustaceans are aquatic arthropods that have two or three body sections, a carapace, and swimmerets. They have five pairs of walking legs and chelipeds used for defense, eating, and grooming.
- Crustaceans hunt for food using their antennae and grind food using mandibles.
- Most crustaceans reproduce sexually through external fertilization, though some undergo internal fertilization.
- Crustaceans impact their biome by feeding on detritus and serving as keystone species for aquatic food chains.

Terms
carapace • swimmeret • walking leg • cheliped • mandible

17.3 INSECTS

- Insects are arthropods that have three body segments, six legs, one pair of antennae, and one pair of compound eyes. Most insects have wings.
- Insects have mandibles used for grinding food. Ground-up food gets broken down further by saliva and travels through the digestive system.
- Females release pheromones to attract a mate. These pheromones are also used to communicate danger or directions to a food source for other insects in a colony.
- Insects begin as a fertilized egg and undergo incomplete or complete metamorphosis to reach adulthood.
- Insects are useful in their roles as pollinators; however, some are known to destroy crops and land. They are also used in medicine and forensic investigations.

Terms
Malpighian tubule • trachea • spiracle • pheromone • metamorphosis • incomplete metamorphosis • nymph • complete metamorphosis • pupa

Arthropods 395

Chapter Reviews—handy statements that sum up the big ideas in each section of a chapter along with a list of each section's key terms

biology

SIXTH EDITION

UNIT 1
LIVING IN GOD'S WORLD

1

THE LIVING CREATION

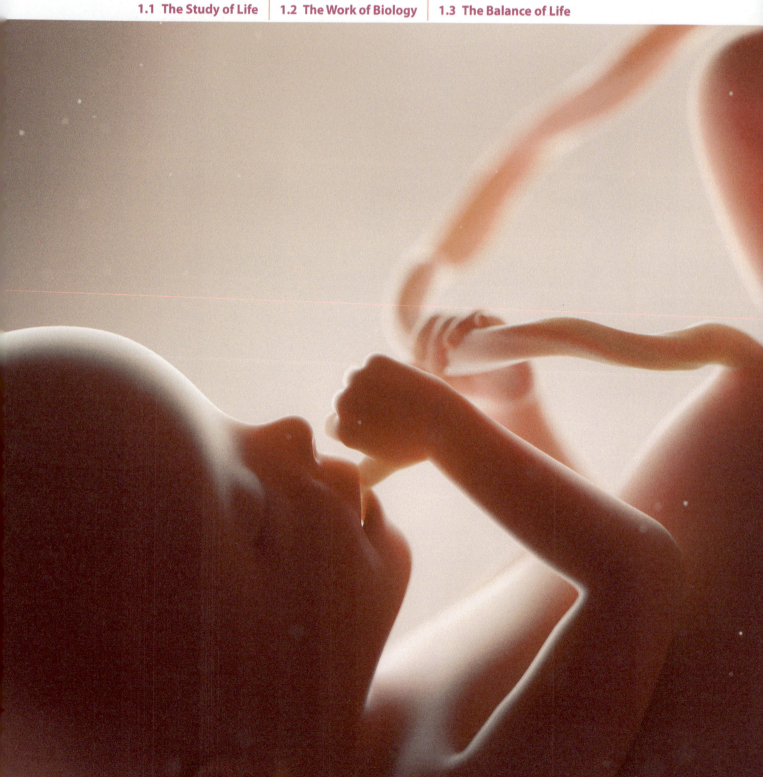

A Little Life

Marina's doctor had told her that the tissue growing inside her had the potential to be a human but wasn't yet. He said that it was her choice to let this fetus grow or to get rid of it. It certainly would be a lot easier to not be pregnant.

But now Marina was at a pregnancy center, plagued with doubts. As she looked at the sonogram, she wept. She saw ten tiny fingers and ten tiny toes, and she felt that an abortion wasn't right. This little life depended on her. Questions flooded her mind. Was she wrong about what she had always thought about life?

1.1 THE STUDY OF LIFE

Worldview and the Big Questions in Biology

Marina isn't alone. Many women seeking abortions are plagued with doubts. They turn to crisis pregnancy centers, and many get sonograms. Almost 80% of women who see a sonogram of their baby continue their pregnancies because they recognize a life when they see one.

People have questions about life. Issues like abortion, genetic engineering, drugs, and evolution fill the headlines, showing that people struggle to answer these questions. God's Word has answers. Looking at science in the light of God's Word helps us understand these issues more clearly. Your journey through this textbook is structured around six big questions that will help you view life on Earth through biblical lenses.

? What makes something alive?

Questions

What is biology?

How does worldview affect the study of biology?

What makes something alive?

How do living things stay alive, grow, and reproduce?

Terms

biology
worldview
Creation Mandate
image of God

1. Where did life come from?
2. How is life designed to function?
3. How does God provide energy for life?
4. How does science work in the real world?
5. What is the balance between preserving Earth's resources and using them to help people?
6. How can I make ethical decisions in the use of biology?

Questions like these are part of the study of **biology**, which literally means "the study of life." But before we get to the big questions, we need to think about an important tool for studying science, including biology, and that is worldview. A **worldview** is a way of seeing and interpreting all aspects of life that arises out of an overarching narrative and that shapes how a person thinks and acts.

To see how worldview colors our perception of the world around us, think about a farmer. When a farmer plants wheat, he believes and expects that wheat will grow in his field. His expectation is not based on blind faith but on observations of previous plantings. He's probably planted wheat for many years and gotten a wheat crop as a result every time. His expectations for future crops are firmly rooted in his observations of past crops. The farmer *doesn't* expect to plant wheat and get a crop of melons as a result. If a farmer couldn't predict what would grow in his field according to what he planted, farming would be a pretty crazy business!

Specific beliefs that make up a worldview are known as *presuppositions*. Most, if not all, people presuppose that the world around them acts in orderly, repeatable, and predictable ways—like when wheat seeds produce wheat plants—but they don't often stop to consider why they believe this or even whether their belief is justified. A *biblical worldview* is grounded in God's Word, the Bible. For Christians, the fact that the creation is orderly makes perfect sense—wheat seeds go in the ground, then wheat seedlings sprout up. The natural laws that we discover at work in creation, which appear to us to be rational and unchanging, are what we would expect from the rational and unchanging God who reveals Himself in Scripture. The God of the Bible is not a god of chaos! In fact, the Bible informs our belief that God made us and our world with good and eternal purposes in mind. That gives us hope for the future!

As sensible as these conclusions seem, they are *not* necessarily conclusions that are naturally arrived at in other worldviews. A *naturalistic worldview* is founded on the belief that our universe is the product of purely natural causes—no Creator God is allowed! Scientific naturalists believe that humans are the result of a random and undirected process of biological evolution. But if this is true, then our rational minds are also the product of random chemical interactions and pure chance. How then could we trust that what our minds perceive about the world is indeed true and reliable? Ultimately, the honest scientific naturalist must conclude that the universe is unknowable and without purpose or meaning. We'll look at other important differences between the two worldviews throughout this textbook.

Everything that you learn in this textbook will help you answer one of the six big biology questions from a biblical perspective. And knowing how to answer questions like these will prepare you for many of the issues that you'll face throughout your life.

Let's get started!

In the Beginning

THE CREATION OF THE EARTH

We live in a world full of life. But where did it all come from? According to Genesis 1, God created the earth and everything in it in six 24-hour days. He created it *ex nihilo*, which literally means "out of nothing," about 7000 years ago according to inferences from biblical records. (See Chapter 10.) His carefully crafted, brand-new world was good, and once He put people in His world on the sixth day, He declared it to be very good.

God created people to fellowship with Him. He also had a job for them to do. The first command God gave to the first man and woman was to fill the earth and have dominion over it (Gen. 1:28). This is often called the **Creation Mandate**. Why would God give us this job? Many creatures are stronger and have sharper senses than humans. Other creatures have amazing skills that humans don't have at all, like the ability to fly or breathe underwater. God asks us to manage the earth because we are a special creation, created in the **image of God** (Gen. 1:26–27), which is the combination of qualities that God has placed in people as a reflection of Himself. The image of God in us includes our personalities and emotions, our abilities to plan and reason, our sense of eternity, and our desire to create, love, and be loved (Eccles. 3:11). One of the most important qualities that come with our unique position in creation is that we are given the choice to either obey the commands that God has put before us or disregard them.

THE FALL OF CREATION

Adam and Eve chose to disobey God. In doing so, they brought a curse on themselves, on the rest of creation, and on their descendants. Since Adam, everyone who has ever lived has been born with a sin nature—an innate desire to disobey God—and the whole earth groans under the affliction of sin (Rom. 8:22). The command to care for the earth still stands, though it is now more difficult under sin's curse. Disease, death, and sin now frustrate our attempts to manage a fallen earth. Biologists feel this frustration as they work to fight decay, disease, and ultimately death. We see the effects of sin at work in a society that accepts evil, such as abortion.

THE REDEMPTION OF CREATION

But God didn't abandon His creation. He provided a way to save people from their sin and restore the world to the good and perfect way that He created it. He promised that from Eve would come a seed that would crush the head of the serpent who had tempted Adam and Eve to sin (Gen. 3:15). And He kept His promise. This Seed, Jesus Christ, came down to the earth as the God-man to perfectly fulfill God's intent for man's dominion (Ps. 8; Heb. 2:6–9).

Jesus came to help people. His acts of mercy—healing people, providing them food, protecting them from harm—also showed His authority over creation and His power to redeem mankind and reverse the effects of the Fall. His ultimate act of mercy was to take the penalty of our sin on Himself by sacrificing His life. Those who repent of their sin and place their faith in the sacrifice of God's Son will be redeemed from the penalty of their sin. One day Christ will come to the earth again and remove the Curse entirely. In the new earth God's people will be restored to rule over creation as He intended from the beginning of time (Rev. 22:5). Until then we are called to shine in a world filled with sadness and sin as lights in a dark place, living out our redemption with good works (Matt. 5:14–16). Acts of kindness, sacrifice, and love should characterize the lives of God's people, who look for the restoration of all things.

WHAT CREATION, THE FALL, AND REDEMPTION MEAN TO BIOLOGY

So why does this biology textbook begin with Creation, Fall, and Redemption? Because biology—the study of life—needs a foundation. Like Marina, we need something more than science to answer the most important questions about life. A biblical worldview provides the necessary foundation.

In the Bible God's narrative of Creation clearly demonstrates that human life has value above all other life because people are made in His image. He wants us to treat other people as the special creation they are. In Matthew 22:39 God gave us one of His greatest commands: to love others as ourselves. Observing this command and the Creation Mandate gives us a biblical approach to any scientific study.

God tells His image-bearers that they are to care for His creation, including children, animals, trees, and everything else He made. And because we are made in His image, we should care for other people above all other parts of creation—including unborn lives. We are more important to God than anything else, and we are responsible for the well-being of His creation.

THE ATTRIBUTES OF LIFE

Cells. All living things are made of at least one cell. As we'll see in Chapter 5, cells are complex and rely on multiple interactions with other cells and the environment. Every cell, tissue, and system in the body contains structure, a specific arrangement of parts.

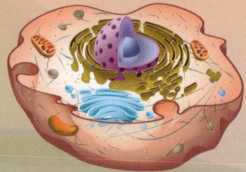

Reproduction. Cells come from preexisting cells; life comes from preexisting life. When our body replaces its cells, the information in the new cells is an exact copy of the information in the old ones. When a baby is conceived, two parent cells come together to form a new person with a new combination of genetic information.

Metabolism. The metabolism of an organism is the sum of all its chemical processes. In other words, it includes all the ways that an organism uses energy. Processes like photosynthesis, respiration, growth, and movement all require energy. These processes are part of an organism's function, the way that the structures in an organism work together.

Organization. Organisms use energy and information in the cells to maintain a high level of order in their bodies. When you eat a burger, it isn't randomly distributed throughout your body; it is methodically broken down into chemicals and then sent to the appropriate cells and systems in your body to be processed. In this way the maximum amount of energy can be removed from the burger. Without this kind of efficient organization your complex body would quickly fall apart.

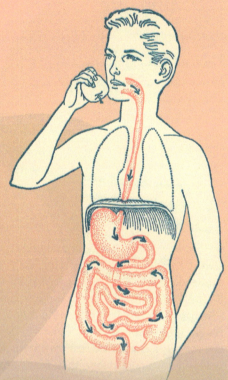

Growth. Outgrowing your pair of running shoes is proof that you are a living creature. All growth involves the assembling of parts to make up an organism. Growth is not only an increase in size; it also includes replacing worn-out or damaged cells.

Response. No organism can exist in a vacuum. All organisms daily interact with and respond to things like temperature, moisture, sunlight, other organisms, and their own bodies. Living things have the ability to adapt, or to change in response to the environment. For example, when a capybara gets hot, it adapts to the heat by either swimming in a marsh or wallowing in the mud.

Defining Life

What exactly is life? For centuries, people have grappled with what makes something alive. From the Greek physician Hippocrates to nineteenth-century biologists, people have tried to describe what they believed was the "vital force" that all living creatures possess. Since then biologists have agreed upon certain attributes that set living things apart from the nonliving world.

Most scientists agree that an organism must have all the attributes described on pages 8–9 to be considered alive. Otherwise, a nonliving thing like a snowflake might be considered alive because it moves and grows in the atmosphere, has a highly organized crystalline structure, and responds to temperature conditions in the environment. But snowflakes are not made of cells and don't reproduce—they clearly are *not* living things!

At some point, no matter how organized an organism's body systems are, worn-out cells die faster than new cells reproduce, and the organism dies. Cell mutations and sickness are consequences of sin that lead to sin's ultimate consequence—physical death. But physical death isn't the end of life for people. God will one day resurrect the bodies of the redeemed, and they will live forever with Christ reigning over the new earth (Rev. 22:5). Those who reject Christ's sacrifice will also be resurrected—to face eternal judgment in the lake of fire (Rev. 20:11–15).

Life is complex and difficult to define! But all living things have a similar design, making life much easier to study in spite of its complexity. This common design points to a loving, all-wise Designer.

Energy and Information

As you read through the attributes of life, you may have noticed two themes. Organisms need energy to build and maintain their structures and functions, and they need information so that their cells know what to build, how to function, and how to create new organisms. Throughout this textbook we will see that every aspect of life comes back to energy and information, and God lovingly provides these for His world.

ENERGY

At some point today you will feel like you've run out of energy. You may begin to feel sluggish, and your stomach may growl and demand food. So, naturally, you eat something. Food is the source of energy for people and many other organisms. Some organisms, such as plants and certain kinds of bacteria, can create their own food using energy from the sun.

Eating food is just one small part of how God supplies energy to living things. There are many ways that energy cycles through organisms and their environments. For example, the light energy of the sun, combined with chemicals taken from the air and soil, is converted to chemical energy in plants. Animals and humans take the chemical energy from plants and convert it into other forms, such as energy for motion, for warmth, and for the production of complex organic compounds. Bacteria and fungi break down waste products, extracting energy from them and returning nutrients back to the soil for plants.

INFORMATION

Once an organism gets its daily dose of energy, how does it know what to do with that energy? All cells contain information that directs the activity of an organism. That information is usually in the form of DNA or RNA molecules. In your body these molecules direct various cells that digest food, feel pain, produce blood, and so on. The information in cells is passed along when either one cell splits in two or two cells join as one. Your body uses this information to give you fingernails, hair, ears, and every other piece of you. Without information to control cells and energy to power them, life would quickly come to a screeching halt.

As we will see in coming chapters, the information required to organize all the functions of an organism is immense and incredibly complex. One gram of your DNA can store roughly 455 exabytes of data—that's 455 *billion* gigabytes of information! God has built into us a code of life with the information needed to carry on life processes. This is a marvelous gift—a gift that the greatest minds in science are still trying to unwrap.

1.1 SECTION REVIEW

1. What is a worldview?

2. Summarize in your own words the creation of the earth, the fall of people and its consequences for the whole earth, and the redemption of the earth and people.

3. What two commands from God can we obey by a biblical approach to science?

4. Copy and complete the table at right to classify the terms or phrases listed below as a structure or a function and relate them to one of the six attributes of life. For example, an egg cell is a structure and is part of reproduction. (*Note:* Not all parts of the table should be filled in.)

   adaptability
   digestion
   arrangement of molecules
   single-celled bacterium
   increasing number of cells

5. A virus can reproduce because it contains genetic material, but it has no metabolism and does not contain cells. Should a virus be considered alive? Explain your answer.

6. How do energy and information in cells relate to each other?

7. How are energy and information used to keep an organism's body organized?

	CELLS	METABOLISM	GROWTH
Structure			
Function			

	REPRODUCTION	RESPONSE	ORGANIZATION
Structure	egg cell		
Function			

THE WORK OF BIOLOGY
The Nature of Science

?

How does worldview affect the way that people study biology?

Questions

What do people base their views of science on?

How reliable is science?

What kinds of questions can science answer?

How does science work in the real world?

Terms
model
theory
law
scientific inquiry
hypothesis
survey

Big Question 4: How does science work in the real world?

In 2007 biologists who had been searching for new species in the mountains of Cambodia for almost a decade made a bizarre discovery: a frog with green blood and bones the color of turquoise. These biologists had discovered over a thousand species during their time in Cambodia, but the characteristics of this frog were unique. These and other biologists asked questions like, "Why does it have green blood and blue bones? Are there more creatures out there like this one?" In today's world of endless information there are still surprises in God's creation.

The more scientists learn about nature, the more questions they have. Sometimes they make breakthroughs that revolutionize the way we think about the universe, such as Galileo's observations of the moons of Jupiter and Pasteur's germ theory. What if biologists discovered something that completely changed the way that we view the fundamental ideas of cell theory? Biology would need to be rethought and reinterpreted within this new framework. In other words, what science claims to know about our universe and the way that it works is subject to change.

Science can be a great tool for understanding God's world and using it wisely, but it's inadequate as a foundation for a belief system. Science isn't capable of dealing with the biggest questions in life: How did we get here? Is there a God? Why is there evil in the world? What happens to us after we die? Only an unchanging and infallible God can give us satisfactory answers to these questions, answers that we find in His Word. God's revealed truth provides us with a sure foundation, for both life and for doing science, that will endure the passing of time.

Science as Modeling

Mr. Stubbs, a seven-foot American alligator that lives in a swimming pool at the Phoenix Herpetological Society, needed a tail. So herpetologists—scientists who study reptiles—used motion-capture equipment to model the motion of alligators in the wild. The motion information was then used to generate a mathematical model,

Mr. Stubbs ▶

which was used to create a similar tail made of rubber that helped Mr. Stubbs function more like a normal alligator. Although his prosthetic tail looked the part, it still wasn't ideal. In 2018 Mr. Stubbs received a new, more high-tech 3D-printed tail. He's still learning to use it, though he will never be able to be released back into the wild. But the herpetologists knew they were making progress when Mr. Stubbs tail-slapped one of his handlers with his new tail. It was a very alligator-like thing to do!

Mr. Stubbs's tail helped more than just him. The science used to develop his prosthesis could be useful for other animals and even for people who have lost limbs. Though Mr. Stubbs's newest tail isn't just like a real one, it works well enough for him. This is an example of the work of science to develop models that help us understand the world around us in ways that help us use it better.

A **model** is a simplified representation of reality that describes or explains something in the world in a workable and useful way. A large part of science is modeling because we learn about the world by examining parts and pieces of it and looking at how those pieces work together. Just as with Mr. Stubbs's tail, our models are not the same thing as reality, but they are workable—they are practical and can be changed when we discover new information. Because models are different from reality and can change as we discover new information, science can't give us final answers.

THEORIES AND LAWS

Models can represent concrete things, such as the plastic model of the human digestive system that many high schools have in their laboratories. Models can also represent abstract ideas, such as the attributes of life mentioned in Section 1.1 or the flow of energy through a cell. Models don't lead to absolute truth, but they can help us visualize, organize, and understand areas of science so that we can use them to make useful predictions.

Two kinds of scientific models that you may already know about are theories and laws.

A **theory** is a model that explains a set of observations. For example, the cell theory explains that cells are the smallest living things and that they are made up of many smaller parts that show amazing structure and function. For centuries scientists observed cells and their actions until they were able to organize the observations into an explanation of how cells work.

A **law** is a model that describes how phenomena relate to each other in a predictable way. The law of gravity describes the relationship between objects on the earth and the earth's gravitational pull. These relationships can often be described mathematically, such as in the equation for the law of gravity.

$$F = G\frac{m_1 m_2}{r^2}$$

And so far, the law of gravity predicts quite well what will happen if you stand under a coconut tree for too long!

There are so many ways to observe the world and model it that there needs to be a process to sort through that information. One way that we can do that is through **scientific inquiry**, sometimes described as the *scientific method*. Scientific inquiry is a logical procedure that helps answer a scientific question.

This method isn't about following a series of steps. Rather, it's a way of thinking that helps people make sense of what they observe. It's part of the larger process of science. Let's take a look back in history to understand how scientists have used scientific inquiry to develop one of the most significant models in biology—the structure of DNA.

Ask a Question.

Biologists have been probing the cell for decades to see how it stores the information of life, beginning with a friar named Gregor Mendel experimenting with pea plants. In 1944 biologists Oswald Avery, Colin MacLeod, and Maclyn McCarty determined that there is a large molecule called *deoxyribonucleic acid* (DNA) that holds this information, and they knew that it was a chain of smaller parts called *nucleotides*. (We'll learn more about DNA in the next chapter.) But they didn't know how these parts fit together to store information in the cell. Biologists were wondering, **"What is the shape of the DNA molecule?"**

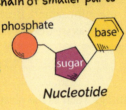

phosphate

base

sugar

Nucleotide

Watson

Crick

Do Research.

Two young biologists, James Watson and Francis Crick, rose to the challenge. They dived into the research, experiments, and data that Avery, MacLeod, and McCarty had collected to find out about DNA and the parts that make it up. On the basis of their results Watson and Crick made a hypothesis: DNA was a spiral, or a helix. A **hypothesis** is a simple, testable statement that predicts the answer to a question, in this instance, the question about DNA's shape. Watson and Crick also learned about some mathematicians who had determined in 1952 that a helix-shaped molecule photographed with x-rays produces an X-shaped image. They also learned about a chemist's hypothesis that DNA was a triple helix. Armed with information and their hypothesis, Watson and Crick decided to assemble a DNA model out of cardboard and wire.

Experiment.

An experiment is simply a way to test a hypothesis. Watson and Crick tried unsuccessfully to assemble their DNA helix. They showed their proposals to groups of scientists, including English chemist Rosalind Franklin. When Watson saw Franklin's x-ray photo of pure DNA crystals, what he saw confirmed their hypothesis that DNA is in fact a helix. Franklin's x-ray photo was the perfect test for Watson and Crick's hypothesis.

Experimenting isn't the only way to gather data for a hypothesis. We can also conduct a **survey**, a sampling of data gathered from an existing group of data. Things like polls and questionnaires are types of surveys. If we wanted to find out the average depth of all the ponds in Ohio, we could directly measure every single pond, or we could measure a large enough sample of ponds to find the average. Scientists also sometimes naturally observe phenomena that they can't reproduce in the laboratory, like an exploding star. These are called *natural experiments*.

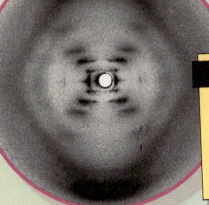

Analyze the Data.

Watson and Crick gathered data from Franklin's x-ray photo to determine how all the parts of DNA fit together. They physically modeled the parts of DNA and worked with them until they developed the double-helix structure that explained how information is copied in a cell.

Publish the Results.

Scientists today publish their findings in science journals, articles, and books. These published experiments often raise new questions in other fields that scientists will try to answer. In 1953 Watson and Crick published their findings in the scientific journal *Nature* for the scientific world to see. Watson and Crick's model of DNA's double-helix structure propelled biology into the field of molecular genetics. For their efforts, Watson and Crick received a Nobel Prize in 1962.

Fig. 2. This figure is purely diagrammatic. The two ribbons symbolize the two phosphate-sugar chains, and the horizontal rods the pairs of bases holding the chains together. The vertical line marks the fibre axis

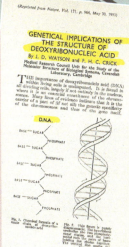

So what is the end result of the scientific method? The result is a model that is workable and that has the ability to make useful predictions. But a model is not the same thing as absolute truth. Models continue to be tested and refined as we gather data and learn more about the world in ways that help us understand and use it better. Viewing and using science this way helps us glorify God by fulfilling His commands.

MINI LAB

Peer Review

The findings that scientists publish in journals are *peer reviewed*, which means that the articles are read and evaluated by other scientists before being published. Peer review adds a layer of accountability to the process of scientific inquiry. Part of the review process is to evaluate whether a procedure has been adequately described. Try your hand at some peer review with this activity.

How can scientists make sure that their results can be repeated? ?

Materials
building materials supplied by your teacher • glue

PROCEDURE

A Obtain ten pieces of building materials from your teacher.

B With your team, design and assemble an "apparatus" for an experiment. Make sure that all ten pieces of material are incorporated into your design.

C Write out a procedure for assembling your apparatus.

D Exchange your procedure with another group.

E Using the procedure provided by the other group, reconstruct their apparatus. When both groups have finished, compare the reconstructed apparatus to its original design.

RESULTS

1. How closely did your reconstructed apparatus mimic the original?

2. How does this activity demonstrate the importance of clear and thorough descriptions?

3. Why is a clear and thorough description of the equipment and procedures used in a scientific investigation an important part of the review process?

1. Read through the following scenarios and decide whether they are observations or interpretations of science.

    a. A couple discovers that their baby will be born with severe disabilities but decides to keep the child.

    b. Some birds fly south for the winter, while others do not.

    c. New-world monkeys have flat noses and old-world monkeys have narrow noses. A biologist discovers a new-world monkey with a narrow nose.

    d. A researcher who develops a new drug determines that children under the age of twelve should not use the drug.

2. Can scientific models provide absolute answers? Explain your answer.

3. Evolutionists say that nature contradicts what the Bible says about the origin of life. Think back to the discussions on observation, interpretation, and modeling. How do you think a Christian could respond to this evolutionary idea?

4. What is the difference between a theory and a law?

5. You've noticed over the past several days that a flock of robins likes to sit in your front yard but never goes over to the neighbor's yard. Using the scientific method, how would you figure out why the robins prefer your yard over the neighbor's?

1.3 THE BALANCE OF LIFE

Big Question 5: What is the balance between preserving Earth's resources and using them to help people?

Preserving and Using Our World

Over seven billion people call this planet home, and hundreds more are born every minute. The United Nations predicts that the world population may surpass 8 billion by 2025 and 10 billion by 2083.

There are many people who view a growing world population as a problem. They say that nature is good and should be kept in its pristine condition. Technology consumes the earth and pollutes precious resources. Their "Mother Earth" must be protected at all costs because they believe that people came from the earth through evolution. This kind of thinking is aligned with a philosophy called **environmentalism**—the idea that the environment, the ultimate cause of man's existence, is fragile and needs to be actively protected. This can lead some who take this position to argue that the environment should be prioritized at the cost of humans.

But Genesis 1:28—the Creation Mandate—tells us that God created people and placed them on the earth to populate it and manage its resources. God also tells us in Matthew 22:37–39 that His greatest commands are to love Him and others, and managing resources is a way that we can do that. Being good stewards of the earth means that we need to preserve some things, keeping their future use in mind. **Conservation** is the preservation and wise use of natural resources. Notice that the word "use" is in that definition. Using resources is part of conservation, though they must be used wisely with a mind for the future.

What should biology be used for?

Questions

Why is conservation important?

How should we use biology?

Why is biology important?

Terms

environmentalism
conservation
biotechnology
bioethics

We can obey the Creation Mandate and the command to love God and others through biblical conservation. These biblical ideas help us keep a balanced view of how we conserve the earth's resources. We should respond to situations that don't seem balanced by asking, "Is this how God wants us to manage the earth?" and "Does this show love for God and His image-bearers?" Throughout this textbook we'll see how conservation can ultimately be used to glorify God and fulfill His commands.

The Application of Biology

Big Question 6: How can I make ethical decisions in the use of biology?

If we are to conserve God's world, how should we use science to meet the challenges of a growing world population? One of the most powerful tools in biology is **biotechnology**, the use of technology to enhance living organisms and their processes. Using biotechnology, scientists have developed more-nutritious grains, prettier flowers, more treatments for disease, and alternative fuel sources that keep our environment cleaner. But scientists also use biotechnology to abort unborn lives, to create "perfect" babies, and to perform gender reassignment surgery.

Because of the potential dangers in biotechnology, it and other tools of science require constant monitoring. One way to carry out such monitoring is through the application of **bioethics**, the study of ethical situations in biology and medicine. Secular bioethics attempts to do this through reason rather than by using God's Word. In some cases secular bioethics unknowingly adopts aspects of a biblical worldview. Thus, not all of its conclusions are contrary to biblical ethics. But secular bioethics is primarily built on the presuppositions of a naturalistic, evolutionary worldview. Since it is not dependent on Scripture, secular bioethics often leads to conclusions that strongly conflict with a biblical worldview. Rather than adopting this system, we need to see issues through the lens of Scripture to discern how we should best use the tools of science.

The Biology Connection

Learning interacts with life. As we have seen in the history of the development of the model of DNA, many disciplines must come together to make progress in our God-given stewardship responsibility.

Seeing the connections between biology and other fields is a way to fulfill the Creation Mandate and love others. A great example of this is prosthetics (see right). The field of prosthetics requires the combined efforts of physicians, engineers, machinists, and physical therapists to produce something that helps thousands of people enjoy their lives again. But the command to love others isn't limited to getting people back on their feet. It also extends to helping people thrive. That's why part of our stewardship should be to continually evaluate the ways that we help people to see whether we can do a better job. For example, neuroscientists and computer programmers joined forces with those in prosthetics to create the cochlear implant, a device that does far more than just amplify sound like an ordinary hearing aid. A cochlear implant creates auditory signals that the brain can process, allowing many deaf people to hear clearly for the first time.

Biology helps us manage the earth and help other people, thinking about a future world in which the unborn lives of today will be the citizens of tomorrow. We must work in biology in ways that connect the disciplines to fulfill God's commands and that ultimately point people to Christ.

1.3 SECTION REVIEW

1. Explain the fundamental difference between environmentalism and conservation.

2. Is conservation the idea that resources should be used as little as possible? Explain.

3. Some environmental groups say that people are overfishing the oceans. Would a Christian fisheries biologist have an obligation to examine this issue? Explain.

4. Why is a worldview important for issues in biotechnology?

5. List two products that are the result of a combination of fields of study. Include the fields of study that were combined.

Use the ethics models on pages 20–22 to answer Questions 6–8.

6. List and briefly describe the three parts of the biblical ethics triad.

7. List and briefly explain what is meant by the four principles of bioethics.

8. What is the chief difference between the biblical ethics triad and the principles of ethics model?

CHRISTIAN ETHICS AND BIOLOGY

The more we know about living things, the better equipped we are to fulfill the Creation Mandate. The study of biology helps us to understand the world that God created for us. But biology also helps us in another way—by giving us information that we can use to help answer hard questions about how to best manage our world and love our neighbors. Sometimes advances in biology introduce us to new, even thornier questions.

Biology, like any science, is a tool that can be used for either good or evil. Applying moral principles to how we live our lives, including our use of biology, is known as *ethics*. Throughout this textbook we will examine instances in which ethics and biology intersect. The way that a person looks at ethics depends on his worldview. Christians have a unique way of looking at ethical questions because their decisions are based on God's Word, the Bible. Obedience to God's Word requires that we consider how our decisions will affect both our neighbors and God's creation.

For each issue that we look at, we will use a model that can be called an *ethics triad* consisting of biblical principles, biblical outcomes, and biblical motivations. The three Bible-based legs of the triad will guide our ethical decision-making.

 ## BIBLICAL PRINCIPLES
What does God's Word say?

God's Image-Bearers. Foundational to our ethical decision-making is the understanding that we all bear God's image. Therefore, we must make decisions out of respect for all people and for their protection (Gen. 1:26–27).

Creation Mandate. God's first commandment to us is to have dominion over the world that He created. Therefore, we must wisely care for God's creation according to this Creation Mandate. We must balance the appropriate use of the world's resources with the needs of people around the world. Nothing belongs to us; we are stewards of God's world (Gen. 1:28).

God's Whole Truth. God's image in people and the Creation Mandate touch on many ethical issues. But other parts of Scripture also give us helpful insights into what God wants us to do. Part of making ethical decisions requires that we understand what His Word teaches. We cannot live any part of our lives separated from God and His Word (2 Tim. 3:15–16).

BIBLICAL OUTCOMES
What results are right?

Human Flourishing. As soon as God created mankind, He blessed him (Gen. 1:28). Throughout Scripture (Ps. 1; Matt. 5) we see that God's desire is for all people to be blessed and to prosper. Jesus came to give us life and to have us live that life abundantly (John 10:10). Our ethical decisions must align with God's will to maximize human flourishing.

A Thriving Creation. Part of our obligation to the Creation Mandate is to ensure that creation thrives (Gen. 1:28; 2:5, 15). God has assigned us an important role as stewards of the world that He created. We must wisely use and develop the earth's resources to ensure that it flourishes.

Glorifying God. Just as Jesus came to glorify God, everything a believer does should glorify God (Matt. 5:16; 1 Cor. 6:20; 10:31). His decisions should show God that he loves and honors Him. This obligation includes every aspect of life: school, work, and play. So it is not enough that his decisions help others or that creation thrives. A Christian's decisions must always give God the honor that He is due.

BIBLICAL MOTIVATIONS
How can I grow through this decision?

Faith in God. The Bible discusses works versus faith (James 2:14–26). The passage concludes that a believer is to live out his faith in God through his works. He is motivated to act because of his faith in God. Good works can stem only from his faith in God (Rom. 14:23; Heb. 11:6).

Hope in God's Promises. In the Bible hope is not something that we wish for; it is something that God has promised. Biblical hope is confident expectation. The promises of God allow a Christian to take action without fear (2 Tim. 1:7–9). Scripture teaches us that God can never lie, so a child of God can act with the assurance that God will follow through on His promises.

Love for God and Others. As stated in 1 Corinthians 13, a believer's greatest motivation for doing right is love. A Christian has the love of God in him, and he does right when he is motivated by love for God and others. John 13:34–35 teaches that love is the outward sign of a transformed life.

THE PRINCIPLES OF BIOETHICS

Bioethics is an ancient study. Hippocrates, a Greek physician, handed down what is known as the *Hippocratic Oath*. The oath bound physicians to do no harm (nonmaleficence) and to help their patients (beneficence). The revolutionary progress achieved in health care in the last century has raised many new questions about how to ethically provide health care. In answer to the growing moral dilemmas, governments and health-care networks have developed a model to enable them to administer health care ethically. Their model is commonly known as the *principles of bioethics*. Four principles are recognized as part of the model:

1. respect for autonomy,
2. nonmaleficence,
3. beneficence, and
4. justice.

RESPECT FOR AUTONOMY

Generally, patients are recognized as having final authority for making decisions regarding their health care. Doctors and other experts should present information and communicate truthfully with a patient in order to help him make an informed, voluntary decision. This principle is the foundation of informed consent—a process of evaluating information and obtaining consent for specific medical decisions and treatments. Health-care professionals must maintain the privacy of such communications unless obtaining permission to do otherwise.

NONMALEFICENCE

You may have heard the expression "do the patient no harm." This principle requires that health-care providers do not purposefully act in any way that might harm a patient. Instead, medical providers must operate in a manner that avoids or minimizes risk. This implies that providers must also be competent, that is, fully trained and authorized to do the kinds of work that they do.

BENEFICENCE

The word *beneficence* comes from the same Latin roots from which we get the word *benefit*, referring to something good, helpful, or advantageous. You can see then that beneficence and maleficence go hand-in-hand: providers are not merely to do no harm—they are also expected to provide a benefit, in so far as they are able.

JUSTICE

Justice refers to treating people equally and encouraging fair and equitable distribution of resources. How can a hospital or health-care system distribute health care fairly? This principle considers several factors:

- individual need,
- the health disparities that exist,
- the effort required to provide the service,
- the potential benefits of the service, and
- the constraints of supply and demand.

CHAPTER REVIEW

Chapter Summary

1.1 THE STUDY OF LIFE

- A worldview is a way of seeing and interpreting all aspects of life and is based on a set of presuppositions.

- A biblical worldview presupposes the truth of God's Word, the Bible. A naturalistic worldview presupposes that our universe is the result of only natural causes.

- The Creation Mandate (Gen. 1:28) is God's command to humankind to fill the earth and have dominion over it.

- Our ability to do the work of biology and use it to exercise dominion is adversely affected by the Fall.

- The proper use of biology includes using it to love and serve others, who are all bearers of God's image.

- Living things are made of cells, are highly organized, use energy, can grow, respond to their environment, and reproduce.

Terms
biology • worldview • Creation Mandate • image of God

1.2 THE WORK OF BIOLOGY

- Scientists work to create models, which are simplified descriptions or explanations of reality.

- Models include theories, which are explanations for sets of data, and laws, which relate phenomena in specific ways.

- Scientific inquiry is a modeling process. As part of the process, a hypothesis is established, then observations are gathered and analyzed.

- The workability of scientific models is evaluated on the basis of results obtained through the inquiry process. Models may be modified or discarded on the basis of new evidence.

Terms
model • theory • law • scientific inquiry • hypothesis • survey

1.3 THE BALANCE OF LIFE

- Environmentalism is a philosophy rooted in scientific naturalism. Environmentalists may elevate the need to protect the earth above the needs of humans.

- Conservation involves the wise use of natural resources with an eye toward safeguarding them for future generations.

- The biblical ethics triad is a model for ethical decision-making that is based on considering what God's Word says, determining what outcomes are right, and identifying how one can grow because of a decision that was made.

- The principles of bioethics are a model for wrestling with ethical questions, especially with regard to health care. The four principles of bioethics are respect for autonomy, nonmaleficence, beneficence, and justice.

Terms
environmentalism • conservation • biotechnology • bioethics

CHAPTER REVIEW
Chapter Review Questions

RECALLING FACTS

1. Define *biology*.

2. Which two commands offer a biblical approach to studying biology?

3. The ability to adapt is part of which attribute of life?

4. What are the two ways that an organism can grow?

5. What do living things obtain their energy from? Where do they obtain the necessary information for growth and reproduction?

6. What does it mean for a model to be workable?

7. What kind of answers can models give us? What kind of answers can they *not* give us?

8. From a biblical worldview perspective, what is the primary difference between environmentalism and conservation?

9. What is the focus of bioethics?

UNDERSTANDING CONCEPTS

10. What is the difference between a biblical worldview and a naturalistic worldview?

11. During a visit to the American Museum of Natural History in New York, you see an exhibit of skeleton fragments. The signs on the exhibit say that these are the remains of a Neanderthal, a relative of modern humans that lived 200,000 years ago. What two presuppositions are revealed in the signs on the exhibit? Which worldview do these presuppositions represent?

12. Relate each of the following descriptions of a structure or process to one of the six characteristics of living things.

    a. A lizard retreats to the shade of a rock ledge during the heat of the day.

    b. A human's stomach lining is replaced every few days.

    c. A pile of leaves decays into mulch.

    d. An onion skin seen under high magnification appears to be made of regularly repeating rectangular structures.

    e. A bacterium divides to produce two new bacteria.

    f. A catfish farm produces 1.0 kg of catfish for every 1.8 kg of feed given to the fish.

13. Make a drawing to show the flow of energy from the sun to a person through natural processes in living things.

14. Wombats don't have tails. Because of this, the offspring of wombats don't have tails either. Is this an example of the flow of energy or of information? Explain your answer.

15. How does worldview affect the way people form models?

16. Describing the eating habits of a new species of frog is an (observation / interpretation) of science. Describing how the frog's eating habits were created by God is an (observation / interpretation) of science.

17. The model of DNA is based on observations and works very well to explain how information is transferred from one cell to another. Why might the DNA model need to be changed one day?

18. How can a Christian fulfill the Creation Mandate and the command to love God and others through conservation?

19. You've been asked to help someone figure out whether an experiment is ethical. What should you base this bioethical decision on?

CRITICAL THINKING

20. Explain why the study of biology must be done within the context of a worldview.

21. A self-driving car is highly organized, uses energy, moves, and can respond to its environment. Is it a living thing? Defend your answer.

22. How would you respond to a person who claims that the purpose of science is to discover ultimate truth? Give an example to support your argument.

23. Choose the sentences in the following example that speak of an activity that can be handled by science and those that cannot.

    (1) Astrobiologists are studying other planets and moons to try to find water on them. (2) They believe that if they find water, they may find evidence of life arising on another planet. (3) These scientists often test their interstellar technology by using it in the most extreme environments on Earth. (4) As a result, they have discovered many fascinating things about hydrothermal vents deep in the ocean and bacteria living under the ice in Antarctica. (5) Astrobiologists think that if they ever find evidence of life on another planet, it will lead them to discover the origin of life on Earth.

24. Design a sample survey to find out how many dandelions are in your city's park.

25. Evaluate this statement: "Cutting down trees for lumber violates the principles of conservation."

2 THE CHEMISTRY OF LIVING THINGS

Socrates put the cup to his lips, knowing that the poison hemlock that it contained would bring him certain death. Accused of corrupting the youth by not recognizing the Greek gods, he chose this punishment rather than exile. But how could a seemingly innocent plant bring on death with just a simple drink?

The odd thing is that the Greeks used this same poison as medicine to relieve nagging coughs, arthritis, and the discomfort of teething in babies! We still use poisons today to cure what ails us. Snake, spider, and scorpion venoms and poisonous plants contain amazing chemicals to fight cancer, cardiac arrest, and stroke.

2.1 MATTER, ENERGY, AND LIFE

Matter

So what is the difference between a poison and a medicine? They are both chemicals that affect the way living things grow and operate. One of the most fruitful applications of studying life is to use chemicals from living things, like the poison from the hemlock, to develop medications that can save and improve people's lives. And lives matter because we are made in God's image.

Living things are deeply affected by matter and the changes that it undergoes. The study of matter and the changes that it experiences is called *chemistry*. **Matter** is anything that takes up space and has mass. We are made of matter and are breathing, using, and eating matter all the time! But we aren't the only ones. All living things—bees, banana trees, chimpanzees—interact with chemicals this way. There is a vital relationship between life and matter.

Atoms are the basic building blocks of matter. They are the smallest possible particles of an element. They have a nucleus that is heavy but extremely tiny compared to the rest of the atom. It is in the nucleus that positively charged *protons* and neutral *neutrons* are found. Whizzing around the nucleus is a cloud of negatively charged *electrons* that seem to occupy different levels, like the orbits of planets around the sun. Electrons in the outermost level are called *valence electrons*, and they are important for interacting with other atoms.

An atom with the same numbers of electrons and protons is electrically neutral. If these numbers are different, then the atom has an electrical charge, making it an *ion*. Extra electrons produce a negative ion, or *anion*. When there are fewer electrons than protons, the result is a *cation*, which has a net positive charge.

A neutral oxygen atom has 8 protons, 8 neutrons, and 8 electrons. There are 6 valence elections in the outer level.

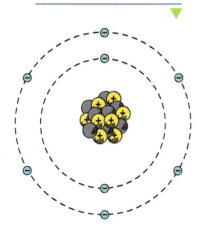

? What is the difference between medicine and poison?

Questions

What is matter, and what is it made of?

What is energy, and how do living things get it?

What is the difference between a physical and a chemical change?

Which kinds of chemical compounds are involved in living things?

Terms
matter
atom
element
energy
temperature
physical change
chemical change
bond
compound
molecule

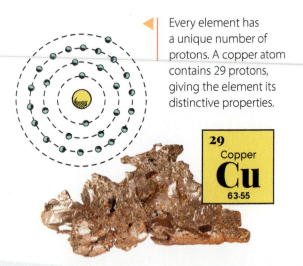

Every element has a unique number of protons. A copper atom contains 29 protons, giving the element its distinctive properties.

29
Copper
Cu
63·55

The attraction of oppositely charged protons and electrons is the force that holds an atom together. Neutrons act like glue, keeping the protons in the nucleus from repelling each other. And it's a good thing, too, or living things would fly apart!

Not all atoms are the same. Atoms are classified as elements by the number of protons in their nuclei. **Elements** are pure substances made of only one kind of atom. For example, chlorine gas, copper nuggets, and hunks of yellow sulfur are each made of only one kind of atom. But living things are made of many different elements. The periodic table of elements shows the different kinds of elements that we know about, including the ones that are commonly found in living things.

Energy

Living things don't just have matter, like a blob just sitting there doing nothing. Living things also have energy. While matter is the material stuff of the universe, **energy** is the ability to do work. Without energy, nothing happens.

Bacteria move with little propellers, whales generate sounds, glowworms make light, and platypuses generate heat to incubate their eggs. These examples illustrate that there are different kinds of energy. When bacteria move, they show *mechanical energy*. When whales make sounds, they generate *acoustic energy*. When glowworms make light, they produce *light energy*. When platypuses generate heat, they produce *thermal energy*. Thermal energy comes from moving particles. The measure of the average speed at which these particles move is the **temperature** of a substance or organism, that is, its relative hotness or coldness. So when we have a fever, the average speed of the particles in our bodies is higher.

Living things need to get energy from somewhere before they can use it. Animals can't make their own energy—they get energy from their food. This energy is stored in the chemicals that make up food—we call it *chemical energy*. Plants, which also do not make their own energy, get energy by transforming energy from the sun. So we see that energy can be changed from one form to another. It's important to note that living things depend on these transformations.

Energy also tends to spread out. Plants concentrate energy in their stalks and leaves, and then other organisms eat the leaves. Plants also release energy as they emit water vapor and carbon dioxide, both of which disperse into the atmosphere. Living things depend on energy's tendency to disperse, a concept known as *entropy*.

Changes in Matter

Just as energy can change from one form into another, so can matter. One of the basic laws of science is that the amount of matter and energy in the universe never changes. This is called the *law of conservation of mass and energy*. The conservation of energy is known as the *first law of thermodynamics*. Living things do not create new energy or matter, but matter can change from one form to another. That is why we observe cycles of matter in the environment, such as the water, nitrogen, and carbon cycles. We will learn more about these cycles in later chapters.

Changes that don't change a substance's identity are called **physical changes**. Icicles melting in the spring, water evaporating from the ocean and condensing to form clouds, and gas bubbles dissolving into a churning mountain stream are all examples of physical changes. Many physical changes can be reversed during other physical changes, but some are irreversible. For example, wheat can be ground into flour, but flour cannot be changed back to wheat.

Sometimes matter changes in a way that completely transforms its identity. When iron rusts, for example, it undergoes a **chemical change**—a change that causes a substance to alter its chemical identity. In the case of the formation of rust, the change occurs as iron particles combine with oxygen particles. Not only is rust a new substance that is totally different from oxygen or iron, but it also has a different physical appearance. When a person digests food to produce glucose, the food undergoes a chemical change. When grass clippings decay to produce compost, a chemical change takes place. Sometimes chemical changes store chemical energy, such as during the digestion of food. Although many chemical changes can be reversed using other chemical changes, others are irreversible.

Another kind of chemical change is nuclear change. Nuclear changes are usually associated with substances that are radioactive. For example, carbon, one of the main elements associated with life, has six protons. But there are different forms, or isotopes, of carbon whose identities are determined by the number of neutrons each has. Carbon-12 is stable, but carbon-14 is radioactive. As carbon-14 undergoes radioactive decay, its atoms transform into nitrogen-14. This fact is what makes carbon-14 dating of fossils possible. We will learn more about this in Chapter 10.

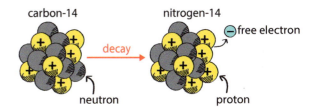

COMPOUNDS AND MOLECULES

When substances made of only one kind of element undergo a chemical change, their atoms react with the valence electrons from other atoms in a way that stores energy. An electrostatic attraction that forms between two atoms is called a **bond**. For example, when iron rusts, a chemical change occurs as the iron forms bonds with oxygen in the air to form rust. Iron and oxygen are also involved in the chemical process that allows our blood to carry oxygen to all the different parts of our bodies. Rust is an example of a **compound**, a pure substance made of two or more chemically combined elements.

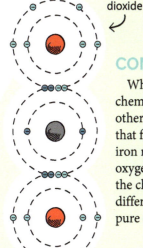

carbon dioxide

IONIC COMPOUNDS

When atoms in the process of bonding give away or receive valence electrons, they transform into ions. One becomes a cation and the other an anion. These oppositely charged ions attract to form *ionic compounds*. The smallest part of an ionic compound is called a *formula unit*. Formula units build on each other to form crystals.

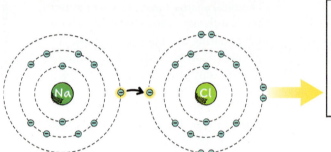

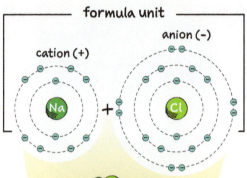

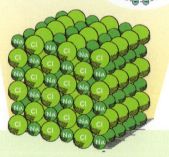

In this example of an ionic compound sodium donates a valence electron to chlorine to form a sodium chloride formula unit that combines with other formula units to form a table salt crystal.

Ionic compounds are important to living things. We use toothpaste containing sodium fluoride (NaF) because our teeth are made of ionic compounds. We drink milk fortified with calcium phosphate [$Ca_3(PO_4)_2$] because our bones are made of ionic compounds. Many domesticated animals such as horses lick salt blocks because salt provides minerals necessary for their biological functions.

COVALENT
COMPOUNDS

Sometimes atoms in the process of bonding share their valence electrons instead of donating them. An example is water, a chemical essential to life. Two hydrogen atoms each share an electron with a single oxygen atom to form a covalent compound. A particle consisting of two or more atoms covalently bonded together is called a **molecule**. Atoms in a covalent molecule typically don't form ions since the numbers of electrons and protons in these atoms are equal. In covalent compounds pairs of shared electrons form the bonds that hold the atoms together. Molecules vary in complexity from simple molecules such as those made of only two oxygen atoms to the immensely complex macromolecules of proteins and DNA that contain hundreds of thousands of atoms. Some of the compounds most essential to life and its processes are mammoth—and some are minuscule.

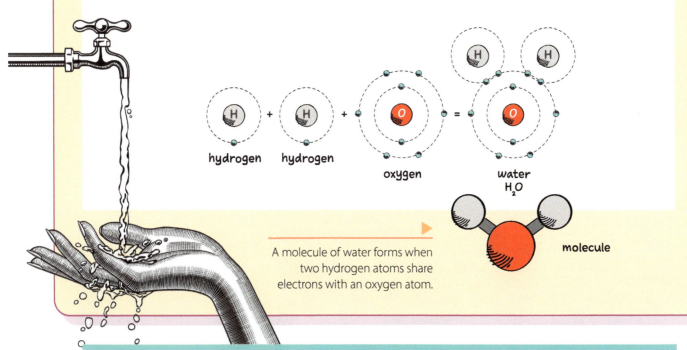

hydrogen   hydrogen   oxygen   water H₂O

molecule

A molecule of water forms when two hydrogen atoms share electrons with an oxygen atom.

2.1 SECTION REVIEW

1. What is the difference between a poison and a medicine? How can we use science to know the difference?

2. Write your own definition of the term *energy*.

3. Give an example of an energy transformation involving a living thing, and state the initial and final types of energy.

4. Create a table that compares physical and chemical changes, including definitions, observable changes, and examples.

5. List at least two differences between ionic and covalent compounds.

6. The illustration below shows the formation of a bond. What kind of compound is forming? How can you tell?

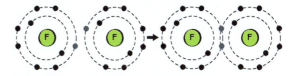

7. Create a chart that shows the relationships between matter, atoms, elements, protons, neutrons, electrons, compounds, and molecules.

How do chemical processes keep life going?

Questions

How does the nature of matter affect life?

What kinds of chemical reactions release or absorb energy?

What makes a chemical reaction speed up, slow down, or not even happen?

Terms

dissolving
acid
base
diffusion
reactant
product
catalyst
enzyme
inhibitor

Physical Changes

Break out your raincoat! It's a rainy day. A fine rain falls from a sky heavy with gray clouds onto a mountainside covered with balsam trees and mountain maples as you hike along a trail. The rain picks up. Water runs down the trail, sending insects scurrying, picking up soil particles, running over limestone rock faces, washing into streams that churn as they swell with rainfall. A rainy day demonstrates several physical changes. We've seen how a physical change alters matter in a way that doesn't change its identity. Let's look at some specific physical changes that are essential to keeping living things alive in this everyday example.

SOLUTIONS

When the water runs down the trail and picks up debris, it forms a mixture. We can see the bits of soil, pebbles, and twigs that get washed into the rainwater. When water runs over limestone deposits, some of the limestone will break up in the water so much that we can't see it. Limestone is the ionic compound calcium carbonate ($CaCO_3$), and it actually breaks up into calcium and carbonate ions mixed in with the water. This forms a solution. A solution is a uniform mixture. The process by which one substance (limestone) is broken up into smaller pieces, usually ions, by another substance (water) is called **dissolving** (illustrated below). The solution contains more water than limestone, so water is the *solvent*, and the limestone, the substance that is dissolved, is the *solute*. The amount of solute packed into a solvent is called the *concentration* of a solution.

  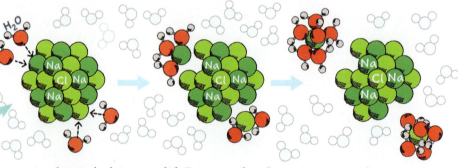

▲ Water molecules tear apart the salt crystal until it is completely dissolved in the solution.

A solute isn't always a solid. For example, when rainwater washes into churning mountain streams, some air dissolves in the water. This is very important for fish, salamanders, and other aquatic animals that rely on dissolved oxygen to breathe. In this case the oxygen is the solute.

Most rainfall is not pure water. It sometimes reacts with pollution, such as sulfuric acid or nitric acid in the air, to form acid rain. An **acid** is a substance that dissolves in water to form hydrogen ions (H^+). A **base** is a substance that dissolves in water to form hydroxide ions (OH^-) or other ions like chloride (Cl^-) or fluoride (F^-) that can accept hydrogen ions. When a strong acid and base react with each other, the hydrogen and hydroxide ions combine to form … water! Acids and bases affect soil and water in the environment as well as digestion and even blood in living things.

Living things live in and use solutions. The atmosphere is a solution of gases, and the oceans are liquid solutions. Later in this chapter we'll see more of water's amazing ability to dissolve and to affect life.

DIFFUSION

Scoop up a cup of pond water and put it under a microscope. What do you see? Lots of little particles (and maybe some critters!) being jostled around. Why are they moving around like that?

Particles are in constant random motion. We've already seen that temperature is related to the average speed of particles in a substance. This motion is caused by collisions between particles, like dust particles in a shaft of sunlight that move around but never seem to settle out of the air. This effect, seen only in particles, is called *Brownian motion*. Brownian motion affects the motion of especially small particles.

Matter, like energy, tends to spread out. If a skunk sprays a wildcat in a windless forest, the stinky smell spreads to fill the area. If an octopus squirts a moray eel with its ink in a quiet tide pool, the ink fills the surrounding water. Both examples demonstrate the dispersion of substances through Brownian motion alone, and not due to wind or water currents, in a process known as **diffusion**. The ability of matter to spread out is another physical change that living things rely on. Later in Unit 2 we'll see just how important diffusion is when we study the processes that make living things grow.

Looking under a microscope is all it takes to see Brownian motion in, well, motion!

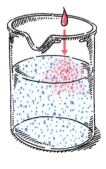 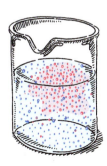 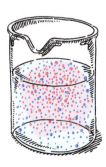 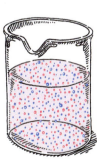

Food dye diffuses through water as a result of Brownian motion.

Chemical Changes

Fire! Smoke! Heat! There's a forest fire in the Mariposa Grove of Yosemite National Park. Smoke fills the air, and waves of heat undulate from the fire. Days later, the mighty sequoia trees in this grove with their fire-resistant bark still rise firm and unmoved, though slightly charred, from the scorched forest floor. It's also important to note that only in the heat of fire will the sequoia cones open to release the seeds of a future grove of mighty sequoias.

A forest fire is a good example of a chemical change because there are visible signs that hint at the unseen molecular changes taking place. We know from Section 2.1 that a chemical change transforms the identity of matter. Molecules in the leaves and wood of the underbrush change or *react* with oxygen in the air to leave charred remains. Some of the most obvious signs of a chemical change include the formation of a gas, a change in color, or the release of heat and light. These all happen in a forest fire, and that's a good thing for the few sequoia saplings that may sprout up days, weeks, or even months later.

CHEMICAL EQUATIONS

Knowing that a chemical change or reaction occurred is just the starting point. What did we begin with? These are the **reactants**. What did we end up with? These are the **products**. Chemists pack all the answers to these questions into an expression called a *chemical equation*. The chemical equation below shows what happens in a forest fire.

$$C_6H_{12}O_6 \, (aq) + 6O_2 \, (g) \longrightarrow 6H_2O \, (g) + 6CO_2 \, (g)$$
$$\text{glucose} \qquad \text{oxygen} \qquad \text{water} \qquad \text{carbon dioxide}$$

The first reactant, $C_6H_{12}O_6$, is glucose, which is the food of plants and is found dissolved in water in the stems of plants. Oxygen is the second reactant. In a forest fire the woody material, pitch, and other flammable substances combust, or burn, in the presence of oxygen to form water vapor and carbon dioxide gas—the products. This process releases lots of thermal and light energy. So you could say that energy is a product also.

$$C_6H_{12}O_6 \, (aq) + 6O_2 \, (g) \longrightarrow 6H_2O \, (g) + 6CO_2 \, (g) + \text{energy}$$

Chemical reactions that release energy are called *exothermic reactions*, where the energy of the products is lower than the energy of the reactants because some of the chemical energy stored in the glucose is not stored in the water vapor and carbon dioxide. Instead, it escapes as heat and light.

But the opposite reaction also happens in a forest. This reaction doesn't happen during a forest fire; rather, it happens when plants make glucose during photosynthesis, and so the reverse reaction takes place.

$$6H_2O \, (g) + 6CO_2 \, (g) + \text{energy} \longrightarrow C_6H_{12}O_6 \, (aq) + 6O_2 \, (g)$$

In this case plants need energy to form glucose, and they get this energy from the sun. This process shows how a plant stores energy and makes its own food and is an example of an *endothermic reaction*, a chemical change in which energy is absorbed and the reactants have less energy than the products (see below).

We can also see the energy necessary to get a reaction started. Both exothermic and endothermic reactions need energy to get jumpstarted. This energy is called the *activation energy* (E_a).

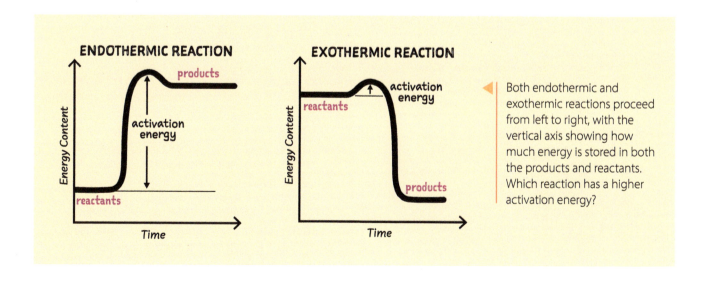

Both endothermic and exothermic reactions proceed from left to right, with the vertical axis showing how much energy is stored in both the products and reactants. Which reaction has a higher activation energy?

CATALYSTS AND ENZYMES

The symbol E_a is such a little symbol, but what it represents makes a big difference for living things. Some chemical reactions necessary to keep something alive and growing have huge activation energies. There is no possible way for these reactions to happen without an immense energy investment. That's good for harmful reactions that we don't want to happen, like spontaneous combustion! But what about reactions that living things need to happen? There is a way to meet the energy requirement. A **catalyst** is a substance that provides an alternate route from reactants to products that doesn't require as much energy to jumpstart. This means that the reaction has a lower activation energy. It speeds up without consuming all the energy in the process. And, as an added bonus, it doesn't affect how much energy is stored or released during the reaction. A catalyst is the key that unlocks a reaction so that it can take place. Where do living things get catalysts? One special class of catalysts, called **enzymes**, naturally occurs in living things. They are large molecules, 10–100 times larger than any molecules scientists have been able to make in the laboratory. And they are super-customized to make processes that are necessary to life happen like clockwork.

Enzymes are responsible for most vital chemical reactions. For example, an enzyme called *RuBisCO* helps plants create glucose during the process of photosynthesis. Scientists have found thousands of enzymes, and each one is specific to a chemical reaction occurring within a living organism. The presence of an enzyme typically makes a reaction happen millions of times faster! Without enzymes these processes would occur too slowly to be of any use to the organism. Enzymes also help living things by lowering activation energy and allowing a chemical reaction to proceed at a lower temperature. Without an enzyme, high temperatures necessary to reach the activation energy would cook the cells and injure or damage tissue in the organism.

So how do enzymes work? Typically, the enzyme has a customized spot where molecules dock in a way similar to the way that a smartphone connects with a charger. This is called the *active site*. The molecule that docks in the perfectly shaped active site is called the *substrate*. The enzyme holds the substrate while it undergoes a chemical reaction, then releases it when the reaction is done, without changing the enzyme.

RuBisCO, which is short for ribulose-1,5-bisphosphate carboxylase/oxygenase, is an enzyme that helps plants convert carbon dioxide in the atmosphere into glucose. Scientists think that it is the most common enzyme on the planet. But that doesn't mean that it's not special—life on Earth depends on this one enzyme!

enzyme

substrate

active site

Some enzymes, such as those that make food spoil, do things that we would rather not happen! Can we counter unwanted effects? An **inhibitor** is a substance used to reduce an enzyme's activity. Inhibitors work by bonding to the enzyme, making it difficult or even impossible for the substrate to bind to the enzyme's active site. Neurotoxins, like atropine in the poison found in nightshade plants, are valuable inhibitors. They are some of the most important medications that doctors use in emergency health care today because they prevent cardiac arrest. A neurotoxin works by blocking an enzyme that helps transmit signals to the heart that trigger cardiac arrest.

Chemically speaking, it's hard to distinguish a medicine from a poison. Both can be catalysts and inhibitors. But the difference in effect between the two can be obvious—whatever helps people is medicine, and whatever hurts them is poison. It takes both a good knowledge of chemistry and a biblical worldview to use medicines the right way. When we take chemicals from plants and animals that are naturally toxic and use them in ways that help people, we are pressing God's world to function in the way that it should.

2.2 SECTION REVIEW

1. Create a T-Chart representing the differences between physical and chemical changes.

2. Identify each of the following as a physical or chemical change.

   a. dew forming on a fern

   b. a dandelion growing

   c. rock being ground into powder by a moving glacier

   d. minerals dissolving into a local stream

   e. acid rain killing a balsam over time

3. Imagine that diffusion didn't happen, that is, that matter never dispersed on its own. How would this affect life? How does diffusion illustrate God's care for life?

4. How does dissolving rely on diffusion?

5. Draw a graph similar to the ones on page 35 for each of the reactions below. Identify the activation energy, reactants, and products. Label each graph as exothermic or endothermic.

   a. H_2O (g) + CO_2 (g) + energy $\longrightarrow$ H_2CO_3 (aq) (the reaction that removes carbon dioxide from blood by forming carbonic acid)

   b. $C_6H_{12}O_6$ (aq) + $6O_2$ (g) $\longrightarrow$ $6H_2O$ (g) + $6CO_2$ + energy (the reaction that turns stored glucose into heat)

6. Consider the image of RuBisCO on page 36. Life on Earth depends on the effectiveness of RuBisCO. It doesn't do its job very fast or very well, making mistakes about 27% of the time, so plants need a lot of it. How would you respond to evolutionists who say that if there were a Creator, He would have made such a key enzyme more efficient?

7. Compare substrates and inhibitors with respect to docking with enzymes.

8. Create an analogy that relates enzymes, substrates, and inhibitors.

9. How can we use chemistry to help living things, especially people, who bear God's image?

2.3 BIOCHEMISTRY

Water

Why can you survive longer without food than without water?

What can you live the longest without? In 2012 a Swedish man found buried under snow in his car went without food for about two months. But that's only because he had an unlimited supply of just one thing that living things need—water. Water is one of those things that we take for granted unless we don't have it. Since our bodies are 65% water, in hot conditions we can become dehydrated after just an hour without it. At most, we can live only three to five days. Not only people, but all life that we know about needs water. Without it we're, well, sunk.

What makes water so important to life? Recall that water is a molecule that has two hydrogens bonded to a central oxygen. This molecule is shaped like a boomerang. The oxygen pulls electrons closer to it than each hydrogen does, so the electron cloud around the molecule is lopsided, though the whole molecule has an overall neutral charge. It forms areas of charge like the poles on a magnet. The electron cloud around the molecule is shifted toward the oxygen, forming an area of negative charge around the oxygen and areas of positive charge around the hydrogen atoms. Because these areas of charge are asymmetrical, water is a **polar molecule**.

The polarity of water makes it a superhero when it comes to dissolving things. Recall that rainwater dissolves oxygen, calcium carbonate, sulfuric acid, and nitric acid in the atmosphere. The cells in our bodies that keep life going rely on water to move nutrients around. Without it, processes that sustain life begin to shut down because cells don't have the materials they need to keep working.

Water's polar structure and boomerang shape also allow it to interact with other water molecules. The hydrogens on one water molecule attract the oxygen on another water molecule. Hydrogens on a water molecule can also attract fluorine and nitrogen on other molecules. We call these intermolecular attractions *hydrogen bonding*. Hydrogen bonding is pretty important in biology. In Chapter 6 we'll learn how hydrogen bonding helps DNA stay in its coiled shape and produces the signature kinks and curls of proteins. The shape of an enzyme is customized to help biochemical reactions, and this shape is dramatically affected by hydrogen bonding.

Because of hydrogen bonding, water molecules attract each other, creating surface tension that can support the weight of a water strider or a basilisk lizard sprinting across its surface. The attraction of water molecules to each other is called **cohesion**. Water's polarity also allows it to attract other substances besides water in a process called **adhesion**, one of the forces that helps water move through conduits in stems and leaves. As long as another substance is polar, water molecules will attract and stick to that substance.

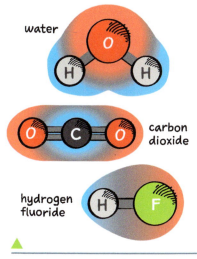

These diagrams show the partial surface charges of covalent molecules. Red areas of the electron cloud have a negative pole, while blue ones have a positive pole. Which molecules are polar?

case study

SHIELD OF ICE

It is wintertime in the woods of Maine. A blanket of snow covers the ice on a pond while more snow flutters down. Wait—there's a mound in the middle of the pond—it must be a beaver lodge.

Beavers spend all winter without breathing the air *outside* their lodges since they have underwater front doors! They fill their time swimming under the ice and lounging in their air-filled lodges. In the still winter air you can noiselessly walk up to the mound over freshly fallen snow and hear the sounds of beavers gnawing on bark from their winter stash of branches that they have stored and retrieved from the mud at the bottom of the pond. On the perimeter of the pond you have a rare sighting of a lynx! You realize with relief that this predator can't get to the beavers, who are protected by a shield of ice that protects their lodge and the pond.

Why does ice float like this? If water behaved like most liquids, the ice would be denser, making it sink to the bottom of the pond. The hydrogen bonds in water make ice less dense than water,

causing it to float. Because of the structure of the bonds in water, it takes a long time for water to undergo a temperature change. It will take many warm days to thaw the beaver's winter shield because so much energy is required for water to change its state. As a result, beavers enjoy a relaxing, long time under the ice!

Questions to Consider

1. What are the benefits of underwater entrances to beaver lodges?

2. Why does the layer of ice that protects a beaver's lodge float?

3. What property of water allows the ice to float and not sink?

Carbon Chemistry

There aren't too many different elements found in living things compared with all the elements we know about. But there is one element that is more significant to life than all the others. Carbon is the element on center stage when it comes to organic compounds, the ones most commonly associated with life. **Organic compounds** are molecules that contain carbon covalently bonded to other elements, typically oxygen, nitrogen, hydrogen, and sometimes certain nonmetals and metals.

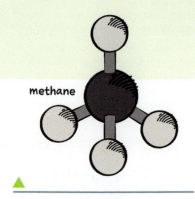

carbon dioxide

methane

Why is carbon so important to life? Carbon usually forms four bonds because neutral carbon has four valence electrons and needs four more to obtain eight electrons, the magic number of stability when it comes to valence electrons. Carbon can bond with itself. It forms a very stable bond with hydrogen and can form single, double, and triple bonds with other elements. It can form long chains or rings. Carbon has an amazing chemical flexibility—scientists know about 50 million carbon compounds! Organic compounds serve many purposes in living things. They hold plants up to keep them from falling over, serving a structural purpose. They act as enzymes, as we've already seen. They can store energy in living things—one reason why people can go without food for so long.

▲

Notice the 3D arrangement of methane and its nonpolar structure.

Carbon Compounds

The simplest carbon compound is methane, CH_4, a compound that microorganisms in the guts of termites often produce as a byproduct. But from there organic compounds can get very big very quickly!

CARBOHYDRATES

Carbohydrates are organic compounds made of carbon, hydrogen, and oxygen. Collectively, carbohydrates refer to sugars, starches, and cellulose. These molecules are the most abundant biological compounds, making up a huge part of our diet, and are the fuel of plants. The exoskeletons of all the insects, crabs, and scorpions on this planet contribute even more to the presence of carbohydrates in living things. Simple carbohydrates are called **sugars**. Glucose is one of the most important sugars for living things, containing six carbons, as we saw earlier in this section. People and animals also use other sugars. Probably the sugar most familiar to you is sucrose, commonly known as table sugar.

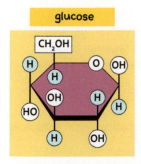

glucose

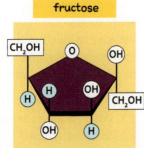

fructose

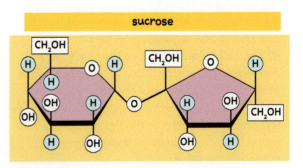

sucrose

▲

Compare the structures of glucose, fructose, and sucrose. Many foods that we eat contain all three of these sugars.

The structures of glucose and sucrose may seem intimidating, but many carbohydrates are even more complex. Carbohydrates can form long chains or *polymers* of six-carbon sugars that coil into a spring shape. Corn starch used for cooking and cellulose in plants are two examples of long, complex carbohydrates.

LIPIDS

Although water mixes with many different substances, it doesn't mix with lipids. This is so because **lipids** are nonpolar molecules. They have two or three long chains that come off a three-carbon base and are made mostly of carbon and hydrogen. It is this characteristic that makes them insoluble in water.

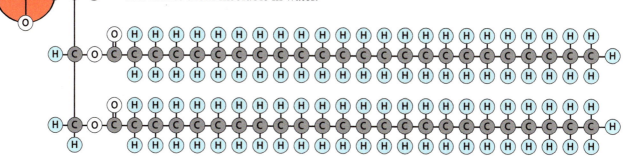

Lipids are incredibly important to biology because living things often use them to store energy. Lipids include substances like fats and oils. When lipids contain a double or triple bond between the carbons in their long chains, they don't have the maximum number of hydrogens, making them *unsaturated fats*. *Saturated fats* have only single bonds. Unsaturated fats like olive oil offer better nutrition than saturated fats like butter.

PROTEINS

Proteins are some of the most important substances in living things. Muscles, hair, blood, and skin all rely on protein chemistry. But proteins are actually polymers of smaller organic compounds. These organic compounds, called amino acids, are much simpler than lipids, and are even simpler than many carbohydrates. **Amino acids** are molecules that have a central carbon bonded to four groups—an amine group (NH_2), a carboxylic acid group (COOH), a hydrogen, and an R-group. The hydrogen in the carboxylic acid group makes them acids. The R-group on the central carbon represents a slot that can be filled by different atoms or groups of atoms. This empty slot allows one amino acid to be different from another. You might have heard of some of them, like tryptophan in turkey meat—the chemical that is often unfairly blamed for making us sleepy after a large Thanksgiving dinner!

amino acid

amine group

carboxylic acid group

R-group

Polypeptides form when amino acids combine to release water through the formation of a peptide bond.

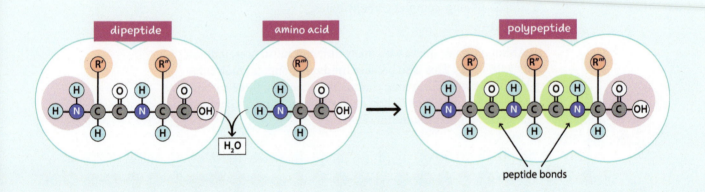

The amine group end of an amino acid can react with the carboxylic acid group of another to release water (above). A polypeptide is a polymer of amino acids linked together this way. Polypeptides can form spirals or helixes held in place by hydrogen bonds, or they can form sheets that look like folded pieces of paper. These structures can combine to form larger and larger structures until we get a customized protein that can function in its own unique way, like the hemoglobin protein that carries oxygen in our blood.

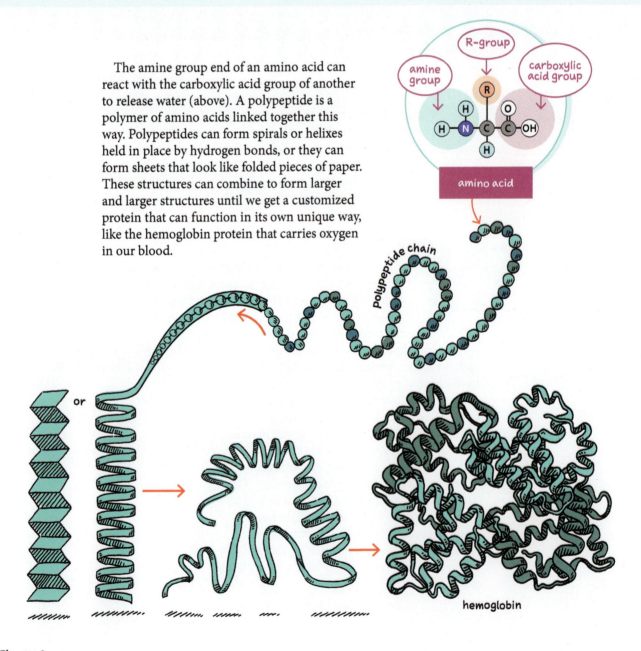

NUCLEOTIDES

Another molecule that is a key molecular building block of life is a nucleotide. **Nucleotides** are made of three parts: a base that has nitrogen in it, a sugar, and a phosphate group (PO_4). The diagram below shows the relationship between these different parts. Notice the word "base" in the description: the base part of a nucleotide can accept hydrogen just like a hydroxide ion can.

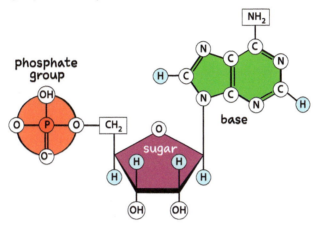

Nucleotides hook together as polymers to form nucleic acids. In Chapter 6 we'll discuss how they form a double strand that is held in a coil by hydrogen bonds, the signature structure for deoxyribonucleic acid (DNA) and ribonucleic acid (RNA). DNA and RNA contain instructions for living things to make proteins. Because proteins are so essential to biological processes, nucleic acids are like the "blueprints" that make each ant, each butterfly, and each person unique. We will learn more about how this happens when we zoom in on the cell in Unit 2.

2.3 SECTION REVIEW

1. Do you think that similarities found in life, like the fact that all life forms that we know about need water, prove that evolution happened? Why or why not?

2. Why is water so important to life? Include examples in your answer.

3. Define *organic compound* in your own words.

4. What is the difference between an organic compound and other kinds of compounds?

5. After comparing glucose and sucrose (see page 40), what do you notice?

6. Give one example of a carbohydrate, protein, and lipid not mentioned in your textbook.

7. What do carbohydrates, lipids, proteins, and nucleic acids have in common? What makes them different?

Starch and Fat Test

We have just learned about the properties and structures of organic compounds, but can we identify them in the foods we eat? Identifying whether foods have fats and starches can actually be quite an easy and fun task. Let's take a look at different foods and try to see whether we can correctly identify them as fats or starches.

How can we spot fats and starches in the food we eat?

Materials
4–6 fat food samples • 4–6 starch food samples • water • large brown paper bag • small cups • permanent marker • iodine

PROCEDURE

Fat Test

A Cut the brown paper bag so that you can spread it out and lay it on a flat, covered surface.

B Using your permanent marker, divide the bag into sections according to the number of food items that you will be testing. Label each section with the name of the food sample being tested in that section. Be sure to include water as your control sample.

C Place each fat food sample into its designated section on the paper. Some of the items will need to be spread onto the paper, while others can simply be placed into their sections. The samples will need to sit for about 15 minutes to allow them to dry. Use this time to make predictions as to whether these items will contain fat.

D After the time is up, remove each food sample. Check whether any oily residue is left behind. An oily residue indicates that the food sample contains fat. Compare your predictions to the results from each trial.

1. Were your predictions correct? What results surprised you?

Starch Test

E Place the small cups on a flat, covered surface according to the number of food samples that you will be using. Label each cup with the name of the sample being tested and place the samples in their respective cups. Be sure to include water in one cup as your control sample.

F Make predictions as to whether the samples will contain starch.

G Put three drops of iodine on each sample. If the sample contains starch, the iodine will turn blue-black. Be sure to record the results.

2. Were your predictions correct? What results surprised you?

GOING FURTHER

3. The substance amylose, present in starch, causes the color of iodine to change from yellow-brown to blue-black. What do you think causes this?

2 CHAPTER REVIEW
Chapter Summary

2.1 MATTER, ENERGY, AND LIFE

- Matter, which makes up everything in the universe, is made of different atoms. Elements are different types of atoms.

- Atoms have a nucleus made of protons and neutrons and are surrounded by negatively charged electrons.

- The different kinds of energy found in living things include mechanical, acoustic, light, thermal, and chemical energy.

- Changes to a substance that do not affect its identity are called physical changes. Changes that transform a substance into a different substance are called chemical changes.

- Ionic compounds form when oppositely charged ions attract one another. Covalent compounds form when atoms share their valence electrons.

Terms
matter · atom · element · energy · temperature · physical change · chemical change · bond · compound · molecule

2.2 THE CHEMICAL PROCESSES OF LIFE

- The process of dissolving occurs when a solute is broken up into smaller pieces by a solvent.

- Diffusion occurs as substances exhibit Brownian motion and disperse throughout a solution.

- Exothermic reactions release energy, resulting in products with lower energy values. Endothermic reactions absorb energy, resulting in products with greater energy values.

- Catalysts are substances that provide alternate routes in reactions and lower activation energy. When a catalyst exhibits undesirable effects, inhibitors may be used to block it from binding.

- Chemistry can be used to help people through the creation of medicines and chemicals for food preservation.

Terms
dissolving · acid · base · diffusion · reactant · product · catalyst · enzyme · inhibitor

2.3 BIOCHEMISTRY

- Water is a polar molecule that is essential for life. Because of its structure, a water molecule exhibits cohesion and adhesion.

- Organic compounds contain carbon covalently bonded to other elements and serve many purposes in living things.

- Carbohydrates are the most abundant organic compounds and are made of carbon, hydrogen, and oxygen.

- Lipids are nonpolar molecules that are insoluble in water and are made of two or three long carbon chains branching off a three-carbon base.

- Proteins are polymers of amino acids, each having a central carbon bonded to an amine group, a carboxylic acid group, a hydrogen, and an R-group.

- Nucleotides are made of a base containing nitrogen, a sugar, and a phosphate group and combine to form nucleic acids.

Terms
polar molecule · cohesion · adhesion · organic compound · carbohydrate · sugar · lipid · amino acid · nucleotide

CHAPTER REVIEW
Chapter Review Questions

RECALLING FACTS

1. What is chemistry?

2. What is energy?

3. What do we call the tendency of energy to disperse?

4. List the five types of energy mentioned in the chapter.

5. In your own words, identify and state the basic law regarding the amount of matter and energy in the universe.

6. What makes up each of the following substances?

    a. elements

    b. covalent compounds

    c. ionic compounds

7. Give an example of a physical or chemical change not given in your textbook.

8. What is the difference between a solution and other mixtures?

9. What do we call the tendency of matter to spread out? What makes this happen?

10. What is usually needed for a chemical reaction to occur in living things?

11. What are two properties of water that make it unique?

UNDERSTANDING CONCEPTS

12. Identify the primary kind of energy released in each of the examples below.

    a. an anglerfish glowing in the ocean depths

    b. an emperor penguin spitting up crop milk to feed its baby

    c. a migrating butterfly

    d. a chirping cricket

    e. a mother kangaroo warming the joey in her pouch

13. Write a brief paragraph that describes the relationship between matter, energy, and life.

14. Is it possible for living things to add new energy to the universe, such as when plants make their own food? Explain.

15. Create a T-Chart that compares ionic and covalent compounds. Include the kinds of elements that make up these compounds, the smallest parts of these compounds, and what makes up the bonds in these compounds.

16. Your stomach contains hydrochloric acid (HCl). What properties does HCl have that are relevant to the stomach environment?

17. Ocean water contains sodium chloride (NaCl), potassium chloride (KCl), calcium chloride ($CaCl_2$), and magnesium sulfate ($MgSO_4$) salts. Identify the solvent(s) and solute(s) in ocean water.

Use the following chemical equation to answer Questions 18–19.

$$C_{12}H_{22}O_{11} + H_2O \xrightarrow{\text{sucrase}} C_6H_{12}O_6 + C_6H_{12}O_6$$
sucrose   glucose   fructose

18. Name the reactants and products in the equation.

19. Notice the named chemical over the arrow—sucrase. This chemical helps sucrose break down into glucose and fructose, a process that takes a long time otherwise. What kind of chemical compound is sucrase? What does it do to the activation energy?

20. The chemical L-arabinose blocks the work of sucrase. It is used in flavorings and artificial sweeteners to prevent the digestion of sugar in people who are trying to lose weight. What kind of chemical compound is L-arabinose?

21. Scientists had to work to create L-arabinose and discover uses for it. How does this scientific work fulfill God's commands?

22. What property of water makes it so effective in dissolving substances? What else does this property help water do?

23. How are organic compounds different from other kinds of compounds?

24. How many bonds can carbon form? What types of bonds does carbon form?

25. Link the following three amino acids into a chain or polypeptide, using peptide bonds.

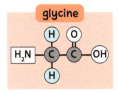

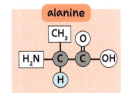

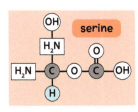

CRITICAL THINKING

26. How are matter and energy related?

27. You are witnessing a salt crystal dissolving on the molecular level. Write a paragraph describing what you see as the crystal dissolves in water.

Use the table below to answer Questions 28–31.

CHEMICAL	DENSITY (g/mL)	COLOR	MELTING POINT (°C)
Water (H_2O)	1.0	clear	0
Salt Water	1.02	clear	0
Ice	0.9	clear	0
Salt (NaCl)	1.6	white	801
Sodium (Na)	0.9	silver	97.7
Chlorine (Cl)	0.003	light green	−10.5

28. Dissolving salt in water is a physical change. How can you prove this from the data in the table?

29. Melting ice to form liquid water is a physical change. How can you prove this from the data in the table?

30. Forming sodium chloride from sodium and chlorine is a chemical change. How can you prove this from the data in the table?

31. If dissolving salt into water to form salt water is a physical change, why is there such a great difference between the data for salt and that of salt water?

32. Study the examples of exothermic and endothermic reactions on page 35. What do you notice? What can you conclude about chemical reactions from these examples?

33. Why do you think living things need water?

34. Do some research on a poison that could be used as medicine, such as yellow scorpion venom, which has the potential to fight cancer. What process must researchers follow to ensure that this drug won't cause more harm than good?

35. Do some research on the Hippocratic Oath, a promise that doctors have historically taken to commit to using their knowledge and skill in medicine for helpful purposes, not harmful ones. Read this oath and evaluate it from a Christian worldview.

36. Choose a career involving chemistry that interests you. Look online for information to write a one-paragraph summary about that career.

Use the ethics box on pages 48–49 to answer Questions 37–40.

37. What authority should we look to when using the biblical triad to examine an issue?

38. Where can we go to get more information about an ethical issue?

39. What should we consider about the authors of those sources?

40. What biblical principles that apply to abortion also apply to other issues as well? List three examples.

USING THE BIBLICAL TRIAD

ETHICS

ABORTION

Recall the abortion story presented in Chapter 1. Marina was faced with a difficult decision: continue her pregnancy or abort the unexpected baby. Her doctor sees the baby as no more than just a bunch of cells and tissue. Is it right or wrong to abort a pregnancy?

ANALYZING ABORTION

In Chapter 1 we introduced a process for making ethical decisions from a biblical perspective. For this chapter we will consider using the biblical triad to analyze the issue of abortion. To help you apply this process, we have provided an outline of ethics questions for you to use in evaluating how to make ethical decisions in biology using the biblical triad.

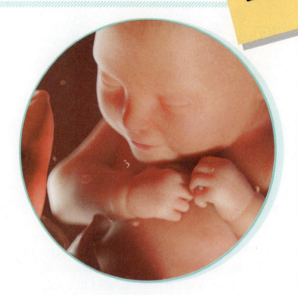

MODELING USING THE BIBLICAL TRIAD
Strategy

The strategy presented here is helpful because it provides a framework that can be used when faced with making ethical decisions. These questions are based on the ethics triad taught in Chapter 1 and are organized into five steps.

1. What information can I get about this issue?

To formulate an appropriate and informed position, we need data and information, so we will need to do some research on the topic before we can evaluate how to ethically approach the issue.

2. What does the Bible say about this issue?

Go directly to God's Word and study the Scripture's teaching that directly applies to the issue presented. Consider the biblical principles of the image of God, the Creation Mandate, and God's whole truth.

3. What are the acceptable and unacceptable options?

Consider the possible options and their outcomes. What will the outcomes for each of these options be? Some options will lead to acceptable outcomes and some to unacceptable outcomes. Compare the outcomes for each option to the biblical teaching and principles that we considered in Step 2. An important aspect of making ethical decisions involves rejecting any options that are inconsistent with biblical principles and outcomes.

4. What are the motivations of the acceptable options?

Once the best possible option is determined, consider how your ultimate decision will affect your growth and maturity as well as that of others. Will this decision promote growth? Is it motivated by faith in God, hope in God's promises, and love for God and others?

5. What action should I take?

Once we have a clear understanding of our position according to the principles of the ethics triad and can justify the action, we need to plan a course of action or form an opinion.

The Issue: Abortion

1. What information can I get about this issue?

Abortion is a medical procedure that terminates a pregnancy through the use of medicine or surgery. Those who approve of a woman being able to choose what is done with her body and who support abortion historically have been called *pro-choice*. Those who argue that life begins at conception and should be allowed to continue developing oppose abortion and are known as being *pro-life*.

2. What does the Bible say about this issue?

The Scripture clearly teaches that a baby inside a woman has the same value as life outside the womb (Exod. 21:22–24). Human life is uniquely important because humans are made in the image of God (Gen. 1:26). God knows us before birth, so our personhood is established before our actual existence (Jer. 1:5; Eph. 1:4). God established the death penalty for murder (Gen. 9:6; Exod. 20:13; 21:12, 14). When we love and protect people, we fulfill the greatest commandment to love God (Matt. 22:37–38).

3. What are the outcomes for each of Marina's possible options?

Marina could choose to continue the pregnancy. Her choices would be to keep the baby or place him for adoption. She could abort the baby. This choice would release Marina from what she thinks are difficult choices ahead of her but is unacceptable because, according to the Bible's teaching, it would intentionally end a human life. It would also have the added consequence of guilt for Marina. This option prioritizes the mother's autonomy at the expense of the baby's life.

4. What are the motivations of the acceptable options?

If we value life the way that God does, we will choose to continue the pregnancy regardless of the outcome. The mother should be motivated to persevere through hardship in order to protect the life of her baby. If we view abortion as the better option, we are motivated by the importance of human choice over the importance of human life. This choice rejects God's command prohibiting murder and His promises to meet our needs.

5. What action should I recommend?

In this case the choice to continue the pregnancy would be in line with Scripture. God values life, and so should we. If a mother is unable to keep the baby, she may choose to place the baby for adoption.

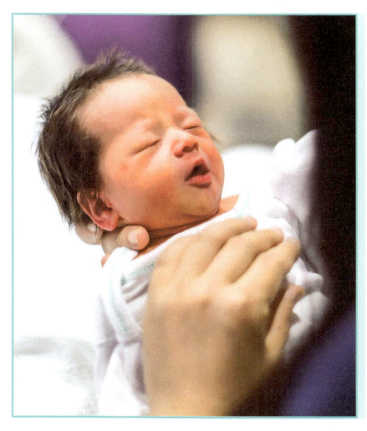

3 ECOLOGY

Extreme Life

We live on a planet that throbs with life to the extreme. Sometimes life takes hold in the most unexpected places—spots on Earth with no sunlight or air, some with oppressively hot or bitterly cold temperatures. Microbes have been found in the deepest ocean trenches, in acid lakes, and even buried deep under Antarctic ice.

Evolutionists look at these forms of life, called *extremophiles*, and wonder whether life like this exists outside of Earth. In fact, in 2024 NASA plans to send a probe to Europa, a moon of Jupiter with an icy surface. They suspect that this moon has liquid water beneath its icy surface and a tectonic structure like Earth's. But what will they find? Will it change the way we think about life on our planet?

3.1 OUR LIVING PLANET
The Biosphere

The search for life originating outside our living planet on places like the moon, Mars, and Europa is called *astrobiology*. It is a hot, new field that is a moving force in NASA's space program. The problem for astrobiologists is that they haven't found life that comes from anywhere other than Earth despite spending millions of dollars developing the finest technology. None at all! So why look for life in space when there is so much around us?

Scientists are looking to extend their beliefs about life on Earth to space. If life and everything in the universe is a product of chance and a big bang, why wouldn't we see life elsewhere in the universe? In their view of biology, evolution is life's designer. The late Carl Sagan, former professor and astronomer at Cornell University, once said, "The universe is a pretty big place. If it's just us, seems like an awful waste of space." Life should be easy to find wherever we look, but that's not what we observe.

God designed life, and He made Earth for life. We see this in the wording and unfolding of the Creation account. The heavens, the earth, the waters, and the stars, sun, and moon are all mentioned in direct connection to the living things that God created. The stars are mentioned almost as an afterthought because life on this planet is the centerpiece of God's creation, though all of creation declares His glory (Ps. 19:1; Rom. 11:36). Earth is the shelter, the haven, the home for God's precious, living creation in the hostile environment of space. When we look around our living planet, we can see evidence of a God who cares and who provides for the living things He loves. The realm of life on Earth, called the **biosphere**, extends from a few kilometers into the atmosphere to a few kilometers into Earth's crust. This thin shell around us is the only place we know of where life can occur.

What makes Earth a good place for life?

Questions

How do ecologists categorize the living and nonliving parts of Earth?

How are living things affected by their environment?

Terms

biosphere
ecology
biodiversity
biome
ecosystem
habitat
niche
abiotic factor
biotic factor
population

Ecology is the study of inter-relationships between organisms and their relation to their physical surroundings. Organisms both affect their environment and are affected by it. Organisms also affect each other. Ecologists explore these relationships as they try to understand the interactions of life and environment. We can use this understanding in ways that help us wisely care for God's good Earth.

Ecology keeps the big picture in mind, rather than focusing on any one type of plant or animal. For example, in the 1800s American bison were hunted to the brink of extinction. Today, herds of bison enjoy protection in state and national parks. Since these herds have grown, ecologists have noticed that the mixed-grass prairies they inhabit enjoy many benefits. Bison graze only on grasses, increasing the possibility of other plants mixing with the grasses, thus providing a healthy balance of nutrients and species. Bison also return nutrients to the soil and shape their environment in ways that benefit prairie dog colonies. These prairie dogs provide food for ferrets, foxes, hawks, and eagles. Abandoned prairie dog homes provide an environment for snakes, lizards, and toads. So we can see that one type of animal can make a big difference in the well-being of many different living things. An area supporting a wide variety of organisms indicates that this corner of the biosphere is ecologically healthy. Ecologists use the term **biodiversity** to refer to the number of species (species richness) in an area. Biodiversity also considers the species evenness, or the proportionality of the populations of species, relative to each other. This variety is related to biological information present in the cells of living things. Genetic variety gives an area of the biosphere a kind of built-in flexibility to deal with changes in the environment such as pests and disease.

Home for Life

When you come home after a busy day out, what makes it feel like home to you? Your family is there. A refrigerator filled with food you like is there. Your bed is there, along with your pets, video games, comfy chair, and all the things you need. Your home is probably in a neighborhood, which is in a city, which is in a state, which is in a country, which is on a continent. Your address has some of this information in it.

Other living things also have homes that house family members, food, and the things they need to thrive. Ecologists give their homes something like an address, dividing up the biosphere into large areas that have a fairly consistent environment and are home to a set of organisms suited to it. These areas are called **biomes**. Biomes can cover large portions of Earth's surface, perhaps up to a continent. For example, tundra is found on land close to the Arctic Circle. Most of Earth's tundra is in the Northern Hemisphere since there is very little land close to the Antarctic Circle in the Southern Hemisphere that will support the tundra biome. We'll learn more about the biomes of the world in the next section.

Within a biome, ecologists investigate **ecosystems**, a limited area smaller than a biome, in which living and nonliving things interact. For example, the Atlantic puffin likes to hang out on rocky cliffs near the Atlantic Ocean. It likes islands, and Iceland is its favorite place—about 60% of all Atlantic puffins live there. The rocky cliffs are the ideal place since they are inaccessible to predators, are close to the puffin's food supply, and are easy places to take off and land. The rocky shores of Iceland are an example of a **habitat**, or a smaller part of an ecosystem that an organism prefers. Puffins prefer the coasts of islands to the inland. They dig burrows under the grasses of these cliffs to line with feathers, twigs, and grasses. The way that puffins live in their habitat is their ecological **niche**. Their niche includes their indirect effect on the habitat, including the erosion of cliffs due to their burrows. Their niche also involves the direct changes that puffins make to their habitat, such as the local populations of fish they eat and the eels they feed their chicks. We might say that the puffin's habitat is *where* it lives, and the puffin's niche is *how* it lives.

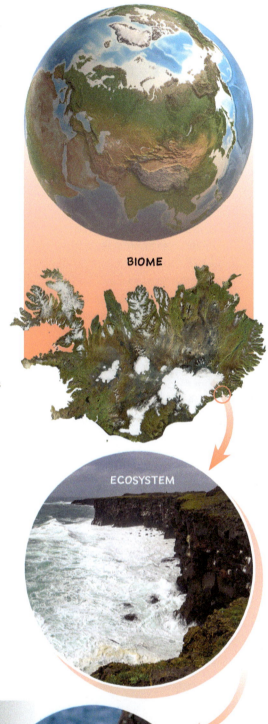

BIOME

ECOSYSTEM

HABITAT

NICHE

case study

THE GREAT BARRIER REEF

Off the coast of Australia lies the Great Barrier Reef, the largest coral reef in the world. Corals are alive, and they attract other living creatures such as sea turtles, clown fish, sea anemones, and crown-of-thorns starfishes. Striped surgeonfish tend to stay near the reef because they have slender, elongated bodies that can slip between cracks in the reef when pursued by predators such as barracuda. Striped surgeonfish are bottom feeders, scooping algae off the ocean floor. But they feed on photosynthetic organisms, so they live only in shallow areas of the ocean. They are territorial, vigorously defending their grazing grounds from other striped surgeonfish and other algae eaters such as damselfish. Male striped surgeonfish are usually surrounded by a harem of females. The striped surgeonfish is important to its ecosystem because it keeps algae and plankton from taking over the area and returns nutrients to the water to nourish other living things.

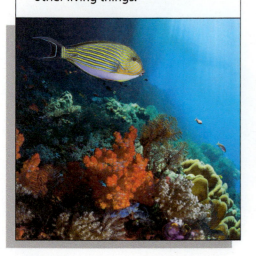

So why do so many puffins make Iceland their home? A variety of factors make it an ideal location for them. Some aspects of their ecosystem—the water, the wind, the rocky cliffs, the cooler temperatures—are elements of the physical environment. These nonliving aspects of an ecosystem are called **abiotic factors**. Other factors—the fish they eat, other puffins, and the seals, foxes, and gulls that prey on them—are living parts of their ecosystem, or **biotic factors**. This would include parasites, like the fleas and ticks that sometimes plague puffins. As part of God's design for our planet, all life—its balance and sustenance—depends on the interaction between abiotic and biotic factors. For instance, abundant nutrients and plenty of summer sunshine in the Northern Hemisphere (abiotic factors) produce the masses of phytoplankton (biotic factor) that are the first link in the arctic food chain that includes puffins. The living things that inhabit the same ecosystem are called a *community*, like the neighborhood that you live in.

abiotic

biotic

Puffins spend most of the time during the year bobbing around on the ocean. But they're not loners. They nest in colonies and often lay eggs all at the same time, and the number of breeding couples in a colony changes every year. They rely on each other to defend the colony from predators. The more the merrier! This is an example of a **population**, a group of organisms of the same species interacting in the same area. Ecologists monitor the populations of animals living in an ecosystem to learn more about the ecosystem and the relationships between biotic and abiotic factors in that environment.

Within a population an ecologist can look at the attributes of the population, such as how many live in a certain area, the ratio of males to females, and so on. This could involve studying individual puffins or puffin couples. The goal would be to learn about the things that individual puffins need, the way they raise their chicks, the way they behave in relation to the colony, and the way individual puffins or groups of puffins interact within the colony. This is an example of how ecologists study individual organisms to learn more about populations, communities, and ecosystems to better care for the place we all call home.

MINI LAB

Who Is in the Community?

Scientists often must make estimates of the size and composition of a community of organisms, such as fish on a coral reef or bromeliads in a rainforest. Many times, only a few of the community members are readily visible. Scientists must figure out ways to accurately sample the entire community, not just those organisms that are easily seen.

How can I determine the makeup of an ecological community?

Materials
opaque bowl filled with colored marbles • sampling tools

PROCEDURE

A Your bowl of marbles represents a community of organisms. Each color represents a different kind of organism within the community. Think of a procedure that you could use to estimate the total number of each color of marble in the bowl, the total number of marbles, and the percentage of the total represented by each color. Write out your procedure on a separate sheet of paper.

B Carry out your procedure. Remember to record your data and show any calculations.

C Record your estimate of the total number of "species" in your community, the total number of individuals of each species, and the percentage of the total population comprised of each species.

ANALYSIS

1. Evaluate your procedure, including its strengths and weaknesses. How could your procedure be improved?

2. Note that you were not asked to count the total number of marbles nor the total of each color. How does this model one of the limitations of real-world science?

3.1 SECTION REVIEW

1. Why do you think NASA is thinking of sending a probe to Europa to search for life rather than to the moon or Mars, both neighbors of Earth?

2. Considering Psalm 19:1, respond to Carl Sagan's statement: "The universe is a pretty big place. If it's just us, seems like an awful waste of space."

3. What part(s) of Earth would the biosphere exclude?

4. The textbook gives the example of your address as an illustration of the relationships between the biosphere, biome, ecosystem, community, population, and individual. Use the analogy of a computer connected to the internet to explain the relationship of these terms, where a letter key on the keyboard represents an individual organism and the internet represents the biosphere.

Use the case study on the Great Barrier Reef on the facing page to answer Questions 5–12.

5. What is the striped surgeonfish's ecosystem?

6. List three abiotic factors in its ecosystem.

7. Name three members of its community.

8. What is the relationship between biotic factors in the striped surgeonfish's ecosystem and its community?

9. What is the striped surgeonfish's habitat?

10. What is its niche?

11. What makes the Great Barrier Reef a good place for the striped surgeonfish?

12. Do you think we could find this fish in other places? Why or why not?

3.2 BIOMES

Climate

Questions

What kinds of biomes are there?

How do mountains and oceans compare with other biomes?

Terms

tundra
deciduous forest
desert
savanna
grassland
coniferous forest
chaparral
tropical rainforest
vertical zonation

Your alarm clock goes off—time to get up. How do you choose what you're going to wear today? Chances are you check the weather. The temperature outside and the presence of anything falling from the sky tell you whether you need to break out the winter coat, umbrella, or sunscreen.

Weather tells you what Earth's atmosphere is like right now, but *climate* is the average weather of an area over an extended period. Figuring out both today's weather and an area's climate relies on temperature data and precipitation (rain, snow, sleet, hail, or freezing rain) data. A meteorologist—a scientist who studies weather—can tell you whether you need a raincoat for the day; a climatologist can tell you what kind of wardrobe you need!

? Why do certain organisms live in certain places?

FACTORS THAT AFFECT CLIMATE

Temperature. From the cold Antarctic to the heat of the tropics, one of the primary factors that affects climate is temperature. Temperatures vary according to both daily and seasonal cycles.

Two climates, two sets of temperature and precipitation data. Notice the axes labels on the bottom, left, and right of the graphs. The bars show precipitation data, and the lines on top of each graph show temperature data.

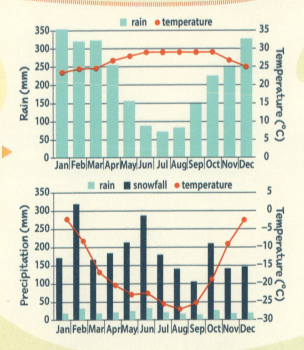

Precipitation. The amount of water that a biome receives is another primary factor. Some regions experience roughly equal amounts of rainfall every month of the year. Others may have very wet winters and drought-like summers. The form in which the precipitation occurs—namely rain or snow—is also a factor.

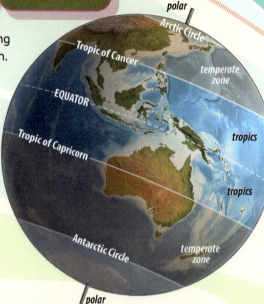

Seasons. The seasons are a direct result of Earth's tilt. Changing seasons often bring changes in temperature and precipitation.

Latitude. The latitude of an area affects how extreme the changes associated with seasons can be. For example, some areas have monsoons associated with changing seasons. In general, warm areas are closer to the Equator and cold areas are closer to the poles.

Elevation. An area's elevation can have the same effect on climate as latitude, affecting the swings of temperature and precipitation.

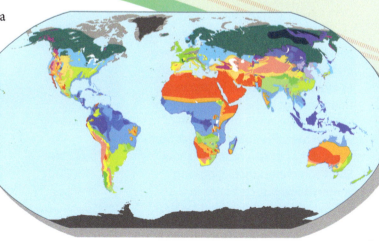

— warm currents — cold currents

Bodies of Water. The presence of large bodies of water affects an area's climate because water changes temperature slowly, moderating how fast air temperatures can change. Warm ocean currents can keep areas that we might expect to be cooler warmer during different seasons. Cold currents have the opposite effect, keeping coastal areas cooler than areas just a few kilometers inland.

Climatologists think of Earth's surface as a system of zones, each of which is based on the temperature and precipitation data of a particular region. In many cases you could examine the plants that live in an area and tell something about the average temperatures and precipitation in that area over a long period of time. The figure at right shows different climate zones of the world. Each climate zone has a different combination of high, medium, or low temperatures and precipitation. Many of these zones have mountains, forests, or bodies of water that create subzones. Also notice that areas of the world that are far apart can have similar climates.

BIOMES OF THE WORLD

Climate zones don't change suddenly as we move from one zone into another. One zone gradually merges into another, with temperatures and precipitation gradually changing. In fact, some living things flourish in places where there is a transition from one kind of climate to another, like the puffins who live on the cliffs along the ocean.

So what effect do climate zones have on life? Climate zones are defined by abiotic factors, but biomes add biotic factors to the mix. Remember that biomes are large areas that have a fairly consistent environment and are home to a set of organisms suited to that environment. Let's take a tour of the major biomes of the world.

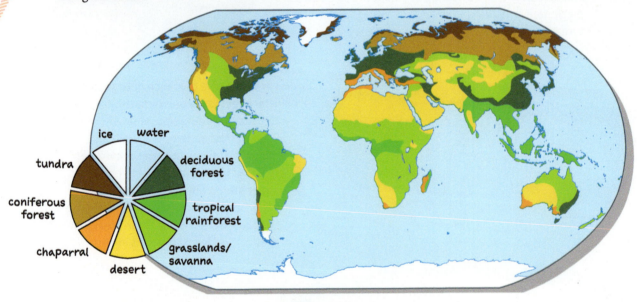

ice water

tundra

deciduous forest

coniferous forest

tropical rainforest

chaparral

grasslands/ savanna

desert

Tundra

The **tundra** has frozen land (called *permafrost*) year-round, with little vegetation except for hardy plants, such as mosses, lichens, and wildflowers, that take root in the top few inches of soil that thaw in the summer. There are few trees. Precipitation varies from a few inches to several feet of snow. Short summers punctuating long winters allow insects buried in the soil to hatch, providing food for birds. Other animals such as caribou and grizzly bears feed during the summer and migrate elsewhere or hibernate to avoid the harsh winter.

Deciduous Forest

Trees, such as oak, birch, ash, and maple, form **deciduous forests** in temperate zones that receive moderate rainfall and have moderate temperatures. These trees experience changes in all four seasons and support an abundance of wildlife, such as cicadas, songbirds, owls, and raccoons.

Desert

Water is precious in the **desert**, a biome that gets less than 25 cm of rain a year. Temperatures can change dramatically from scorching daytime temperatures to chilly nights, though overall, deserts have high temperatures. Some deserts are cold, occasionally getting snow. Organisms in this biome, such as cactuses, succulents, reptiles, insects, and spiders, survive by conserving water. Most animals, such as rattlesnakes and kangaroo rats, are more active at night.

Savanna

Savannas are the realm of lions, wildebeests, and baobab trees. These biomes have widely spaced trees that dot a rolling landscape of grasses. Savannas have moderate to high temperatures year-round and moderate but seasonal precipitation. The rainy season brings 100–150 cm of rain, relieving parched plants, animals, and the ground, and reducing the fires that can break out during the dry season.

Grasslands

Rippling fields of grasses give the **grassland** biome its name. This temperate biome doesn't get enough rainfall to support trees (25–75 cm per year), but moderate rainfall and temperatures do support grasses that are crucial to the many herbivores that live here, such as wild horses, pheasants, and gazelles.

Coniferous Forest

Evergreen and fragrant, the **coniferous forests** have two seasons each year, with cooler temperatures and moderate precipitation. These conifers change the soil and block out the light for other plants, making the conifers the kings of the forest. Hemlocks, firs, and redwoods support the other life here, including elk, wolverines, moose, pileated woodpeckers, and rabbits.

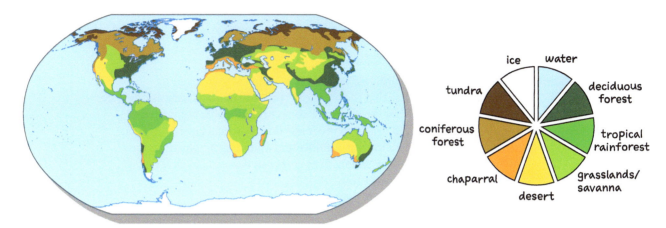

ice
water
tundra
deciduous forest
coniferous forest
tropical rainforest
chaparral
grasslands/savanna
desert

Chaparral

The scrubby **chaparral**, like the grasslands, receives moderate amounts of precipitation. Cool, wet winters alternate with warm, dry summers. Droughts and fires are common. Chaparral plants typically have tough, waxy leaves to retain moisture and thorns to deter browsers like deer and antelope. Although chaparral covers the least amount of Earth's surface, it has a large number of plant species—an estimated 20% of all plants!

emergent layer
40 m
canopy layer
25 m
understory layer
10 m
forest floor

Tropical Rainforest

Dripping vegetation, warm, humid air, and an amazing year-round biodiversity categorize the **tropical rainforest**. Rainfall exceeds 250 cm per year, supporting an abundance of life, such as lemurs, sloths, pythons, geckos, and cacao trees. The forest is so tall that there are layers of growth, beginning with the top, or emergent layer, then the canopy; these two layers get the most light and rainfall. The understory beneath the canopy is home to poison dart frogs and birdwing butterflies. The lowest and shadiest level is the forest floor.

Vertical Biomes

In the tropical rainforest biome description on the preceding page you will notice that the types of plants and animals vary as one proceeds down from the emergent zone through the canopy, then to the understory, and finally to the forest floor. Vertical layers within a biome form because the temperature and amounts of light and rainfall can change with elevation. Ecologists call this ecological layering in a biome **vertical zonation**.

MOUNTAIN BIOME

snow line

alpine tundra

subalpine forest

montane forest

oak woodland

Another biome that experiences vertical zonation is the highland or mountain biome. For example, Mount Kilimanjaro in Tanzania starts in the African savanna. As hikers travel to its volcanic summit, they pass through a tropical rainforest, grassland, moorland, an alpine desert, and an ice cap. Other mountains typically have a deciduous zone, a coniferous zone, and a tree line beyond which there is insufficient rainfall and temperatures are too low for trees. Beyond this timberline is an area of ice and snow. Gaining 300 m in elevation usually produces about a 6 °C change in temperature, the equivalent of traveling 500 km toward a pole.

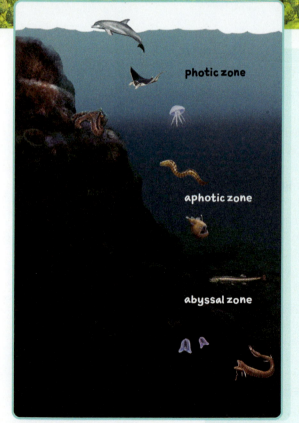

photic zone

aphotic zone

abyssal zone

AQUATIC BIOMES

A third biome that experiences vertical zonation is the aquatic biome. Oceans cover about 71% of our planet, and life in the oceans depends on light. Water temperature is driven by sunlight; in turn, temperature affects how much oxygen can dissolve in the water—a useful thing if you're a fish! In the ocean, light is also essential for plants and algae that depend on photosynthesis. Because of these factors, different animals and plants live in different places in the ocean. *Benthic organisms*, such as kelp, crabs, and snails, live on the ocean floor. *Pelagic organisms*, such as plankton, squid, sharks, and whales, swim or drift with the currents. As we can see on the left, benthic and pelagic organisms that need light live in the photic zone, the thin upper layer of the world's oceans. Benthic and pelagic organisms that don't need light can live in the aphotic and abyssal zones where there is little or no light. Life looks very strange and alien down there!

When we step back and look at all the different biomes of the world, we can see an amazing tapestry of environments that are home to many different living things. In the quilt of life, the interactions of organisms and ecosystems show a variety and care that point to a caring and creative Designer.

3.2 SECTION REVIEW

1. What is the difference between a climate zone and a biome?

2. Why do some biomes have very few trees?

3. List some abiotic factors that affect a biome.

4. Copy the table below on a separate sheet of paper and fill it in. Label temperature and precipitation as low, moderate, or high for each biome. Use the map of climate zones on page 57 and the biomes map on page 58 to identify each biome's climate zone.

5. Which biome would the climate graph on the right match? Give three examples of things that could live in this biome.

6. Why do vertical zones form in an ecosystem?

7. How do naturalists explain order, design, and function in the biosphere? How should a Christian respond to this order?

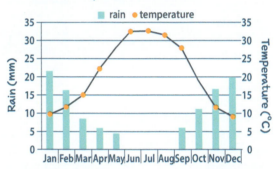

BIOME	TEMPERATURE	PRECIPITATION	CLIMATE ZONE	EXAMPLES OF ANIMAL LIFE
Tundra				
Deciduous Forest				
Desert				
Savanna				
Grassland				
Coniferous Forest				
Chaparral				
Tropical Rainforest				

Producers and Consumers

Questions

Where do living things get energy?

How are living things influenced by each other?

How likely is it that a web of living things could exist on another planet?

Terms

producer
consumer
competition
detritus
decomposer
food web
ecological pyramid
symbiosis
parasitism
mutualism
commensalism
neutralism
amensalism

When pioneering climatologist Wladimir Köppen began thinking about climate zones and biomes around the world in 1884, he looked at one thing: plants. Plants give us a lot of information about the abiotic factors in an environment, and they also form the base of support for a whole network of life.

A plant is an example of an organism that can make its own food through photosynthesis. (We'll study how it does this in Chapter 7.) Plants and algae are the most common food manufacturers, what biologists call **producers** (or *autotrophs*).

Almost all other organisms get their energy from the energy that plants capture from the sun. Since these other organisms can't produce their own food but instead consume the producers, they are called **consumers** (or *heterotrophs*). There is a flow of energy in an ecosystem that involves different types and sources of energy.

Let's consider an African safari. In the African savanna the grasses and trees are the producers, forming the foundation of the biotic community in this ecosystem. Giraffes graze from apricot and mimosa trees and sometimes from grasses. They are *herbivores*—animals that get their food from plants. Lions and cheetahs also lurk in the grasses of the savanna, preying on mongooses, wildebeests, gazelles, and even giraffes. They are *carnivores*—animals that get their food from other animals. Leopards join the pack of *predators* on the African savanna, hunting guinea fowl and impalas at night and dragging their prey up into a tree to consume later. Impalas, guinea fowl, and even monkeys are the leopard's *prey*. And hyenas and vultures clean up everyone's meaty leftovers, making them *scavengers*. Carnivores are either predators or scavengers; often they are both. If there is a limited supply of meat in an ecosystem, the populations of carnivores may experience intense **competition**, which occurs when organisms try to use the same resource. This can also happen during the dry season as water holes in the savanna begin to shrink.

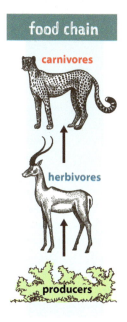

food chain

carnivores

herbivores

producers

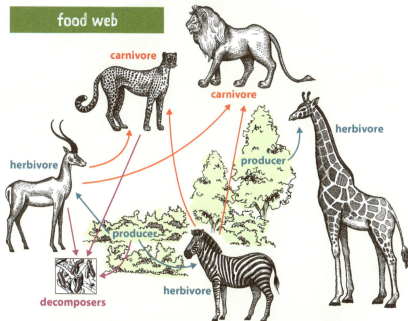

food web

carnivore

carnivore

carnivore

herbivore

producer

herbivore

producer

herbivore

decomposers

Side-striped jackals also roam the African savanna, feeding on insects and springhares during the dry season and eating fruits like apricots in the wet season. These jackals are *omnivores*—animals that eat both other animals and plants. The African savanna is home to aardvarks, which eat a kind of cucumber that grows underground. They also like to invade termite mounds and lap up the inhabitants with their long tongues.

Termites are tiny, but they use the power of the colony to reshape and renew the savanna landscape. They create mounds that offer burrows for animals and act as termite high-rise apartments. They also feed on dead plants, wood, and even animal dung—the waste products of plants and animals, called **detritus**. Termites recycle the nutrients in these substances to enrich the soil as they make their mounds. Plants can grow here in the future, or termites can return nutrients to the biome when other animals like aardvarks and mongooses eat them. Organisms that break down nutrients are called **decomposers**. Termites get a little help breaking down detritus from microbes that live in their guts.

Energy in a biome passes from one organism to another in a sequence known as a *food chain*. For example, apricot trees grow in the savanna, feeding giraffes who feed lions. Food chains need producers, consumers, and decomposers to work. But as we've also seen in the savanna, the food chains of different animals can overlap, creating a web of life, or a **food web**. Food chains are connected in a food web.

food pyramid

carnivores

herbivores

producers

Another way to demonstrate the flow of energy in an ecosystem is to use an **ecological pyramid** that illustrates a food chain. Each layer, or *trophic level*, represents a different type of organism in the food chain, and each depends on the level below it for energy, though these pyramids don't cover all the energy gained or lost in an ecosystem. The bottom layer is usually the producers, followed by herbivores, omnivores, and carnivores. The bottom layer is broadest because there are more producers in that part of the food chain, representing a greater amount of living matter, or *biomass*. The healthier an ecosystem is, the more biomass it supports.

As we continue to track the energy through an ecosystem, we note that producers make food from the sun's energy. With each step in the food chain, or each trophic level, there is a transfer of energy from one organism to another. Ecosystems that are full of life abound in producers, allowing the ecosystem to support several trophic levels on the ecological pyramid.

Symbiosis

White rhinoceroses, giraffes, and hippopotamuses graze in the rustling fields of the savanna. Some gray birds flit between these grazing animals, appearing to hitch a ride like mini tourists on a safari. These animals interact with each other within their habitat, a phenomenon that ecologists call **symbiosis**. Let's consider several types.

Oxpeckers pick off and eat ticks that burrow into the tough hides of giraffes, rhinoceroses, and hippopotamuses and possibly weaken them. They are actually doing the large animals a favor! The ticks exhibit **parasitism**—they are organisms that depend on a much larger animal, called a *host*, for food. Over time, parasites weaken their hosts. The oxpeckers are predators of the tick, helping out the host animals and themselves. A mutually beneficial relationship between several members of an ecological community is called **mutualism**. It's a win-win situation for everyone—except the tick!

The rhinos tramp and munch their way through the savanna, raising up a cloud of insects. Cattle egrets, a type of bird, trail the rhinos, scooping up the insects. The egrets illustrate a kind of symbiosis known as **commensalism**: they benefit from the rhinos, but the rhinos are neither helped nor harmed by the egrets.

Symbiosis is usually a good thing, but nature is not always kind. Interactions between organisms can be neutral or injurious to one or both of them. For example, the black mamba, one of the most venomous snakes in Africa, shares a neighborhood with the oxpecker, but it isn't directly affected by the oxpecker. This relationship is called **neutralism**. But grazing herbivores could startle a black mamba hiding in the grasses. It can then be spotted and picked off by a watching mongoose. The grazing herbivores are not affected by the black mamba, but the snake is definitely impacted by the herbivores! This relationship is called **amensalism**—one organism is injured while the other is unaffected.

We've also seen how animals must kill for nourishment or compete with each other for limited resources, especially dwindling water holes during the dry season. Relationships between organisms can change quickly in the environment, such as when two lions produce offspring but eventually abandon a sick or injured lion cub during a famine, drought, or predator attack. The African savanna can be a harsh and unforgiving place.

The struggle for survival on Earth often seems cruel. Does this bother you? It should! The cruelty we observe in nature stirs in us a sense that something isn't quite "right" about our world—a cursed and fallen world where sin has marred God's creative order. All of nature— each biome, ecosystem, community, and organism—waits for the time when He will restore His fallen creation and death will be no more. Indeed, at that time the Bible tells us that "the wolf also shall dwell with the lamb, and the leopard shall lie down with the kid; and the calf and the young lion and the fatling together; and a little child shall lead them" (Isa. 11:6). But until that time, we see that God graciously provides for His creatures.

If life were ever found on another planet, would those organisms also have to cope with things like predation, parasitism, and death? So far science cannot answer such questions because no life on other planets has yet been found. All our observations up to the present time have supported what we see in Scripture—that Earth is a unique address in our universe, a place specially designed to support an abundance and diversity of life.

3.3 SECTION REVIEW

1. Give an example of a producer and a consumer.

2. Give an example of an herbivore, omnivore, and carnivore from the deciduous forest biome.

3. What do you think a *detritivore* is? What is another name for this kind of organism?

4. Create a food chain of the African savanna, including lions, savanna grass, and giraffes.

5. Draw an ecological pyramid from the food chain that you created in Question 4.

6. Label each trophic level in your ecological pyramid from Question 5 as a producer, herbivore, omnivore, or carnivore.

7. Where do omnivores fit on an ecological pyramid?

8. Sketch out a food web for the African savanna including the following organisms: savanna grasses, apricot trees, giraffes, wildebeests, lions, cheetahs, termites, anteaters, and side-striped jackals.

9. How would you respond to a person who says, "The natural world shaped by evolution is a brutal place where animals prey on other animals. How could a loving Creator make a world like this?"

10. Complete the table below to show the symbiotic relationships between two different organisms. For each organism, indicate whether the relationship is helpful (+), harmful (−), or neutral (0). In the last column give an example of a pair of organisms that illustrate the indicated type of symbiosis. The first row has been done for you as an example.

SYMBIOSIS	ORGANISM 1	ORGANISM 2	EXAMPLE
Competition	−	−	lion/cheetah
Predation			
Parasitism			
Mutualism			
Commensalism			
Neutralism			
Amensalism			

3 CHAPTER REVIEW
Chapter Summary

3.1 OUR LIVING PLANET

- The biosphere consists of all parts of Earth where life exists.

- Ecologists study the relationships between organisms and between organisms and their environments, including both biotic (living) and abiotic (nonliving) factors.

- For study, the biosphere can be divided into a nested hierarchy of smaller regions, including biomes, ecosystems, and habitats.

- Biomes are large areas of the planet defined primarily by climate.

- Ecosystems are limited portions of biomes where particular kinds of organisms live and interact. A habitat is that portion of an ecosystem utilized by a particular organism.

- The manner in which a particular organism lives and interacts with its environment is its niche.

Terms
biosphere • ecology • biodiversity • biome • ecosystem • habitat • niche • abiotic factor • biotic factor • population

3.2 BIOMES

- The characteristics of a biome are primarily determined by yearly patterns of temperature and precipitation. Additional influences include seasons, latitude, elevation, and proximity to large bodies of water.

- Tundras experience long, cold winters and short summers. They are characterized by the presence of permafrost.

- Deciduous forests are dominated by broadleaf trees such as oak or maple. These occur in regions of moderate temperature and rainfall.

- Deserts are characterized by a lack of precipitation. Deserts may be either hot or cold.

- Savannas are grasslands with widely spaced trees. They typically have moderate to high temperatures year-round, but precipitation is seasonal.

- Grasslands are temperate biomes that do not receive enough annual precipitation to grow trees.

- Coniferous forests are dominated by cone-bearing trees such as pines and firs. These occur in regions with moderate precipitation and often long, snowy winters.

- Chaparrals are regions with mild, wet winters and warm, dry summers. Many chaparral plants have adaptations to survive drought and fire.

- Tropical rainforests occur in areas with high temperatures year-round and high amounts of precipitation. Rainforests have greater biodiversity than other biomes.

- Some biomes change with increasing elevation or depth, exhibiting vertical zonation.

Terms
tundra • deciduous forest • desert • savanna • grassland • coniferous forest • chaparral • tropical rainforest • vertical zonation

3.3 WEB OF LIFE

- Organisms can be categorized as either producers, which make their own food, or consumers, which obtain energy by consuming other organisms.

- Consumers can be further divided into those that eat primarily plants (herbivores), those that prey on other consumers (carnivores), and those that break down nutrients in dead organic material (decomposers).

- The flow of energy in an ecosystem can be modeled by food chains, food webs, and ecological pyramids.

- Different kinds of organisms can interact with one another in a variety of ways, collectively referred to as symbiosis.

Terms
producer · consumer · competition · detritus · decomposer · food web · ecological pyramid · symbiosis · parasitism · mutualism · commensalism · neutralism · amensalism

Chapter Review Questions

RECALLING FACTS

1. Why is astrobiology such a popular new field in science?

2. Place the following terms in order of increasing land area: *biome, biosphere, ecosystem,* and *habitat*.

3. What factors limit the extent of the biosphere?

4. What are the primary factors that determine the type of biome that will exist in a particular area?

5. Which type(s) of biomes have uniformly moderate to high temperatures throughout the year?

6. List the biomes of the world that have low precipitation (rain, snow, hail, sleet).

7. Why does vertical zonation occur on a tall mountain?

8. What kinds of organisms form the basis of an ecological pyramid?

9. Honeybees get food from apple trees and also pollinate them, which enables the trees to produce fruit and seeds for propagation. What kind of symbiosis is this?

10. The color of a tree frog perfectly matches that of the tree leaves among which it lives, helping the frog escape the attention of hungry predators. What type of symbiosis exists between the tree frog and the tree?

UNDERSTANDING CONCEPTS

11. Do ecologists ever study individual organisms? Explain your answer.

12. Why is biodiversity a sign of a healthy ecosystem?

13. An ecologist is studying a group of emperor penguins in winter, each of them including a father sheltering a solitary egg on Antarctic ice while the mother is out fishing. Is the ecologist studying an ecosystem, community, population, or individual? Explain.

case study

TIDE POOL ECOLOGY

Near Acadia National Park in Maine, waves buffet the rocky coasts. In this place where land and water meet, living things abound because of plentiful sunlight, clean water, and a good supply of nutrients. An added benefit for the animals that live here is all the nooks and crannies to hide in.

But life isn't always easy for the constantly battered starfish, sea anemones, snails, crabs, and mussels that call this place of transitions home. When the tide is out, they can be left high and dry for creatures to snack on. Animals find a way to cling onto the rocks to endure periods of exposure. Some fish called killifish live in areas of the tide pool where there is always standing water. These fish can gulp air from the surface when the tide is low and there is no fresh supply of oxygen in the water. They can also jump out of the pool and flop across land to find a new pool!

Mussels, snails called limpets, and barnacles in the tide pool snack on algae and larval sea creatures called plankton. Mussels and limpets are eaten by starfish and crabs. These crabs could be the next meal for herring gulls or otters.

Use the case study above to answer Questions 14–22.

14. Name three abiotic factors in the tide pool.

15. Name three biotic factors in the tide pool.

16. Is a tide pool an example of a biome, ecosystem, or habitat? Explain.

17. How do you think temperature would affect the tide pool?

18. How do you think rainfall would affect the tide pool?

19. Do you think that tide pools might have their own small version of a vertical zonation? Explain.

20. Create a food chain from the information in the case study using algae, limpets, crabs, and otters.

21. Use your food chain from Question 20 to make an ecological pyramid. Label the trophic levels with *producer*, *herbivore*, *omnivore*, and *carnivore*.

22. Draw a food web for the tide pool including mussels, barnacles, algae, plankton, crabs, limpets, starfish, herring gulls, and otters.

23. Using the major lines of latitude, locate the temperate biomes and climate zones on the globe at right.

24. Using the major lines of latitude, locate the tropical biome and climate zones on the globe at right.

25. The coniferous forest biome has sufficient precipitation for many trees. Why does it support few grasses?

26. Why do the things that live and grow on a mountain change with elevation?

27. How do you think freshwater aquatic biomes are different from oceanic aquatic biomes?

28. Create a food chain and an ecological pyramid that include a producer, herbivore, omnivore, carnivore, and decomposer from the same biome.

CRITICAL THINKING
Use the graphs below to answer Questions 29–36.

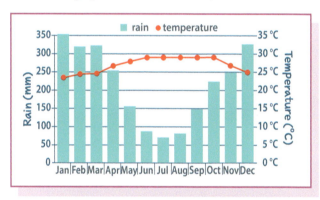

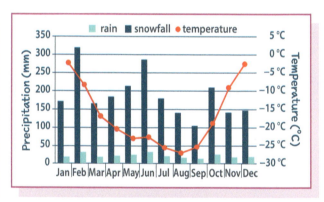

29. What kind of climate does Graph 1 represent: tropical, temperate, dry, or polar? How do you know?

30. Is this climate zone in the Northern or Southern Hemisphere? How do you know?

31. Do you think that Graph 1 represents a climate that can support a high biomass and great biodiversity? Why or why not?

32. What kind of climate does Graph 2 represent: tropical, temperate, dry, or polar? How do you know?

33. Is this climate zone in the Northern or Southern Hemisphere? How do you know?

34. Do you think that Graph 2 represents a climate that can support a high biomass and great biodiversity? Why or why not?

35. Do you think that Earth in its originally created form had the biomes that it has now? Why or why not?

36. A friend tells you that because there are an estimated one billion trillion stars in the observable universe, the chance that life exists on other planets must certainly be high. How would you respond to your friend's assertion? Use Psalm 8:3–9 to narrow your answer by considering man's place in creation.

4 INTERACTING
WITH THE BIOSPHERE

The Butterfly Effect

Can a butterfly floating on the breeze in Bolivia affect the weather in Mississippi? That's the idea behind the butterfly effect—that small changes in a system can have drastic effects. The name comes from the metaphor of a butterfly flapping its wings, altering the air patterns around it. It was thought that these tiny changes in the atmosphere, compounded with others, eventually create a hurricane. The butterfly effect can apply to many different phenomena, not just weather. It's mathematical and predictive, and it is just one of the many models that try to explain the interactions that take place within the biosphere.

4.1 SUSTAINABILITY
The Cycles of Matter

Why doesn't the earth run out of oxygen?

Hurricanes, of course, aren't caused by butterflies, but small changes in the environment do add up. When we take a step back and look at the world, we find that small changes can be a big deal. One such example is the decline in bee populations. There are around 20,000 species of bees. These essential pollinators help provide us with the food we need for survival. But why is the bee population declining? Some say that it's because of habitat loss and disease; others say pesticide use and climate change are having an impact. Regardless of the reason, it's important for us to be good stewards of our world by avoiding careless actions that harm bees and cause the population to decline further. Small changes can have a big impact.

The good news is that we don't always need to feel responsible to ensure that populations increase or decrease. We don't have to worry about cockroaches or mice taking over the world or fast-growing weeds choking out all other plants. **Sustainability** is the ability of the biosphere to maintain its balance indefinitely, keeping life forms in check. God has built checks and balances into Earth to maintain the planet's productivity and diversity and prevent us from being overwhelmed as we try to manage the populations that really do need our help in a fallen world. One way that Earth has been designed to sustain itself is through the cycling of matter. Water, oxygen, carbon, nitrogen, and phosphorus are constantly absorbed by organisms and released back into the environment. The movements and interactions of the matter in the biosphere, geosphere (Earth's crust), and atmosphere are collectively known as the **biogeochemical cycle**.

Questions

How does matter flow through an ecosystem?

Why do some organisms survive in an environment, and others do not?

How can we tell how populations change?

How does the environment recover after a disaster?

Why are ecosystems predictable and orderly?

Terms

sustainability
biogeochemical cycle
nitrogen fixation
population density
limiting factor
exponential growth
carrying capacity
primary succession
pioneer species
secondary succession
climax species

The Water Cycle

Water vapor in the atmosphere cools and becomes liquid again through the process of *condensation*. These tiny droplets of water or ice crystals form clouds.

Water from places like rivers, lakes, oceans, or rain-filled ditches *evaporates*, also entering the atmosphere.

As water vapor in the atmosphere reaches saturation, it begins to fall to Earth as rain, sleet, hail, or snow in a process called *precipitation*.

Plants absorb water from the ground through their roots. Once plants use the water for their life processes, the water exits through the plants' leaves by *transpiration* and enters the atmosphere as water vapor.

Precipitated water either goes back to plant roots, evaporates back into the atmosphere, or becomes runoff water that percolates into the water table.

Water from groundwater, a system of underground reservoirs, seeps out of the ground and into springs, lakes, oceans, and underwater vents.

The Oxygen & Carbon Cycles

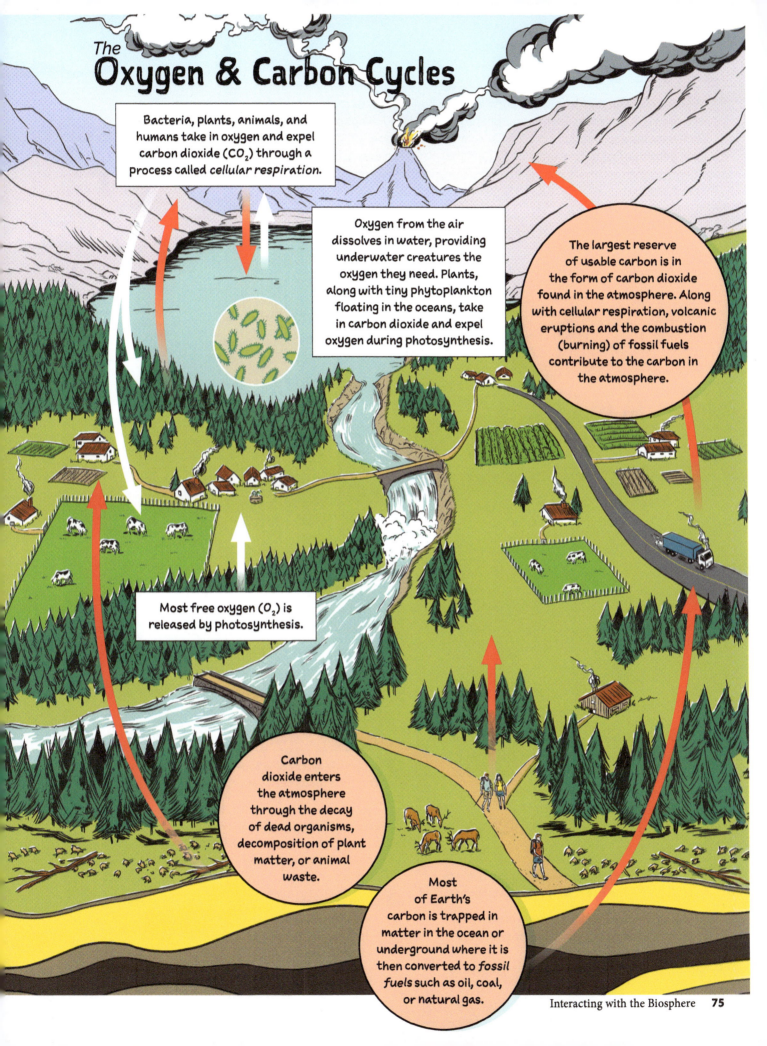

Bacteria, plants, animals, and humans take in oxygen and expel carbon dioxide (CO_2) through a process called *cellular respiration*.

Oxygen from the air dissolves in water, providing underwater creatures the oxygen they need. Plants, along with tiny phytoplankton floating in the oceans, take in carbon dioxide and expel oxygen during photosynthesis.

The largest reserve of usable carbon is in the form of carbon dioxide found in the atmosphere. Along with cellular respiration, volcanic eruptions and the combustion (burning) of fossil fuels contribute to the carbon in the atmosphere.

Most free oxygen (O_2) is released by photosynthesis.

Carbon dioxide enters the atmosphere through the decay of dead organisms, decomposition of plant matter, or animal waste.

Most of Earth's carbon is trapped in matter in the ocean or underground where it is then converted to *fossil fuels* such as oil, coal, or natural gas.

Interacting with the Biosphere   **75**

The Nitrogen Cycle

Proteins are essential for cells, and protein production requires nitrogen. Nitrogen is abundant in Earth's atmosphere, but most living things can't use it directly. Instead, through a process called **nitrogen fixation**, atmospheric nitrogen must first be converted into a form that plants can absorb and use.

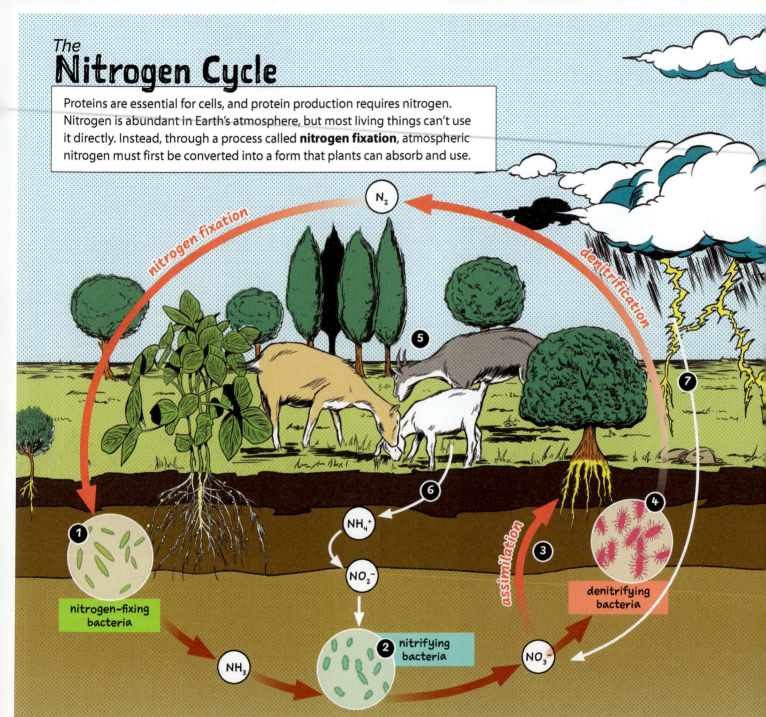

1. Free-living nitrogen-fixing bacteria in the soil and others in the roots of legumes (e.g., beans, peanuts, and soybeans) fix atmospheric nitrogen as ammonia.

2. Nitrifying bacteria convert ammonia to nitrate.

3. In assimilation, plants take in nitrate through their roots and use it for protein production.

4. During denitrification, bacteria convert some nitrate to nitrogen gas, which reenters the atmosphere.

5. Animals—and people too—get their nitrogen from the foods they eat.

6. Animals excrete nitrogen in their metabolic wastes. Bacteria break down some of this waste as ammonium. Other bacteria convert ammonium to nitrite; still others convert nitrite to nitrate.

7. Some atmospheric nitrogen is fixed as nitrate during lightning strikes. The nitrate dissolves into water droplets and falls to earth in precipitation.

The Phosphorus Cycle

Phosphorus is essential for life. It promotes growth in animals and plants, forms genetic material (DNA and RNA), and is needed for structure and energy transport in cells.

Weathering events, such as rain or erosion, cause rocks to release phosphate compounds and other minerals in soil and water.

Animals that eat plants excrete waste and eventually die and decay, releasing phosphates back into the soil. Phosphates that have been deposited into aquatic systems are absorbed by algae or are consumed by organisms.

Phosphates are distributed throughout the soil and water and are absorbed by plants.

Bacteria in the soil break down phosphates through a process called *mineralization.* Eventually, phosphorus is buried and absorbed by rocks or works its way into underground reservoirs and the cycle continues.

Population Growth and Biodiversity

Sometimes after a heavy rain, fertilizers from a farm wash into a nearby lake—your grandpa's favorite fishing spot! The fertilized lake quickly turns murky green. All that green looks incredibly healthy, but when you and your grandpa arrive at the lake for a fishing trip, the fish aren't biting. If you could peer into the water, you would see why—there aren't any fish in the lake!

Microscopic algae are causing the lake to turn green. The algae have a high **population density**, which is the number of individuals in a certain amount of space. The fertilizers in the runoff flowing into the lake add excess nutrients in a process known as *eutrophication*. Algae growth is accelerated in response to the excess nutrients. These increased nutrients result in algae blooms. More algae growing also means more algae die. As dead algae and other organisms drift to the bottom of the lake, the population of decomposers increases. More decomposers means more demand for oxygen, lowering the amount of available oxygen. Reduced oxygen levels are dangerous to fish. The

problem is worse at night when photosynthetic organisms like algae don't produce oxygen. But they still use oxygen for respiration, reducing oxygen levels even more. Oxygen levels may plunge below what fish can tolerate, resulting in a die-off. And of course, the dead fish make the decomposition problem worse. Grandpa's lake has a lot of nutrients but not enough oxygen to support all the organisms living in it. This kind of situation illustrates what is known as a **limiting factor**—a factor that limits the growth of a population.

Limiting factors cause populations to grow or shrink in two ways. The first involves the ratio of birthrate to death rate. When the birthrate within a population is greater than the death rate, a population grows. When the birthrate is less than the death rate, the opposite is true. Sometimes the ratio holds steady, but the size of the population still fluctuates. This is because population size also changes when outside individuals join the population (*immigration*), or members of the population leave (*emigration*).

BIODIVERSITY

Nothing looks nicer than a lush green lawn. It takes a lot of work to create a uniform field of grass—you may have observed one of your neighbors on his hands and knees, picking out each weed that dares to pop up. Why is so much effort required to keep just one kind of plant growing in a lawn? Because Earth is designed to support diverse populations.

As populations vary, the biodiversity of an area changes. If algae kill all the fish populations in a lake, then the insect populations that the fish once ate may grow. Insect-eating birds from other areas may discover this new insect buffet and move closer to the lake. Fish-eating bird populations will leave and find a new lake. This ripple effect is the result of one population—the algae—becoming unbalanced. A change in biodiversity can be natural, or it can result from four major human factors: habitat loss, overuse of resources, introduction of a species to a new area, and overuse of chemicals such as fertilizer. Though these types of situations can't always be avoided, it is part of our responsibility to understand how humans might be affecting the growth and diversity of populations.

Exponential Growth

Limiting factors don't fully explain why a population can't continue to grow until it takes over the planet. Think about a tropical rainforest, filled with water, warmth, and nutrients. In theory, a population living there could grow endlessly. This rapid rate of growth is known as **exponential growth**. A square foot of algae in a pond turns into two square feet, then four, then eight, then sixteen. In a tropical rainforest two orchids in a tree that reach maturity in a month could exponentially grow to 4000 orchids covering the tree in a year. But you'll never find a tree with that many orchids on it because every population has a carrying capacity.

Carrying capacity is the maximum number in a population that an area can sustain. For example, orchids depend on tall trees to help them get closer to the rainforest canopy and reach sunlight. If orchids started multiplying and exceeded their carrying capacity, their tree would quickly be overwhelmed by the orchids and die. A dead, crumbling tree would drop its orchids on the ground, and since the orchids depend on the tree's height to get sunlight, the orchids would soon die. Algae in the fertilizer-tainted pond have temporarily exceeded the pond's carrying capacity as well. As the algae use up the fertilizer and the nutrient levels return to normal, algae will begin to die off because of the competition for declining resources.

Therefore, a more realistic model for growth is *logistic growth*, which accounts for the limiting factors and carrying capacity of a population. But even logistic growth can't perfectly predict how a population will change because of the huge number of factors that interact in the biosphere.

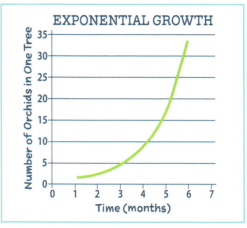

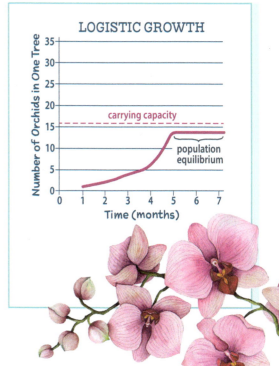

FERRETING OUT THE GROWTH RATE

The black-footed ferret has come back from the grave not once, but twice. This member of the weasel family was thought to have gone extinct prior to 1964, but in that year a remnant population was discovered in South Dakota. Some of those ferrets were caught and used in an attempt to establish a captive breeding population. The attempt ultimately failed. When the last wild South Dakota ferret died in 1974, followed by the death of the last captive ferret in 1979, the black-footed ferret was once again thought to be extinct.

But in 1981 a rancher discovered another remnant population in Wyoming. Scientists studied this population, but diseases were quickly taking their toll on the ferrets. In 1987 the remaining ferrets were taken into captivity. This tiny remnant, 18 ferrets in all, became the founding stock of a successful captive breeding program, whose offspring were eventually released back into the wild as part of a carefully managed ferret recovery plan initiated by the US Fish and Wildlife Service. Starting with the reintroduction of 42 young ferrets in 1991, the recovery plan hoped to have 1500 adult ferrets in the wild by 2023 and 3000 by 2043. By 2009 there were an estimated 448 adult ferrets back in the wild. Were the scientists on track to reach their goal of 1500 ferrets in 2023?

To estimate the growth rate of a population, wildlife biologists use the equation $P = P_i(1 + r)^t$, where P is the final population, P_i is the initial population, r is the rate of growth, and t is the time interval between the two population values. Using 18 as the value for t would give the biologists the growth rate of the ferret population for the eighteen years between 1991 and 2009. In the wild approximately 40% of young ferrets survive to reproductive age, so we will assume that about 16 of the original 42 survived and reproduced.

$$448 = 16(1 + r)^{18}$$

We divide both sides by 16.

$$28 = (1 + r)^{18}$$

We take the 18th root of both sides.

$$1.20 = 1 + r$$

We solve for r.

$$r = 0.20$$

The growth rate is usually expressed as a percentage, so $0.20 \times 100\% = 20.0\%$. This means that during the first eighteen years of the recovery plan, the adult ferret population would be expected to increase by an average of 20% each year. At that rate, the ferret population should have reached the target population of 1500 adults by 2016, well ahead of the 2023 target date. But by 2012 the adult ferret population had declined to 274. What happened? Think through the following questions to see whether you can figure out what was going on with the ferret growth rate.

1. Use the population growth equation to calculate the growth rate of the ferrets between 2009 and 2012.

2. What are some possible reasons why the ferret population declined between 2009 and 2012?

3. Calculate a new long-term (from 1991) ferret population growth rate using the 2012 population of 274 adult ferrets.

4. Which of the two growth rates that you calculated do you think is a more reliable estimate: the rate based on 2009 population data or the one based on 2012 data? Explain.

5. On the basis of your 2012 growth rate, will the ferret population still reach the 2043 target of 3000 adult ferrets?

Modeling nature and trying to predict the course that nature will take are difficult tasks. So the plan to increase the black-footed ferret population isn't a failure. It's a workable plan that can be adjusted as ecologists learn more about black-footed ferrets and the environment in which they live.

Pioneering Ecosystems

When Mount St. Helens blew its top in 1980, the ash from the erupting volcano could be seen in eleven surrounding states. The scorching heat from the pyroclastic flows that raced down the side of the mountain destroyed everything in its path. Once the volcano was spent, one side of it was nothing but bare rock. Fast forward to today, where the bare rock has been replaced with a thriving community of plants and animals. How does a devastated ecosystem bounce back like that?

Change in a biotic community over time like what happened around Mount St. Helens is known as *succession*. Succession that begins with a barren landscape is **primary succession**. This is the type of succession that Mount St. Helens underwent—in several places around the mountain there was nothing but ash and hardened lava covering the ground with no soil left for plants to grow in. As birds flew over these areas, leaving nutrient-rich droppings and seeds behind, a few plants began to reappear. Insects and other small creatures moved into the crevices of the lava rocks, and as they died, their bodies decomposed to form another rich layer of organic material. Plants and animals that are the first to return to an area are known as **pioneer species**. They tend to be short-lived as other species come along and replace them.

Succession in a disturbed landscape that already has soil is **secondary succession**. This type of succession follows primary succession. It also takes place when a farm field or lawn is abandoned. Seeds enter the field's habitat and grow unchallenged. The new kinds of vegetation in the field attract new animals. Soon, small trees begin growing, offering shelter to even more kinds of plants and animals. If left alone for about a hundred years, the habitat will become an old-growth forest filled with **climax species**—stable, long-lived species that mark the end of succession.

Mount St. Helens

pioneer species

Through primary succession, the debris fields around Mount St. Helens began to host little populations of plants after just a few years.

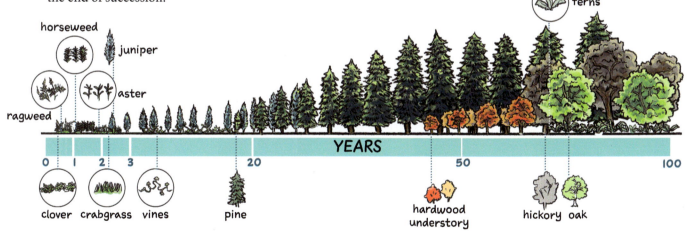

ferns

horseweed

juniper

aster

ragweed

YEARS

0   1   2   3   20   50   100

clover   crabgrass   vines   pine   hardwood   hickory   oak
                                                             understory

Succession from grassland to a climax forest in the southeastern United States

Succession doesn't necessarily follow a step-by-step path from primary to secondary succession. In the case of the Mount St. Helens eruption, the pioneer species and climax species were often growing at the same time. Eventually, if the area is not disturbed again, the climax species will take over, and the pioneer species may no longer grow there.

Finding Order in Chaos

The cycles of matter through the biosphere and the delicate balance between populations are very complex systems that can appear chaotic and random at times. A system is a part of the universe under study. Scientists identify an imaginary boundary around a system to analyze it and the factors that affect it. For example, we can define a population of puffer fish as a system while we study them. Everything outside this population is outside the system that we have defined.

The butterfly effect is sometimes viewed as a random process that serves as an engine for life to evolve from nonliving material. An evolutionist can view the butterfly effect this way because it does show us that we cannot completely model nature. But does a model that appears chaotic really prove that we are the products of chaos and that we will evolve into something else in the distant future?

The answer to this question has three parts. First, human models are limited because the systems they attempt to model are incredibly complex. There are thousands of tiny variables that our finite models can't handle.

Second, as scientists have improved their models, they have discovered that the world is more orderly and predictable than it might appear. For example, in the "boom and bust" cycle of populations, a population seems to randomly swell (the "boom")

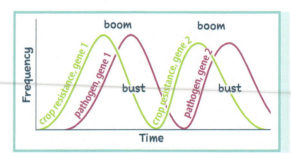

and then quickly shrink (the "bust"). Ecologists are beginning to create models that help explain this roller-coaster cycle. And limited as they are, the models are still useful. In one case farmers found that a particular fungus caused their crops to bust. So they planted a new variety of fungus-resistant crops. The crops would boom until another fungus attacked the crops and caused them to bust again. After studying the boom-and-bust model of crops and fungi, plant breeders developed varieties that resist several different kinds of fungi.

Third, a person's view of chaos in nature ultimately reflects his worldview. Where a naturalistic worldview interprets randomness, chaos, and even cruelty as evidence for evolution, a biblical worldview sees results of the Fall. The lack of order we see doesn't mean that we evolved from less-ordered creatures. We can still see God's order and design, but it has been broken because of sin.

4.1 SECTION REVIEW

1. How do biogeochemical cycles contribute to Earth's sustainability?

2. How does water cycle throughout the biosphere?

3. After a plant absorbs nitrogen from the soil, what process needs to happen before its nutrients can be used? Explain why this process must happen.

4. If a lake has turned green and the fish are dying, what has happened? What can be done to keep the fish alive?

5. Sea urchins love to eat algae. When the population density of algae becomes too high, the population density of sea urchins increases to over 10/m². According to the graph on the right, in which years were the algal populations too high?

6. Research and write about two ways to encourage biodiversity.

7. What is the difference between primary succession and secondary succession?

8. Fireweed is a common pioneer species that grows after a forest fire. Do you think that a climax species like a birch tree can ever grow at the same time as fireweed? Explain your answer.

9. How will people with a biblical worldview regard the complexity of ecosystems? How does this view of nature affect their understanding of models?

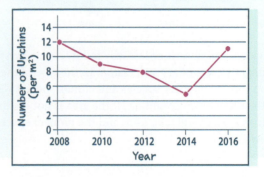

Predation and Populations

While predator and prey conflicts are not always easy to witness, they are important in maintaining balance within an ecosystem. Because of the carrying capacity of an area and the natural predation that occurs, populations rarely grow out of control. This is just another example of how God has structured our world to sustain biodiversity and stability. Let's take a look at a common predator and prey relationship—the owl and the mouse—and see whether we can determine the limiting factors that impact population size.

How do predator and prey interactions affect population size?

Materials
80 beans (per group) • 4 pipe cleaners (per group) • chalk (or tape) • large, flat surface (desk or floor)

PROCEDURE

A To determine the mouse population's environment, use chalk or tape to mark off a 1 m × 1 m square on the table.

B To represent the mouse population, scatter 80 beans randomly inside the area.

C Bend each of the 4 pipe cleaners into a 10 cm × 10 cm square. These will represent the owls.

D Toss the first owl within the mouse population. All of the "mice" within the square should be considered "eaten" and removed. Copy the table below onto a sheet of paper and record the number of mice eaten for the first owl on Day 1. Continue with the other three owls and record the number of mice eaten on Day 1. Each owl must eat and be removed before the next owl can feed.

E In order to stay alive, each owl must eat 4 mice within 3 days. If any owl eats less than 4 mice, the owl dies and must be crossed out on the table.

F Follow the same procedure to capture mice for the remaining 9 days, recording the number of mice eaten by each owl for each day.

ANALYSIS

1. Describe any observed trend in the mouse population.

2. Describe any observed trend in the owl population.

3. Identify the factors that affect the mouse and owl populations.

4. List any factors that were not a part of the lab activity but which could affect the mouse and owl populations.

GOING FURTHER

G Sometimes different species of plants or animals may be significantly reduced because of invasive species. Research this topic by doing an internet search for invasive species. Be sure to give examples and state how the ecosystem is affected.

NUMBER OF MICE EATEN

OWL	DAY 1	DAY 2	DAY 3	DAY 4	DAY 5	DAY 6	DAY 7	DAY 8	DAY 9	DAY 10
1										
2										
3										
4										
Total Mice Left										

THE HUMAN NICHE

Views of Ecology

We hear terms like *ecological footprint* and *climate change* all the time on social media, but what do they mean? These terms are scientific. People of all worldview perspectives are concerned about their ecological footprint and climate change. But many who are most concerned with these things have a worldview that does not acknowledge God as the Sustainer of Earth or people as His image-bearers. Let's define some of the common claims in ecology and use the Bible to bring these views of ecology into focus.

GREENHOUSE GASES

Many of the compounds that move through the biogeochemical cycles are **greenhouse gases**—gases in the atmosphere that trap heat. Greenhouse gases are necessary for life because they keep Earth warm and absorb harmful radiation from the sun. The most common greenhouse gases are carbon dioxide, nitrous oxide (N_2O), methane (CH_4), water vapor (H_2O), ozone (O_3), and gases that contain fluorine such as chlorofluorocarbons (CFCs). Many scientists believe that as these gases build up in the atmosphere and trap heat, they cause Earth to heat up in a process called *global warming*. They often say that human activities, such as burning fossil fuels or wood, increase our *carbon footprint*, or the amount of greenhouse gases that people release when they burn carbon-containing fuels. Some scientists project that Earth's temperature will soon increase as we burn more carbon-containing fuels and cut down more carbon-absorbing trees.

Are humans ruining the world?

Questions

Why should we take care of God's creation?

How does evolution affect how people view ecology?

How is ecology connected to other fields of science?

Terms

greenhouse gas
ecological footprint
climate change
invasive species
bioremediation

Evaluation

Temperature data suggests that Earth's temperature has risen about half a degree over the past fifty years, a very small change. Predictions of a great temperature increase in the future that are based on a small increase in the past are projections—not observations—and thus may not be reliable.

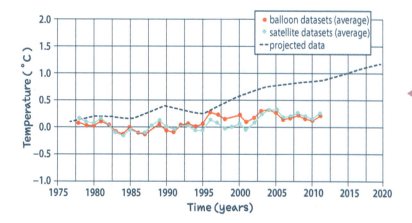

Comparing temperatures observed with weather balloons and satellites to temperatures predicted with models over three decades. Notice how much higher the predicted temperatures are than the actual temperatures.

Though human industry does produce greenhouse gases, natural processes are responsible for most of the greenhouse gases released into the atmosphere. When greenhouse gases reach a certain level in the atmosphere, bacteria, algae, and plants flourish, taking in these gases and removing them from the atmosphere. This doesn't give people license to carelessly release nitrogen and carbon into the environment—part of fulfilling the Creation Mandate is to manage the byproducts of human industry. We should monitor the amount of greenhouse gases that we produce, improve our technology to produce less of them, and use trees wisely. But we can be assured that the fear of destroying our world with greenhouse gases is unfounded. God designed His earth to recycle its byproducts, including greenhouse gases.

ECOLOGICAL FOOTPRINT

How we humans relate to Earth's carrying capacity is sometimes called our **ecological footprint**. This idea was developed in 1992 as a measure for how many resources people use and how quickly the planet can replenish those resources. According to the ecological footprint idea, 0.021 km² of land and water supports one person, but on average each person requires 0.027 km², meaning that humans are using up Earth's resources faster than it can sustain itself and releasing greenhouses gases in the process. But is this true? Is our ecological footprint too big?

Evaluation

Ecological footprint calculations are generally faulty on a large scale for two reasons. These calculations struggle to consider the complexity of modeling human interactions with the environment. A single calculation cannot accurately capture each factor involved. Because these calculations are based on current research and data, they also don't take into account future technology or additional collected data. Even as the human population has increased exponentially—in the last sixty years world population has doubled—we haven't exponentially increased the amount of land that we use because gains in technology have helped us to grow more food in less space.

Though ecological footprint calculations are not completely accurate, they can point out places where resources are being mismanaged. For example, large swaths of rainforest have been cut down and replaced with crops. People need these crops to eat, but when nothing else is planted except for one kind of crop plant, the soil quality becomes poor. A mix of crop plants and rainforest plants would increase the area's biodiversity and therefore its soil quality. Knowing the effect on the environment on a local scale can motivate us to recycle our trash and use water wisely. But it is important to know that when it comes to calculating global footprint, many ecologists admit that footprint models are limited because of the many factors that must be considered.

Earth itself fights against increasing levels of nitrogen and carbon. Though we should assess our impact on the land and its resources and correct our actions if needed, God has given us a world that can handle our footprint.

The ecological footprint model compares an area's used natural resources—cropland, fish resources, forests, developed land, and pastureland—with the amount of natural resources that it generates.

CLIMATE CHANGE

Scientists have observed fluctuating temperatures, such as a warm decade leading to a cooler one. And changes such as melting glaciers are hard to ignore. These are part of **climate change**, which is the change of global temperatures and weather patterns over time. Once again, most scientists blame people for Earth's shifting climate and its effect on plant and animal populations.

Evaluation

Most of the fears surrounding climate change are based on melting ice and snow, rising ocean levels, and increased rainfall. These changes are observable, and some people see them as the first steps toward Earth's destruction. According to a biblical worldview, our world has gone through at least one major change in its climate during its approximately 7000-year history. The Flood, followed by the Ice Age, drastically changed Earth, but in God's grace the planet has rallied to replenish itself. And changing climate might be a sign that Earth is still recovering from the Flood. Also, God promised us after the Flood that seedtime and harvest, cold and heat, summer and winter, and day and night would continue without fail (Gen. 8:22). So we can be confident that melting glaciers and changing temperatures, though they may change the landscape and populations, will never drastically change the patterns of nature that God has designed for Earth.

Over the decades that scientists have studied climate change, they have found that it is an extremely complex issue. Because of this complexity, no wholly satisfactory explanation has yet been found to account for it. One theory is that the sun's activity is responsible for it. Some scientists have suggested that El Niño, a global weather event that occurs every few years, affects climate. But other scientists continue to propose theories that we humans cause climate change and should be the ones to fix it before we destroy our planet.

We are charged by God to learn as much as we can about Earth's changes, model them, and act wisely on what we learn. If we can wisely control our impact on climate, then we should do so. While we must be careful not to get caught in the hype of popular theories that seek to leave out God's promises and His control over Earth, it is equally important to remember that there is a balance between human flourishing and mitigation efforts. Although there is an increase in greenhouse gases with increased technology, the growth in technology has actually revolutionized human existence, significantly reduced infant death rates, and propelled upward the average lifespan.

Melting glaciers, such as in the Northwest Passage, have opened shipping routes, cutting shipping costs and connecting continents in a whole new way.

EXTINCTION RATES

The pygmy three-toed sloth is on the verge of extinction. It lives on several islands off the coast of Panama. Locals harvest trees from the islands to build their homes, destroying the sloth's habitat in the process. Many evolutionary biologists view the demise of the pygmy three-toed sloth as an example of the ever-growing decrease in the number of species. And they accuse Earth's dominant species, humans, for this decline. These biologists assume that millions of years ago, before humans evolved, only ten species went extinct each year. They also maintain that Earth today is losing thousands of species each year, and that there were many more species in the distant past—over a billion species—than there are now (almost 1.5 million cataloged species).

These assumptions lead people to think that increasing extinction rates will wipe out diversity in a few decades. These biologists have concluded that the biggest reason for all this extinction is the growth of the human population. Is this true?

Evaluation

If we use God's Word as our authority, then we know that God created the kinds of creatures that He wanted on Earth, and they quickly filled every corner of the globe. Then God sent the Flood about 4000–5000 years ago, causing a mass extinction of life. We see evidence in the fossil record of the great variety of life before that worldwide event. After the Flood, all land animals "in whose nostrils was the breath of life" died (Gen. 7:22). Only plants and animals that could survive in the water lived through the flood.

Through Noah God preserved each kind of creature so that they could repopulate the world. Therefore, a biblical worldview assumes that the number of species has been increasing since that time, and that seems to fit current observations. Researchers are discovering many new species every year!

Because of their naturalist worldview, evolutionists interpret the massive extinction that they see in the fossil record as part of a normal cycle of extinction in Earth's history. So they believe that today's species are beginning another extinction cycle. They then inflate the actual extinction rate to make it higher than the rate of species discovery. What is the actual extinction rate? In the past 500 years about 1000 species have gone extinct. That's about two species every year. The number may be higher than that, but scientists haven't observed thousands of species going extinct every year.

Should we care about killing off a species? Some argue that if the building of a home destroys the habitat of an endangered species like the pygmy three-toed sloth, the home should not be built. Others argue that sloths don't matter and that people should build whatever they want. A more reasonable view is that the answer lies somewhere in between. We should not willfully destroy a species if possible, but neither should we let plants and animals run the planet—that's not their role. Our responsibility is to watch over threatened species, fill the world with more of God's image-bearers, and provide for people above all other creatures.

Connecting to Ecology

Issues in ecology branch out and touch many areas of life, and it takes more than ecologists to solve ecological problems. One example of different groups coming together to solve a problem has to do with **invasive species**, organisms that move into a habitat where they are not native and then compete with native species for resources. A case in point is the Burmese python, a snake native to Southeast Asia, which is popping up more and more in parts of Florida. The pythons were introduced to the Florida environment by a python breeder whose shop was destroyed in 1992 by Hurricane Andrew and by people who bought them from pet stores and then released them once they got too big. These giant snakes are thriving in Florida's Everglades, but they're disrupting the native species. Wildlife biologists knew that they were no match for the growing numbers of pythons, so they began to work with other people, such as researchers at a local natural history museum, park rangers, hunters, volunteers, government agencies, and even the police to combat the snakes. Together these groups are working to solve an invasive problem that affects Florida's ecosystems and economy.

Another ecological problem—oil spills—is one that we may have found a solution for. **Bioremediation** is a technique that uses organisms to remove or neutralize hazardous materials. In 2010 the largest oil spill in US history released millions of gallons of oil into the Gulf of Mexico. People used physical barriers called booms to manually gather the oil floating on the surface as well as chemicals to break down the oil, but it was still spreading throughout the Gulf. So scientists deployed their last hope: booms filled with fertilizer and oil-munching bacteria and fungi. The fertilizer gave the microbes enough nutrients to help them multiply, and the rest of their nutrition came from the oil. Groups from around the country donated materials for the booms, and volunteers helped scientists construct them and get them afloat. There's still oil along the Gulf coast, but the microbes are becoming more and more effective in cleaning up what remains.

The command to care for our world can sound simple, but carrying it out is complicated. For every wise decision we make about ecology, we often make a foolish one. But that doesn't mean that we should give up. Though managing our responsibility is a difficult balancing act between preserving resources and using them to help people, we can learn how to face these tough ecological situations with God's wisdom.

4.2 SECTION REVIEW

1. What are two ways that bacteria help keep Earth's processes in check?

2. What do ecological footprint calculations fail to consider?

3. Why will rising ocean levels that are due to climate change never destroy life on Earth?

4. People with a naturalistic worldview say that species are generally (growing / declining). Those with a biblical view say that species are generally (growing / declining).

5. A swampy patch of land next to a town has become a haven for mosquitoes. The townspeople want to drain the swamp to get rid of the pests, but the area is protected by the government because it's home to an endangered frog. What might be some solutions for this problem?

6. Do some research and write a paragraph on an example of bioremediation being used to combat an invasive species. Include at the end of your paragraph whether the effort solved the problem or created a new one. What other areas of science did ecologists draw from to solve this problem?

7. How could a high-school student in Florida team up with wildlife biologists to help control the Burmese python population?

CREATURES AND CLIMATE CHANGE

Introduction

Several years ago it was reported that malnourished baby sea lions off the coast of California were coming onshore in droves. Many wondered whether this was related to a change in climate. What could have caused this odd behavior?

Task

Choose a migrating animal, such as a specific bird, butterfly, or sea creature, and try to determine whether there's a connection between the animal's behavior and climate change. Then decide whether humans should be part of the solution for helping it get back on its feet.

Procedure

1. Once you've chosen an animal, research to find out the following information about it:

   - typical migration patterns, eating habits, and habitat preferences
   - new or changing migration patterns, eating habits, and habitat preferences
   - possible causes for the changing behaviors
   - ways that people are helping it recover

   Pay attention to words such as *predicted*, *projected*, *forecasted*, and *potential*. Many statistics that you'll find about the effect of climate change on animals may be projections into the future and not observed data.

2. Create a map that compares the animal's migration pattern in the past (ten to twenty years ago) and its migration pattern now. Include on your map a summary of the research that you did.

3. Suggest a possible way that your animal can be helped. For example, there are now several sea lion rescue groups in California that nurse baby sea lions back to health and then release them. There are also new fishing restrictions to help increase the fish populations that sea lions like to eat.

4. Include a few sentences about how worldview might affect a person's interpretation of data. Base your sentences on the two questions below.

   - Why do many scientists believe that changing migration patterns are caused by humans?
   - What is the biblical role of people when it comes to fulfilling the Creation Mandate?

Conclusion

It makes sense that as Earth's climate changes, animal populations change too. We should try to help struggling animals as much as we can because they are resources that are under our care. We should also keep in mind that "the earth is the Lord's" (Ps. 24:1) and that He is in control of every change.

4 CHAPTER REVIEW
Chapter Summary

4.1 SUSTAINABILITY

• The water, oxygen, carbon, nitrogen, and phosphorus cycles are biogeochemical cycles that maintain balance in the biosphere.

• Population density and limiting factors contribute to changes in the growth and biodiversity of populations.

• The carrying capacity of an area may be reached or exceeded as population growth occurs.

• Primary succession of an area takes place when barren land begins to produce vegetation. Secondary succession follows primary succession and occurs in areas already containing soil.

• Models in nature are useful yet limited. They are helpful in predicting outcomes and evaluating the growth and decline of populations.

Terms
sustainability • biogeochemical cycle • nitrogen fixation • population density • limiting factor • exponential growth • carrying capacity • primary succession • pioneer species • secondary succession • climax species

4.2 THE HUMAN NICHE

• As God's image-bearers, man is responsible for taking care of Earth and balancing available resources.

• Views of human activity affecting greenhouse gases, ecological footprint, climate change, and extinction rates are influenced by worldviews.

• When relating ecological issues to human activity, historical and biblical evidence must be considered.

• Various methods of bioremediation can be used to address ecological problems.

Terms
greenhouse gas • ecological footprint • climate change • invasive species • bioremediation

Chapter Review Questions

RECALLING FACTS

1. What is sustainability?

2. What are the five major biogeochemical cycles?

3. List two ways that water leaves Earth's surface and enters the atmosphere.

4. Which two biogeochemical cycles are most closely tied together? Why are they linked?

5. Which organisms give off oxygen? Which give off carbon dioxide?

6. Identify the three ways that nitrogen can be made usable for plants.

7. Describe how nitrogen enters the body.

8. How does lightning contribute to the nitrogen cycle?

9. If a population's death rate is higher than its birthrate, the population will (grow / shrink).

10. If a population's emigration rate is higher than its immigration rate, the population will (grow / shrink).

11. What is the name given for the species of plants that began to populate the area around Mount St. Helens after its eruption?

12. What is the biblical role of people in sustaining the living and non-living components of the environment?

13. What is responsible for most of the greenhouse gases in the atmosphere?

14. According to a naturalistic worldview, why is climate change such a big concern?

15. On what assumptions do evolutionists base their extinction rates for species?

16. Why are invasive species harmful?

17. List two branches of science that may be involved in cleaning up an oil spill.

UNDERSTANDING CONCEPTS

18. Describe the flow of carbon through the biosphere.

19. Draw a simple diagram showing the flow of nitrogen from a peanut plant to a rabbit and back to a peanut plant.

20. Explain why excess amounts of phosphorus that leach into waterways from the fertilizers used on farms can be harmful to life.

21. A population of bacteria in a petri dish will grow rapidly and then begin to shrink even if the bacteria are well fed. What causes the bacterial population to shrink?

22. Give an example of God's orderly and predictable design in the way that succession takes place in an abandoned field.

23. When a scientist finds that she cannot accurately model the behavior of a herd of elephants, should she assume that their behavior is too chaotic to model? Or could it be that there may be more variables to include in her model? Explain your answer, including how worldview affects the assumption that you choose.

24. Some scientists predict that if the climate continues to change, we may no longer have seasons. What is wrong with their reasoning?

25. Should Christians ignore climate change data? Explain your answer.

26. Why should we be concerned about the amount of carbon that our industries release into the atmosphere?

27. Why is it difficult to figure out how quickly species form today?

28. What is bioremediation and how does it help fulfill the Creation Mandate?

CRITICAL THINKING

29. Social media has been buzzing with news about the increasing degradation of our planet. Many scientists have claimed that constant pollution, deforestation, and overuse of resources will destroy it. Explain how this claim displays bias. How should those with a biblical worldview view this issue?

30. How might an artist who frequently draws birds be able to help researchers track down a new species of bird?

31. People around the world are working on creative ways to use invasive species, such as turning kudzu vines into biofuel. Do some research and write a paragraph about one of the interesting ways that people are using an invasive species.

Use the table below and the following information to answer Questions 32–39.

The table shows the land area covered by colonies of monarch butterflies that migrate to Mexico in the winter.

MONARCH COLONIES MIGRATING TO MEXICO

YEAR	AREA (ACRES)	YEAR	AREA (ACRES)
1993	15.39	2004	5.41
1994	19.30	2005	14.60
1995	31.16	2006	16.98
1996	44.95	2007	11.39
1997	14.26	2008	12.50
1998	13.74	2009	4.74
1999	22.16	2010	9.93
2000	9.46	2011	7.14
2001	23.13	2012	2.94
2002	18.63	2013	1.65
2003	27.48		

32. How do you think ecologists collected this data?

33. Why would ecologists use land area instead of actually counting the monarch butterflies?

34. Graph the table data, placing the year on the *x*-axis and the land area in acres on the *y*-axis. Use a graphing calculator or a spreadsheet program. Then find the equation that best fits the trend line on your graph.

35. How does the equation of your trend line compare to the equation used in the case study on page 80?

36. Monarchs make their winter colonies in the fir and pine trees in the mountains of Mexico. They have one of the longest migrations of any insect, gathering energy for their migration by drinking from milkweed along the way. According to the table, what is happening to the number of monarchs making it to their winter hibernating grounds?

37. List two reasons why monarch butterflies are struggling to migrate.

38. What can ecologists do to help the monarch butterfly population recover?

39. How do the actions that you suggested in Question 38 align with the Creation Mandate in Genesis 1:28?

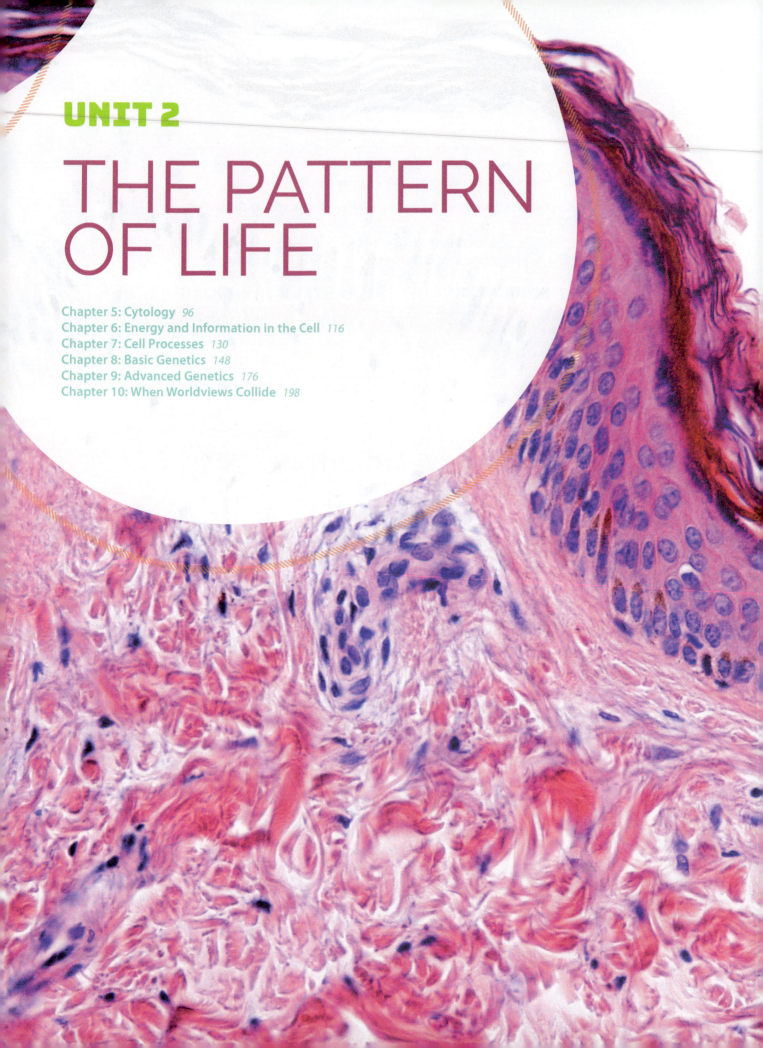

UNIT 2

THE PATTERN OF LIFE

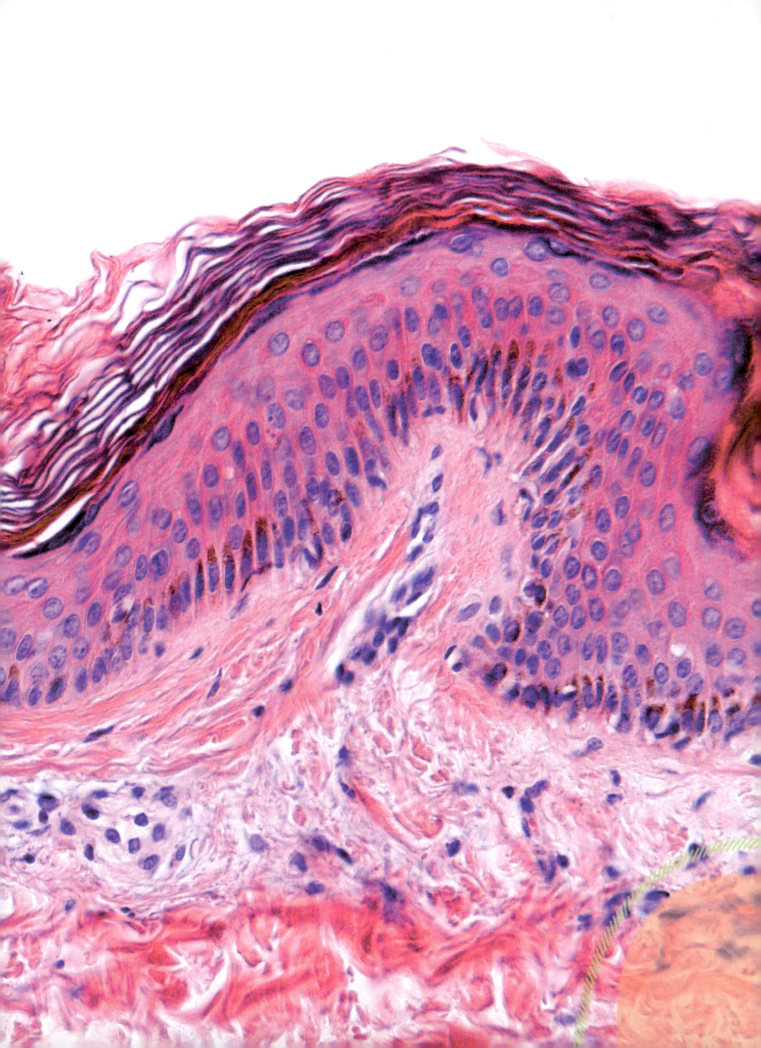

5 CYTOLOGY

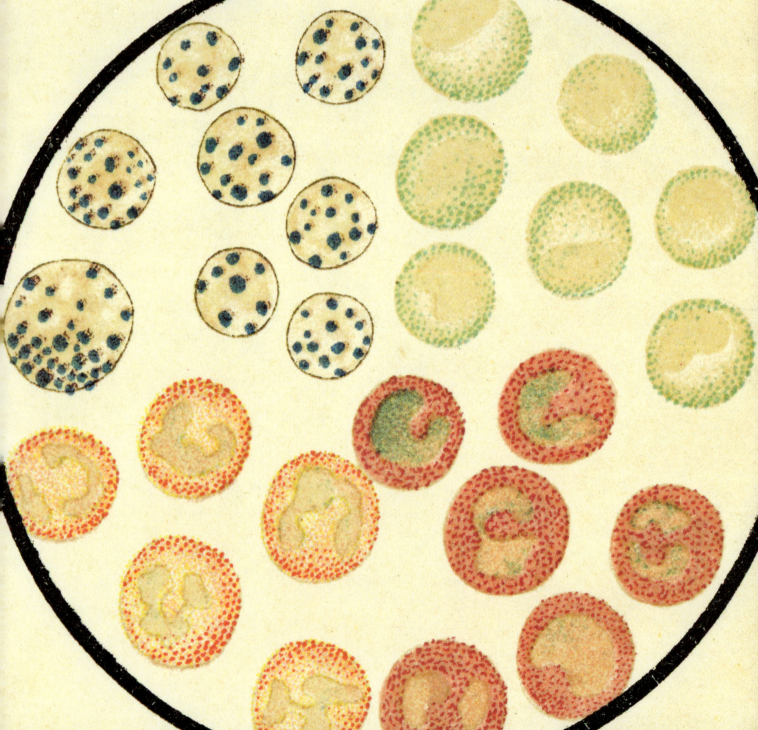

Mending a Broken Heart

In 2009 science fiction became science fact. A medical center in Los Angeles cultured stem cells from the hearts of several patients who had suffered a heart attack. The stem cells were then grown in a laboratory, and the healthy batch of new cells was reinserted into each patient's heart. The new cells immediately went to work in each patient, healing the portions of the heart that had been damaged during the heart attack.

A year after the experiment, the patients' damaged heart tissue had regenerated. This was a medical breakthrough that had the potential to help thousands of people. But the LA medical center wasn't finished. Later, they worked to infuse stem cells with iron so that the cells could be dragged to injured areas of the heart with a magnet, mending hearts better and faster and helping thousands more. Even with the fantastic technology that the medical center had already developed, they were highly motivated to keep pushing the boundaries of medical science. This drive is in line with God's commands to love others and have dominion over creation.

5.1 THE STRUCTURE AND FUNCTION OF CELLS

Modeling the Cell

The idea of the cell, the basic unit of life, hasn't always existed. When biologists first peered into their newly invented microscopes in the 1600s to study plants and animals, they weren't exactly sure what they were looking at. What they saw didn't fit with their current theories of spontaneous generation, the formation of a living thing from nonliving materials, and preformation, the idea that eggs and sperm contain miniature people that simply increase in size after conception. Being able to see an actual cell under a microscope dispelled those theories.

Over the centuries biologists have assembled the tiny puzzle pieces of the cell to create the picture of the cell that we have today. And the discoveries haven't ended—new puzzle pieces are still turning up! In the future, scientists may unveil a never-before-seen part of the cell and completely change how we understand cells. Until then, the ideas we have about cells are proving useful in predicting how they function.

It wasn't until 1665 that Englishman Robert Hooke used the word *cell* to describe the tiny box-like structures that he saw in a piece of cork under a microscope. This was just the beginning of *cytology*—the study of cells. In 1682 Dutch scientist Antonie van Leeuwenhoek discovered that blood is made of cells. He also studied the cells of sperm, bacteria, and larvae under his microscope, which led him to two ideas: each cell had a central part, and living cells didn't arise from nonliving materials or contain mini versions of adult organisms. The model of the cell was taking shape.

What does a cell do?

Questions

Where did the idea of cells originate?

How are the cells in various organisms similar? How are they different?

How do the parts of a cell work together?

Terms

cell theory
unicellular organism
colony
multicellular organism
tissue
organ
organ system
organelle
eukaryote
prokaryote
cytoplasm
cytoskeleton
flagellum
nucleoid
capsule
ribosome

(continued)

Robert
Brown

Almost one hundred fifty years went by before scientists revisited this model. In 1831 British botanist Robert Brown named the central cell part that Leeuwenhoek and others had described—he called it the *nucleus*. Just a few years later the ideas of four young German scientists came together to complete the model of the cell. In 1837 botanist Matthias Schleiden and zoologist Theodor Schwann met for dinner to discuss an idea that they had both been thinking about: that all plants and animals are made of cells. Then in 1855 Robert Remak discovered that cells divide, producing new cells. Three years later Rudolf Virchow expanded on the ideas of Schleiden, Schwann, and Remak by declaring that all cells come from preexisting cells.

Years later the work of these scientists and others after them led to the development of the modern **cell theory**:

1. Cells are the structural and functional units of all living things.

2. Cells come only from other preexisting cells.

These two simple statements contain a lot of information. Ideas such as heredity (the flow of information), metabolism (the flow of energy), and chemical composition (the makeup of cells and their products, information, and energy) are all included in these statements.

The cell theory is a very reliable and predictable model that people use every day to understand cells. But since cells are big enough to observe under a microscope, why do we need a model for them? Biologists still have to explain what they observe about how cells handle energy and information and interact with their environment. As scientists continue to study cells and their environment, cell theory has changed. And one day it may be replaced completely. That's okay because a theory is useful to scientists only when it can explain what we have already observed, make predictions, and be modified to account for new observations.

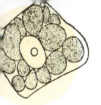

Organizing with Cells

Our bodies are made of an estimated thirty-seven trillion cells that are working together to keep us going. Exactly how these cells work together is one way that biologists classify organisms. An organism that is made of only one cell is called a **unicellular organism**. Bacteria and protozoans are typically unicellular, as are several algae and fungi. A unicellular organism can fully function without help from other cells.

Some unicellular organisms work together in a colony. Within a **colony**, single cells take on specific tasks, such as reproducing, moving the colony, and removing wastes, but individual cells can break away from the colony and function on their own.

A **multicellular organism** is made of two or more cells that depend on each other to function. Some algae and fungi are multicellular, and so are all plants, animals, and humans. Most cells in a multicellular organism are highly specialized and cannot function without the support of other cells.

For example, blood cells have a completely different role than skin cells do, but one can't function without the other.

Why are cells so small? The genetic information inside a cell governs its size. But even if that information were turned off, a cell would still stay tiny. Why? Because the cell's surface area must be comparable to its volume.

The nutrients that a cell needs enter the cell through its surface. Whenever the surface area of an object increases, its volume increases even more, creating more volume inside a cell than can be taken care of by the cell's surface. To illustrate this, consider that the average cell has a surface area of $314 \ \mu m^2$ and a volume of $524 \ \mu m^3$, a ratio of 1:1.7. The largest cells have a surface area of $0.031 \ mm^2$ and a volume of $0.52 \ mm^3$, a ratio of 1:16.8. So a large cell's volume is 16.8 times greater than its surface area. That volume requires a lot of nutrients, so the surface of a large cell is a very busy place and would shut down completely if the cell were any bigger.

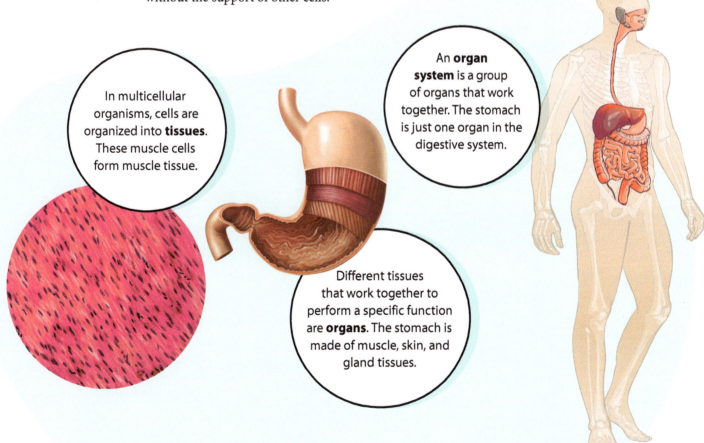

In multicellular organisms, cells are organized into **tissues**. These muscle cells form muscle tissue.

An **organ system** is a group of organs that work together. The stomach is just one organ in the digestive system.

Different tissues that work together to perform a specific function are **organs**. The stomach is made of muscle, skin, and gland tissues.

The Cell's Parts and Purposes

There's another way that organisms can be classified other than as unicellular or multicellular, and that is by the kinds of parts they have. The parts inside a cell are called **organelles**, which literally means "little organs." Many organelles are surrounded by a membrane, and an organism with cells that contain membrane-surrounded organelles is a **eukaryote**. Humans, animals, plants, fungi, and protists are all eukaryotic organisms. An organism whose organelles lack surrounding membranes is a **prokaryote**. A bacterium is an example of a prokaryotic organism.

STRUCTURE & FUNCTION in CELLS

cytoplasm
consists of everything within the cell membrane except the nucleus; contains cytosol, the fluid in which the organelles are suspended; has molecules used for building structures in the cell

cytoskeleton
a system of fibers in the cytosol that helps maintain the cell's shape and provides protein motors and a track to move substances around the cell in a process called *cytoplasmic streaming*

Typical Bacterial Cell (Prokaryote)

flagellum (pl. flagella)
an extension of the cytoskeleton; usually only one or a few on a cell; propels the cell through its environment using a protein motor

nucleoid
found in prokaryotes; contains the genetic material for the cell; floats freely in the cytosol

capsule
found in bacteria outside the cell membrane and cell wall; protects the cell; contains water to keep the cell from drying out, often making it feel slimy

ribosome
found in both prokaryotes and eukaryotes because it's not surrounded by a membrane; contains proteins and RNA; lines up amino acids to make proteins; either attached to the endoplasmic reticulum (ER) or floating in the cytosol

smooth endoplasmic reticulum (smooth ER) transports compounds around the cell and helps maintain the cell's shape; processes fats and breaks down toxic substances in liver cells

rough endoplasmic reticulum (rough ER) transports compounds around the cell and helps maintain the cell's shape; is studded with ribosomes and processes the proteins made by ribosomes

chloroplast
a type of pigmented plastid; found in plants and algae; converts light energy from the sun into chemical energy

Typical Plant Cell (Eukaryote)

leucoplast
a type of colorless plastid; found in plants and algae; stores starches, lipids, and proteins

granum (pl. grana) found inside chloroplasts; made of stacks of thylakoids, which contain the green pigment *chlorophyll* to carry out photosynthesis

cell wall
found in plants, fungi, algae, and bacteria; provides strength and rigidity to the cell; contains pores so that materials can pass through it

central vacuole
found mostly in plants; stores water, salts, sugars, and proteins; maintains *turgor pressure*, the water pressure inside the central vacuole that keeps the cell rigid

nucleus (pl. nuclei)
found in eukaryotic cells; "control center" of the cell; controls the actions of the cell and contains its genetic material

nucleolus (pl. nucleoli)
contains RNA and proteins; area of the nucleus where ribosomes are assembled

chromatin
the genetic material of the nucleus; contains DNA, RNA, and proteins

cilia (s. cilium)
extensions of the cytoskeleton; often cover an entire cell or a portion of a cell; shorter than flagella; propel the cell through its environment; move particles past the cell, such as moving mucus out of the lungs, into the throat, and down to the stomach

Typical Animal Cell (Eukaryote)

cell membrane
surrounds each cell of both prokaryotes and eukaryotes; protects the cell and allows certain materials to move through it; contains proteins that perform several different functions for the cell

lysosome
a type of vacuole; found in human, animal, and animal-like cells (protozoans); contains digestive enzymes that digest food, kill bacteria and viruses, and recycle old cell parts; moves to the cell membrane by cytoplasmic streaming to release wastes outside the cell

centrosome
found in animal and human cells; builds parts for the cytoskeleton

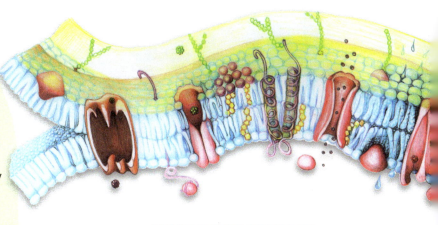

lipid bilayer
a structure based on lipids having a hydrophilic (water-loving) end and a hydrophobic (water-fearing) end, thus arranging themselves into two layers: the hydrophilic ends facing the watery environments inside and outside the cell, and the hydrophobic ends facing each other to escape the water

Golgi apparatus
the "post office" of the cell; receives substances from the ER and packages them into membrane sacs called *vesicles*; sends vesicles either to places within the cell or to the cell membrane to deliver their cargo outside the cell

mitochondrion (pl. mitochondria)
the "powerhouse" of the cell; transforms energy from sugars into usable energy for the cell; has an inner membrane that contains folds (*cristae*), allowing for more surface area to increase energy output and more proteins to be embedded in the membrane; may have many or few cristae depending on how much energy a certain type of cell needs

WHY STUDY CELLS?

The model of cells is still developing. For example, the vesicle in plant cells that makes unripe fruit taste bitter was discovered inside chloroplasts as recently as 2013. But why bother improving the cell theory? Does it really help us to fulfill the Creation Mandate and love others?

It was while studying cells that researchers in Los Angeles discovered a way to turn stem cells from a patient's bone marrow into heart cells. To do so, they had to tinker with each protein in the stem cell and see how the cell reacted. Once they figured out which proteins were essential for changing a bone cell into a heart cell, they began testing their discovery. This led to clinical trials where heart attack patients could benefit from this groundbreaking research. Patients were able to go back to work, walk farther without getting tired, ride a bike, and even play the trumpet again.

Non-Christian researchers do their work for the sake of improving people's lives, just as Christian researchers do. But a Christian has a bigger picture: to love people as Christ does, working for their benefit so that they might see Him. Christians can also view cell research as a way to obey God's command to manage the earth. If studying the tiniest and most insignificant parts of cells can save and improve people's lives, then we are bound by God's commands—managing the earth and loving others—to study the lifesaving possibilities that are still locked inside the cell.

Scaling Up a Cell

Take another look at the cell diagram on page 100. Sure, it looks like a cell, but how well does it represent a real cell? Is the number of each of the different kinds of parts correct? Are their sizes accurately portrayed? Does the diagram convey a sense of how small a cell really is? Let's think about the potential difficulties involved in making scale models of cells, that is, models whose parts are made in the proper proportions relative to each other. We'll do this by thinking about something that probably everyone has experience with—plastic building bricks!

How well do models represent real cells?

Materials
metric ruler • metric tape measure • calculator

PROCEDURE

A Let's start by modeling the innermost bits of a cell—its DNA. The human genome, found in most human cells, contains roughly 6.4 billion base pairs. Let one 2 × 4 building brick represent one base pair.

1. If each brick is 16 mm wide (the base pairs in our model will stack along their long sides), what will be the total length of our brick genome? State your answer in meters.

2. Do you see an issue with your answer from Question 1 with regard to modeling the DNA of an entire cell using building bricks? Explain.

B So maybe DNA base pairs are too small to model with building bricks. Maybe we should try using the bricks to model something larger, like an entire structure made of cells, such as an index finger. Let's try it!

3. Use your ruler to measure the length of your index finger in millimeters and record your value.

4. If a skin cell is 30 μm long (over 88 000 times larger than a DNA base pair), about how many are there along the length of your finger?

5. If you were to use 2 × 2 bricks (16 mm × 16 mm) to represent each cell along the length of your finger, how long would the entire model of your finger be? Report your answer in meters.

C To get an idea of how large your model would be, use the tape measure to measure out the length you reported for Question 5. You may need to go outside to do this!

6. Summarize what you have learned from this activity about the difficulties of modeling to scale cells, cell parts, and structures made of cells.

1. Which part of the cell theory refers to the flow of energy through a cell? to the flow of information through a cell?

2. (True or False) As discoveries were made that couldn't be explained by spontaneous generation, scientists came up with an updated version of the spontaneous generation model.

3. Copy and fill in the concept map on the right on your own paper to complete the information on the different levels of organization of cells.

4. How are organelles in a prokaryotic cell different from those in a eukaryotic cell?

5. Since ribosomes are found in bacterial cells, do ribosomes have a membrane around them? Why or why not?

6. Some diseases cause people to become extremely tired. They don't seem to have any energy. What organelle might be damaged in their cells? Research and describe a few treatments that are being used to help people who have this damaged organelle.

7. Would a muscle cell that works hard every day have many or few cristae in its mitochondria? Explain.

8. How does a biblical worldview direct a Christian researcher's work?

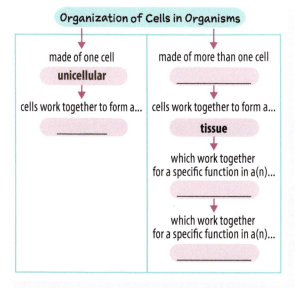

Organization of Cells in Organisms

made of one cell
unicellular
cells work together to form a...
_____

made of more than one cell
_____
cells work together to form a...
tissue
which work together for a specific function in a(n)...
_____
which work together for a specific function in a(n)...
_____

5.2 THE CELL ENVIRONMENT
Balance in the Cell

It's always exciting to see a package that you ordered show up on your doorstep. You rip the box open and happily pull out the book, shoes, or gadget that you've been waiting for. Have you ever thought about the massive effort required for something you purchased a thousand miles away to arrive at your house? It takes the coordination of huge shipping centers to process incoming and outgoing items and to ensure that your new shoes end up at your house and not your neighbor's.

That's how your cells work—they are the shipping centers of the body. They not only send products around the body; they also make the products. That's a lot of work for the tiniest parts of your body! To do this job every day, your cells must maintain **homeostasis**, which is the internal balance of a system that keeps conditions stable.

Another term used to describe this balance is *dynamic equilibrium*: the balance of changes and motions in a system. For example, a shipping center has enough space in its warehouse to hold 10,000 items; if the warehouse has fewer than 3000 items, the shipping center will close. Every day, items are shuttled into and out of the warehouse, and the warehouse manager makes continual adjustments to maintain the warehouse's dynamic equilibrium of more than 3000 items but fewer than 10,000.

? Could my goldfish survive in the ocean?

Questions

How does a cell maintain balance in different environments?

How does a cell send and receive materials?

Terms

homeostasis
positive feedback
negative feedback
osmosis
hypertonic solution
hypotonic solution
isotonic solution
passive transport
facilitated diffusion
active transport

In a cell, things like temperature, nutrition, and illness can affect homeostasis, so the cell is designed to constantly adjust to maintain an equilibrium. Under ideal conditions cells may not have to work that hard. These conditions are in the optimal range for cells. But if conditions change, cells leave the optimal range and enter the range of tolerance, a status that they can handle with some extra work. If you get too hot or cold, your body does its best to stabilize itself and get back into the optimal range. When you're hot, you sweat to remove heat from your skin. When you're cold, you shiver to generate heat. But when conditions go beyond what cells can handle, called the *limit of tolerance*, the cells die. Sunburn and frostbite are the result of cells reaching the limit of their tolerance.

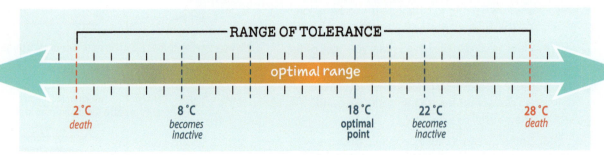

RANGE OF TOLERANCE

optimal range

| 2 °C | 8 °C | 18 °C | 22 °C | 28 °C |
| death | becomes inactive | optimal point | becomes inactive | death |

GETTING FEEDBACK

How do cells regulate the speed of processes or even start and stop them when needed? Signaling occurs through two important processes: positive feedback and negative feedback.

Positive feedback occurs when a substance involved in a cellular process causes the process to speed up. For example, when you breathe in air, the oxygen must move from your lungs to all the cells of your body. Remember from Chapter 2 that each type of protein has its own unique shape that enables it to do its job. Red blood cells contain a protein called *hemoglobin*, which has four subunits that can bind with oxygen molecules. In their natural shape, the subunits have only a low attraction for oxygen, so it would take a long time for oxygen molecules to bind to all four subunits. But when oxygen enters the bloodstream from the lungs, an oxygen molecule will usually bind to at least one of the subunits. This makes the other three subunits change shape so that they can quickly bind to oxygen molecules, guaranteeing that every hemoglobin molecule carries a full load of oxygen to the cells.

Negative feedback occurs when a substance produced by a cellular process causes the process to slow down or stop. For example, mitochondria provide cells with usable energy in the form of a molecule called *ATP*. One of the many enzymes involved in this process is *PFK*. When the cell needs energy because the number of ATP molecules is low, PFK swings into action. But as the number of ATP molecules rises, they begin binding to PFK, deactivating it. Thus, the product (ATP) brings the process to a halt until the cell begins to run low on ATP again.

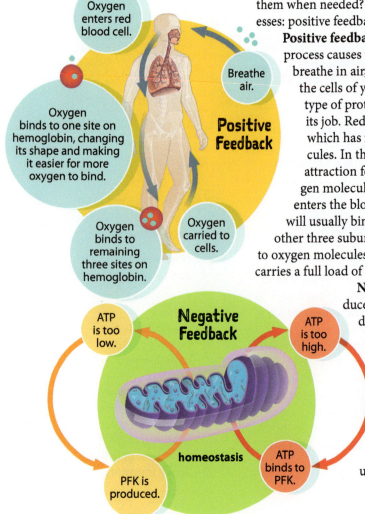

Oxygen enters red blood cell.

Breathe air.

Oxygen binds to one site on hemoglobin, changing its shape and making it easier for more oxygen to bind.

Positive Feedback

Oxygen binds to remaining three sites on hemoglobin.

Oxygen carried to cells.

ATP is too low.

Negative Feedback

ATP is too high.

PFK is produced.

homeostasis

ATP binds to PFK.

Solutions and the Cell

Cells are always in some sort of solution. The air we breathe and walk through and the water we drink and swim in are solutions that affect cells. We experience the chemistry of solutions when we put lots of salt on our French fries. We end up feeling thirsty because the cells in our mouth and our digestive system become depleted of water. The same thing happens when a sailor gets lost at sea and drinks ocean water out of desperation. He'll be dead in a few days! But why?

Remember from Chapter 2 that the amount of solute dissolved in a solvent is called the concentration of a solution. Water is the solvent both in the ocean and in a person's cells. Ocean water contains a lot of solutes, being about a 3.5% salt solution. It is a more concentrated solution than a person's cells are, which are about a 0.9% salt solution. Also think back to the discussion in Chapter 2 about diffusion—the spreading of solutes throughout a solution until they reach equilibrium. But a person's cells are surrounded by a cell membrane to keep its contents in and everything else out. So something other than diffusion has to take place to move water into and out of our stranded sailor's cells.

The process that takes place is **osmosis**, which is the diffusion of a solvent through a semipermeable membrane, a membrane that allows certain substances to pass through it but not others. The cell membrane is a semipermeable membrane. When ocean water contacts the thirsty sailor's cells, the water in his cells travels through the cell membranes toward the greater concentration of solutes.

semipermeable membrane

ocean water with solutes

water

02:00

04:00

osmosis

This leaves the cells depleted of water and triggers a negative feedback response of "I'm thirsty!" so that the cells can reach equilibrium again. Instead, he drinks more ocean water. Poor guy!

Ocean water is known as a **hypertonic solution**, a solution that is more concentrated than the cytoplasm of a cell. This difference in solute concentration causes water to flow out of the cells through the semipermeable membrane. When cells are in a solution that is less concentrated than their cytoplasm, they are in a **hypotonic solution**. For example, when cells are in pure water, there is more solute inside the cells, so the pure water passes through the cell membrane and into the cells, causing them to swell. When cells and the solution that they are in have the same concentration of solutes, they are **isotonic solutions**. Your body works to keep most of its solutions isotonic.

Transport across the Membrane

PASSIVE TRANSPORT

When we drop a tea bag into a mug of hot water, what happens? The water enters the tea bag, mixes with the ground tea leaves, and then exits as a much tastier substance. We can speed up the process by using hotter water or by stirring, but ultimately the process of tea diffusing out of the tea bag happens without any help. This is known as **passive transport**—the movement of molecules across a membrane without the use of chemical energy. The particles' own natural motion causes them to move from an area of high concentration to an area of low concentration, and cells don't have to expend any energy to get these particles across their membranes.

> **Four factors determine whether a particle can be passively transported across a cell membrane.**

1. *Particle Size*—Very small particles like water, oxygen, carbon dioxide, and nitrogen can pass through the pores of a cell membrane easily.

2. *Particle Shape*—Even if a particle is small enough, it may still be unable to pass through an opening in the cell membrane. A square particle will not fit into a round hole!

3. *Particle Polarity*—Charged regions within a particle can let it in or keep it out. Oil and water don't mix because oil is nonpolar (contains symmetrical charges) and water is polar (contains asymmetrical charges). Likewise, the outer surfaces of a cell membrane's lipid bilayer are polar (hydrophilic), but the inner surfaces are nonpolar (hydrophobic). The nonpolar surfaces of a cell membrane repel polar substances, but the membrane allows nonpolar molecules to pass through easily. Only very small polar molecules like water can pass through the membrane using passive transport. Also, ions with their unbalanced charges cannot pass through the membrane passively.

4. *Membrane Composition*—The lipid bilayer contains different proteins and different-sized openings that affect what can get through a cell membrane. For example, because liver cells process toxins that enter the body, the openings in their cell membranes are very large to accommodate the different toxin particles that the liver cells need to absorb. But in the membrane around a liver cell nucleus, only particles that present a specific chemical "ticket" can pass through.

We've already discussed two types of passive transport—diffusion and osmosis. A third type is known as facilitated diffusion. Glucose, the sugar that your body uses for energy, is a large polar molecule. Yet in some cells it can pass through the cell membrane without the cell expending any energy. How is that possible? In the process of **facilitated diffusion**, molecules move through the cell membrane with the help of transport proteins. These proteins respond to the concentrations of the solutions inside and outside the cell. When the concentration of glucose is higher outside the cell than inside, transport molecules automatically shuttle the glucose molecules into the cell without using any chemical energy.

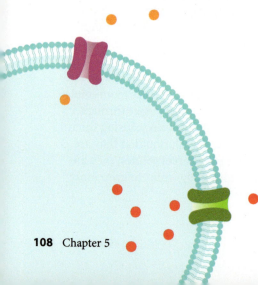

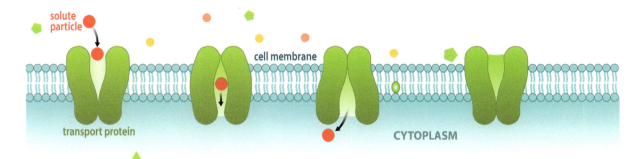

In facilitated diffusion the pressure of the higher concentration of particles outside the cell forces a particle into the transport protein. The protein changes shape to move the particle through the membrane. Then the protein releases the particle into the cell and returns to its original shape.

ACTIVE TRANSPORT

Whether a human is fighting a strong ocean current or a salmon is trying to swim upstream, it takes energy to move against the flow of water. This is a lot like **active transport**, the movement of molecules across a membrane using chemical energy. A cell often needs to move materials from an area of lower concentration to an area of higher concentration. How can the cell actively use energy to do that?

Molecules such as sodium ions attach themselves to transport proteins on the cell membrane, like what happens in facilitated diffusion. But the transport proteins don't automatically shuttle the ions into the cell. The cell must apply some energy to the transport protein before the protein will move the sodium ion across the membrane.

Cells use two other kinds of active transport to handle really big particles. Instead of moving a particle across its membrane, cells can use energy to simply engulf the particle they want, a process known as *endocytosis*. During endocytosis a molecule that comes in contact with the outside of the cell membrane is then surrounded by the membrane. The membrane forms a sack around the molecule and then pinches off to become a vesicle.

Whenever you cut your finger, white blood cells rush to the wound to engulf any bacteria that may have entered your body. Without endocytosis, you wouldn't last very long in the battle against bacteria.

The opposite of endocytosis is *exocytosis*, which occurs when a vesicle fuses with the cell membrane to release its contents outside the cell. When our cells produce proteins, they are shuttled from the endoplasmic reticulum to the Golgi apparatus to be packaged into a vesicle. Once packaged, the proteins are sent to the cell membrane. Here, the vesicle fuses with the cell membrane, opens up, and releases the proteins into the solution around the cell for other cells to pick up and use.

Endocytosis engulfs particles; exocytosis releases them.

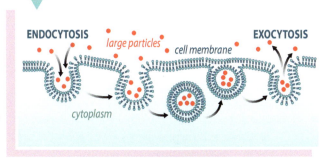

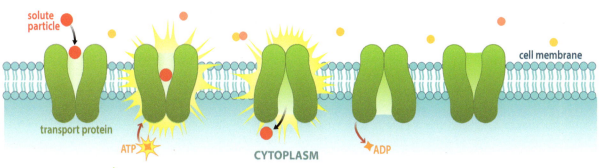

Active transport requires energy to move molecules across the cell membrane from an area of lower concentration to one that is higher. Notice the energy that comes in to activate the protein before the protein will change its shape.

PASSIVE TRANSPORT AND KIDNEY STONES

Doubling over in pain, Jamal felt like someone was stabbing him with tiny knives in his lower back. Until the pain began, he was having a good day—he had a hard but fun soccer practice and had eaten a steak for dinner. His parents rushed him to the ER, where a physician quickly diagnosed the pain as a kidney stone. Jamal's mom mentioned to the ER physician that she had also had a few minor kidney stones recently. Could their cases be related?

Kidney stones can form in several different ways, one of which is through a defect in the transport proteins that carry amino acids from kidney cells to the blood, a condition called *cystinuria*. Amino acids enter the kidneys through active transport but exit the kidneys through passive transport. This means that if the transport proteins that carry the amino acids out are defective, the acids will build up inside kidney cells. One amino acid in particular, cystine, can build up so much that its molecules crystallize together to form stones in the kidneys.

The ER physician told Jamal that once he passed the kidney stone, he could bring it in to be tested. Two painful days later, Jamal passed the stone, and he dropped it off at the hospital laboratory. The laboratory analyzed it and found that it was made of cystine crystals. Armed with this knowledge, Jamal and his mom went to their family physician, who told Jamal that he had cystinuria, an inherited condition. The physician then told him a few ways to manage his newly discovered condition—drink plenty of fluids to help dilute the amino acids in the kidneys, especially when he's active. And eat fewer amino acids—that means less protein.

5.2 SECTION REVIEW

1. When you catch a cold, your body fights the cold virus and tries to eliminate it as quickly as possible. Why does your body react this way to a virus?

2. Two of the most important parts of our blood are red blood cells and plasma. They are usually isotonic. What would happen to the red blood cells if the plasma became hypotonic?

3. (True or False) Osmosis occurs when a solvent moves from an area of lower concentration to an area of higher concentration.

4. Even though sodium ions are tiny, they cannot cross a cell membrane using passive transport. Why not?

5. Sketch out a cell membrane that shows the four factors that can affect a particle as it tries to pass through a cell membrane.

6. Why is facilitated diffusion a type of passive transport?

7. Why are endocytosis and exocytosis considered kinds of active transport?

Refer to the case study above to answer Questions 8–12.

8. Although cystinuria is an inherited condition, some people never know they have it until something triggers it. Judging by the treatments that Jamal's physician suggested, what do you think triggered Jamal's kidney stone?

9. What kind of passive transport is used by the kidneys to move amino acids into the blood? How do you know that?

10. Even though Jamal's mom had never been diagnosed with cystinuria, she decided to follow the physician's advice herself to drink more fluids and eat less protein. Her kidney stone problem quickly diminished. Without testing her kidney stones, how can you know that she has cystinuria?

11. Why do you think that a condition like cystinuria cannot be cured at this time?

12. How would searching for a cure for Jamal's cystinuria show obedience to God?

5 CHAPTER REVIEW
Chapter Summary

5.1 THE STRUCTURE AND FUNCTION OF CELLS

- Typical for scientific modeling, the cell theory was developed over time through the work of multiple scientists.

- The cell theory states that cells are the structural and functional units of all living things and that cells come only from other preexisting cells.

- All living things are made of cells, though they may be unicellular (single-celled), colonial (multiple cells that can each live independently), or multicellular (composed of many cells).

- Most cells in multicellular organisms are specialized for certain tasks and are organized into tissues, organs, and organ systems.

- Every cell contains a multitude of specialized organelles that carry out the different functions needed to live, grow, and reproduce.

- Eukaryotic cells have membrane-bound organelles and a membrane-bound nucleus; prokaryotic cells do not.

- Cellular models are still works in progress. Studying cells can help Christians obey the Creation Mandate and show love for others.

Terms
cell theory • unicellular organism • colony • multicellular organism • tissue • organ • organ system • organelle • eukaryote • prokaryote • cytoplasm • cytoskeleton • flagellum • nucleoid • capsule • ribosome • smooth endoplasmic reticulum • rough endoplasmic reticulum • chloroplast • leucoplast • granum • cell wall • central vacuole • nucleus • nucleolus • chromatin • cilium • cell membrane • lysosome • centrosome • lipid bilayer • Golgi apparatus • mitochondrion

5.2 THE CELL ENVIRONMENT

- All cells must work to maintain homeostasis, a condition of dynamic equilibrium with their environment.

- The further outside its optimal range the environment for a cell is, the harder it must work to maintain homeostasis. If homeostasis cannot be maintained, the cell will die.

- Cells use feedback mechanisms to maintain homeostasis. Positive feedback occurs when a substance involved in a cellular process causes the process to speed up. Negative feedback occurs when a substance produced by a cellular process causes the process to slow down or stop.

- A cell's semipermeable membrane allows water to easily diffuse in or out. In an isotonic solution the rates of inflow and outflow are balanced. In a hypertonic solution a net outflow occurs, resulting in dehydration. In a hypotonic solution a net inflow occurs, causing swelling.

- Cells have different means for transporting substances through their cell membranes. The method required for a particular substance depends on the substance's particle size, particle shape, particle polarity, and the composition of the cell's membrane.

- Passive transport, either diffusion or facilitated diffusion, does not require the cell to expend energy and always occurs along the concentration gradient between a cell's interior and exterior environments.

- Active transport methods (transport proteins, endocytosis, and exocytosis) require energy and can move substances against a concentration gradient.

Terms
homeostasis • positive feedback • negative feedback • osmosis • hypertonic solution • hypotonic solution • isotonic solution • passive transport • facilitated diffusion • active transport

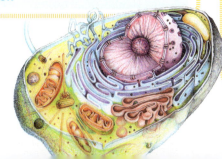

5 CHAPTER REVIEW
Chapter Review Questions

RECALLING FACTS

1. What is the difference between the cells in a colony and those in a multicellular organism?

2. Which kind of cell contains organelles with membranes around them?

3. List three structures that are found in all types of cells.

4. List the three organelles that cells use to move themselves or to move other substances.

5. Place the following plant organelles in order from smallest to largest: chloroplast, granum, thylakoid.

6. Which organelle is most responsible for making proteins?

7. What do you think makes the rough endoplasmic reticulum look rough?

8. What are cilia and flagella made of? What function do they both have in cells?

9. Which two organelles in an animal cell are responsible for moving materials to the cell membrane to be released outside the cell?

10. Which organelle is sometimes thought of as the "post office" of the cell? the "powerhouse" of the cell?

11. Which passively travels across the cell membrane more easily: a polar or a nonpolar molecule?

12. Large molecules can enter a cell passively using _____ or actively using _____.

UNDERSTANDING CONCEPTS

13. Suppose it's the year 2185 and scientists have discovered that something smaller than the cell is the foundation of all living things. How would this discovery affect the cell theory?

14. Do unicellular organisms have tissues? Why or why not?

15. Your study partner says that prokaryotic cells differ from eukaryotic cells in having no genetic material. Is your partner correct?

16. Draw a eukaryotic cell and include the following organelles: cell membrane, centrosome, cytoplasm, Golgi apparatus, lysosome, mitochondrion, nucleolus, nucleus, ribosome, rough endoplasmic reticulum, and smooth endoplasmic reticulum. Label the organelles in your drawing.

17. How is keeping a seesaw from touching the ground an example of a dynamic equilibrium?

18. When we don't get enough sleep, our bodies produce high levels of stress hormones to deal with the lack of sleep. Which range of conditions does this example represent? Explain.

19. What is the difference between positive and negative feedback?

20. How are solutions and feedback signals in cells related?

21. How could you turn a hypertonic solution like ocean water into a hypotonic solution relative to the cell?

22. If you add too much fertilizer to your houseplant's water, the plant will die. Why do you think this happens?

23. How are diffusion and osmosis different? How are they similar?

24. How is active transport different from facilitated diffusion? How are they similar?

25. Since passive transport doesn't use any energy, how does a particle move from the outside to the inside of a cell?

26. Considering the case study on page 110, why do you think amino acids can enter the kidneys but sometimes not exit?

CRITICAL THINKING

27. Why should Christians study cells?

28. When a virus enters a plant cell, it hijacks the cell's nucleus and inserts its own genetic material into it to make more virus particles. This leads to the cell filling with virus particles and eventually dying. Thinking through how each organelle works, how do you think the virus's genetic material takes over and leads to cell death?

29. Hair cells come from the same tissue layer that brain cells do. Because of this, researchers have begun to reprogram hair cells to turn them into brain cells in order to study them. How could this research be used to fulfill God's command to love others?

30. Many single-celled organisms that live in a freshwater environment have an organelle called a *contractile vacuole*. This vacuole continually pushes water out of the cell. Why do these organisms need this organelle?

Use the information in the ethics box on pages 114–15 to answer Questions 31–34.

31. What is the major flaw in the conclusion about abortion reached by secular bioethics?

32. How does the recommended action (Question 6) compare with the recommended action at the end of the biblical ethics triad in Chapter 2?

33. The principles of bioethics have secular origins, and the biblical ethics triad is purposefully Christian. Do the answers that these two models produce always differ? Explain.

34. What will lead a person to come to wrong conclusions when using the principles of bioethics?

5 CHAPTER REVIEW

ETHICS

USING THE PRINCIPLES OF BIOETHICS STRATEGY

In Chapter 2 we analyzed an ethical issue—abortion—using the biblical ethics triad with five questions that guided our analysis (pp. 48–49). Here we will demonstrate the process of ethical decision-making using the principles of bioethics presented in Chapter 1 (p. 22). For this we will look at six questions. Note that the first and last questions are the same as those used with the biblical ethics triad.

1. What information can we get about this issue?

Just as with the biblical ethics model, to formulate an appropriate and informed position we need data and information, so we will need to do some research on the issue before we can evaluate how to ethically approach it.

2. How are the people involved in this issue respected and given the freedom to choose?

This addresses the principle of respect for autonomy. Compare the different options available and formulate a response that shows respect for an individual's freedom to choose or have a loved one choose for him if he is unable to choose for himself.

3. How are individuals protected from harm or injury?

Now we move to the principle of nonmaleficence. We must consider whether individuals are harmed or injured by action taken or action withheld.

4. How are people helped by action taken?

Looking at the principle of beneficence, as part of protecting people from harm, we must also consider what option provides the best care to individuals.

5. How is the action taken just and fair?

Lastly, we must consider the principle of justice. To do justice, we must determine which of the options that we have considered provides equal treatment and fair distribution of the necessary goods and services to individuals.

6. What action should we recommend?

Once we have a clear understanding regarding the option that provides an individual with the best possible care, we can form a final position according to the principles of bioethics and justify the actions that we believe need to be taken.

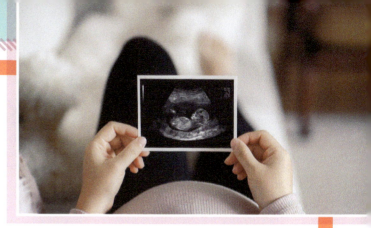

ISSUE: ABORTION

1. What information can we get about this issue?

This question is the same for both models, so the answer with regard to this issue will be the same as well. Abortion is a medical procedure that terminates a pregnancy through the use of medicine or surgery.

2. How are the people involved in this issue respected and given the freedom to choose?

Women have the right to make decisions regarding their own health-care needs. Additionally, parents or guardians have the responsibility for making health-care decisions for minors. Expectant parents have the responsibility to make decisions that affect their unborn children. Ultimately, in the case of abortion, the mother's autonomy supersedes the unborn child's autonomy and trumps the father's input. A common secular position is "My body, my choice."

3. How are individuals protected from harm or injury?

There are many reasons that a woman may seek an abortion. Sometimes the pregnancy is the result of rape or incest, and the woman is looking to move past the traumatic experience. From a secular position, many would argue that allowing her to have the abortion will allow her to move past the trauma and begin the healing process. In other cases, the pregnancy creates a threat to the physical well-being of the mother. Secular doctors would argue that they should protect her life even at the risk of death for the infant. They would argue that this is similar to other medical conditions for which the treatment could cause the death of the unborn child. In many cases, the decision is financial or based on the mother's perceived ability to care for the child. Again, a secular doctor would support the abortion in order to prevent harm to the mother's mental health that may result from the stress of carrying, birthing, and raising the child. While it is recognized that abortion definitely harms the unborn child, the needs of the mother are given first priority. As with many medical procedures, abortions include an element of risk to the mother, but advocates argue that these risks necessitate access to safe abortions in a medical facility.

4. How are people helped by the action taken?

Secular doctors would argue that abortion allows the victims of trauma, resulting in pregnancy, to emotionally move past that event. They would also argue that in cases where abortion is considered medically necessary to protect the life of the mother, the mother is helped. In cases when abortion is done to relieve the stress of an unwanted pregnancy, many believe that ending the pregnancy frees the mother from financial strain, the emotional distress of caring for a child alone, or other difficult circumstances. Of course, the aborted child receives no benefit at all.

5. How is the action taken just and fair?

Justice is about the fair and equitable distribution of services. Pregnancy is a normal and necessary part of life, and in the vast majority of cases, especially in our modern world, pregnancy poses very little health risk. Abortions are therefore elective procedures; that is, they are not normally required to preserve the life of the mother. Justice in this case would mean providing access to abortion in the same manner as for other elective procedures.

Many people with a secular worldview would argue that justice becomes a factor in countries that have limited medical resources, especially since conditions in those countries tend to make pregnancy riskier for the mother. In developed countries the only issue is whether everyone has equal access to medical services.

6. What action should I recommend?

If we consider only the principles of bioethics, abortion should be considered a viable alternative to an unwanted pregnancy. This is especially true in cases in which the mother's life is in danger and the baby has little chance of survival or in cases where the pregnancy was the result of a traumatic event. All of this hinges on prioritizing the needs of the mother.

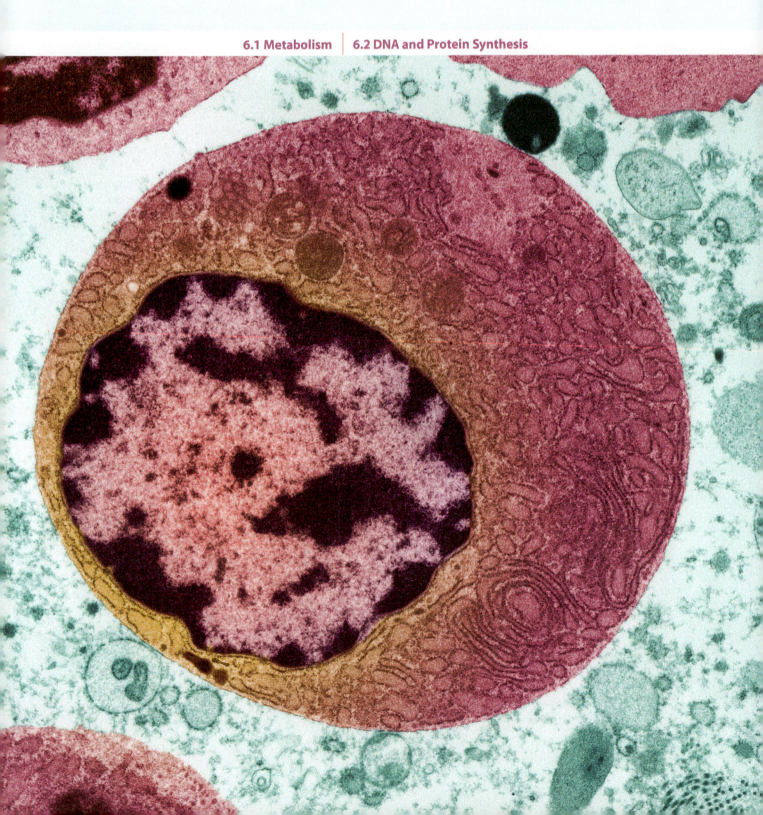

6 ENERGY & INFORMATION IN THE CELL

6.1 Metabolism | **6.2 DNA and Protein Synthesis**

Molecular Machines

Conveyor belts, robotic arms, transport vehicles, and machines with working gears—you expect to see these in an auto factory assembly line that kicks out a new car every minute. You don't expect them to be microscopic molecular machines inside the cells in your body, but they are.

A cell's microtubules work like conveyor belts, and its ribosomes, making proteins, act like robotic arms. The cell also contains kinesin, a transport vehicle. And bacterial flagella function as molecular machines with working gears. We learned about some of these machines in Chapter 5; now let's look at some new evidence of God's intelligence, design, and order in the energy and information in the cell.

6.1 METABOLISM

Energy and Living Things

Staying alive is hard work! In Chapter 2 we learned that living things rely on energy found in their food. Chemical energy gets locked away and stored so that an organism can have it when it needs it.

Organisms that produce their own food are called *autotrophs*. Some living things—called *heterotrophs*—can't make their own food, so they need to get their energy from another living thing that has stored chemical energy. Herbivores, omnivores, carnivores, and parasites are all heterotrophs. Autotrophs are producers and heterotrophs are consumers.

Energy is like money. Autotrophs convert one kind of "money" into another kind by doing a currency exchange, similar to exchanging US dollars for currency in another country. Autotrophs transform energy from the environment into a different form. Likewise, heterotrophs get energy by taking it from another organism. We can save energy or spend it, but we can't use it any faster than we can get it—there aren't any energy credit cards! The law of conservation of energy reminds us that we can't create energy. Even autotrophs rely on energy transformations.

Energy and the Cell

When living things want to store energy to use later, where do they store it? An organism's cells are its "energy bank." Organisms must get energy from their food or from light and store it in cells through chemical processes called **anabolism**. Living things then use stored energy to carry on life processes through other chemical reactions—**catabolism**. Anabolism combines small molecules to make larger ones that store energy, while catabolism breaks down large molecules into smaller ones to release energy. Anabolism is like building a tower with blocks; catabolism tears down the tower to use the blocks for something else.

The energy released during catabolism, usually in the form of heat, helps make chemical reactions happen in organisms by providing the activation energy needed for life-sustaining reactions to proceed. All the processes that gain, save, and spend energy in an organism are together known as **metabolism**—the sum of the chemical processes that an organism uses to get energy from food to grow and function.

What powers a cell?

Questions

Where does an organism's energy come from?

Where does a cell's energy come from?

Terms

anabolism
catabolism
metabolism
adenosine triphosphate (ATP)
adenosine diphosphate (ADP)

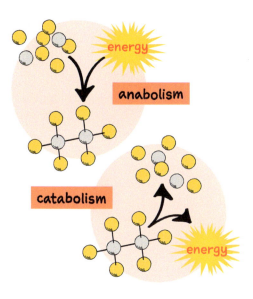

For metabolism to work efficiently, homeostasis within the cells of an organism is required. Cells maintain homeostasis by balancing energy, waste, and their internal environment.

If energy is like money, then its currency is **adenosine triphosphate (ATP)**. We would be dead without this molecule! As we can see in the image below, there are three parts to an ATP molecule. Two parts remain the same during chemical reactions. But the third part, the phosphate tail, is responsible for releasing and storing energy. All three phosphate groups are charged, which means that they push against each other. This is an unstable molecule. ATP is like a compressed spring ready to expand or a dollar bill waiting to be spent.

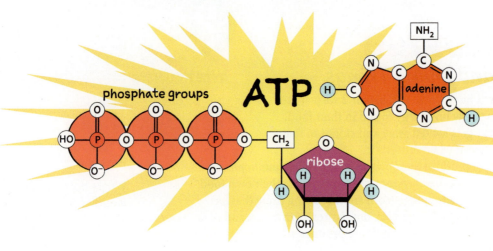

Here's how it works. Anabolic processes take small molecules in food, like glucose molecules, and use them to form large molecules, such as starches and fats. Energy for long-term use in the form of chemical energy is held in the many bonds of these large molecules. But there is too much energy stored there for cells to use all at once, like having a twenty dollar bill and needing only a dollar to buy a drink. If these bonds were all broken at the same time, the energy released would destroy the cell.

Instead, cells store energy in ATP, ready for immediate use. These are manageable packets of energy, or "pocket change," for the cell. When cells need energy, ATP reacts with water to chop off one of the phosphate groups, producing **adenosine diphosphate (ADP)**, one phosphate group, and energy for the cell to use.

Muscles can use their entire reserve of ATP molecules in one minute, so what happens when there's no more ATP to power a muscle? Our bodies very capably recycle ADP. In a working muscle cell ten million ATP molecules are used and regenerated every second! The large molecules provide the energy to regenerate ATP. A phosphate group is linked to the ADP molecule, and once again, energy is stored in the bond, ready to be released when the cell needs it to maintain homeostasis. God's provision for life goes all the way down to the molecular level.

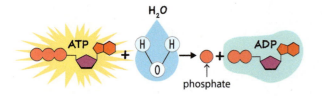

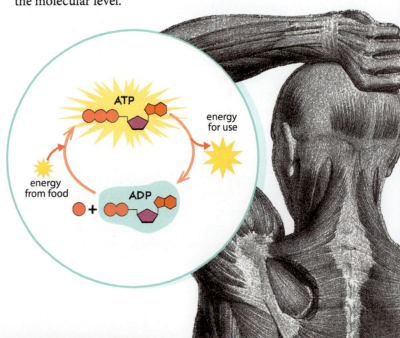

SIGNATURE IN THE CELL

You pop a little chip into your computer, and documents and photos appear. Where did they come from? The information on that chip is stored in a code. Someone designed the chip that stored that code, and you created and saved the information.

We've been learning about the code that is built into cells, codes that can create the cellular machinery in the form of motors (flagella), dump trucks (Golgi bodies), and highway systems (microtubules). We don't question the intelligent design of the memory chip in our computers. So why do people question the intelligent design of cellular machinery? Both are the result of code.

Stephen Meyer is a scientist and philosopher and is the director of Discovery Institute's Center for Science and Culture, the headquarters of the intelligent design (ID) research community. He wrote the book *Signature in the Cell*, in which he argues that information always has an intelligent source. In his book he demonstrates that DNA contains information—a remarkably complex kind of information. Therefore, DNA (and life itself) must come from an intelligent source.

Stephen Meyer

Introduction

You are a writer for a Christian scientific journal, and you are writing a review of Meyer's book, a classic on ID.

Task

Your task is to write a 150–200 word summary of his book, examining his claims and analyzing them from a biblical perspective.

Procedure

1. Do a search on the title of the book, reading summaries and comments about it. Consider visiting the book's website. If you feel ambitious, you could consider reading parts of or all of the book.

2. Do some internet research on ID.

3. Do the same search on the Answers in Genesis website.

4. Plan, write, and revise your review.

5. Get another student or adult to proofread your review, and ask your teacher to publish it on a class website.

Conclusion

The existence of God and His involvement in and care for creation extend to the molecular level. When we look at the code for creation, we are driven to see God's "signature in the cell."

6.1 SECTION REVIEW

1. Do both heterotrophs and autotrophs have metabolic processes? Explain.

2. Explain how the sources of energy for heterotrophs and autotrophs are different.

3. Why can't organisms just use energy as soon as they get it without storing it in ATP?

4. How do ATP molecules store energy for the cell?

5. Explain what an organism does with the energy that it takes in from food or light.

6. Identify the structural difference between ATP and ADP.

6.2 DNA AND PROTEIN SYNTHESIS

?

What guides a cell's growth and function?

Questions

Where does a cell's information come from?

How does this information guide a cell's growth and function?

Terms
deoxyribonucleic acid (DNA)
base pair
replication
mutation
ribonucleic acid (RNA)
transcription
translation
anticodon
protein synthesis
codon

The Code of Life

If energy is like money and ATP is its currency, then the cell is a whole Wall Street of buyers and investors! Though it seems chaotic, there is a carefully ordered flow of energy and information within the cell. Cells know just what to do to get energy out of storage and how to use it in metabolic processes to sustain life and maintain homeostasis. As the energy for life extends down to the molecular level, so also the information of life is coded in molecules in the cell. The molecule that is responsible for this is **deoxyribonucleic acid (DNA)**. DNA is a nucleic acid made of a chain of nucleotides that stores information. This information is used to direct the metabolic processes of the cell and will also be copied and transmitted when a cell reproduces.

DNA

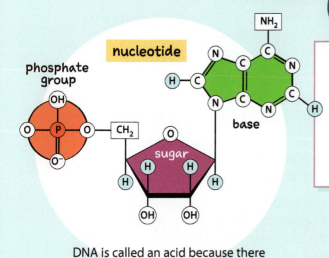

nucleotide

phosphate group

base

sugar

Nucleotides have three parts—a nitrogen-containing base, a sugar, and a phosphate group. The phosphate and sugar in each kind of DNA nucleotide are the same, but the nitrogen-containing base is different.

DNA is called an acid because there is a place on each nucleotide that can give away a hydrogen ion.

There are four different types of bases, often represented with letters in the code of life. These bases are adenine (A), thymine (T), guanine (G), and cytosine (C).

thymine

adenine

guanine

cytosine

The bases on adenine and thymine as well as those on guanine and cytosine lock together through the attractions of hydrogen bonds, represented by dotted lines at right, to form the "rungs" of the DNA ladder. These pairs of nucleotides are called **base pairs**.

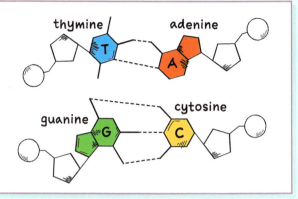

base pairs

The two strands of DNA look like a twisted ladder, which scientists call a *double helix*.

The phosphate group on each nucleotide links sugars on different nucleotides. The base dangles off the sugar. This means that the two strands of the DNA "ladder" are made of alternating phosphates and sugars, and the "rungs" are made of nitrogen-containing bases.

DNA Replication

DNA can be copied when an enzyme called *helicase* pulls apart the base pairs to unzip the DNA.

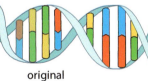

DNA polymerase

copied DNA strands

helicase

original DNA strand

copied DNA strands

Another enzyme called *DNA polymerase* shuttles in nucleotides that match each unzipped part of the DNA. When the strands have been completely unzipped and matching nucleotides are linked together, two copies have been created in a process called **replication**. Mistakes in the replication process can cause **mutations** when a nucleotide is mistakenly inserted, deleted, or skipped altogether. The cells have repair enzymes to recognize and repair mistakes added during the replication process. This is a powerful evidence of design since, without these enzymes, mutations could have disastrous effects on an organism. However, some mutations persist, usually causing trouble for organisms.

BUILDING PROTEINS

Cells make copies of DNA for two reasons: to get a cell ready to divide—so that the new cell will have the information it needs to continue life processes—and to make proteins.

Proteins are the machines that help a cell do its job. DNA is like the blueprint for a skyscraper, and proteins are like the dump trucks, bulldozers, and cranes that turn the blueprint into an actual building. Enzymes are proteins that fulfill a specific purpose to make metabolic processes happen in living

things. As we saw in Chapter 2, proteins are chains of smaller molecules called *amino acids*. These large chains fold into complex shapes that help a protein do its job.

So how does a cell use DNA to make proteins? The cell makes a copy of a section of DNA, but this copy has only one strand with a few modifications. This single-strand nucleic acid is called **ribonucleic acid** (**RNA**). This copying process takes place in the nucleus and is called **transcription**.

TRANSCRIPTION

Transcription happens in the nucleus.

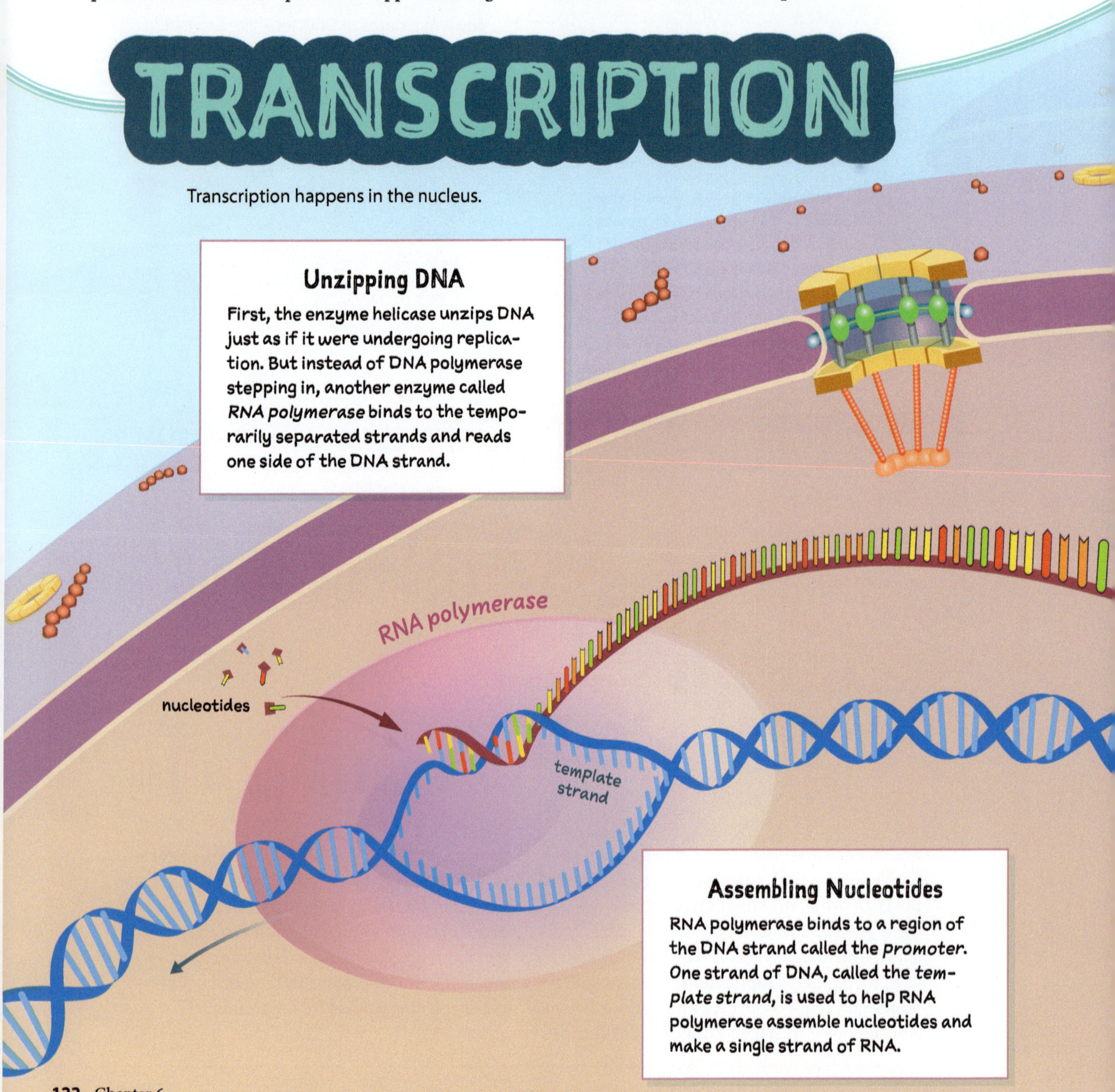

Unzipping DNA

First, the enzyme helicase unzips DNA just as if it were undergoing replication. But instead of DNA polymerase stepping in, another enzyme called *RNA polymerase* binds to the temporarily separated strands and reads one side of the DNA strand.

RNA polymerase

nucleotides

template strand

Assembling Nucleotides

RNA polymerase binds to a region of the DNA strand called the *promoter*. One strand of DNA, called the *template strand*, is used to help RNA polymerase assemble nucleotides and make a single strand of RNA.

Leaving the Nucleus

Once the edits are complete, the newly transcribed strand of *messenger RNA (mRNA)* leaves the nucleus through a *nuclear pore* and goes into the cytoplasm. In the cytoplasm, cellular machines "read" the code from the mRNA and assemble amino acids into proteins through a process known as **translation**.

Editing RNA

Before RNA can be made into a protein, it must have portions of it cut out or put back together.

Introns are sections of the RNA transcript that do not contain information about the protein being produced. They are cut out of the RNA before it is sent from the nucleus to the cytoplasm to make proteins.

Exons are sections of the RNA that code for proteins. They are included in the final version of RNA transcript that is sent out of the nucleus to be translated into proteins.

nuclear pore

intron

exon

intron

exon

mRNA strand

Contrasting DNA and RNA

In RNA strands, thymine is replaced by uracil (U). Also, the sugar part of the nucleotides in RNA is different than the one in DNA, having one extra oxygen atom.

↓ DNA   ↓ RNA

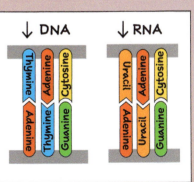

TRANSLATION

Translation happens in the cytoplasm.

Amino Acids Transferred

Once in the cytoplasm, mRNA is ready to begin decoding. But first, another type of RNA, *transfer RNA (tRNA)*, picks up amino acids in the cytoplasm. On one of the four prongs of tRNA are three unattached bases called **anti-codons**, which align with codons on the mRNA strand to properly position amino acids for the new protein chain being built. The tRNA is assembled and released from the nucleus just as mRNA is.

Amino Acids Bonded by Ribosomes

The cellular machine that links the amino acids delivered by tRNA is called a *ribosome*. It contains another type of RNA called *ribosomal RNA (rRNA)* that is used to make proteins. While at the ribosome, tRNA matches the amino acids to the coded mRNA.

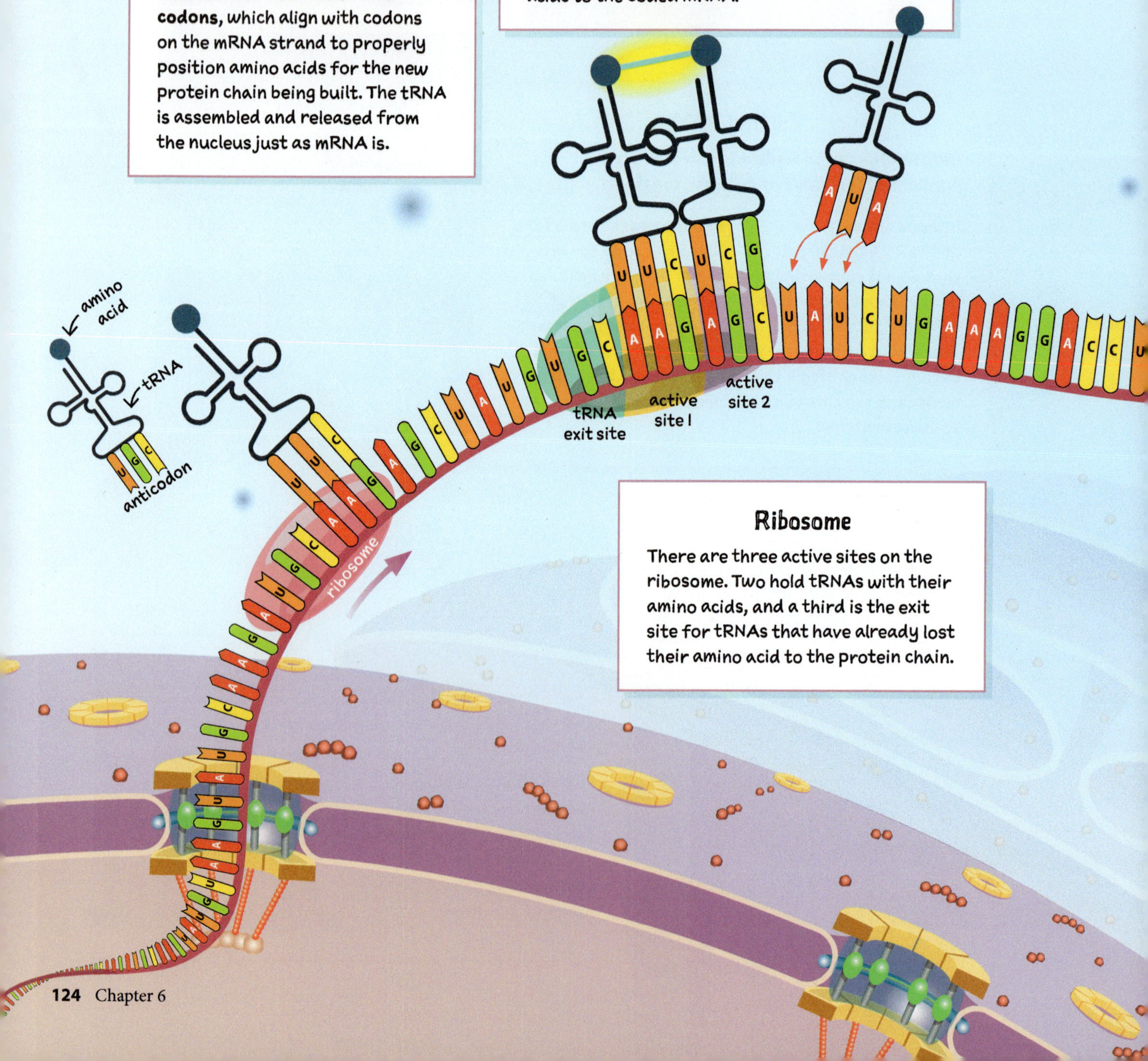

Ribosome

There are three active sites on the ribosome. Two hold tRNAs with their amino acids, and a third is the exit site for tRNAs that have already lost their amino acid to the protein chain.

amino acid

tRNA

anticodon

ribosome

tRNA exit site

active site 1

active site 2

Proteins Created

When two anticodons are lined up with their corresponding codons, they dock, and a *peptide bond* forms between their amino acids in a process called **protein synthesis**. As the protein builds, it forms its own signature folds and turns on the basis of chemical interactions.

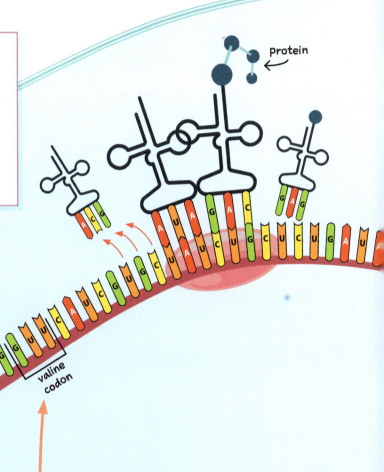

protein

valine codon

Codons Read

Like using letters to form words, three consecutive nucleotides form **codons**, which are triplets of nucleotide bases in the RNA transcript that provide the code for an amino acid during translation. The combination of bases in a codon codes for one of twenty possible amino acids to add onto a protein chain. For example, in the codon GUU the first letter represents guanine, the second is uracil, and the third is uracil. That codon codes for valine, which is an amino acid that can be used to build a protein.

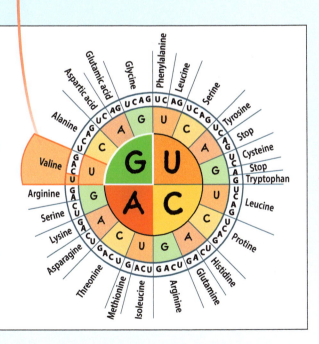

MAKING PROTEINS

A tRNA carries an amino acid to the ribosome. If its anticodon fits with the codon, it docks to form base pairs with the messenger RNA strand. Protein synthesis always begins at a *start codon*. In eukaryotes, start codons always code for methionine.

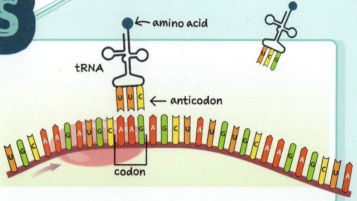

A second tRNA carrying a matching anticodon docks. This loosens the first tRNA's grip on the amino acids. These amino acids form peptide bonds with the addition of energy from ATP.

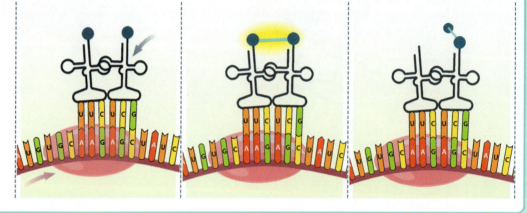

The ribosome moves down the mRNA to read a new codon, and the first tRNA that has transferred its amino acid to the protein chain exits so that another can be added. The process repeats until the ribosome reaches a *stop codon*.

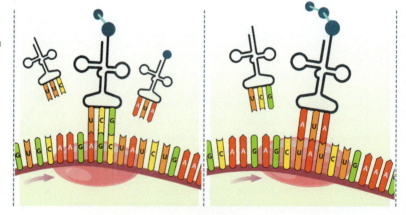

The protein forms its own signature folds and turns as it is manufactured.

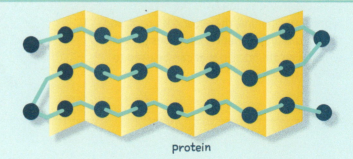

protein

MINI LAB

Modeling DNA and RNA

When we consider what RNA actually is, it may be easier to think of it as a working copy of DNA. RNA has many functions, but synthesizing proteins is especially important because proteins help determine an organism's characteristics.

How can we use DNA to model RNA?

Materials
model-building supplies, such as toothpicks, beads, string, pipe cleaners, or modeling clay

PROCEDURE

A With a partner, make a plan for building models of DNA and RNA using the provided supplies. Be sure to have your teacher approve your plans before you begin modeling.

B Begin creating the model of DNA first. This will be your template for creating your RNA model.

C Create your RNA model from your DNA template according to the process shown on pages 122–23.

D Present your final models to the class.

CONCLUSION

1. Evaluate your models. How well do your models represent DNA and RNA?

2. What are your models' limitations?

3. Use your models to demonstrate transcription and translation. Be sure to use the bolded terms defined in your textbook to explain your demonstration.

GOING FURTHER

4. Explain why the models that you created are limited in their ability to explain the true structure of DNA and RNA.

6.2 SECTION REVIEW

1. How does DNA contain information?

2. In your own words, describe the DNA replication process.

3. What happens when there is a mistake in the transcription or translation process? What do we call this phenomenon?

4. Compare DNA and RNA. You may choose to use a graphic organizer like a T-Chart (see Appendix B).

5. How does the information in DNA help a cell grow and function?

6. Name the three types of RNA and give their purposes.

7. Create a process map relating the terms DNA, RNA, protein, amino acid, transcription, and translation. Show which processes happen in the nucleus and which happen outside the nucleus (see Appendix B).

8. After learning about the energy and information in the cell, how would you respond to the quote below by Richard Dawkins, a biologist and one of the world's most outspoken atheists?

> *"We are machines built by DNA whose purpose is to make more copies of the same DNA. This is exactly what we are for. We are machines for propagating DNA, and the propagation of DNA is a self-sustaining process. It is every living object's sole reason for living."*

Energy and Information in the Cell   **127**

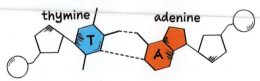

6.1 METABOLISM

- Cells within an organism manage metabolism to maintain homeostasis.

- ATP is the molecule within cells that temporarily stores energy.

- When cells need energy, ATP is converted to ADP by releasing a phosphate. ADP is converted back to ATP with the addition of a phosphate, and energy is stored in the bond.

Terms
anabolism • catabolism • metabolism • adenosine triphosphate (ATP) • adenosine diphosphate (ADP)

6.2 DNA AND PROTEIN SYNTHESIS

- DNA is a nucleic acid, made of nucleotides, that stores information. The information it contains directs metabolic processes and is copied for cell reproduction.

- RNA is a single-stranded copy of DNA used to make proteins.

- Transcription occurs in the nucleus of the cell and is the process in which DNA is copied to make RNA.

- Translation occurs in the cytoplasm and is the process in which proteins are created.

- DNA must be replicated and used to make RNA and allow for protein synthesis to occur.

Terms
deoxyribonucleic acid (DNA) • base pair • replication • mutation • ribonucleic acid (RNA) • transcription • translation • anticodon • protein synthesis • codon

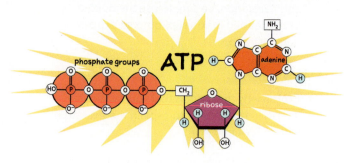

Chapter Review Questions

RECALLING FACTS

1. How are the two types of metabolic processes related?

2. Why do living things need both anabolic and catabolic processes?

3. Where is energy stored in ATP?

4. What must happen for ATP to be turned into ADP?

5. What smaller units make up DNA?

6. Which parts of a nucleotide form the backbone and rungs of the DNA double helix?

7. What holds a base pair together?

8. What unzips DNA for replication?

9. List three ways that RNA is different from DNA.

UNDERSTANDING CONCEPTS

10. Describe what must take place when an animal eats in order for food to be turned into energy.

11. Draw a process diagram that shows the relationship between ATP, ADP, energy, and a free phosphate group.

12. Explain how a protein results from DNA.

13. If there is a mistake that occurs during DNA replication, how can cells fix the problem?

14. Compare transcription and translation.

15. Explain the role of DNA and RNA in protein synthesis.

16. Relate the roles of codons and anticodons.

17. Describe the three types of RNA involved in making proteins.

CRITICAL THINKING

18. Using the codon circle chart on the right, decipher the code below and name the three amino acids that it represents.

GGU GGA UGG

For Questions 19–21, use the sequence of bases from the single strand of DNA shown below.

GGAATCGGACTATGGGT

19. Make the complimentary strand of DNA.

20. Make the complimentary strand of DNA, but include one mutation.

21. Make the corresponding strand of RNA.

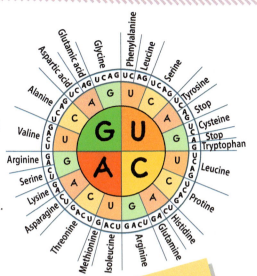

CRISPR TECHNOLOGY

ETHICS

ISSUE

CRISPR (Clustered Regularly Interspaced Short Palindromic Repeats) is a highly studied piece of medical technology that has rocked the world of medicine and science. It has impacted many areas of research and has opened the doors to potential preventive and specialized medical treatments. This technology has showed promising effects, but many people have expressed ethical concerns regarding its use.

TASK

A friend has recently researched the topic of CRISPR and has stated that this technology is going to change humanity as we know it. Not understanding much of CRISPR yourself, you decide to learn more about it by researching the topic for a biology ethics essay that was recently assigned. For the assignment, your teacher has asked that you apply the biblical ethics triad to prepare an essay stating your position on the topic.

1. *What information can I get about this issue?*

2. *What does the Bible say about this issue?*

3. *What are the acceptable and unacceptable outcomes of supporting CRISPR technology?*

4. *What are the motivations of the acceptable options?*

5. *What action should I take?*

PROCEDURE

1. Begin planning and do the appropriate research needed for your topic. Do an internet search for CRISPR and CRISPR technology ethics. Use your Bible as a reference for a biblical perspective.

2. Draft and revise your essay. Be sure to properly cite your sources.

3. Have a teacher, peer, or both proofread your essay and provide feedback.

4. Publish your essay.

CONCLUSION

The purpose of CRISPR must be considered when debating the use of CRISPR. Regardless of the purpose, we are obligated to look to God's Word for wisdom and direction on what constitutes loving others with the use of medical technology. We must be careful—what may seem loving for one person could actually harm future generations.

7 CELL PROCESSES

7.1 **Photosynthesis** | 7.2 **Cellular Respiration and Fermentation**

Liquid Sunshine

You put the nozzle of the gas pump into your car's gas tank and squeeze the lever. Out comes liquid sunshine. Sound impossible? Well, it might not be as far-fetched as it sounds. Researchers connected to companies like ExxonMobil Corporation have worked with different species of tiny marine algae to create biofuels that are usually made from crops like corn, soybeans, and sugarcane. We won't have to use land or fresh water to raise the algae, and this biofuel won't consume crops that could be used to feed people and animals. But how do algae convert energy from sunlight into fuel for your car? It happens through a cell process called *photosynthesis*.

7.1 PHOTOSYNTHESIS
Harvesting Light

A kelp forest sways in the undulating current of an ocean bay. Its leaflike colonies bask in the sunlight that fills the green waters with a warm glow. The same process that grows kelp also helps algae turn sunlight into biofuel. What you might not expect is that 70%–80% of the oxygen that we breathe comes from algae in the oceans! Rain-forests produce the other 20%–30%, but they cover only about 2% of the earth's surface, though 50% of the organisms on our planet make their home there. Compare that with the oceans, which cover 71% of the earth's surface. You can see why algae are important to the rest of life in the world.

So why don't we make our own energy like algae do when we go out into the sun? It would definitely save on grocery bills! The reason is that we don't have the right information or organelles in our cells to pull off this feat. It takes specially designed cells and special enzymes to turn sunlight into food through the process of **photosynthesis**. Without photosynthesis widespread life on Earth wouldn't be possible. In Chapter 2 we studied the chemical reaction that describes the process of photosynthesis. We model it with the chemical equation shown below.

$$6H_2O\ (l) + 6CO_2\ (g) + \text{light energy} \xrightarrow{\text{chlorophyll}} C_6H_{12}O_6\ (aq) + 6O_2\ (g)$$

water   carbon dioxide   glucose   oxygen

Just as there are two parts to the word *photosynthesis*, meaning "light" and "forming," there are two stages to the process of photosynthesis. The first involves capturing the energy from sunlight, called the **light-dependent phase**. The second involves changing that energy into a form that an organism can use by making special chemicals; this takes place during the **light-independent phase**. These two phases of photosynthesis take place in different parts of a cell's chloroplasts.

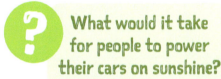
What would it take for people to power their cars on sunshine?

Questions

Why is photosynthesis so important?

How can I model photosynthesis?

How does a cell generate its own energy?

Do plants make their own food at night?

What happens to photosynthesis when factors in the environment change?

Terms

photosynthesis
light-dependent phase
light-independent phase
stroma
thylakoid
chlorophyll
electron transport chain
Calvin cycle

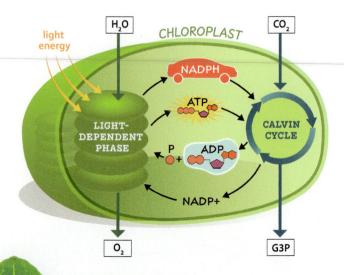

As you study the following images that show the process of photosynthesis, make sure that you look for two pairs of chemicals to see what's happening to the energy in these reactions. We already know about one pair of chemicals—ADP and ATP. When ATP is produced, the process stores energy. When ADP is produced, the process releases energy.

The other pair of chemicals to keep your eyes on is NADP+ (nicotinamide adenine dinucleotide phosphate) and NADPH. NADP+ is like an electron shuttle, and when it captures a hydrogen ion and two electrons, it forms NADPH. Without these two pairs of molecules, photosynthesis wouldn't happen at all.

PHOTOSYNTHESIS

Light-Dependent Phase

During the first phase of photosynthesis, the *light-dependent phase*, cells gather energy from light. In plants, water comes from the roots and carbon dioxide comes in through the tiny openings called *stomata* (s. stoma) on the undersides of leaves. As a byproduct of photosynthesis, stomata also release the oxygen that we breathe.

light

stoma

carbon dioxide

oxygen

stroma

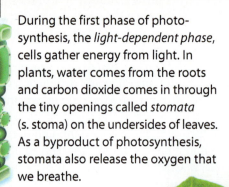

Sunlight enters the chloroplasts in plant cells, passing through the clear fluid in the chloroplasts called **stroma**.

thylakoids

In the interior of the chloroplasts are *grana* (s. granum), stacks of green sacks called **thylakoids**. Each thylakoid is a membrane-bound compartment.

Chlorophyll appears green and makes plants and algae look green because it absorbs all the other colors of light and reflects green light.

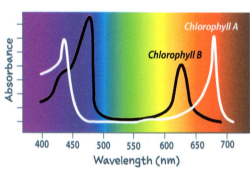

Thylakoid membranes contain chlorophyll embedded in strategic places. **Chlorophyll** is a chemical, a *pigment* that absorbs energy from sunlight. It uses sunlight to energize electrons, kicking off photosynthesis. Without chlorophyll photosynthesis doesn't happen!

❶ When chlorophyll reacts with light, it splits water molecules to produce high-energy electrons and hydrogen ions (H^+). These electrons move through a series of protein molecules, like water over a waterwheel, in a system called the **electron transport chain** (shown below). Energy from the electron transport chain increases the H^+ concentration in the thylakoid and decreases the H^+ concentration in the stroma.

❷ The hydrogen ions are released into the stroma through a protein called *ATP synthase* that turns ADP into ATP.

❸ Hydrogen ions and electrons entering the stroma bond with NADP+ to form NADPH.

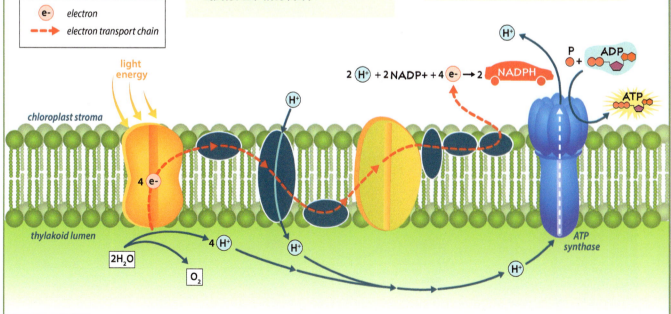

PHOTOSYNTHESIS

Light-Independent Phase

Now the second phase of photosynthesis begins. It does not rely on sunlight, thus the name *light-independent phase*. NADPH and ATP from the electron transport chain drive the Calvin cycle in order to make food.

In the **Calvin cycle** (shown below) ATP provides energy to turn carbon dioxide and water into glucose. NADPH provides hydrogen ions and electrons to keep this cycle going. Note the number of carbons at each step in the Calvin cycle, showing how carbon dioxide with one carbon becomes glucose with six carbons.

❶ Carbon dioxide from the air that enters the stomata of the leaves is the key ingredient for the Calvin cycle. A molecule called *ribulose 1,5-biphosphate* (RuBP) will serve as the carbon receptor.

❷ An enzyme called *RuBisCO* links a carbon dioxide molecule with a molecule of RuBP, forming a very unstable six-carbon intermediate. The intermediate quickly splits into two three-carbon intermediates called *3-phosphoglyceric acid* (3PG). During the reduction phase a two-step reaction converts 3PG to G3P, glyceraldehyde-3-phosphate.

RuBisCO

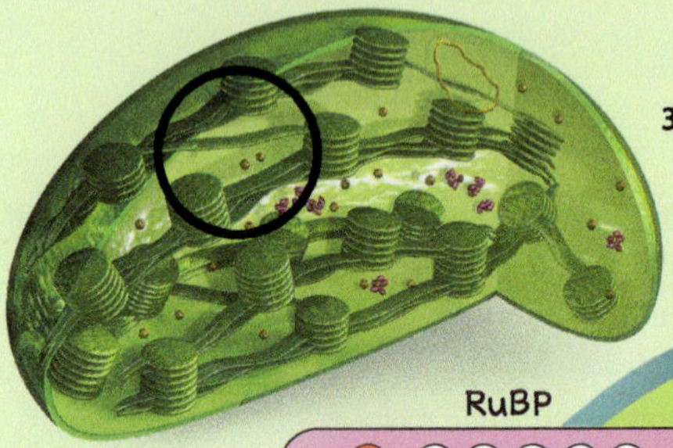

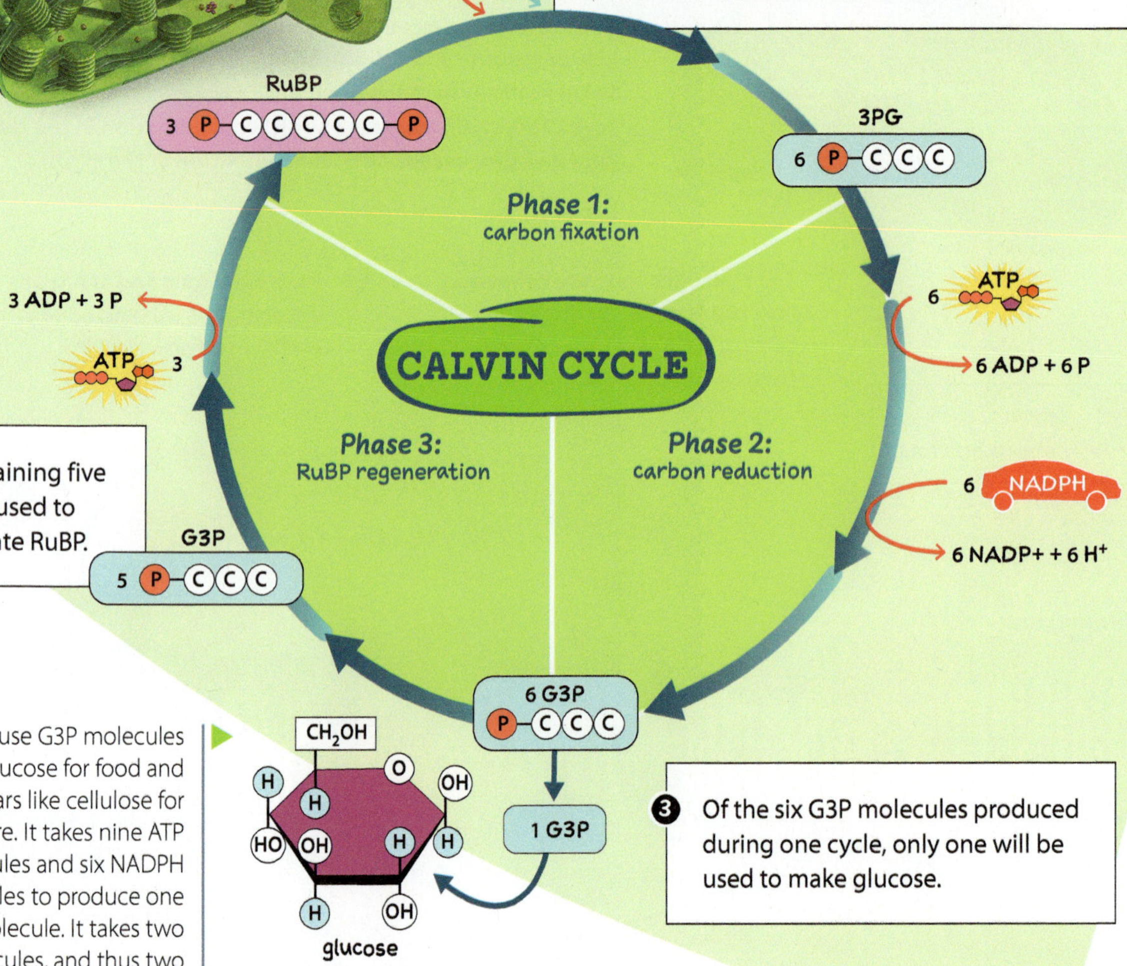

$3\ CO_2$

$3\ H_2O$

RuBP
3 P C C C C C P

3PG
6 P C C C

Phase 1:
carbon fixation

6 ATP

$6\ ADP + 6\ P$

$3\ ADP + 3\ P$

ATP 3

CALVIN CYCLE

Phase 3:
RuBP regeneration

Phase 2:
carbon reduction

6 NADPH

$6\ NADP+ + 6\ H^+$

❹ The remaining five G3P are used to regenerate RuBP.

G3P
5 P C C C

6 G3P
P C C C

1 G3P

❸ Of the six G3P molecules produced during one cycle, only one will be used to make glucose.

Cells use G3P molecules to make glucose for food and other sugars like cellulose for structure. It takes nine ATP molecules and six NADPH molecules to produce one G3P molecule. It takes two G3P molecules, and thus two turns of the cycle, to make one glucose molecule.

CH₂OH
H
O
H
OH
HO
OH
H
H
H
H
OH

glucose

Environment and Photosynthesis

Recall in the photosynthesis diagram (p. 132) that plants and other photosynthetic organisms need light, carbon dioxide, and water to make glucose. Plants usually have enough carbon dioxide and light to undergo photosynthesis. But in some biomes, like the desert and chaparral, water can be scarce. So how do plants cope to maintain homeostasis? Most plants are called *C3 plants*, which means that they use the Calvin cycle to produce a three-carbon precursor of glucose—G3P. G3P molecules can also be used to create starch, cellulose, sucrose (table sugar), and fructose for plants.

But when it gets hot and dry and plants open their stomata to take in carbon dioxide, they lose water. When there's not enough water, the Calvin cycle can't make G3P molecules. Instead, it *uses up* oxygen and *produces* carbon dioxide, opposite of the way that the cycle usually works. This process is called *photorespiration*—a kind of survival mechanism for plants, not normal operating procedure!

To get around photorespiration, some plants, called *CAM plants*, open their stomata only at night when it is cooler so that they can take in carbon dioxide without losing water to their environment. Instead of creating G3P, they create other compounds that can be made into glucose. But these processes are not as energy efficient as the Calvin cycle. They cost more ATP molecules to make glucose. Succulents like pineapples, cacti, and orchids are CAM plants.

Another option is for plants to just continue the Calvin cycle with less carbon dioxide. Instead of creating a three-carbon G3P precursor to glucose, these plants create a four-carbon molecule, though it requires more energy. Because of this, these plants are called *C4 plants*. Corn and sugarcane are examples of C4 plants, though both operate like C3 plants within limits when conditions are more favorable for the Calvin cycle to produce G3P.

Now let's step back and see how photosynthesis affects the biosphere. Light energy is the only significant source of energy in many ecosystems. Plants can absorb only about 34% of the total light energy available to them. The rest is reflected, lost as heat, or composed of wavelengths that are too long to be absorbed. Of the absorbed light energy, only a tiny fraction is actually stored as chemical energy in sugar. Of course, plants use nearly half of that sugar for their own metabolism. In the end, of all the light energy available to plants, only about 4% ends up being stored as biomass. It's amazing that the rest of the biotic community can thrive on such a small percentage of the energy from the sun. Ultimately, the sun is responsible for sustaining most life on Earth.

MORE THAN JUST ENERGY

So far we've looked mostly at how plants provide energy for themselves and for consumers. But have you ever taken a moment to think about all the amazing things that plants provide for us other than food and oxygen? Are you wearing cotton or linen clothing? That came from plants. The flavoring in your vanilla ice cream and the cocoa in the chocolate syrup that you pour over it—those also came from plants. The lumber used to build your house came from trees, which are plants, of course. Even the aspirin that you take for a headache is derived from a chemical found in plants. All these products, and many more besides, are available to us because of the incredible process of photosynthesis. And researchers continue to find new ways to use plants to help us exercise wise dominion over God's creation.

Will that dominion include using more of the biofuels mentioned at the beginning of this chapter? Possibly. The International Energy Agency is projecting biofuels to provide a quarter of the world's transportation fuel needs by 2050; in 2019 that figure stood at only 3%, and the 2050 target is looking unrealistic. Scientists are still learning how to get the most out of photosynthesis.

7.1 SECTION REVIEW

1. According to Genesis 1:29–31, how does photosynthesis fit into God's design for His creation?

2. Diagram the chemical reaction of photosynthesis, using red circles to symbolize oxygen, white circles to symbolize hydrogen, and black circles to symbolize carbon. Use the chemical reaction to determine how many molecules you need.

3. Why are plants and algae green?

4. God designed stroma to be a clear fluid. Explain what could happen if stroma were opaque to light.

5. List the two stages of photosynthesis and indicate where they take place.

6. Create a process map that shows the different steps of photosynthesis. Make sure to account for all reactants, catalysts, and products in the chemical equation for photosynthesis. Also make sure that you show the flow of energy in this process. (See Appendix B.)

7. Under what conditions will a plant resort to photorespiration?

8. How can people exercise good and wise dominion over the process of photosynthesis for God's glory?

7.2 CELLULAR RESPIRATION AND FERMENTATION

Cellular Respiration

We already know that people and animals eat to get energy from their food. In the last section we saw how plants and other photosynthetic organisms make food from the water, carbon dioxide, and sunlight in their environment. Both producers and consumers get energy from energy-containing molecules in their food through a different process—cellular respiration. It's a meal of molecules!

These energy-containing molecules are glucose or other sugars, lipids, and even proteins. **Cellular respiration** produces energy in the form of ATP that can be used to fuel the processes of life for both producers and consumers.

Cellular respiration is similar to the chemical process of photosynthesis but in reverse. Photosynthesis takes place in the chloroplasts of photosynthetic organisms; cellular respiration occurs in the mitochondria, and most eukaryotic cells have these organelles.

? How do cells free up chemical energy from sugar?

Questions

How does a cell use energy?

What is the role of oxygen in cellular respiration?

How does cellular respiration help in the cycling of matter and energy through the environment?

Terms

cellular respiration
glycolysis
citric acid cycle
anaerobic process
fermentation

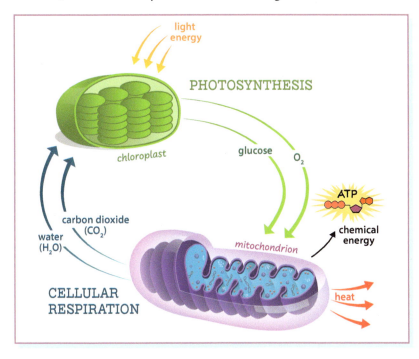

The energy from photosynthesis is stored in the bonds of glucose. When plants need energy for cell processes, they break down glucose in a process known as **glycolysis** to release the energy and form ATP. When herbivores and omnivores eat plants, they get the glucose and other energy-containing molecules that plants store. This means that other organisms besides plants can also carry out glycolysis.

Cellular respiration is very much like burning fuel. It usually requires oxygen, so it is an aerobic process. We'll see later on that fermentation can occur without oxygen in what are called *anaerobic processes* (see page 142), but less energy is produced.

There are three phases of aerobic cellular respiration: glycolysis, the citric acid cycle, and the electron transport chain. Glycolysis occurs in a cell's cytoplasm, and the citric acid cycle and electron transport chain take place in the mitochondria. The **citric acid cycle** takes smaller molecules produced by glycolysis and uses them to form citric acid. The cycle then gradually takes atoms from this six-carbon molecule to produce ATP, NADH, and carbon dioxide. The main function of the citric acid cycle is to harvest high-energy electrons to feed into the electron transport chain. This process occurs in the mitochondria as the last phase of cellular respiration. See the chemical reaction for cellular respiration below.

$$C_6H_{12}O_6\ (aq) + 6O_2\ (g) \longrightarrow 6H_2O\ (g) + 6CO_2\ (g) + energy$$

glucose   oxygen   water   carbon dioxide

Notice the three parts of cellular respiration: glycolysis in the cytosol, the citric acid cycle in the matrix of the mitochondrion, and the electron transport chain in the inner membrane of the mitochondrion.

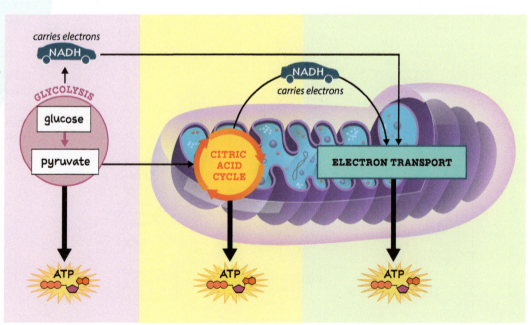

CELLULAR RESPIRATION

Phase 1: Glycolysis (p. 138)

Pyruvate

Phase 2: Citric Acid Cycle

The citric acid cycle happens in the fluid of the innermost part of each mitochondrion, called the *matrix*. The cycle releases carbon dioxide, which can be exhaled or simply allowed to diffuse back into the environment, and energy, which an organism can use. Overall, the citric acid cycle produces one molecule of ATP each time around. It takes pyruvate from glycolysis to drive the citric acid cycle so that it can transport high-energy electrons to the electron transport chain that will eventually produce large amounts of ATP.

1 Pyruvate from glycolysis is moved into the mitochondrion, where one carbon breaks off, forming carbon dioxide and a two-carbon coenzyme called *acetyl-CoA*. A molecule of NADH is also produced during this step.

2 Acetyl-CoA combines with a four-carbon molecule to form citric acid, which has six carbons like glucose.

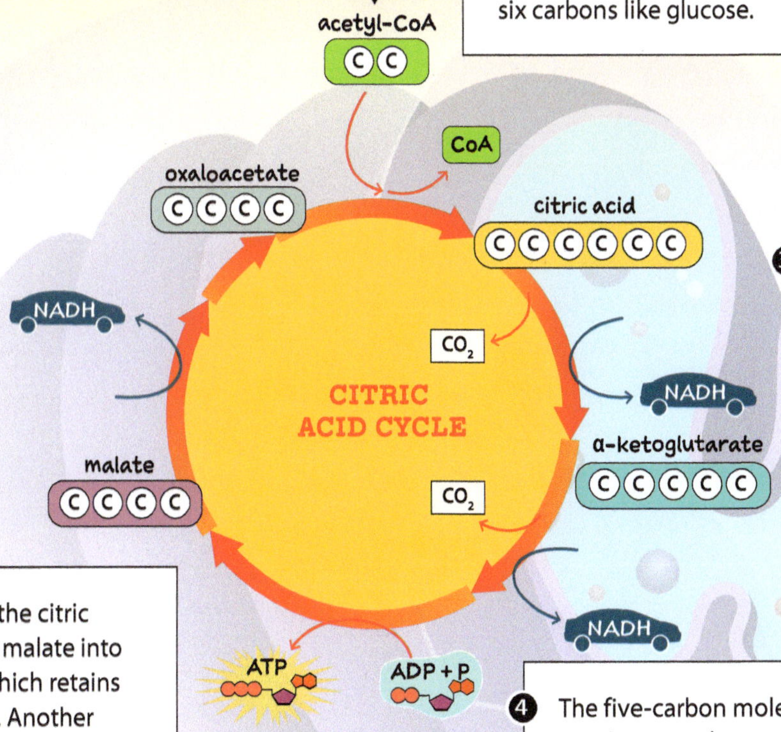

acetyl-CoA
C C

oxaloacetate
C C C C

CoA

citric acid
C C C C C C

NADH

CO$_2$

CITRIC
ACID CYCLE

NADH

α-ketoglutarate
C C C C C

3 Citric acid loses a carbon to carbon dioxide, generating one molecule of NADH in the process and a five-carbon molecule called *α-ketoglutarate*.

malate
C C C C

CO$_2$

NADH

ATP

ADP + P

5 The last step of the citric acid cycle turns malate into oxaloacetate, which retains all four carbons. Another molecule of NADH is also produced. Oxaloacetate will combine with more pyruvate from glycolysis to start the cycle all over again.

4 The five-carbon molecule from the previous step loses another carbon to carbon dioxide, generating another NADH molecule and one molecule of ATP. Through a series of intermediate steps, it forms a four-carbon molecule: malate.

CELLULAR RESPIRATION

continued

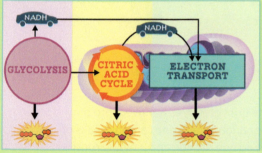

Phase 3: Electron Transport Chain

The electron transport chain in the mitochondria uses the NADH of the citric acid cycle to produce large amounts of ATP. Compare this figure with the electron transport chain in photosynthesis on page 133.

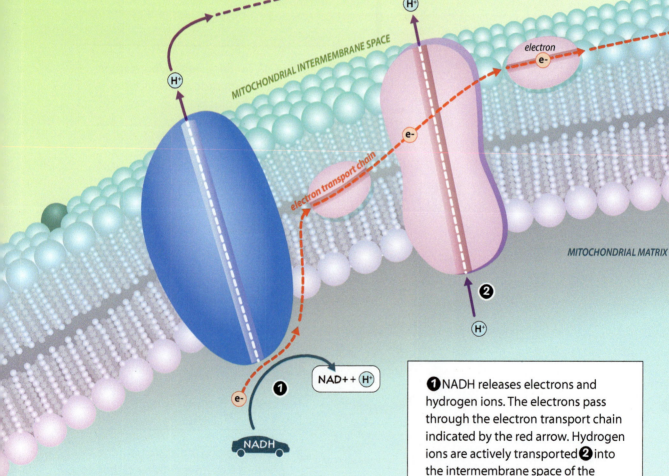

❶ NADH releases electrons and hydrogen ions. The electrons pass through the electron transport chain indicated by the red arrow. Hydrogen ions are actively transported ❷ into the intermembrane space of the mitochondrion because of the difference in charges between these two areas.

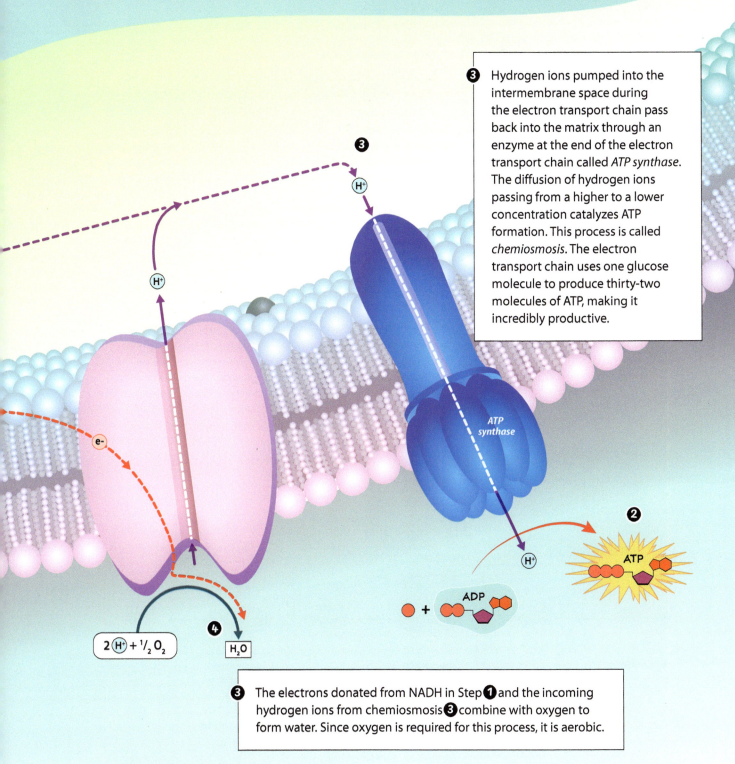

3 Hydrogen ions pumped into the intermembrane space during the electron transport chain pass back into the matrix through an enzyme at the end of the electron transport chain called *ATP synthase*. The diffusion of hydrogen ions passing from a higher to a lower concentration catalyzes ATP formation. This process is called *chemiosmosis*. The electron transport chain uses one glucose molecule to produce thirty-two molecules of ATP, making it incredibly productive.

3 The electrons donated from NADH in Step **1** and the incoming hydrogen ions from chemiosmosis **3** combine with oxygen to form water. Since oxygen is required for this process, it is aerobic.

Glycolysis happens in the cytosol and doesn't require oxygen. It gets all the materials ready for the citric acid cycle. Glycolysis uses several different enzymes to jumpstart the different steps that break down glucose into the three-carbon molecule pyruvate, producing two ATP and two NADH electron carrier molecules in the process. Again, tracking these molecules in the preceding diagrams will help you understand the flow of energy in the process.

So what does the whole process of aerobic cellular respiration provide for living things? Glycolysis produces two ATP molecules, the citric acid cycle produces two ATP molecules, and the electron transport chain produces thirty-two ATP molecules, giving us a grand total of thirty-six ATP molecules. This process has about a 40% efficiency rate, with the rest of the energy released as heat. Even this thermal energy is still useful to keep us warm. But this is one of the most efficient energy processes we know of. Consider that the engines in our cars are only about 20% efficient. The cell is a marvel of engineering!

Fermentation

In two of the three steps of cellular respiration—the citric acid cycle and the electron transport chain—oxygen is necessary, making this *aerobic cellular respiration*. But sometimes oxygen isn't available, as in the sediments of a tidal mudflat or in the salted brine covering a batch of fermenting sauerkraut. This can even happen in our bodies when we exercise strenuously. Does that mean that necessary cellular processes can't occur in these places to give energy?

Though cellular respiration can't happen in these situations, other processes can provide energy in a way that doesn't require oxygen. These are called **anaerobic processes**. Without these our muscles would shut down when they run out of oxygen! Releasing energy by anaerobic processes requires two steps—glycolysis and **fermentation**. Since glycolysis doesn't require oxygen, fermentation can use glycolysis just as aerobic cellular respiration does. Not so with the citric acid cycle and electron transport chain since both require oxygen. Fermentation happens in the cytoplasm. We saw that the electron transport chain is the most productive part of aerobic cellular respiration, so we would expect that fermentation would produce many fewer ATP molecules.

There are multiple kinds of fermentation that produce two ATP molecules during the glycolysis step. During the second step of fermentation NADH is recycled back into NAD+ so that glycolysis can continue. Without this the cell would die.

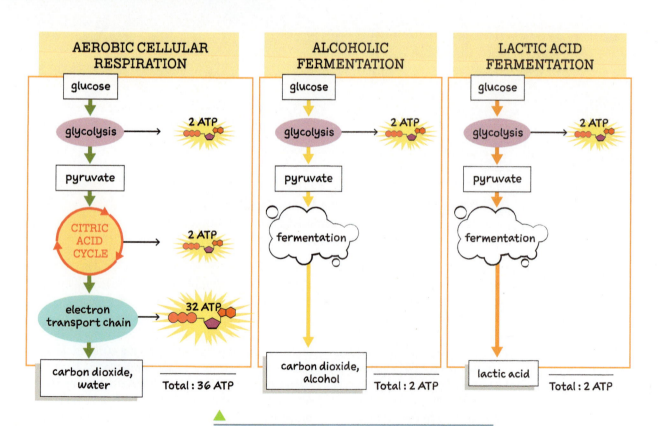

Compare the energy yields of aerobic cellular respiration, alcoholic fermentation, and lactic acid fermentation.

In *alcoholic fermentation* pyruvate from glycolysis is changed into alcohol using NADH. Yeast, a fungus, uses alcoholic fermentation to process sugar. Bread rises and alcoholic beverages are bubbly because of the release of carbon dioxide in alcoholic fermentation. See the chemical reaction below.

$$pyruvate + NADH \longrightarrow ethanol + CO_2 + NAD+$$

Bacteria use another type of fermentation called *lactic acid fermentation*. This is the type of fermentation used to make sauerkraut. Bacteria that feed on the sugars in the cabbage leaves produce pungent sulfur compounds as byproducts. You might be put off by the smell of sauerkraut, but bacteria also help make less-polarizing cheese and yogurt. In one step during lactic acid fermentation, pyruvate forms lactic acid as it reacts with NADH.

$$pyruvate + NADH \longrightarrow lactic\ acid + NAD+$$

The reactants are the same as in alcoholic fermentation, so what makes the difference? Different organisms or parts of an organism have environments that are favorable to different enzymes. These enzymes allow the different types of fermentation to take place and produce different products. Lactic acid fermentation happens in our muscles when we exercise vigorously. Notice that this doesn't produce any carbon dioxide.

We can see from the equations above that both alcoholic and lactic acid fermentation produce NAD+, which can be recycled back into NADH. This electron-carrying molecule is necessary to keep glycolysis going, producing energy. Without fermentation, glycolysis stops.

case study

HYDROTHERMAL VENTS

It may surprise you to learn that not all life on Earth depends on photosynthesis. As early as 1890, Russian microbiologist Sergei Winogradsky proposed that some kinds of bacteria could obtain their energy from inorganic chemical compounds such as hydrogen sulfide (H_2S). Such forms of metabolism later became known as *chemosynthesis*.

In the mid-1970s scientists discovered the existence of hydrothermal vents on the ocean floor. The water gushing from these vents, heated by magma in the earth's crust, ranges in temperature from 60 °C to 464 °C—far above the 2 °C temperature of the surrounding seawater! You'd think that nothing could live in the inky-black, scorching waters there, right? Wrong! In 1977 scientists aboard a deep-sea submersible discovered thriving communities of marine organisms living near the vents. What could form the basis of a food web in a region devoid of the light needed for photosynthesis?

Thermal vent consumers depend on chemosynthetic bacteria, not plants, to pull nutrients from the water, including hydrogen sulfide. These bacteria grow in mats and are eaten by other organisms such as shrimp, clams, and giant tube worms that can reach over 2 m in length. The community even includes predators, such as crabs and eelpouts. Even for the seemingly most inhospitable places on our planet, God has designed life that can do more than just survive—it thrives!

Cell theory is a model—it is a simplified representation of reality that describes or explains the cell in a workable and useful way. We may be used to thinking of force, energy, and atoms as being parts of science that we can't directly observe. But the cell is another black box in science. Scientists can't slice open a cell and watch it work, but they can experiment with cells and watch how they respond and make inferences about the processes that cause these responses.

Cell processes—metabolism, DNA replication, protein synthesis, photosynthesis, cellular respiration, and fermentation—are incredibly complex. Depending on the food supply and environment, cells can adjust to run different processes to maintain homeostasis. The interconnectedness and flexibility of cell processes leave us baffled. Scientists are still working to grasp the intricacies of the cell with models that are simplified representations of this wonderfully designed unit of life. Though we know much about the cell, it remains an element of mystery. Even as we learn more, we uncover more questions waiting to be answered.

7.2 SECTION REVIEW

1. Why does the phrase "burning carbs" or "burning fat," used in connection with exercise, make sense in light of what you have learned about cellular respiration?

2. Fill in the process map below, relating photosynthesis, cellular respiration, water, carbon dioxide, oxygen, glucose, and ATP.

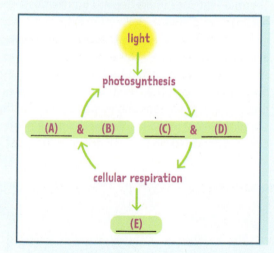

3. Why must glycolysis happen at the beginning of cellular respiration?

4. What is the difference between aerobic cellular respiration and fermentation? Include in your contrast the reactants, parts, and products.

5. How many molecules of ATP are produced during each phase of aerobic cellular respiration?

6. Why does fermentation take place at all if it produces only a small amount of energy?

7. What is the particular benefit for life in general being able to produce energy through anaerobic cellular respiration?

8. Compare the two types of fermentation.

9. How would life on Earth be different if no aerobic cellular respiration were possible?

10. Do research on the human metabolic map. Find the citric acid cycle, which should look like a circle. Where is it on the map? What does this map show you about the processes in the cell?

11. What goes through your mind as you look at the metabolic map from Question 10?

Use the case study on page 143 to answer Questions 12–13.

12. Compare chemosynthesis and photosynthesis.

13. Earlier in the chapter you read that the sun is responsible for sustaining most life on Earth. Is this true even for hydrothermal vent communities? Explain.

MINI LAB

The Effect of Temperature on Fermentation

If you have ever made homemade bread, you may recall the instructions telling you to mix yeast with warm water. Yeast produces alcohol (ethanol) and carbon dioxide gas as it metabolizes the sugars in flour. The carbon dioxide makes the bread rise, and the alcohol, which evaporates during cooking, contributes to the bread's characteristic taste. But is warm water really the best temperature for yeast? Let's find out!

Materials
large test tubes (3) • disposable pipettes (3) • washers (6) • scissors • tape • marking pen • yeast-sugar solution • cold water, room-temperature water, warm water

PROCEDURE

A Use the tape and marker to label the three test tubes "cold water," "room-temperature water," and "warm water." Then fill each test tube about three-quarters full with water of the indicated temperature.

B Using the pipettes, obtain three samples of the yeast-sugar solution provided by your teacher. Each sample should fill the squeeze bulb of the pipette about half-full. Be sure to hold the pipettes with their tubes facing upward so that the solution doesn't spill out.

C Cut the tube on each pipette so that about 4–5 cm is left above the squeeze bulb.

D Place two washers over the tube of each pipette; then gently place one pipette, bulb down, into each of the test tubes. The water in the test tube should completely cover the pipette; if it doesn't, add some more water of the proper temperature to the tube.

E Observe! The production of bubbles indicates that the yeast is busy fermenting the sugar in the yeast-sugar solution.

ANALYSIS

1. Rank the three yeast samples from slowest to fastest rate of fermentation. On what basis did you determine your ranking?

2. Do your results confirm the wisdom of the bread-making instructions?

3. On the basis of your results, write a general statement about the relationship between temperature and yeast fermentation rates.

GOING FURTHER

4. If warm water is better for yeast, then surely hot water must be better, right? Why do you suppose the instructions for neither the bread nor this lab activity called for hot water?

7.1 PHOTOSYNTHESIS

- Photosynthesis plays a critical role in God's creation by converting energy from sunlight into stored chemical energy. This energy is then available for a plant's own needs and for the needs of consumers.

- Plants, through photosynthesis, also remove carbon dioxide from the atmosphere and produce the oxygen needed for cellular respiration.

- The reactants in photosynthesis are carbon dioxide and water, and the products are glucose and oxygen.

- Photosynthesis occurs in two phases: the light-dependent phase and the light-independent phase.

- During the light-dependent phase the green pigment chlorophyll uses energy from sunlight to split water molecules and generate hydrogen ions and free electrons, both of which enter the electron transport chain. The end results of this process are NADPH and ATP.

- In the light-independent phase the Calvin cycle generates G3P, which cells use to produce glucose and other sugars.

- The rate of photosynthesis can be affected by environmental factors such as temperature and the availability of water.

Terms
photosynthesis · light-dependent phase · light-independent phase · stroma · thylakoid · chlorophyll · electron transport chain · Calvin cycle

7.2 CELLULAR RESPIRATION AND FERMENTATION

- Cellular respiration transfers the energy from glucose to ATP. The process may be either aerobic or anaerobic.

- Aerobic cellular respiration consists of three phases: glycolysis, the citric acid cycle, and the electron transport chain.

- Glycolysis takes place in the cytosol of a cell and produces pyruvate and two ATP molecules.

- The citric acid cycle takes place inside mitochondria and produces NADH and another two ATP molecules.

- The electron transport chain, which also happens inside mitochondria, uses free hydrogen ions and electrons released from NADH to generate another thirty-two ATP molecules. Carbon dioxide and water are byproducts of the process.

- Anaerobic cellular respiration includes alcoholic fermentation and lactic acid fermentation. These processes can be done in the absence of oxygen but do not produce as much ATP as aerobic cellular respiration (only two ATP molecules for each type of fermentation).

Terms
cellular respiration · glycolysis · citric acid cycle · anaerobic process · fermentation

Chapter Review Questions

RECALLING FACTS

1. What is the role of NADP+ and NADPH in cellular processes?

2. What is the role of ATP and ADP in cellular processes?

3. Write the overall chemical reaction for photosynthesis, giving the names of the reactants, products, and catalysts.

4. Overall, the process of photosynthesis changes _____ energy into _____ energy.

5. Why must the light-dependent phase of photosynthesis happen in the thylakoid membrane?

6. What cycle is at work in the light-dependent phase of photosynthesis? in the light-independent phase?

7. What environmental factors affect photosynthesis?

8. How do some plants adjust for environmental conditions that are not favorable to photosynthesis during the day?

9. Give the three steps of aerobic cellular respiration.

10. Which steps of cellular respiration are aerobic?

11. Which steps of cellular respiration and fermentation are anaerobic?

UNDERSTANDING CONCEPTS

12. How is photosynthesis part of God's good design for life in light of Genesis 1:29–31?

13. Your study partner says that photosynthesis happens only in the leaves of plants. Is she correct? Explain.

14. You're proofreading a friend's science report. In it he says that chlorophyll absorbs light from the green portion of the visible light spectrum. How would you respond?

15. Is it correct to say that plants never need oxygen from the environment because they make their own during photosynthesis? Explain.

16. Which of the following show(s) exercising good and wise dominion using photosynthesis?

    a. increasing food production with artificial sunlight

    b. using algae to produce biofuel

    c. using plants to produce illegal drugs

    d. planting trees in a city to improve air quality

17. Create a flow chart for aerobic cellular respiration, showing the three steps and the number of ATP molecules produced in each step. Find a way to show where each of the three steps happens.

18. Think about a person who is exercising more than normal. Why does fermentation even happen in the muscles since it doesn't produce very much energy? Is this a beneficial or harmful thing?

19. Botulism is a potentially fatal illness caused by toxins produced by the bacterium *Clostridium botulinum*. The most common causes for botulism are improperly preserved foods in sealed containers and colonization in infant digestive tracts. What do those two environments have in common? What does that tell you about the type of cellular respiration used by *Clostridium botulinum*?

20. Discuss the role that cellular respiration plays in an organism's environment.

21. How is God's care for His creation seen in the cellular processes that you have learned about in this chapter?

CRITICAL THINKING

22. Chemists can take a leaf extract and rub it onto a special kind of paper called *chromatography paper*. Some of the color rubs off on the paper, leaving a green smudge. If they put the edge of the paper in a solvent like acetone, the acetone will travel up the paper, carrying the leaf extract with it and separating it into different colors. What do you think this indicates about the pigments in leaves?

23. What do you think is the difference between an organism that is an *obligate anaerobe* and one that is a *facultative anaerobe*? Per your answer, which kind of organism is the *Clostridium botulinum* bacterium from Question 19?

Use the graph below to answer Questions 24–27.

Cabbage goes through a fermentation process to produce sauerkraut with its characteristic taste and smell. This process happens naturally if the cabbage is kept at a temperature of about 18 °C. The graph shows how long this process takes.

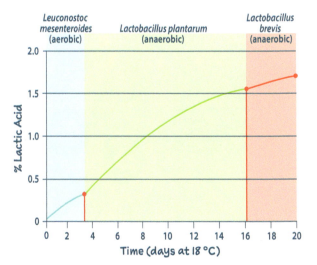

24. The *y*-axis shows how much lactic acid is in the sauerkraut. How can producers tell when sauerkraut is ready?

25. What do you think *Leuconostoc mesenteroides*, *Lactobacillus plantarum*, and *Lactobacillus brevis* are?

26. Why does sauerkraut undergo lactic acid fermentation and not alcoholic fermentation?

27. Identify other foods that go through a similar process as sauerkraut.

8 BASIC GENETICS

Mighty Microrobots

When we think of robots, generally we think of intelligent machines, capable of doing some type of work. But what about robots that we can't even see? Is that even possible? You might be surprised to know that tiny robots, called *microrobots*, are being researched in mice to develop ways to deliver treatment to cancer cells in the brain. Scientists created these microrobots with a nanoparticle gel containing magnetic iron beads and a cancer drug. This gel allows the microrobots to be controlled by magnetic fields. To gain access to the brain, the microrobots are coated with *E.coli* bacteria.

Cells that protect our bodies—*neutrophils*—gobble up these bacterial treats and travel via the bloodstream to cancer hot spots in the brain for treatment delivery.

Although this research is currently limited to mice, the hope is that the technology can soon be used for treating brain cancer in humans. Like most research, it takes many years before technology such as this can be used for treating disease. In the meantime, we can continue to use what we learn about our bodies to love and serve others through the use of medicine.

8.1 CELL DIVISION
Chromosomes and Genes

Each cell in the human body contains DNA. This is the cell's genetic material and is found in the nucleus as part of the nucleus's *chromatin*. Strands of DNA in the chromatin are tightly coiled around proteins called *histones*. Just before a cell divides, the chromatin condenses to form structures called **chromosomes**. Think of a chromosome as a spool of thread: DNA is the thread, and the histones are the spool that the DNA wraps around. One chromosome can contain between 500 and 2000 of these spools of DNA.

Many of the sections of DNA within a chromosome are **genes**, which code for a specific protein to create a certain characteristic in an organism. Genes are the basic units of heredity. Almost everything about us is a result of our genes—eye color, height, blood type, natural muscle size, and so on are all coded in our genes and spooled together in our chromosomes. Genes might also code for proteins that suppress tumors, direct the growth of nerves in the brain, or help cells with RNA transcription.

STRUCTURE OF CHROMOSOMES AND GENES

Just before a cell divides, its chromosomes duplicate and form an X shape. The duplicated chromosome is made of two identical halves, called **chromatids**, which are joined together at the **centromere**. Centromeres play a crucial role during cell division because they not only keep chromatids together but also help break them apart at the right time so that the information in chromosomes can be divided between two new cells.

? How long does it take for my skin cells to completely change?

Questions

How does information get from one cell to another?

How do cells contribute to an organism's growth and reproduction?

What are the differences between mitosis and meiosis?

Terms

chromosome
gene
chromatid
centromere
sex chromosome
autosome
diploid organism
haploid organism
cell cycle
interphase
mitosis
cytokinesis
meiosis
gamete
zygote

KINDS OF CHROMOSOMES

There are two kinds of chromosomes. **Sex chromosomes** determine whether an organism will be male or female. They also carry a few genes that determine other traits. In humans the two sex chromosomes are X and Y. The X chromosome is significantly larger than the Y, and so it carries a lot more genetic information. Children inherit one sex chromosome from each of their parents, giving them a combination of either XX, which codes for a female, or XY, which codes for a male.

Autosomes are chromosomes that *do not* determine an organism's sex. They contain most of an organism's genes. Humans have forty-four autosomes that pair up to form twenty-two sets. Unlike sex chromosomes, the two autosomes in a pair are about the same size.

Chromosomes aren't aligned in the nucleus in a way that they can be counted easily, so scientists sort out the chromosome pairs, arrange them in size order, and take a picture of them under a microscope. The resulting image of the ordered chromosomes is called a *karyotype*.

A human karyotype showing the twenty-two pairs of autosomes and one pair of sex chromosomes for a total of twenty-three pairs. Notice the size difference between the two sex chromosomes.

▼

diploid organism

haploid cell

CHROMOSOME PAIRS

The pairs of autosomes in the karyotype shown on the previous page are *homologous*, from the Greek words meaning "same speak." Both chromosomes contain the same genes that code for a set of characteristics. Your mother gave you one of the chromosomes in the pair, and your father gave you the other. Note that sex chromosomes are not considered homologous, though they're still a pair of chromosomes.

Organisms that have homologous pairs of chromosomes are **diploid organisms**. You are diploid, your dog is diploid, and so is the oak tree in your front yard. But not all organisms have chromosomes from each of their parents. They may have just one set of chromosomes and so are **haploid organisms**. Diploid organisms often have a few haploid cells in their bodies. Eggs and sperm are haploid cells that, when joined, form a new organism that is diploid.

COUNTING CHROMOSOME SETS

The number of different chromosome sets that an organism has in its cells is indicated by a number and the abbreviation n. So a diploid cell is $2n$, and a haploid cell is simply n. In humans $n = 23$, so $2n = 46$. Keeping track of n is important for organisms that may have more than two chromosome sets, a condition called *polyploidy*. For example, wheat can be $2n$, $4n$, or $6n$, depending on the species.

DIPLOID NUMBER OF CHROMOSOMES IN SOME COMMON ORGANISMS

Organism	Number	Organism	Number
adder's tongue fern	1260	kangaroo	16
chicken	78	mosquito	6
chimpanzee	48	shrimp	92
dog	78	watermelon	22
human	46	wolf	78

The Cell Cycle & Meiosis

Believe it or not, a month ago you were living in a different skin. How is that possible? Skin cells die and are replaced every three to four weeks through the **cell cycle**, which is the process of cell growth that leads to cell division for many cells. The cell cycle is responsible for an organism's growth and repair, and it faithfully transfers the genetic information from the one cell that formed you at conception to every cell thereafter in your body, keeping all your cells genetically identical.

The Cell Cycle

The cell cycle is broken into three stages: **interphase**, **mitosis**, and **cytokinesis**.

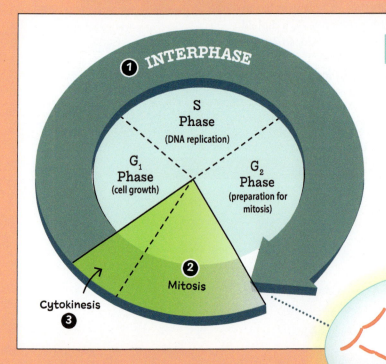

❶ INTERPHASE

Interphase is the phase in which the cell spends most of its time. During interphase the cell grows, and its nuclear DNA is duplicated. At the end of interphase, the tetraploid (4n) cell with its extra organelles, proteins, and chromosomes is ready for the four phases of mitosis and cytokinesis.

❷ MITOSIS

Prophase

In the cytoplasm, centrosomes (p. 102) are carried to opposite ends of the cell by protein motors. The centrosomes then assemble fibers from the cytoskeleton into a structure called the *mitotic spindle*, which will align and separate chromosomes in later phases. In the nucleus the chromatin begins to coil into visible chromosomes, the nucleolus is disassembled, and the membrane around the nucleus breaks down to release the chromosomes.

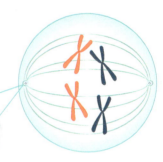

Metaphase

Metaphase begins as the fibers of the mitotic spindle attach to proteins on the centromere of each chromosome. These fibers pull the chromosomes to align them along an imaginary line called the *equatorial plane*.

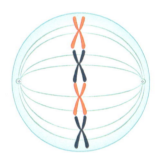

Anaphase

The centromeres, which hold the two chromatids of each chromosome together, send out a signal that splits the centromeres in half. The mitotic spindle then pulls the chromatids toward the centrosomes.

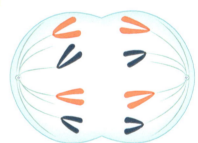

Telophase

The chromatids are now their own chromosomes, and the nucleus begins to re-form around them. In the nucleus the nucleolus reassembles, and the chromosomes uncoil to become chromatin again. In the cytoplasm the mitotic spindle disassembles. Cytokinesis also begins.

❸ CYTOKINESIS

While mitosis is finishing, the cell divides the contents of its cytoplasm. The cell also forms a *contractile ring* between the two nuclei, which pinches the cell in two. There are now two genetically identical cells, each with its own organelles, proteins, and diploid set of chromosomes. There are no visible chromosomes during cytokinesis and interphase.

Meiosis

During sexual reproduction, **meiosis** happens instead of mitosis to form haploid gametes. Just as in mitosis, a tetraploid (4*n*) cell about to enter meiosis has doubled its number of chromosomes along with its cytoplasmic proteins and organelles.

MEIOSIS 1

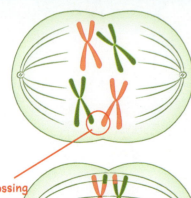

crossing over

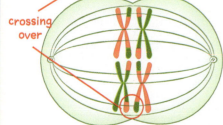

❶ *Prophase I*

Once the membrane around the nucleus has dissolved, each chromosome joins with its homologous chromosome to form a tetrad. Chromosomes in a tetrad exchange genes in a process called *crossing over*. The centrosomes migrate and the spindle forms.

❷ *Metaphase I*

The spindle fibers align tetrads on the equatorial plane.

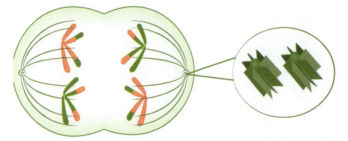

❸ *Anaphase I*

The spindle separates each tetrad and pulls whole chromosomes toward the centrosomes instead of individual chromatids.

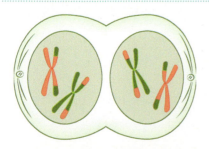

❹ *Telophase I and Cytokinesis*

Each side of the cell now has a haploid set (n) of chromosomes. A nucleus re-forms around each set, but the chromosomes don't uncoil. The cytoplasm is divided, and the cell is pinched in two.

MEIOSIS 2

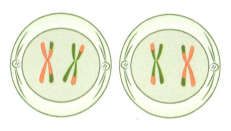

❺ Prophase II

Right after the cells divide, a spindle re-forms in each cell to begin meiosis II. The chromosomes do not duplicate.

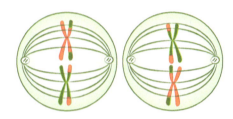

❻ Metaphase II

The spindle fibers align chromosomes on the equatorial plane.

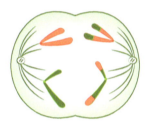

❼ Anaphase II

The chromosomes' centromeres split, and the spindle pulls the chromatids toward the centrosomes.

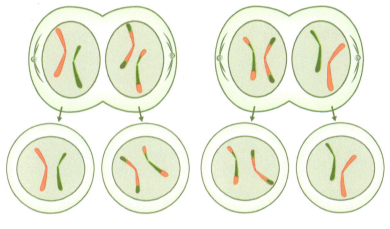

❽ Telophase II and Cytokinesis

A nucleus re-forms around each haploid set of chromosomes. Because of the chromosomes' crossing over, the four cells are genetically different from each other and from the original cell.

Mitosis creates the trillions of identical diploid cells that are in the body. But the body also contains some haploid cells known as **gametes**, which are cells that can unite to form a zygote. A **zygote** is a diploid cell that is genetically different from either parent. The zygote divides by mitosis as it grows and develops before being born. In humans the two kinds of gametes are egg cells and sperm cells. In a diploid organism haploid gametes are produced through meiosis, the reduction of a cell's chromosome number from diploid to haploid. Just as in mitosis, a tetraploid (4*n*) cell about to enter meiosis has doubled its number of chromosomes, proteins, and organelles. Unlike mitosis, the cells go through two divisions, which are called *meiosis I* and *meiosis II*.

MITOSIS	**MEIOSIS**

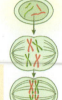

- leads to genetically identical cells
- generates two diploid cells after one cell division
- has homologous pairs of chromosomes that are separate
- is the basis of *asexual reproduction*, a process that produces offspring that are genetically identical to their parents

- leads to genetically different cells
- generates four haploid cells after two cell divisions
- has homologous pairs of chromosomes that are connected to form tetrads
- is the basis of *sexual reproduction*, which occurs when two haploid cells join to form a genetically different diploid zygote

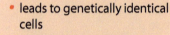

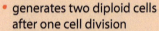

Similarities

- Both are a form of cell reproduction.
- Both begin with a tetraploid cell.
- Meiosis II is almost the same process as mitosis.

8.1 SECTION REVIEW

1. Relate DNA to genes and chromosomes.

2. How many homologous pairs of chromosomes do you have?

3. What is the difference between the size of sex chromosomes and autosomes?

4. In which stage of the cell cycle is the cell preparing itself to divide?

5. A single muscle cell can contain dozens of nuclei. Which stage of the cell cycle is the cell *not* completing? How do you know?

6. During which phase of mitosis do the centromeres split?

7. During which phase of mitosis does the mitotic spindle assemble? In which phase does it disassemble?

8. In which phase of meiosis is genetic information exchanged between chromosomes?

9. Why is meiosis split into meiosis I and II?

10. Give two differences between anaphase I and anaphase II in meiosis.

11. Which type of cell division turns a diploid cell into a haploid cell?

12. What is the difference between asexual and sexual reproduction?

Gregor
Mendel

8.2 THE INHERITANCE OF TRAITS

Mendel's Genetics

In a small eastern European abbey during the 1850s a monk named Gregor Mendel (1822–84) was making scientific history. Gardeners had known for centuries how to breed pea plants to get the kinds of peas or flower colors that they wanted, but no one knew the mechanism that caused a trait to be handed down to the next generation until Mendel discovered it. Though he would never realize it during his lifetime, Mendel's pea plant experiments in his garden behind the abbey revolutionized the way scientists thought about **heredity**, the passing of traits from one generation to another.

Mendel published his ideas and experiments on pea plants in a scientific journal, but his ideas were so advanced for that time that scientists couldn't make heads or tails of them. So his work lay forgotten until about sixteen years after his death when a few botanists unknowingly repeated his experiments. As they researched the groundbreaking results of their experiments, they discovered that Mendel had gotten the same results decades earlier. His work was finally getting the attention it deserved, and he is now considered the Father of Modern Genetics.

So what were the results of Mendel's experiments that changed the way people thought about heredity? They can be boiled down to three principles: heredity, segregation, and independent assortment.

Why do some children look more like one parent than the other?

Questions

How is a cell's genetic information expressed?

Why do some offspring look different from their parents?

How can we predict the traits that an organism may have?

Terms
heredity
dominant trait
recessive trait
allele
homozygous organism
genotype
phenotype
heterozygous organism
Punnett square
monohybrid cross
dihybrid cross
pedigree
incomplete dominance
codominance
multiple alleles
polygenic inheritance
sex-linked trait
carrier

PRINCIPLE OF HEREDITY

Mendel bred thousands of pea plants together and traced their traits through each generation. He learned that each parent plant contributed what he called a *factor* for each trait, giving the offspring two factors for that trait. For example, two tall parent plants could each give their offspring a tall factor, resulting in tall offspring. But sometimes, two tall parents would produce short offspring. So what happened to the tall factors?

Mendel explained that since an organism has two factors for each trait, one of those factors must win out over the other one. The trait that is expressed is the **dominant trait**. This doesn't mean that the losing factor disappears. It still codes for the **recessive trait**, the trait that isn't expressed. Mendel showed this by breeding a tall plant that was purebred (a plant that always produced tall plants) with a short plant that was purebred (a plant that always produced short plants).

GENOTYPE	PHENOTYPE
TT—homozygous (purebred) *Tt*—heterozygous (hybrid)	tall—dominant trait
tt—homozygous (purebred)	short—recessive trait

Since tallness in pea plants is a dominant trait, it is represented by an uppercase *T*. Since shortness is considered recessive, it's given a lowercase *t*. These individual letters represent **alleles**, which are the different forms that a gene can have. Alleles are the factors that Mendel observed. For example, the tall trait (*T*) is just one allele for the gene that controls height. A purebred organism, also known as a **homozygous organism**, has two of the same alleles for a trait. A purebred tall plant has two tall alleles, *TT*, and a purebred short plant has two short alleles, *tt*. These letters help us to visualize an organism's **genotype**, or genetic makeup.

Notice in the diagram below that the purebred tall and short pea plants first produce only tall plants. Though the parent plants (P_1) and offspring (F_1) both have the same **phenotype**, or physical expression of a trait, they have different genotypes—in other words, they're all tall but their alleles are different. *TT* tall plants are purebred, and *Tt* tall plants are hybrid; that is, they contain two different alleles. Organisms with two different alleles for a gene are **heterozygous organisms**.

When two F_1 offspring breed, they produce another generation of plants (F_2). When the *Tt* alleles of F_1 combine, they can produce purebred tall (*TT*), hybrid (*Tt*), and purebred short (*tt*) plants as the F_2 offspring.

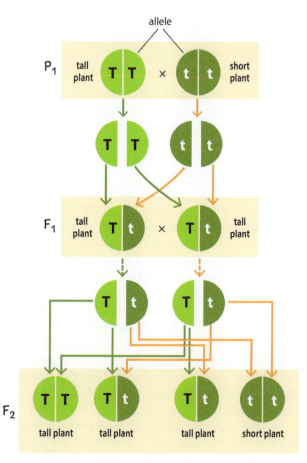

PRINCIPLE OF SEGREGATION

After studying his pea plants, Mendel decided that the factors that determine a trait must separate, or segregate, into the reproductive cells of the plants. Though he didn't know it, he was describing the process of meiosis, which forms haploid gametes that contain combinations of alleles that are different from the parents' combinations of alleles.

PRINCIPLE OF INDEPENDENT ASSORTMENT

Mendel didn't study just the height of his pea plants; he also studied the color and shape of peas. He noticed that each of these traits segregated independently during reproduction. In other words, the trait for yellow peas (*Y*) or round peas (*R*) didn't always accompany the trait for tallness (*T*); some tall plants had green (*y*), wrinkly (*r*) peas, and other tall plants had green, round peas. The traits mixed and matched freely.

Each chromatid of a chromosome carries an allele for each trait. There's no particular way that the chromosomes align on the equatorial plane during meiosis, so the chromatids, and therefore the alleles, are recombined in the gametes in new ways.

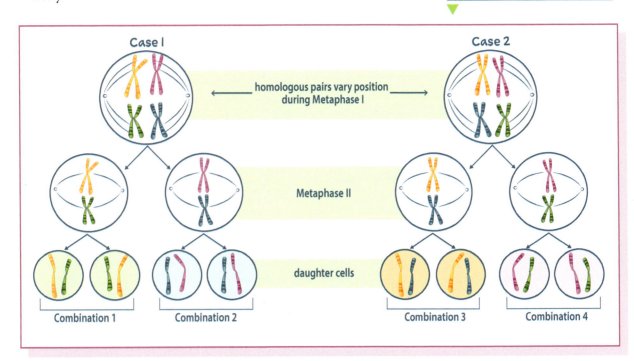

The independent assortment of traits is what creates variation from organism to organism. It's the reason why you may have your mom's eye shape, your dad's eyelashes, and an eye color that doesn't match anyone else's in your family.

Punnett Squares

Sorting out which alleles end up in which offspring can quickly become confusing. So Reginald Punnett (1875–1967), an avid student of Mendel's ideas of genetics, came up with a simple solution. He drew little squares to show the different ways that alleles can segregate. Known as **Punnett squares**, these diagrams are still helpful today for understanding how parents' alleles can be mixed in their offspring.

Reginald Punnett

The image on the right shows a Punnett square that demonstrates how albinism can be passed from a mother and father to their children. Albinism is a recessive trait, and when it is expressed, a person's body fails to produce melanin, the pigment that gives color to skin, hair, and eyes.

In the Punnett square the trait is represented with an *M* for normal melanin and an *m* for no melanin. A homozygous genotype of *mm* codes for albinism, showing that it's a recessive trait. The horizontal row of alleles is the father's, and the vertical column is the mother's. This Punnett square shows that both parents are heterozygous for albinism (*Mm*). When the parents' alleles are hypothetically crossed (*Mm × Mm*), known as a *cross*, we can see the possible combinations of alleles that might appear in their children.

	Father (normal)	
	M	m
M	**MM** normal melanin	**Mm** normal melanin
m	**Mm** normal melanin	**mm** albino

Mother (normal)

These are just possibilities, however. These heterozygous parents might have ten children and never have a child with albinism. What a Punnett square can show you, along with the possible combinations of alleles, is the probability of a particular combination. Notice how three of the four combinations of alleles in the Punnett square above include the dominant trait for normal melanin. So each of these parents' offspring has a 75% chance of inheriting the dominant trait. There's only a one-in-four chance (25%) that a particular child of theirs will have albinism. About 1 in every 20,000 children inherits the recessive trait of albinism (*mm*).

Tracking Traits

What if you want to track more than just one trait at a time, such as the color *and* shape of pea pods? The Punnett square on the previous page deals with a **monohybrid cross**, a cross that tests only one set of alleles. Crossing two sets of alleles results in a **dihybrid cross**. The Punnett square shows a dihybrid cross for pea pod color and shape. A parent plant that is purebred for the two dominant traits (green and inflated pea pods—*GGII*) is crossed with a parent plant that's purebred for the two recessive traits (yellow and constricted pea pods—*ggii*).

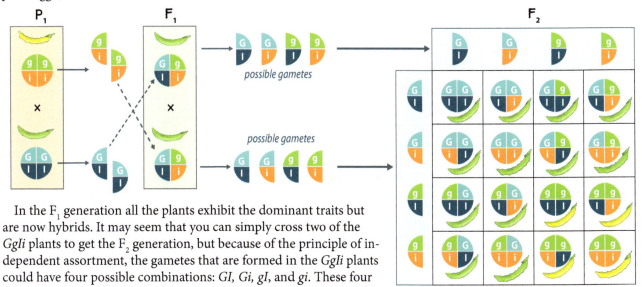

In the F$_1$ generation all the plants exhibit the dominant traits but are now hybrids. It may seem that you can simply cross two of the *GgIi* plants to get the F$_2$ generation, but because of the principle of independent assortment, the gametes that are formed in the *GgIi* plants could have four possible combinations: *GI*, *Gi*, *gI*, and *gi*. These four possible gametes are crossed to produce the F$_2$ generation.

It's fairly simple to use a cross to track traits in pea plants because they can have hundreds of offspring in a short amount of time. But what about organisms that produce only a few offspring? Or what if we have a question about a trait from our family's history? For these situations we can use a **pedigree**, a family history that charts a hereditary trait through past generations. Pedigrees show the phenotypes of traits as they are passed down from one generation to the next. Tracing phenotypes can help scientists figure out the genotypes of a family's ancestors and predict the genotypes of current family members. Notice the pedigree of fruit flies below.

In fruit flies red eyes are dominant and brown eyes are recessive.

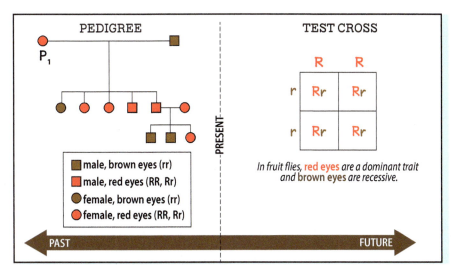

Remodeling Genetics

Poppies were the go-to flower in Britain during the early 1900s. Everyone had to have those vibrant red blooms. Red poppies became so admired that they inspired one of the most famous poems written about World War I, "In Flanders Fields," and now they serve as a remembrance for Britain for that war.

But the highly praised red poppy didn't stay red for long. Amos Perry, the owner of a popular plant nursery in England, stumbled across a pink poppy in the middle of a bed of red poppies. Intrigued, he collected the pink poppy and began selling its pink offspring. But not everyone was happy with his pink poppies—a customer told Perry that one of the pink poppies he had bought was flowering white. Perry quickly apologized and replaced the poppy for the customer, and then he scooped up the rascally white poppy and created yet another breed of poppies.

Gregor Mendel's ideas don't fully explain the red-to-pink-to-white change, so is there something else going on with genes? The following five types of gene inheritance are models that deviate from Mendel's model and help explain many traits, including color in poppies.

INCOMPLETE DOMINANCE

The traits that Mendel studied in pea plants are either dominant or recessive. But in some organisms, both alleles for a trait are expressed at the same time, creating a third phenotype. This blending of alleles is known as **incomplete dominance**. In poppies a red flower has the genotype *RR*. A white flower is *rr*. Using Mendel's principle of heredity, we might think that the genotype *Rr* would give a red-flower phenotype, but it actually creates a blend of red and white—a pink flower.

Since neither color is truly dominant, an uppercase *C* (for color) is used to represent flower color, and lowercase superscripts are used for the different colors. So a red poppy is $C^r C^r$, a white poppy is $C^w C^w$, and a pink poppy is $C^r C^w$. Crossing two pink poppies can result in pink, red, or white poppies.

In incomplete dominance the two alleles for a trait (red and white flowers) combine to produce a heterozygous genotype that is neither dominant nor recessive (pink flowers).

When two pink poppies are crossed, they may produce more pink poppies, but there's also a 25% chance for each that they'll produce either white or red poppies as well.

Incomplete dominance also occurs in some human traits. For example, someone with a normal cholesterol level may have a genotype of $C^n C^n$ (*C* is for cholesterol and n is for normal). Someone with very high cholesterol may have a genotype of $C^h C^h$, while someone with a hybrid genotype of $C^n C^h$ may have a cholesterol level that's higher than normal but not as high as that of a $C^h C^h$ person.

CODOMINANCE

Sometimes both inherited alleles are expressed in what is known as **codominance**. Think of a king and queen coreigning, both expressing their commands at the same time—this is codominance.

Some breeds of chickens exhibit codominance in their feathers. Since both alleles for feather color are dominant, they are both represented with an uppercase letter. When a white chicken ($C^W C^W$) mates with a black chicken ($C^B C^B$), an alternating pattern, or variegation, of black and white ($C^W C^B$) is produced.

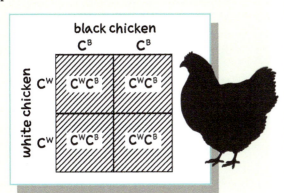

black chicken

	C^B	C^B
C^W	$C^W C^B$	$C^W C^B$
C^W	$C^W C^B$	$C^W C^B$

white chicken

MULTIPLE ALLELES

So far we've discussed traits that have two alleles, but some traits have **multiple alleles**. One of the most familiar traits with multiple alleles is human blood type. Blood type can contain the alleles *A*, *B*, or *O*. People can express only the two blood type alleles received from their parents, so no one can have all three. Traditionally, the two dominant alleles, *A* and *B*, are represented with an uppercase I, and the recessive allele, *O*, with a lowercase i.

Genotype	Phenotype
$I^A I^A$, $I^A i$	Type A
$I^B I^B$, $I^B i$	Type B
$I^A I^B$	Type AB
ii	Type O

Type B Blood

Type A Blood	I^B	i
I^A	$I^A I^B$	$I^A i$
i	$I^B i$	ii

If one parent is heterozygous for blood type A and the other is heterozygous for blood type B, their children have the potential for any blood type. The "i" that is used as the symbol for blood type comes from the word *"immunoglobulin,"* which is another name for an antibody in blood that latches onto disease-causing substances in the blood.

POLYGENIC INHERITANCE

Some traits have more than one gene that controls them. **Polygenic inheritance** involves two or more genes working together to express a trait. Many human traits are the result of several genes working together, creating a greater variety of expressions than a single gene could. The wide spectrum of skin color is an example. We have several genes, possibly dozens, that code for the amount of melanin that our cells produce. Environmental factors such as UV rays also play a part. But let's consider just three genes for skin color. Very light skin is a recessive trait (*aabbcc*), and very dark skin is dominant (*AABBCC*). A completely heterozygous genotype, *AaBbCc*, creates a medium-color phenotype. The more dominant alleles in the skin color genotype, the darker the skin will be.

SKIN COLOR

	ABC	ABc	AbC	aBC	Abc	aBc	abC	abc
ABC	AABBCC	AABBCc	AABbCC	AaBBCC	AABbCc	AaBBCc	AaBbCC	AaBbCc
ABc	AABBCc	AABBcc	AABbCc	AaBBCc	AABbcc	AaBBcc	AaBbCc	AaBbcc
AbC	AABbCC	AABbCc	AAbbCC	AaBbCC	AAbbCc	AaBbCc	AabbCC	AabbCc
aBC	AaBBCC	AaBBCc	AaBbCC	aaBBCC	AaBbCc	aaBBCc	aaBbCC	aaBbCc
Abc	AABbCc	AABbcc	AAbbCc	AaBbCc	AAbbcc	AaBbcc	AabbCc	Aabbcc
aBc	AaBBCc	AaBBcc	AaBbCc	aaBBCc	AaBbcc	aaBBcc	aaBbCc	aaBbcc
abC	AaBbCC	AaBbCc	AabbCC	aaBbCC	AabbCc	aaBbCc	aabbCC	aabbCc
abc	AaBbCc	AaBbcc	AabbCc	aaBbCc	Aabbcc	aaBbcc	aabbCc	aabbcc

There are eight possible gametes that can be produced in each of two people with a heterozygous genotype (*AaBbCc*) for skin color. Crossing the gametes in a Punnett square shows that their children are much more likely to have medium skin than either dark or light skin.

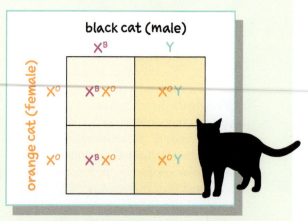

An orange female cat (X^O and X^O) crossed with a black male cat (X^B and Y) can produce only calico females and orange males.

SEX-LINKED TRAITS

Most of our traits are determined by our autosomes. But a few are controlled by the sex chromosomes and are called **sex-linked traits**. Most sex-linked traits are determined by the X chromosome. Why? The X chromosome is much larger than the Y (look back at page 150 to see how much larger), and so it carries a lot more information. The color of calico cats is a great example of a sex-linked trait. Almost all calico cats are female. The calico color contains a mix of orange and black fur. Both colors are carried only on the X chromosome and are expressed codominantly. A female cat can be $X^B X^B$ (black), $X^O X^O$ (orange), or $X^B X^O$ (calico). A male cat can be $X^B Y$ (black) or $X^O Y$ (orange).

A common sex-linked trait in humans is red-green colorblindness. The allele for this trait is found on the X chromosome. Normal vision is dominant (X^N), and red-green colorblindness is recessive (X^n). Men are much more likely than women to inherit colorblindness because they need only one recessive allele on their X chromosome ($X^n Y$) to express the trait. Women must have the recessive allele on both of their X chromosomes ($X^n X^n$) to be colorblind.

Notice in the Punnett square on the left that any of the daughters of the colorblind man and his homozygous, normal wife will be heterozygous, carrying one allele for normal vision and one for colorblindness. People who carry an allele but don't express it are called **carriers**. A carrier daughter can't have colorblindness because it's a recessive trait, but she could potentially pass it along to her sons, making them colorblind, or pass it along to her daughters, making them carriers.

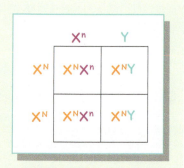

A colorblind man ($X^n Y$) won't have any colorblind children if his wife is homozygous for normal vision ($X^N X^N$), but his grandchildren could be colorblind.

Since people with red-green colorblindness can't tell the difference between the colors in a traffic light, some cities have altered the colors or changed the shape of the stoplight itself.

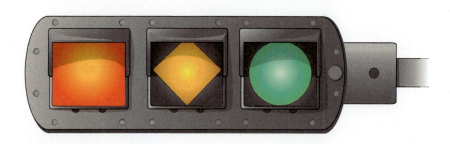

EXPERIMENTING WITH ANIMALS

Research companies often try to avoid testing their products on animals, but for potentially dangerous products it's not always avoidable. Before treatments like stem cell transplants or new medications are ever used on a person, they're tested extensively on animals. Every research laboratory must follow a set of rules for taking care of the animals, making sure that they're treated as humanely as possible. But many of them still die as a result of the treatments and diseases that they're given. Could there be a better way to create treatments that are safe for people without sacrificing so many animals?

For thousands of people who suffer from diabetes, animal research has been a lifesaver. Before the 1920s diabetes was a death sentence. But two physicians began experimenting on dogs and discovered that diabetes is caused by the faulty production of insulin. They were able to isolate insulin and offer it to people with diabetes as a treatment. Today we don't have to rely on animals to produce insulin because medical researchers have found ways to produce insulin artificially.

A more recent move away from using animals is in asthma and allergy research. The chemicals in asthma inhalers and allergy pills are usually tested on animals, but a new procedure is gaining in popularity—taking a few cells from a person's respiratory system and testing the medications on those cells. This procedure might be the first step toward eliminating animal studies for many kinds of medications.

A darker side of animal research is manipulating animals' genes in such a way that it permanently disfigures them. In the late eighteenth century a sheep with a crippling form of dwarfism was prized because it was unable to jump over low fences. This trait was bred into an entire line of sheep that became known as Ancon sheep (*Ovis aries*). These sheep became extinct in the nineteenth century because breeding for the genetic disease of dwarfism made them incredibly weak. Many research animals today have their genes manipulated on the molecular level so that they can be cloned, produce human diseases, grow human organs, or create bizarre combinations of species that don't survive. The purpose of any animal research should be to benefit people or the animals themselves; it shouldn't be used as a playground for creating animals for no useful reason.

The Bible gives us a balanced view of how to use animals. They were used daily as sacrifices to show the Israelites their need for the permanent Sacrifice that would one day come, but the sacrificed animals weren't wasted. Meat that was left over after certain sacrifices fed the priests (Lev. 7:34). In Exodus 20:10 God includes animals in the Sabbath day rest, and in Deuteronomy 25:4 God reminds His people to take care of the animals that they have. This command is repeated in 1 Corinthians 9:9 and 1 Timothy 5:18, applying not only to animals, but to anyone who serves another. We should also care for animals that we may not value as highly as others (Matt. 10:29). Our pets aren't the only animals that deserve to be treated well!

We live in a world marred by sin and suffering, and because of that our stewardship is less than ideal. But that doesn't mean that we should treat animals with little or no respect. Human lives have more value than animal lives (Matt. 10:31), but we should continue to improve our technology and research techniques so that we can minimize the harm that we bring to the animals in our care.

CRUELTY FREE
NOT TESTED ON ANIMALS

Predicting Genotypes

Do you ever wonder why you look the way you do? The answer to this question lies within your genes. Up to this point you have learned that genes combine to create your physical features. The dominant genes that are present are the traits that are physically expressed. Using a Punnett square, let's take a look at a common trait to see the effects of expression when different genotypes are crossed.

How can I predict an offspring's traits?

Materials
paper • pencil

PROCEDURE

A Determine the genotypes of two parents who are heterozygous for dimples, which is a dominant trait (*D*). The recessive trait is no dimples (*d*).

B Draw a Punnett square by placing each allele of one parent in the top row and each allele of the other parent in the left side column.

C Within the boxes of the Punnett square write the genotypes of the offspring that result when the parents are crossed.

D Create another Punnett square which crosses a homozygous dominant parent with a homozygous recessive parent.

ANALYSIS

1. What are the genotypes for each heterozygous parent?

2. What are the possible genotypes for the offspring?

3. What is the genotype ratio of the offspring?

4. What are the genotypes for a homozygous dominant and homozygous recessive parent?

5. What are the possible genotypes for the offspring?

GOING FURTHER

6. Summarize the phenotypes that may possibly occur from these crosses.

7. What cross(es) would give a higher likelihood of the recessive phenotype being expressed?

1. In gerbils solid-colored fur is a dominant trait (*C*), and multicolored fur is recessive (*c*). If a gerbil has the genotype *Cc*, what is its phenotype?

2. Briefly state the differences between Mendel's three genetic principles.

3. Can we identify the genotype of a person with normal melanin? Why or why not?

4. In many plants variegated leaves (leaves with stripes and other color variations) are a recessive trait, and solid-colored leaves are dominant.

   a. Choose a letter and list the two alleles using that letter.

   b. Using a Punnett square, show the possible genotypes that result from a cross between a plant with variegated leaves and a plant with heterozygous solid leaves.

   c. List their possible phenotypes.

5. Jake has albinism (*mm*), but no one else in his family does. How did his parents pass the trait down to him? Draw a Punnett square to help explain your answer.

6. If Jake wanted to trace the trait of albinism to see where it appears earlier in his family's history, what could he do?

7. Eye color is a result of polygenic inheritance—two alleles on one gene and two alleles on another. One gene codes for either brown (*B*) or blue (*b*). The other codes for dark green (*G*) or light green (*g*). Brown is dominant over green. Someone with the homozygous genotype *bbgg* has very blue eyes; someone with the homozygous genotype *BBGG* has very brown eyes. Can a couple who both have brown eyes have a child with blue eyes? Support your answer with a Punnett square.

8. If a black-feathered chicken were bred with a white-feathered chicken, and their offspring's colors were a result of incomplete dominance, what color would their offspring's feathers be? Why?

9. Using the information and the Punnett square that you created in Question 7, determine whether brown-eyed parents can have a green-eyed child. If so, what are the genotypes for green eyes? Is it more likely that these parents would have a blue-eyed child or a green-eyed child?

10. Is type AB blood an example of incomplete dominance or codominance? Explain.

11. Fabry disease is a recessive genetic disorder that leads to the buildup of lipids throughout the body. More men than women have this disease and usually show severe symptoms. Occasionally women have it, and they have either mild or severe symptoms. Using this information about Fabry disease, what two types of gene inheritance do you think are involved? Explain.

8.3 GENE EXPRESSION

Activating Genes

Every gene we have can be found in almost every cell of our bodies. That means that the genes that code for eye color are found not just in the eyes—they're also in the stomach, kidneys, ears, feet—they're everywhere! So how does the body know to express eye color genes only in the eyes?

Gene expression is the process in which genetic information is activated to make a genetic product. The journey from DNA to proteins that we studied in Chapter 6 is part of the process of gene expression. A genetic product could be an obvious trait, like the proteins that produce eye color, or a trait that we'll never see, like the enzymes that help digest materials in the small intestine. Two main factors—DNA molecules and the environment—control gene expression.

?

Why are stem cells so useful?

Questions

How are genes and cell development related?

How does the environment affect gene expression?

How do genes determine a cell's function?

Should we use animal genes to benefit people?

Terms

gene expression
promoter
TATA box
cell differentiation
homeotic gene
stem cell
embryo

THE ROLE OF DNA

During the transcription of DNA little sequences of DNA called **promoters** come before each gene and tell the *RNA polymerase* molecule where to bind to begin copying the gene into a strand of mRNA. Promoters are found in all eukaryotic cells and accompany every gene in a cell. One common promoter, called the **TATA box**, binds to a group of *transcription proteins* that are attached to RNA polymerase. Through the different combinations of promoters and proteins, genes are turned either on or off in a cell.

DNA also contains sequences called *enhancers* and *silencers*. These can bind to the promoters in front of genes and act as switches, telling the RNA polymerase whether the gene needs to be turned on (enhanced) or turned off (silenced). A promoter is told what to do by enhancers and silencers. Think of a promoter as a car that's on but not moving yet. It requires an enhancer (the gas pedal) to go or a silencer (the brake) to stop.

But the enhancers and silencers don't act alone. Just as the RNA polymerase molecule and transcription proteins bind to promoters, proteins called *regulatory proteins* bind to enhancers and silencers. Regulatory proteins are the actual link between the enhancers and silencers and the promoters. Going back to the car analogy, regulatory proteins are like your foot stepping on the gas or brake pedal. When your eye cells are expressing eye genes, they are stepping on the gas when eye genes appear in the DNA strand and hitting the brakes when other genes appear. Once the right genes are turned on and transcribed to mRNA, the *introns* (noncoding sections of DNA) are removed, and the *exons* (coding sections of DNA) are stitched together and are ready to leave the nucleus to be translated into a protein.

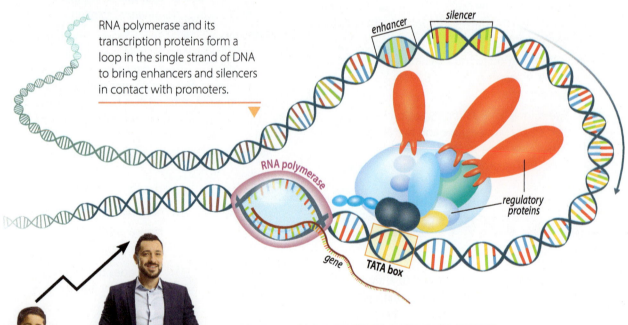

RNA polymerase and its transcription proteins form a loop in the single strand of DNA to bring enhancers and silencers in contact with promoters.

enhancer

silencer

RNA polymerase

regulatory proteins

gene

TATA box

THE ROLE OF THE ENVIRONMENT

Transcription can be affected by things other than promoters and proteins.

Hormone-activated traits. A person's gender can affect gene expression. For example, when a child reaches puberty, the body sends out a wave of hormones (chemical messengers) to tell the body how to begin preparing for adolescence. These hormones bind to the DNA strand during transcription and activate genes that have never been active before. Because males have different levels of hormones than females do, different genes are activated, resulting in two very different developments between the sexes.

Light and temperature. The fur color of a Himalayan rabbit changes depending on the climate of the region it lives in. When the rabbit is warm, its genes produce a uniform coat color. When it is cold, its extremities—ears, nose, feet, and tail—turn a much darker color than the rest of its coat. Sometimes a Himalayan rabbit that lies against the side of a cold wire cage will develop dark lines of fur where the cage was touching, creating "grill marks" on the rabbit.

One of the reasons we get sleepy when it's dark outside is due to gene expression. As light levels drop, our genes send out a signal to produce melatonin, the hormone that regulates sleep patterns. Being in a brightly lit area all day and night will alter the body's production of melatonin and therefore its sleep cycle.

Chemicals. Chemicals such as drugs and pollutants can radically alter gene expression in babies that are still developing. If a pregnant woman takes certain medications, inhales or consumes pesticides, or touches certain parasites during her pregnancy, the chemicals can disrupt transcription and create defective genes throughout her baby's cells. On the other hand, researchers are looking for ways to use chemicals to turn off genes that code for disorders. The hope is that these chemicals might one day lead to a better treatment for cancer, cystic fibrosis, sickle cell anemia, and the scores of other genetic disorders that we have to deal with in our fallen world.

▲

A Himalayan rabbit that was raised below room temperature (below 73 °F, or 23 °C) will have darker ear, nose, and paw coloration than one raised above room temperature.

Genes and Development

So how does a tiny, one-cell zygote turn into an organism with trillions of cells that each do something different? Zygotes go through the process of **cell differentiation**, which changes cells so that they can perform more specialized tasks.

While a baby is still a single-cell zygote or a several-cell embryo, its cells are *totipotent*; that is, they have the potential to turn into any kind of cell. A cell from a baby during the totipotent stage can become a liver cell, eye cell, bone cell, or any other kind of cell. After a few cell divisions, the cells become *pluripotent*, still able to produce many kinds of cells, but not all kinds.

A special set of genes in a developing baby's cells, known as **homeotic genes**, are working to produce each kind of cell that he will need. They make sure that a person looks like a human, that a cow looks like a cow, and so on. Once that's done, the baby's cells are no longer totipotent, and the genes that were responsible for cell differentiation are switched off.

These amazingly changeable cells are called **stem cells**—cells that have not yet differentiated into specialized cells. The cells described in the previous paragraph are called *embryonic stem cells* because they appear during the zygote and embryo stages of development. An **embryo** is the early stage of an unborn young of a multicellular organism. But there are a few cells that don't differentiate in a fully developed organism. These are called *somatic stem cells*, or adult stem cells.

When a one-cell zygote divides, it produces a slightly specialized cell and another stem cell. As the cells continue to divide, they become more specialized, and fewer stem cells are produced.

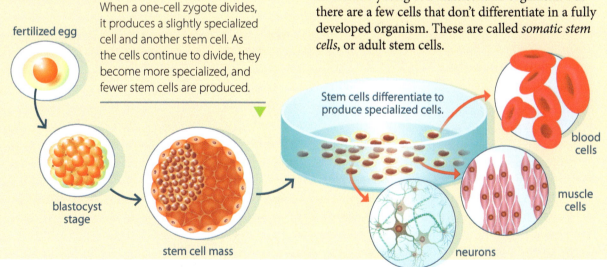

fertilized egg

blastocyst stage

stem cell mass

Stem cells differentiate to produce specialized cells.

blood cells

muscle cells

neurons

▲

Surgeons collecting bone marrow

Physicians have been using somatic stem cells for years to help people with bone cancer or other diseases that damage bones. These people receive a bone marrow transplant from another person because bone marrow contains somatic stem cells. Once the donor cells have been transplanted and are established, they begin differentiating, producing healthy cells to replace the damaged ones. And stem cells are found not only in bone marrow; researchers are finding more and more places in the human body that contain them. They hope to harness the versatility of these cells to create treatments and cures for even more diseases.

8.3 SECTION REVIEW

1. How does a cell in the spleen know how to express spleen genes? Use the terms *enhancer, promoter, regulatory protein, RNA polymerase, silencer,* and *transcription protein* in your answer.

2. Gestational diabetes occurs when a pregnant woman has high levels of blood sugar. Cells in the pancreas usually keep blood sugar levels normal, but hormones produced during pregnancy deactivate the cells. After pregnancy, hormone levels return to normal, and so do the pancreas cells. Is this an example of DNA or the environment controlling gene expression? Explain your answer.

3. What would happen if there were only enhancers and no silencers in a strand of DNA?

4. Which has the greater potential to turn into any kind of cell—a totipotent or a pluripotent cell?

5. If a teenager inhaled a lot of chlorine at the pool, which of his cells would be affected? If an embryo were exposed to chlorine, which of his cells would be affected?

6. What is the difference between an embryonic stem cell and a somatic stem cell?

7. What is the balance between caring for animals and using them for medical research? Justify your answer with a Bible reference.

8. Give two problems associated with research using human embryonic stem cells and two ways that somatic stem cells provide solutions to those problems.

8 CHAPTER REVIEW
Chapter Summary

8.1 CELL DIVISION

- Chromosomes are structures that contain DNA and supporting proteins. Genes are the sections of DNA within chromosomes that code for specific proteins.

- Sex chromosomes determine whether an organism will be male or female. Autosomes are chromosomes that do not determine an organism's sex.

- The cell cycle has three stages: interphase, mitosis (prophase, metaphase, anaphase, telophase), and cytokinesis.

- Mitosis generates two diploid cells, occurs during asexual reproduction, and goes through only one cell division.

- Meiosis generates four haploid cells, occurs during sexual reproduction, and undergoes two cell divisions.

Terms
chromosome • gene • chromatid • centromere • sex chromosome • autosome • diploid organism • haploid organism • cell cycle • interphase • mitosis • cytokinesis • meiosis • gamete • zygote

8.2 THE INHERITANCE OF TRAITS

- Gregor Mendel introduced the principles of heredity, segregation, and independent assortment.

- Dominant traits are traits that are expressed, and recessive traits are those that are not.

- Monohybrid crosses are used to predict the genotypes of offspring using one set of alleles. Dihybrid crosses use two sets of alleles to predict genotypes.

- Incomplete dominance, codominance, multiple alleles, polygenic inheritance, and sex-linked traits are all examples of non-Mendelian genetic inheritance.

Terms
heredity • dominant trait • recessive trait • allele • homozygous organism • genotype • phenotype • heterozygous organism • Punnett square • monohybrid cross • dihybrid cross • pedigree • incomplete dominance • codominance • multiple alleles • polygenic inheritance • sex-linked trait • carrier

8.3 GENE EXPRESSION

- An organism's genes are found in most of its cells.

- Genes can be turned "on" or "off" by the actions of promoters, transcription proteins, enhancers, and silencers.

- Different genes found within DNA can be expressed in response to their environment.

- Hormones, light, temperature, and chemicals may all affect gene expression.

- Stem cells are cells that have not differentiated into specialized cells. Somatic stem cells are undifferentiated adult stem cells.

Terms
gene expression • promoter • TATA box • cell differentiation • homeotic gene • stem cell • embryo

8 CHAPTER REVIEW
Chapter Review Questions

RECALLING FACTS

1. List the following terms from least complex to most complex: chromosome, DNA, gene.

2. Describe two differences between sex chromosomes and autosomes.

3. The haploid number of chromosomes (*n*) for wheat is 7. How many sets of chromosomes does wheat have if it has forty-two chromosomes total?

4. The green-spotted pufferfish has forty-two chromosomes. Its eggs and sperm each have twenty-one chromosomes. What is its haploid number of chromosomes? What is its diploid number?

5. Using the numbers from Question 4, how many chromosomes would you suggest the offspring of green-spotted pufferfish have? How many homologous pairs of chromosomes do they have?

6. During which part of interphase does a cell make a copy of its chromosomes?

7. How do cells become genetically different from each other during meiosis?

8. How is cytokinesis after mitosis different from cytokinesis after meiosis I?

9. Gametes divide by _____, and zygotes divide by _____.

10. Heterozygous is to hybrid as homozygous is to _____.

11. A(n) _____ tries to predict future generations; a(n) _____ traces past generations.

12. What is the difference between incomplete dominance and co-dominance?

13. Multiple allele inheritance involves more than two _____ coding for the same trait. Polygenic inheritance involves two or more _____ coding for the same trait.

14. During what process of protein synthesis are genes turned on or off?

15. Which type of stem cells can be harvested and used without killing the donor of the cells?

UNDERSTANDING CONCEPTS

16. Sketch out the four phases of mitosis and cytokinesis. Include in your sketch the location of the nucleus (whether it's in the cell during a phase or not), the way that the genetic material is stored (chromatin, chromosome, or chromatid), and labels for each phase.

17. Sketch out the eight phases of meiosis and cytokinesis. Include in your sketch the location of the nucleus (whether it's in the cell during a phase or not), the way that the genetic material is stored (chromatin, chromosome, or chromatid), and labels for each phase. Also label the phase at which the diploid cell becomes haploid.

18. Explain how mitosis and meiosis begin with tetraploid cells and end with diploid cells, haploid cells, or both.

19. Of Mendel's three principles of genetics, which deals with the movement of a trait into the reproductive cells of an organism? Which deals with the dominance of a trait? Which deals with each trait moving separately, regardless of how other traits move?

20. Green cucumbers are dominant over orange cucumbers. Show a cross between two green cucumbers that could produce orange cucumbers. First, list the genotypes of the two parent cucumbers and draw a Punnett square to show the cross. Second, identify the genotype(s) of the offspring that would be orange.

21. In a dihybrid cross, why must we figure out all the possible combinations of gametes before crossing them to produce an F_2 generation?

22. Why is Mendel's model of genetics not the only model of inheritance?

23. Wavy hair is the result of one allele for curly hair and one allele for straight hair. Is wavy hair an example of incomplete dominance or codominance? Draw a Punnett square to show the cross between someone with curly hair and someone with straight hair to show the likelihood of producing wavy-haired children.

24. In Texas Longhorn cattle a red cow bred with a white bull produces offspring with a roan coat—cattle with both red and white fur. Are roan cattle an example of incomplete dominance or codominance? Draw a Punnett square to show the cross between a red cow and white bull to show the likelihood of producing roan cattle.

25. Is the polygenic inheritance of skin color a type of incomplete dominance or codominance? Explain.

26. Using a Punnett square, show a cross between a colorblind woman and a man with normal vision. Identify the possible phenotypes of their children.

27. Copy the flow chart below, which shows how genes are expressed. Fill in your copy by associating the correct term with each lettered blank.

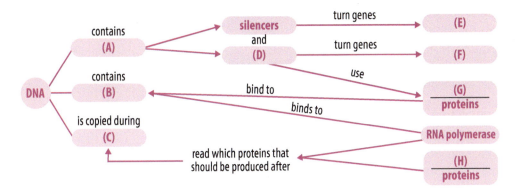

28. Explain how DNA relates to traits that organisms express.

29. If it's cool outside, a turtle will lay eggs that hatch into mostly male turtles. If it's warm due to sun exposure, the turtle will lay eggs that hatch mostly into females. What two environmental factors of gene expression are at work here?

30. A group of researchers reprogram certain genes in fruit flies, creating legs in the place of wings on the flies. Which genes are the researchers reprogramming?

CRITICAL THINKING

31. A peanut farmer discovers a special trait in one of his peanut plants: it's resistant to rootworms, a pest that eats peanuts before they're harvested. Which method of reproduction—asexual or sexual—would be better to use to make sure that this trait stays in future peanut plants? Explain.

32. A person has type O blood even though neither of his parents do. Explain what blood types his parents could have.

33. Explain how a non-colorblind male and non-colorblind female can have a male offspring who is colorblind.

Use the case study below to answer Questions 34–36.

34. What is the unique function of Hox genes?

35. Explain why evolutionists claim that these genes may be evidence of common ancestry between animals.

36. What is the weakness in the evolutionary claims about Hox gene mutations?

case study

HOPE IN HOX GENES

Hox genes are very important genes in the body. In fact, without them, who knows where all our body parts would end up! These highly specific genes are responsible for the arrangement and position of body structures seen within organisms. Historically, evolutionists have believed that evolutionary change must be slow and gradual, taking place through a long series of very small changes. But there is little evidence for this type of change in the fossil record. Hox genes have been hailed as a possible vehicle of rapid change from one type of animal body plan to another. It has been suggested that mutations in these genes may induce sudden, large-scale changes. But various laboratory studies have shown that when these genes mutate, the resulting changes in body plan tend to make organisms less fit for survival. The mutation-induced changes would be fatal to such organisms in the wild. If these mutations cannot produce even small beneficial changes, much less turn a fish into a reptile, should this theory still be supported?

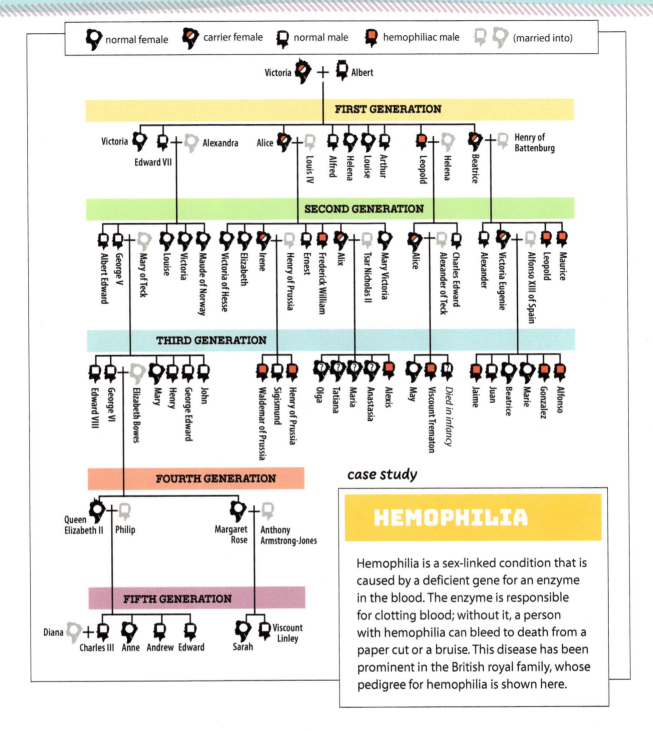

Legend: normal female · carrier female · normal male · hemophiliac male · (married into)

Victoria + Albert

FIRST GENERATION

Victoria · Edward VII + Alexandra · Alice + Louis IV · Alfred · Helena · Louise · Arthur · Leopold · Helena · Beatrice + Henry of Battenburg

SECOND GENERATION

George V / Albert Edward + Mary of Teck · Louise · Victoria · Maude of Norway · Victoria of Hesse · Elizabeth · Irene + Henry of Prussia · Ernest · Frederick William · Alix + Tsar Nicholas II · Mary Victoria · Alice · Charles Edward / Alexander of Teck · Alexander · Victoria Eugenie + Alfonso XIII of Spain · Leopold · Maurice

THIRD GENERATION

Edward VIII · George VI + Elizabeth Bowes · Mary · Henry · George Edward · John · Waldemar of Prussia · Sigismund · Henry of Prussia · Olga · Tatiana · Maria · Anastasia · Alexis · May · Viscount Trematon · Died in infancy · Jaime · Juan · Beatrice · Marie · Gonzalez · Alfonso

FOURTH GENERATION

Queen Elizabeth II + Philip · Margaret Rose + Anthony Armstrong-Jones

FIFTH GENERATION

Diana + Charles III · Anne · Andrew · Edward · Sarah + Viscount Linley

case study

HEMOPHILIA

Hemophilia is a sex-linked condition that is caused by a deficient gene for an enzyme in the blood. The enzyme is responsible for clotting blood; without it, a person with hemophilia can bleed to death from a paper cut or a bruise. This disease has been prominent in the British royal family, whose pedigree for hemophilia is shown here.

Use the case study above to answer Questions 37–40.

37. Using the pedigree, determine whether hemophilia is a dominant or recessive condition.

38. Use a Punnett square to predict the genotypes of Victoria and Albert's children. What percentage of them should have been homozygous for the normal allele of the clotting enzyme gene? How does this percentage compare with the actual number of normal genotypes among the couple's children?

39. Victoria and Albert had a son Leopold with hemophilia. In the 1800s most people with hemophilia didn't live long enough to marry and have children, but Leopold did. How did he and his wife Helena have a child with a completely normal genotype? Use a Punnett square to support your answer.

40. Today there are medications for people with hemophilia that allow them to lead normal lives. How does this affect the probability of the deficient enzyme allele being passed down from generation to generation?

9 ADVANCED GENETICS

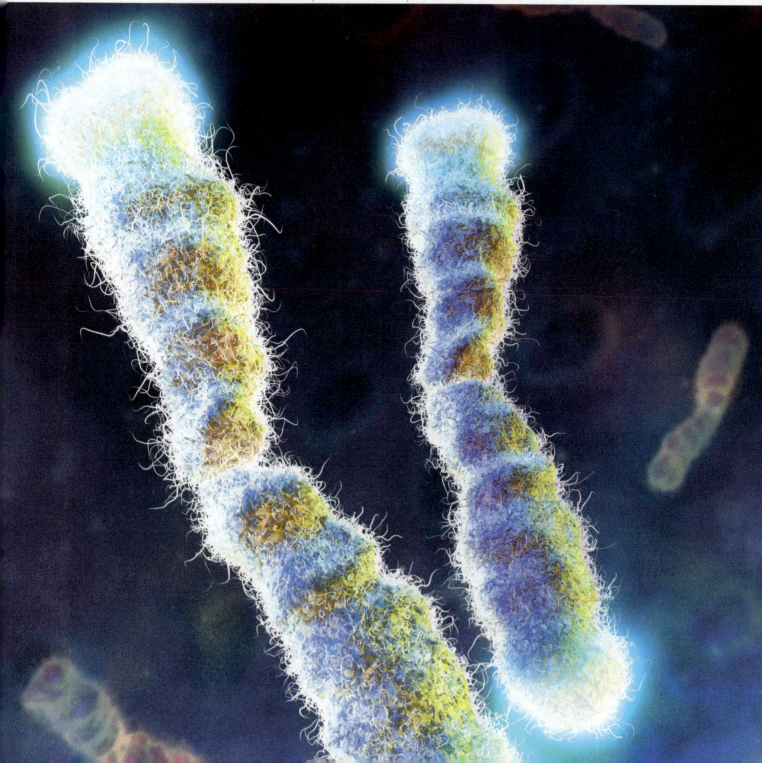

Food for the Future

In 1960, agriculture appeared to be struggling to feed a growing world population. The International Rice Research Institute developed a new strain of rice called *IR8*. Rice production soared. The Philippines had a self-sufficient rice harvest for the first time in years. The Vietnamese called IR8 Honda rice because bumper crops allowed them to buy new motorcycles. Indian parents named their children after the rice. This crop and many others produced in laboratories led to the Green Revolution, the period between the 1930s and the 1960s during which food production around the world grew dramatically because of advances in science and technology. The Green Revolution may have saved a billion people from starvation around the world.

By 2050 there may be nine billion mouths to feed, and we need a new Green Revolution. Biotechnology—the use of technology to enhance living organisms and their processes—may be the power tool that helps us fight against world hunger. Geneticists are working to produce crops that resist drought, pests, and herbicides to produce greater harvests with less land. How can we responsibly use this science to help the most people?

9.1 POPULATION GENETICS

Genetic Variability

When researchers developed IR8, they were using selective breeding to tease out desirable traits from a rice population. IR8 was so successful because it had a short stalk and produced abundantly without falling over. Since then researchers have developed more cultivars, or varieties, of rice and other crops using selective breeding and hybridization. These cultivars are bred to taste and produce better than the original.

The sum of all the alleles that all individuals of a population of organisms can conceivably possess is called its **gene pool**. If a farmer plants a whole field of IR8 rice, he is taking advantage of the combination of alleles that are unique to IR8. This can be dangerous, however, as pests, viruses, and changing environmental conditions can wipe out an entire field of plants. But if farmers plant their fields with a variety of cultivars and strains of the same organism, the same field can represent a greater genetic diversity. That strategy creates a lower risk of losing the crops to pests and poor environmental conditions that result in plant disease.

God created the world full of living things that have the potential for amazing variety. Just think about all the different kinds of people that came from Adam and Eve! When you look around at all the people at the mall, you see a perfect example of variation. **Variation** in a population is the range of genotype differences between individuals from the same gene pool.

As people are born and die or move to or away from an area, there is a constant incoming and outgoing of genetic information in the population. This ebb and flow of genetic material in a population is called **gene flow**. Some alleles are more common than others. So if you were to count the number of people with blue, brown, and green eyes that passed you in ten minutes at the mall, you would be considering **allele frequency**—how often an allele shows up in a gene pool. For example, in Scotland about 57% of the population has blue eyes.

? Why does 57% of the population of Scotland have blue eyes?

Questions

Why do organisms in a population change?

What limits genetic change in a population?

Terms

gene pool
variation
gene flow
allele frequency
plasmid
genetic equilibrium
genetic drift
bottleneck effect
founder effect

Frequency of Eye Color

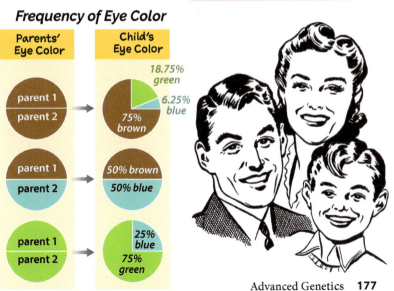

Parents' Eye Color	Child's Eye Color
parent 1 / parent 2	18.75% green / 6.25% blue / 75% brown
parent 1 / parent 2	50% brown / 50% blue
parent 1 / parent 2	25% blue / 75% green

Advanced Genetics **177**

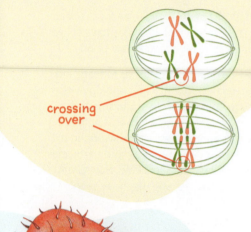

crossing
over

plasmid

prokaryote
(bacterial cell)

Where does all this variety come from? One source of genetic variation is the different types of alleles that combine when two parents reproduce. Another source of genetic variation happens during meiosis, when chromatids can switch information to form new combinations of alleles, called *crossing over*. Also, single-celled organisms such as bacteria can exchange genes through **plasmids**—small, circular DNA molecules that function like chromosomes—without the process of reproduction. This is called *lateral gene transfer*.

In Chapter 6 we saw that a cell can make errors when copying or decoding DNA. These errors—mutations—can be passed down to introduce variation and new traits into a population, though most mutations aren't passed on. The number of mutations sustained by an organism or a gene pool is called its *mutation load*. A population's mutation load is directly related to its *genetic load*, a measure of the health and fitness of a population. A heavier mutation load means a heavier genetic load and a less healthy population since mutations are usually harmful to a population. We'll look more closely at mutations in the next section of this chapter.

Genetic Limits

In large populations with thousands of individuals the allele frequency is stable. Geneticists call this condition **genetic equilibrium**. For example, we wouldn't expect the blue-eyed trend in Scotland to change unless there was a mass exodus of a large segment of the population or an influx of a new group. But we can't assume, just because blue-eyed people are the majority in Scotland, that blue eyes are a dominant allele.

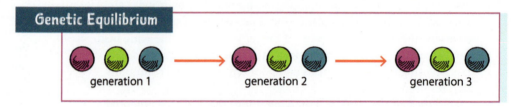

Genetic Equilibrium

generation 1   generation 2   generation 3

At the start of the 1900s, when the study of genetics was just beginning and Mendel's work had been rediscovered, scientists began wondering how gene pools and populations might be affected over time. One of the questions scientists wanted to investigate was, "Will the dominant forms of an allele eventually replace the recessive ones?"

Two researchers working independently of each other were able to demonstrate that dominant alleles will not automatically replace the recessive alleles for a population in genetic equilibrium. The *Hardy-Weinberg principle* was named in honor of the two men who investigated this idea.

Genetic equilibrium is the theoretical result of the Hardy-Weinberg principle. No population is ever truly at equilibrium because it is constantly changing. Its size, individuals, and mutation load are constantly in flux. But what happens when an allele randomly becomes more common in a gene pool? This doesn't usually happen in large populations, but it could happen in small ones like those living on islands. Geneticists call a change in the allele frequency that is based on random events **genetic drift**.

What kinds of random events can cause genetic drift? When a large portion of a population dies, for example, observation of the survivors may reveal that the gene pool and allele frequency have changed dramatically. This is called the **bottleneck effect**. In another phenomenon known as the **founder effect**, a small population moves to a new area and the allele frequency changes as this smaller group starts a new population.

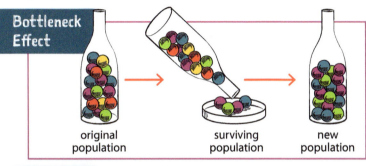

Bottleneck Effect

original population → surviving population → new population

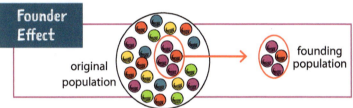

Founder Effect

original population → founding population

Evolutionists see genetic drift as being a major player in natural selection. The fact that genetic drift happens, however, doesn't prove evolution. Genetic drift usually represents a loss of variation and information in a population. This becomes evident when we compare the original population with the new populations after the bottleneck effect or the founder effect, shown in the image above. No one has ever observed genetic drift creating a new kind of animal. People who believe God's Word can see the variety within a kind of animal, such as butterflies, as an example of genetic drift without the need for large-scale change. Variety in nature is a fingerprint of God's creativity and order.

9.1 SECTION REVIEW

1. Suggest three factors that affect a gene pool.
2. What are three sources of genetic variation in a population?
3. How does mutation load affect the variability of a population?
4. What is the difference between gene flow and genetic drift?
5. Evolutionists look to genetic drift as a main agent of evolution to produce different species. From a biblical standpoint, why is this view incorrect?

Use the case study above to answer Questions 6–9.

6. Describe some of the variations in the orange population.
7. Is it possible for the allele frequency to hold constant despite citrus greening? Explain.
8. Is the orange population in Florida in equilibrium? Explain.
9. What kind of genetic drift is the orange population experiencing?

The spirit bear is a black bear that is white because of two mutations in the gene for fur color.

9.2 MUTATIONS

DNA, Chromosomes, and Mutations

So how do geneticists breed plants like the miracle rice IR8? The key to unleashing the traits that we need in plants is locked up in their DNA. DNA is information, and information comes only from a designer. The wording of the Creation Mandate in Genesis 1:28 is closely linked with investigating how we can use plants and the DNA they contain to provide a growing world population with food.

GENE MUTATIONS

In Chapter 6 we studied the process by which a cell replicates DNA and uses it to produce proteins. This copy of DNA provides information for the cell to carry on life-sustaining processes such as photosynthesis, cellular respiration, and protein synthesis. It is also used to prepare for cell growth and reproduction through mitosis or meiosis. By now you should think of a gene as a section of DNA that can be decoded to create a certain protein. As we explored in Chapter 7, these genes produce certain recognizable characteristics in an organism. Your genes, and thus your DNA, make you special, different from everyone else.

But, as we saw in Chapter 6, mistakes in the process of decoding, copying, and using DNA sometimes occur. We use the term *mutation* to represent a genetic error that is produced when a nucleotide base in a section of DNA is added, deleted, or skipped in either replication or decoding. A mutation is like a misspelling in the words and sentences of DNA nucleotide bases. Some mutations happen randomly, and some are caused by a **mutagen**, a physical or chemical substance that changes genetic structure. A mutagen could be any toxin in the form of a virus, radiation, or a chemical (e.g., cigarette smoke, pollutants, dyes).

Why is cancer bad?

Questions

How do changes in DNA affect chromosomes and genes?

How can one mutated cell affect other cells?

How does a mutation show itself in an organism?

Terms
mutagen
point mutation
frameshift mutation
telomere
nondisjunction
polyploidy
tumor
cancer
carcinogen

The results of a mutation depend on whether a nucleotide base was added, deleted, or skipped and where on the DNA chain the mistake happened. When just one nucleotide base changes in the DNA chain, we call it a **point mutation**, whether it is substituted, added, or deleted. Point mutations can have a domino effect depending on what they involve. Mutations in gametes or in cells that produce gametes are called *germ mutations* and can get passed down to the next generation. Mutations in cells that don't make gametes are called *somatic mutations*; these are not passed down. Germ mutations affect every cell in the offspring produced; somatic mutations affect only a few cells of the original organism. When a nucleotide base is swapped out for another one, a point mutation may not be harmful to the organism. But things can get thornier when one is added or deleted, especially if it is a germ mutation.

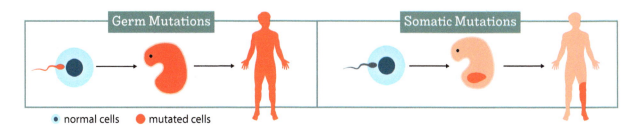

Consider the diagram illustrating protein synthesis shown below. If a nucleotide base is added or deleted from the chain, what happens to the codon? The base shifts the whole codon, producing a **frameshift mutation**. This means that when RNA is coded from the DNA, it will probably make a very different protein than it would have before the mutation since the codons and the amino acids they code for will be much different. The protein will have a different shape than the one it was intended to have, and this will probably cause problems for an organism.

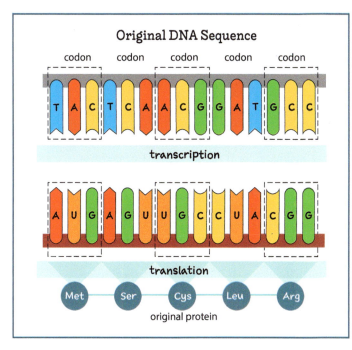

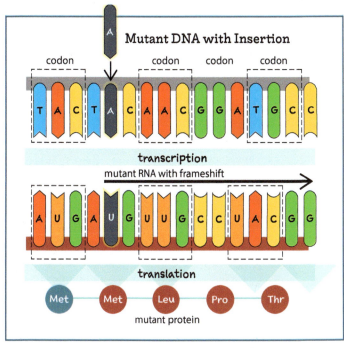

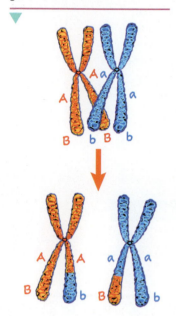

Crossing over occurs when pairs of chromosomes exchange genetic information.

CHROMOSOMAL MUTATIONS

Sometimes mutations affect more than just genes. A mutation may involve rearranging a chromosome in a way that also affects a large portion of the DNA sequence. Crossing over is a normal way for chromosomes to exchange information. But **telomeres**, which are the tips of chromosomes, don't usually participate in crossing over. They mark the end of a chromosome. An enzyme called *telomerase* ensures that the telomeres remain fairly stable during meiosis. If a mutation that stops the production of telomerase occurs, the telomeres get shorter and shorter, leading to cell death. If a mutation causes telomerase to be produced continually, the telomeres mutate, which can lead to uncontrolled cell division.

Sections of chromosomes can experience changes. Occasionally, a segment of a chromosome is left out after it separates during cell division. This complete loss of a segment of the chromosome is called a *deletion*. The chromosome piece usually winds up outside the nucleus, where it disintegrates. This mutation can create serious problems for the next generation if valuable genetic material is missing. Also, a segment of a chromosome may break off and reattach in the same position, but upside down. This is called an *inversion*. Inversions are less likely to cause serious conditions because the genetic material is still present on the chromosome, though in a different order.

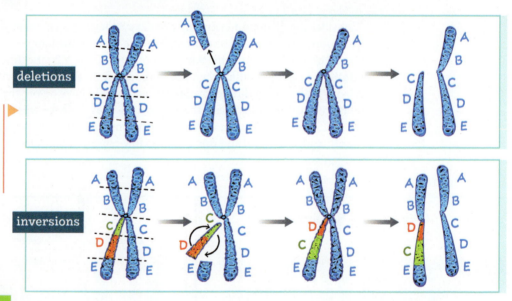

Both deletions and inversions are chromosomal mutations. In the first, the chromosomal fragment fails to attach. In the second, the chromosome fragment reattaches in an inverted position.

deletions

inversions

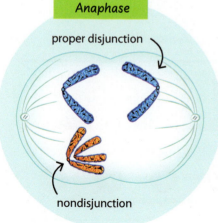

Anaphase

proper disjunction

nondisjunction

Sometimes homologous chromosomes or sister chromatids in a cell don't fully separate during anaphase I or II in meiosis. The gamete mutation is then propagated through mitosis. This failure of the chromosome to separate is called a **nondisjunction**. A nondisjunction leads to gametes that either have extra chromosomes or not enough. This means that haploid gametes produce something other than a normal diploid zygote ($2n$). Depending on which anaphase of which process the disjunction occurs in, they could produce a *monosomy* ($n - 1$) with one less chromosome or a *trisomy* ($n + 1$) with one extra chromosome. A nondisjunction doesn't change the entire set of chromosomes of an organism. It simply adds or subtracts a chromosome from an otherwise normal set.

But when gamete cells do not divide properly, especially during metaphase I in meiosis, a diploid gamete can unite with a normal haploid gamete to produce a *triploid* (3n) zygote with three whole sets of chromosomes. Two diploid gametes sometimes unite to produce a *tetraploid* (4n) zygote with four sets of chromosomes. Cells such as these with three or more full sets of chromosomes exhibit **polyploidy**. Trisomy in cells that are normally diploid is usually lethal, but there are a few exceptions. Most people born with three sex chromosomes can have a relatively normal life without any major health issues, though some may have impaired fertility. Babies that develop trisomy in autosomes usually die, but people with three of Chromosome 21 often live a fairly typical life. This latter condition, known as Down Syndrome, causes intellectual disability and developmental delays throughout a person's life. The effects of the condition vary from person to person. Medical advances and therapy have helped people with Down Syndrome enter the workforce and live very fulfilled lives. They are known for being affectionate and personable.

Polyploidy can be used in agriculture to produce triploid grapes, oranges, and other fruits. Because they lack seeds, triploids must be reproduced asexually. Many valuable crops are polyploids—corn, wheat, cotton, grapes, alfalfa, bananas, and potatoes. Though man didn't breed these polyploids, their survival is due completely to man's cultivation. We have used the larger sizes and genetic flexibility of polyploids to grow them in different areas to feed a growing world population. Without good and wise dominion over polyploids, people could starve.

seedless fruit (triploids)

VARIOUS CHROMOSOME NUMBER CHANGES			
Euploidy *(complete chromosome sets)*	haploid (*n*)	X X X	gametes, mosses, algae, fungi, bee drones
	diploid (2*n*)	XX XX XX	humans, animals, most plants
	triploid (3*n*)	XXX XXX XXX	seedless plants (watermelons, bananas)
	tetraploid (4*n*)	XXXX XXXX XXXX	Irish potatoes, alfalfa, some rare plants
	tetraploid (2*n* + 2*n*)	XX XX XX XX XX XX	corn, wheat
Aneuploidy *(missing or extra chromosomes)*	monosomy (2*n* - 1)	XX XX XX	various types of plants, Turner syndrome in humans
	trisomy (2*n* + 1)	XX XXX XX	various types of plants, Down and Klinefelter syndromes in humans

A large bump on a tree called a *burl* is a plant's version of a tumor. These tumors are rarely harmful to the tree because cell walls keep abnormal cells from spreading. Woodworkers prize burls for making bowls with decorative grains.

Mutations and Cell Growth

We've seen the different types of mutations that can happen both in genes and chromosomes. But what about their effect on cell growth? A single mutated cell can significantly affect an organism when it multiplies. Just as genes code for different phenotypes in an organism, there are specific genes that govern the cycles of cell growth and reproduction and the proteins that the cells use to make this happen. When a mutation affects these genes, mutated cells can begin to divide uncontrollably, resulting in an abnormal mass of cells called a **tumor**. Anything that has cells can develop a tumor, even plants and animals. *Benign* tumors don't spread to other parts of the body. An example of a common benign tumor is a polyp in the large intestine. But tumors can also be *malignant*, which means that they may *metastasize*, or spread to other parts of the body. Malignant tumors can cause major parts of an organism to shut down, preventing them from fulfilling their purpose in life-sustaining processes for an organism.

intestinal tumor

healthy blood

blood with leukemia

Cancer

Cancer causes one out of every four deaths in the United States, and anyone can develop it. Malignant tumors are one form of **cancer**, which is an unrestrained growth of abnormal cells in the body. But not all cancers form tumors. *Leukemia* is a form of cancer that involves spreading abnormal blood cells in the circulatory system. This kind of cancer is harder to treat because the abnormal cells can be more widely distributed in an organism.

Because cancer is such a widespread disease, the search for effective treatments is a priority for science. The main goal of treating cancer is to destroy abnormal cells in ways that do the least amount of damage to healthy cells. The best ways to kill abnormal cells are to poison, burn, or cut them out. *Chemotherapy* uses chemicals to poison and disrupt cell processes in abnormal cells so that they can't divide, killing the cells and shutting down their reproduction. *Radiation therapy* uses high doses of energetic radiation to burn cancer cells in ways that slow down or stop their growth completely. Doctors can use *surgery* to cut out clumps of abnormal cells that have formed a tumor. For many cancer patients a combination of these therapies is used to kill cancer cells while hurting as few normal cells as possible.

Most people know that, to lower their risk of certain kinds of cancer, they shouldn't smoke or stay in the sun too long. Substances like radiation, tobacco smoke, and certain chemicals that increase our cancer risk are called **carcinogens**. Many carcinogens are also mutagens, though some work by disturbing cellular metabolism. What can people do when their own genetics is fighting against them?

In 1989 a biochemist working late one night had the idea that cancer might have a genetic component. To test his idea, he removed a suspected cancer gene from a tumor sample and replaced it with a normal gene. The cancer cells in the tumor immediately stopped growing. This discovery was an amazing breakthrough, but it created lots of unanswered questions about how to best treat cancer.

Researchers are working to develop an ethical gene therapy treatment that targets cancer cells. They've also come up with a solution using a virus and the body's own immune system. Because cancer is so common in our world, doctors and researchers are on a quest to produce treatments that don't require poisoning, burning, or cutting to improve quality of life for patients who suffer from this disease. To do this, they use the tools of genetic engineering, our topic in the next section.

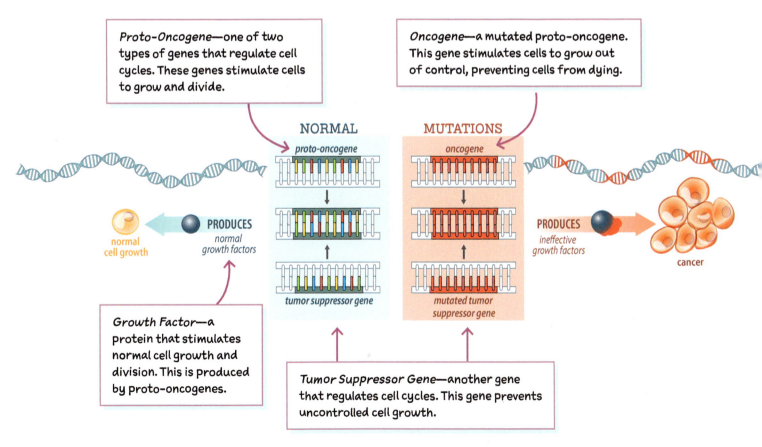

Proto-Oncogene—one of two types of genes that regulate cell cycles. These genes stimulate cells to grow and divide.

Oncogene—a mutated proto-oncogene. This gene stimulates cells to grow out of control, preventing cells from dying.

NORMAL

proto-oncogene

tumor suppressor gene

PRODUCES
normal growth factors

normal cell growth

MUTATIONS

oncogene

mutated tumor suppressor gene

PRODUCES
ineffective growth factors

cancer

Growth Factor—a protein that stimulates normal cell growth and division. This is produced by proto-oncogenes.

Tumor Suppressor Gene—another gene that regulates cell cycles. This gene prevents uncontrolled cell growth.

Point Mutations

As we've seen, a point mutation's potential for harm depends on which kind of mutation it is and where it occurs in a DNA strand. Let's model a mutation and see what happens!

Materials
none

How does a point mutation affect an organism?

PROCEDURE

Start with the following DNA nucleotide sequence (three codons).

GGG CAC CTT

A Assume that the nucleotides are numbered 1–9 from left to right. Use a random number generator to select which of the nucleotides will be affected by a mutation.

B Use the random number generator to determine whether the affected nucleotide will experience a substitution mutation (1), deletion (2), or addition (3). Assume that an addition, if it occurs, will insert a new nucleotide *after* the selected nucleotide.

C If the mutation is a substitution or addition, use your random number generator to determine whether the new nucleotide is adenine (1), cytosine (2), guanine (3), or thymine (4). Ignore any result for a substitution that yields a like-for-like substitution, such as adenine for adenine—that's not a substitution, obviously!

ANALYSIS

1. What amino acid sequence does the original nucleotide sequence code for? Refer to the codon circle on page 125, and don't forget to transcribe the DNA first!

2. What is the new nucleotide sequence produced by your mutated DNA sequence?

3. What amino acid sequence will be produced by the new sequence?

4. Discuss the implications of your mutated DNA strand with respect to translation.

5. Compare your result with those of your classmates. What can this comparison tell you about the nature of mutations?

9.2 SECTION REVIEW

1. Use the DNA base sequence ATT CGA TAG GCT to demonstrate the three types of point mutations.

2. What is the difference between a gene mutation and a chromosomal mutation?

3. What is the major difference between a point mutation and a chromosomal mutation?

4. Which is less likely to be harmful to a chromosome, a deletion or an inversion? Explain.

5. What determines whether a nondisjunction will create monosomy or trisomy?

6. Using what you have learned, indicate what specific kinds of mutations would be most harmful to a population of organisms.

7. Give examples of the ways that a mutation can be expressed in an organism.

8. Why do tumors form in organisms?

9. A doctor has just done a biopsy of a mole on a patient's arm and found that it is cancerous. What treatment(s) would he recommend? Justify your answer.

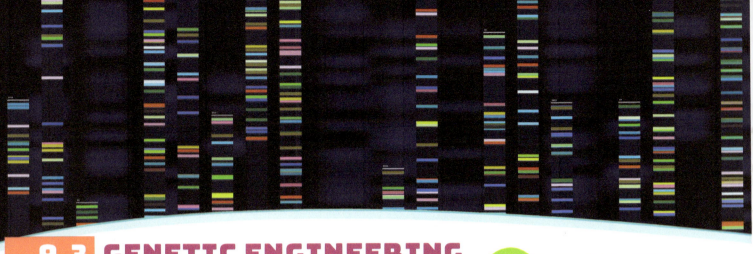

9.3 GENETIC ENGINEERING

Studying DNA

In 1984 Francis Collins began writing an encyclopedia. This encyclopedia wasn't about famous people in history or countries of the world—it was all about the human genetic code. Collins was head of the Human Genome Project, a scientific study to identify and list all human genes, the phenotypes they express, and the variations in these phenotypes. This was no small task, since recent analysis suggests that humans have almost 47,000 different genes! This project, which claimed completion in 2003, was the largest team effort in biology in history. But even with such a concerted effort there remained gaps in the human genome that had not been sequenced. It took another eighteen years of work to fill in the gaps.

Francis Collins

Genomics is the study of **genomes**, or an organism's full set of genetic information coded in its DNA. Scientists have studied the genomes of a variety of organisms, ranging from the almost 5 million base pairs in *E. coli* bacteria to the 3.1 billion base pairs in human DNA.

Researchers have found that the genomes of two different people share 99.9% of the exact same information. This is not surprising since most life processes in people are exactly the same. About 1.5% of the human genome contains codes for making proteins. Some of the functions of the remainder are known, such as codes that control genes, and some consist of repetitive sequences. But the exact function of much of the nonprotein-coding DNA is not yet known. Still, mapping the human genome has yielded concrete benefits. Through this work scientists have been able to trace some genetic disorders back to their DNA roots.

Reading and Manipulating DNA

We can determine which genes cause certain characteristics in organisms, but is there a way to change those genes to bring out a new characteristic? **Genetic engineering** is the process of deliberately manipulating the genes within an organism in ways other than natural processes like reproduction. It involves changing an organism's DNA by inserting new genetic material. Genetic engineering is a blossoming field of biology with real potential to solve significant real-world problems.

? How can DNA be used to find out who committed a crime?

Questions

How are genes separated, catalogued, and rearranged?

How do scientists change DNA?

How can we use genetic engineering to better exercise dominion over God's world?

Terms

genome
genetic engineering
restriction enzyme
sequencing
polymerase chain reaction
recombinant DNA
genetically modified organism (GMO)
transgenic organism
gene therapy
DNA fingerprinting

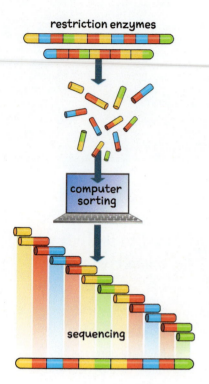

restriction enzymes

computer sorting

sequencing

READING DNA

To begin studying an organism's genome, researchers must isolate DNA from gametes of that organism. DNA molecules are too long to work with easily, so they must be divided into pieces. Geneticists use a **restriction enzyme**—an enzyme used to cut DNA in specific places into pieces for study. An uncombined base pair at the end of a snipped piece of DNA is called a *sticky end* since it is available for pairing.

Then comes the work of **sequencing** the DNA—determining the order of nucleotides within it. Geneticists use an automated sequencing machine to help them sequence billions of base pairs. In addition, computers aid geneticists by comparing sections of the genome to find areas that overlap. They can use this information to figure out the original sequence of the DNA. The team on the Human Genome Project used this process to map the human genome.

MANIPULATING DNA

Genetic engineers often need to make multiple copies of a section of DNA before working on them. They use the cell's own natural process of DNA replication to generate millions of copies of DNA in just a few hours. The process of making multiple copies of a gene under study is called the **polymerase chain reaction**. This process involves several steps, beginning with separating the DNA strand to be copied by heating it up. After the DNA cools, complementary primers (short chains of base pairs) are added and bond to the sticky ends to jump-start replication. DNA polymerase takes over the normal replication process, beginning at primers to produce two copies. This cycle of heating and cooling is repeated until enough copies of the gene are produced.

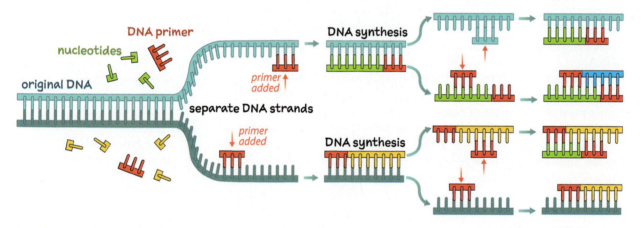

Geneticists can introduce genes into strands of DNA to produce customized DNA. Sometimes they can connect smaller pieces of DNA containing genes that code for certain phenotypes that they want an organism to express. These smaller pieces of DNA are combined with original DNA from an organism to form a customized DNA called **recombinant DNA**.

Making Recombinant DNA

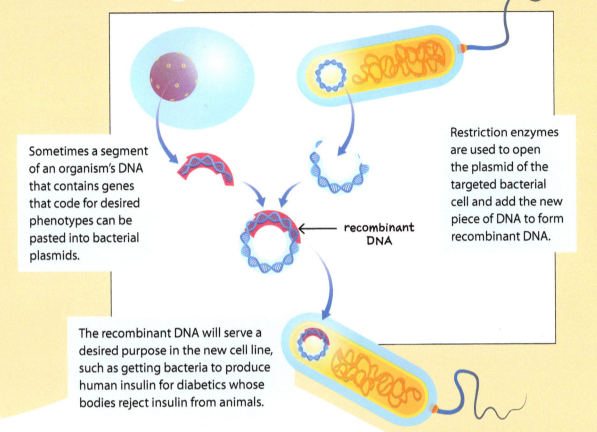

Sometimes a segment of an organism's DNA that contains genes that code for desired phenotypes can be pasted into bacterial plasmids.

Restriction enzymes are used to open the plasmid of the targeted bacterial cell and add the new piece of DNA to form recombinant DNA.

recombinant DNA

The recombinant DNA will serve a desired purpose in the new cell line, such as getting bacteria to produce human insulin for diabetics whose bodies reject insulin from animals.

Using Engineered DNA

GENETICALLY MODIFIED ORGANISMS

We've already seen some of the applications of biotechnology, such as detecting disease and producing more useful crops that feed more people and use land efficiently. Genetically engineered plants are just one kind of **genetically modified organism** (GMO), which is an organism that has undergone genetic engineering. Geneticists can use recombinant DNA to introduce genes from other living things of the same species or of different species. Organisms that contain genes that have been introduced from other kinds of organisms are called **transgenic organisms**.

Monsanto Company, a leading biotechnology company, has developed many of the world's genetically modified crops. Scientists from this company found plants near an herbicide factory that had developed the ability to resist the herbicide.

genetically modified convict cichlid

They were able to isolate the herbicide resistance gene and introduce it into crops using bacteria. Now farmers can apply herbicides that kill weeds without harming the crops planted in their fields.

Most of our corn, sugar, and soybeans are genetically modified crops. Genetically engineering a plant can have the same effect that breeding has had for centuries, but in quicker and more efficient ways and with a more reliable outcome. At this time we are unsure of the long-term effects that using and eating genetically modified organisms will have on our environment, culture, and health. But geneticists and botanists are working to produce genetically modified crops that won't rely as heavily on chemicals such as fertilizers, herbicides, and pesticides.

Gene Therapy

Another way that scientists can use genetic engineering is to change a gene in an organism to relieve a genetic disorder or disease. This kind of genetic engineering is called **gene therapy**.

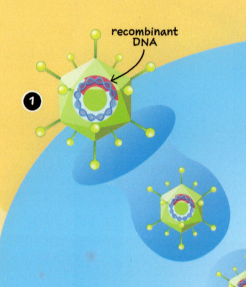

recombinant DNA

1

Gene therapy relies on a microscopic vehicle (known as a *vector* **1**) to deliver recombinant DNA. Vectors can be bacteria or viruses called *bacteriophages* that attach to certain bacteria or cells. Scientists use vectors to transfer new genetic material to an organism in order to change its DNA.

2

cell nucleus

After this new genetic information has been inserted **2** , the therapeutic DNA produces the desired effect. Sometimes therapeutic DNA produces proteins needed to reverse a disease or counteract a problem.

Though gene therapy sounds foolproof, sometimes the treatment doesn't last for long periods of time or even work at all. Gene therapy is never a permanent solution. Often this requires treatment that patients need to continue for the rest of their lives. An organism can reject the vector completely, sometimes causing serious side effects. It's also possible that the gene therapy fails to be effective because geneticists don't completely understand the disease and all the genes that are related to it, so the therapeutic DNA is insufficient to reverse the disease even if it is successfully introduced. Scientists are still experimenting with gene therapy in clinical trials.

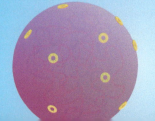

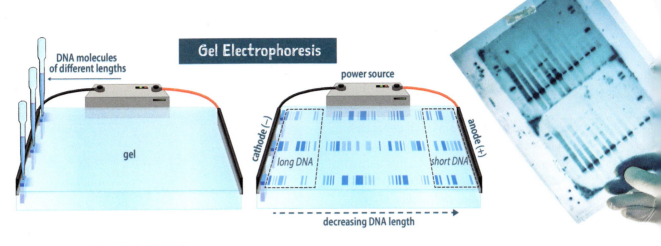

Gel Electrophoresis

DNA molecules of different lengths

power source

gel

cathode (−)

long DNA

short DNA

anode (+)

decreasing DNA length

DNA FINGERPRINTING

Another application of genetic engineering that is especially helpful in forensics and archaeology is to use samples of DNA to identify individuals since no two people have the exact same DNA (identical twins are an exception). The process of using fragments of DNA to identify individuals is called **DNA fingerprinting**. Since 1987, when DNA fingerprinting was first used in forensics, hundreds of court cases have been decided with the help of DNA fingerprinting, clearing the innocent and incriminating the guilty. Another important use is identifying lost family members or the remains of people killed in natural disasters or accidents. DNA fingerprinting can also be used to diagnose inherited disorders in newborns, such as cystic fibrosis, hemophilia, sickle cell disease, Huntington's disease, and some types of Alzheimer's.

The earliest forms of DNA fingerprinting developed in the 1980s involve a process called *gel electrophoresis*. Fragments of DNA created by restriction enzymes are applied to a gel-covered plate attached to a power source. When an electrical current is applied, the DNA fragments are sorted as they travel down the plate, with the longest strands traveling slower and staying closer to the power supply while the shortest fragments migrate farthest away. Bands of different lengths of DNA are revealed by staining the finished gel. Today, DNA fingerprinting looks for unique patterns in an individual's DNA to identify them using this process.

Stepping back, we can see that God has given us an amazing resource coded in the DNA of all living things. We can exercise good and wise dominion over the field of genetics to heal diseases, serve out justice, and provide food for people with a mind for the future citizens of the world. Our Creator is the only source of the remarkable arrangement of information in DNA. Origen, an early church father who died in AD 254, said, "I cannot understand how so many distinguished men have supposed [the world] to be uncreated, that is, not made by God himself the Creator of all things, but in its nature and power the result of chance." (*On First Principles*, 2.1.4.)

9.3 SECTION REVIEW

1. Do you think that the Human Genome Project was a worthy endeavor? Why or why not?

2. Draw a diagram that shows how a gene can be transferred from one plant to another.

3. Give an example of a genetically modified organism.

4. List and explain four ways that DNA can be manipulated.

5. What are three practical ways to use the tools of genetic engineering?

6. How would you respond to an evolutionist who supports his belief in evolution with the fact that 98% of the DNA in chimpanzees and humans is the same? (The genomes of both have been sequenced.) Consider doing a keyword search with the words "human chimpanzee DNA comparison" to help you form your response.

7. Researchers hope to use genetically modified orange trees that use two genes from spinach to give oranges immunity to citrus greening. The goal is to rescue the orange population in the United States without the heavy use of pesticides. What things should scientists, farmers, and orange juice producers consider if they decide to grow genetically modified oranges?

FIGHTING DROUGHT WITH GENETICS

One of the major problems with genetically engineered seed is that the crops they produce are designed for modern farms. This translates into crops that rely on automated watering, regular fertilizing, and applications of pesticides and herbicides. Also, the desirable traits of genetically modified plants are not passed on to the next generation of seeds, so farmers need to purchase seeds year after year. This makes it difficult for farmers in underdeveloped countries who rely on rainfall and harvest their own seed for the next year's planting. Recently, Monsanto Company has been making an attempt to produce strains of crops specifically for underdeveloped countries, since they are the people who could benefit the most from biotechnology. The crops of small local farmers feed a larger percentage of the world's population than industrial farms, so it makes sense to develop seeds to help the neediest people in the world.

Introduction

You are an American farmer working for the Kenya-based African Agricultural Technology Foundation. You are working with Monsanto Company to produce water-efficient corn to help small, drought-stricken Kenyan farms fight hunger.

Task

Your task is to develop a general plan to use biotechnology to produce and distribute drought-resistant corn that will allow farmers to use their harvested seed and use a minimal amount of chemicals on their fields. You will communicate your plan in a ten-slide presentation.

Procedure

1. Do some research on selective breeding.
2. Research the process and the tools used to genetically modify plants.
3. Plan your presentation and collect any photos or videos you need, giving proper credit. Make sure that you think of how to distribute the new seed.
4. Create your presentation and show it to another person.
5. Deliver your presentation to your class or family.

Conclusion

Several foundations, such as the Bill and Melinda Gates Foundation, have funded the collaboration between Monsanto Company and the African Agricultural Technology Foundation to produce a strain of corn called *DroughtTEGO® corn*. Scientists have used biotechnology to develop this corn to provide the most food for the people who desperately need it. This is a good example of how we can wisely use biology to help God's image-bearers, with a mind for the future.

9.1 POPULATION GENETICS

- The sum of all the possible alleles that exist in a population of organisms constitutes a gene pool. Variation in the pool describes the population's range of genotypic differences.

- Gene flow occurs as individuals enter or leave a population. Gene flow can affect the allele frequency of a population, that is, the percentage of a population that carries a particular allele.

- Large populations tend to remain in genetic equilibrium. Genetic drift tends to occur in small, isolated populations such as happens when a large part of a population dies off (bottleneck effect) or a small part of a population moves to a new area (founder effect).

Terms
gene pool • variation • gene flow • allele frequency • plasmid • genetic equilibrium • genetic drift • bottleneck effect • founder effect

9.2 MUTATIONS

- Mutations are random changes in the DNA sequence of an organism and may be either natural or the result of the effects of a mutagen. Only germ cell mutations can be passed to offspring.

- When just one DNA nucleotide is changed, a point mutation occurs. This may be either an insertion, deletion, or substitution. Insertion and deletion mutations have more potential for harm because they cause a frameshift that affects how the affected sequence is read during transcription.

- Chromosomal mutations affect larger sequences of DNA either between chromosomes (crossing over) or within them. Mutations within chromosomes may take the form of lost segments (deletions) or sections detached and reattached in reverse order (inversions).

- Nondisjunction during meiosis results in aneuploid germ cells with more or fewer chromosomes than normal. The effects of aneuploidy vary depending on both the kind of organism affected and the chromosome that is lost or duplicated.

- Tumors occur when mutations take place in the genes that regulate cell growth. Benign tumors do not spread, but malignant tumors (cancers) have the potential to metastasize and spread to other parts of an affected organism

Terms
mutagen • point mutation • frameshift mutation • telomere • nondisjunction • polyploidy • tumor • cancer • carcinogen

9 CHAPTER REVIEW
Chapter Summary

9.3 GENETIC ENGINEERING

- Genomics studies the genomes of organisms, that is, their entire set of genetic information.

- Advanced microbiology techniques have made possible the sequencing of entire genomes. The information gleaned from sequencing DNA can be used in genetic engineering, the deliberate manipulation of an organism's genes. GMOs are organisms whose DNA has been genetically modified.

- Genetic engineering can produce transgenic organisms whose genes contain recombinant DNA, which is DNA that has had the gene for a desired trait inserted into it.

- Practical applications of genetic engineering include gene therapy to treat disease and DNA fingerprinting to aid in solving crimes or identifying genetic disorders.

Terms

genome · genetic engineering · restriction enzyme · sequencing · polymerase chain reaction · recombinant DNA · genetically modified organism (GMO) · transgenic organism · gene therapy · DNA fingerprinting

Chapter Review Questions

RECALLING FACTS

1. What are some of the advantages of using biotechnology to improve crops?

2. Identify three ways to add genetic information to increase variations in a population.

3. What can change the allele frequency in a gene pool?

4. What is the difference between germ mutations and somatic mutations?

5. Draw a chromosome and label the telomeres.

6. Why do tumors form in organisms?

7. Are tumors always cancerous? Explain.

8. How are biologists and doctors working to fight cancer to save the valuable lives of people?

9. How is a transgenic organism created?

10. Describe four ways that scientists manipulate an organism's genetic material.

UNDERSTANDING CONCEPTS

11. A survey of students in your school reveals that about 36% of the students have detached ear lobes. Is it correct to conclude that detached earlobes is a recessive trait? Explain.

12. Are large or small populations most susceptible to genetic drift? Explain.

Use the information below to answer Questions 13–14.

Some strains of rainbow trout have more black spotting than others. Suppose a strain of heavily spotted rainbow trout is introduced into a lake whose resident rainbow trout are lightly spotted. During the ensuing years, anglers catch some trout with heavy spotting, some with light spotting, and some with levels in between. Many years later a disease strikes the lake's trout population. Although all the trout are susceptible to the disease, the fatality rate for the heavily spotted trout is 90%, while virtually 100% of the lightly spotted trout die. When the population recovers, only heavily spotted trout are caught.

13. Assuming that spotting is determined by a single gene, explain what happened when the heavily spotted trout were introduced.

14. Is this recovered trout population an example of gene flow or genetic drift? Explain.

Use the original sequence of DNA below to answer Questions 15–17.

AAG CTG ATG

15. Identify the following sequences of RNA that are transcribed from the DNA above as correct (no mutations), as a point mutation, or as a frameshift mutation. (*Hint:* You may want to begin by writing the correct mRNA on your paper.)

   a. UUC GAC UUC

   b. UUC CGA CUA C

   c. UUC GAC UAC

   d. UUG GAC UAC

16. What does each of the groups of three nucleotide bases in Question 15 represent?

17. What makes the frameshift mutated strand of mRNA different from the other strands of mRNA?

18. Do all mutations create problems in organisms? Explain.

19. How are chromosomal mutations different from gene mutations? How are they similar?

20. Explain how the same nondisjunction event during meiosis I can produce either of two different kinds of polyploidy.

21. Draw a Venn diagram to illustrate the relationships between mutagens and carcinogens. Use one circle to represent each. Also use the words *mutation* and *cancer* in your diagram.

22. Selective breeding produces organisms with desirable genomes. Is this an example of genetic engineering? Explain.

23. Are all genetically modified organisms plants? Explain.

24. Can DNA fingerprinting be used to tell the difference between identical twins? Explain.

CRITICAL THINKING

25. How will the genetic load of a population of organisms change over time? How will this affect the genetic variability within the population?

26. During the Human Genome Project researchers initially found that 97% of the human genome did not appear to code for any specific genes and appeared to serve no function. Compare this with a typical prokaryotic genome, which has 20% noncoding DNA. People began to refer to this as "junk DNA" and to assert that this was proof of human evolution. How would you respond to this claim?

27. How do DNA, genomes, and genetic engineering point to a Master Creator?

Use the case study below to answer Questions 28–30.

28. Scientists thought that the reason children in early studies developed leukemia was because the vector had inserted its recombinant DNA close to proto-oncogenes in the host cell. How might this have caused leukemia?

29. How did doctors solve this problem?

30. Some people argue that diseases like SCID show that either God doesn't exist, or, if He does, that He is unconcerned for His creatures or unable to do anything about their condition. How would you respond to this from a biblical worldview?

case study

GENE THERAPY FOR SCID

Scientists are experimenting with gene therapy on bubble babies, children born with a disease called *severe combined immunodeficiency* (SCID). Their immune systems are so weak that the only way they can survive is to live in a germ-free plastic bubble, limiting both their life expectancy and their quality of life. SCID can be caused by roughly twenty defective genes that produce different types of the disease. Early efforts in gene therapy to treat this disease resulted in four young children developing leukemia, with one of them dying, though the approach did cure some children born with the disease. Now doctors are enlisting the aid of an unlikely gene therapy ally: the AIDS virus. Using the AIDS virus as a vector, doctors were recently able to successfully treat forty-eight of fifty children. While this new technique has provided patients with a working immune system in the short-term, questions remain about how long the cure will last. Doctors remain hopeful, but only time will tell.

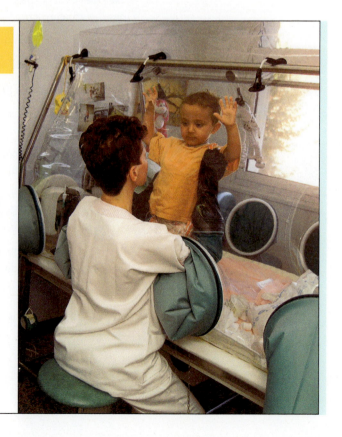

31. What are some of the drawbacks of genetically modified organisms, especially crops?

32. A civil court is trying to determine the father of a child, with two men both claiming to be the father. Both want custody of the child. The judge orders DNA fingerprinting to be done, using DNA from the mother, the child, and both men. On the basis of the gel electrophoresis samples below, who do you think the father is? Explain.

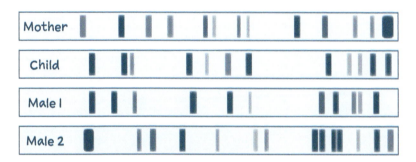

Use the graph below to answer Questions 33–36.

The average height of men and women in the United States has remained relatively constant over the past decade. It is also one of the highest in the world, along with that of Europeans. The graph shows a sampling of male and female height in the United States in 2009.

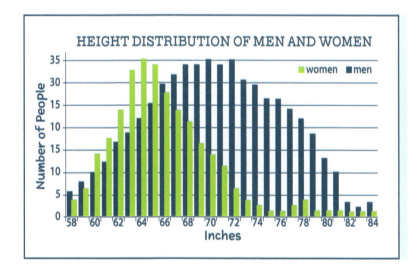

33. What is the average height of men in the United States?

34. The shortest people in the world are usually from underdeveloped countries like Tanzania and Guatemala. Create a sketch of what you think the graph would look like for Nepal, where average adult male height is 4 ft 11.5 in.

35. Do you think human height is affected only by genetics? Explain.

36. What would happen to average height if immigration from South America to the United States increased dramatically? What is this phenomenon called?

10

WHEN WORLDVIEWS COLLIDE

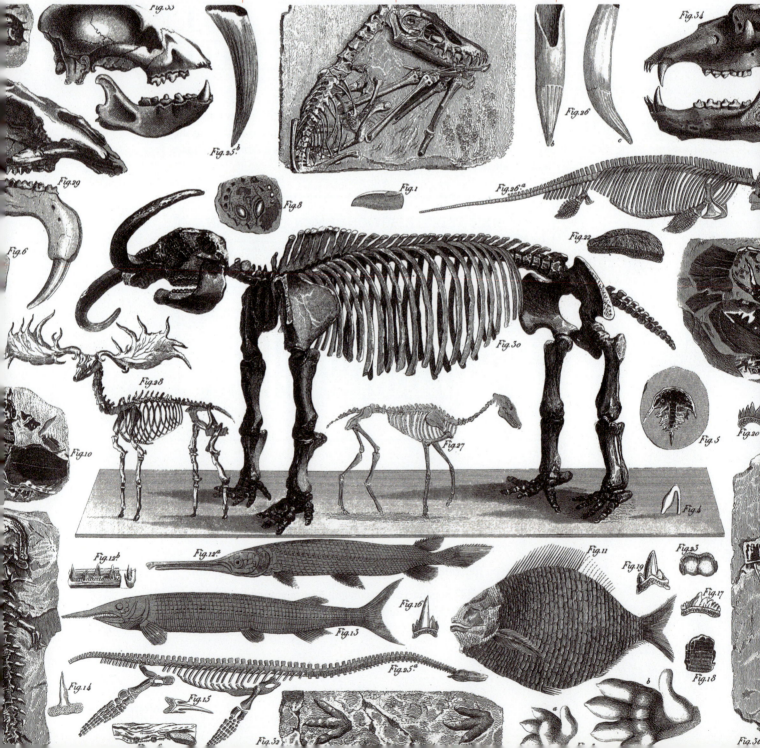

The iDINO Project

Tyrannosaurus, Stegosaurus, Triceratops—people love dinosaurs. But all that is left of them is just old, hard bones, right? Actually, no. Scientists in the Creation Research Society are investigating *Triceratops* fossils from Montana that hold a surprising mystery: original, unfossilized tissues that contain intact cell structures, proteins, and blood vessels. Bones that are 66 million years old are not supposed to have such things! But that is what scientists across the board have observed.

This growing body of data began to draw the world's attention in 2005 with a study of soft, original tissues in a *Tyrannosaurus rex* fossil that revealed evidence of red blood cells. Creation scientists participating in a study of soft dinosaur tissue called the *iDINO Project* want to uncover more evidence that confirms that dinosaur fossils are really not that old at all.

10.1 DARWIN'S THEORY OF EVOLUTION

The Origins Question

Living things can change. This fact should surprise no one since evidence of such change is all around us. In fact, there are probably examples of changed organisms in your own house. Do you own a Persian cat, a corgi, or a fantail goldfish? Each of these pets is quite different from its wild cousins. Sweet corn from the grocery store and hybrid tea roses in the garden are likewise very different from their wild forebears. We could cite thousands of similar examples.

But how do living things change and to what extent? That is the big question, and how one answers that question depends on one's world-view. Creationists and evolutionists arrive at very different answers, even though they start with the same evidence. It all hinges on interpretation. To understand why the two sides see things so differently, we need to delve into a little history. Let's start with a look back at one of mankind's most significant voyages of discovery.

AN IDEA TAKES SHAPE

Young Charles Darwin would probably not have struck anyone as a man likely to change history. By the summer of 1831 he had tried his hand first at studying medicine, then at preparing for the ministry, neither of which appealed to him. What did interest him was natural history—studying organisms in their natural environments—and collecting beetles. Darwin was particularly inspired by one of his professors, botanist John Henslow. It was Henslow who recommended Darwin as a naturalist to serve aboard HMS *Beagle*, a survey ship. In the rigid social hierarchy of the Royal Navy at that time, fraternization between officers and common sailors was frowned upon. Captains often took men of their own social rank aboard ship as companions, and such was the arrangement between Captain Robert FitzRoy of the *Beagle* and Darwin.

Charles Darwin

Was evolution Darwin's idea?

Questions

What is evolution?

How did Darwin come up with the theory of evolution?

What is "descent with modification"?

Who else contributed to the evolution model?

Terms

biological evolution
artificial selection
uniformitarianism
comparative anatomy
homologous structures
vestigial structure
fossil
descent with modification
common ancestor
adaptation
natural selection

Marine surveying is slow work. The leisurely pace of the voyage—nearly five years—allowed Darwin plenty of time to go ashore. There he made many observations, which he meticulously recorded in his journals, and collected specimens of plants and animals. But it would be a mistake to assert that Darwin's experiences aboard the *Beagle* alone led to the idea of evolution. As is often the case, the truth is more complicated than that.

BEFORE DARWIN

The Concept of Transformation

Today when we use the words **biological evolution**, they are typically understood to mean the idea that organisms can slowly change over time into other kinds of organisms. But the idea that organisms can *transform*, as the concept was called at the time, was not original to Darwin himself. People had already known for centuries that domesticated animals could be bred to have certain desirable characteristics, or traits. Different kinds of dogs, for example, could be bred to be large guard dogs, nimble herding dogs, or small, docile companion dogs. When people do the choosing of which animals to breed to fix a desired trait in a population, it is called **artificial selection**. People also recognized that selection was inherently a *conservative process*; that is, it conserved some traits in a population while eliminating others. Consider again an example from the dog world. All the dog breeds we know of, from huge Neapolitan mastiffs to tiny chihuahuas, have been selectively bred from an ancestral dog *kind* (see Section 10.4). In the present, no amount of breeding chihuahuas with other chihuahuas can ever reproduce the original dog kind; the genes for large size have been lost in the chihuahua population, while the genes for small size have been conserved.

C. 300 BC
Aristotle, along with many ancient Greeks, believes that living things do not change over time, an idea called the *fixity of species*. Aristotle classifies organisms on his "ladder of life."

1555
Pierre Belon publishes his *Natural History of Birds*. His comparison of human and bird skeletons is regarded as one of the earliest examples of comparative anatomy.

The ROAD to EVOLUTION

Even before Darwin, some scientists had recognized that organisms in the wild could transform over time as well. They also believed that such change could be *creative* in nature, that is, producing new kinds of organisms. Darwin's own grandfather, Erasmus Darwin, even wrote about the idea in a 1794 book. What these scientists didn't know at the time, though, was *how* such change might occur. One of the best-known of these pre-Darwinian theories is that of the *inheritance of acquired characteristics*, widely held but most closely associated with the French naturalist Jean-Baptiste Lamarck. This theory stated that organisms could pass on to their offspring traits that had been acquired during the organism's lifetime.

The Concept of an Ancient Earth

Around the same time that men like Lamarck were considering ways that organisms might change, other scientists were noting that the earth itself had apparently changed over time. One such scientist, Scottish geologist James Hutton, was a chief contributor to this idea. Hutton noticed that some of the exposed cliffs in Scotland consisted of vertical layers of sedimentary rock. He concluded that these rock layers must have formed by multiple cycles of horizontal deposition, followed by vertical uplift. Thus, the cliffs could not be a permanent and unchanging feature as many people of the time supposed them to be. In addition, Hutton believed that great amounts of time were necessary to account for the cycles of deposition and uplift. He concluded that the earth must be old, not young. Hutton had no difficulty in reconciling his discoveries with his worldview. As a deist, he believed that God had created the world but afterward allowed it to be governed by the laws of science alone.

Exposed Scottish cliffs (Isle of Skye)

1785

The first part of James Hutton's *Theory of the Earth* is read before the Royal Society of Edinburgh. Hutton recognized that many of Earth's features, such as seaside cliffs in Scotland, were not original creations, but rather the result of secondary processes such as sedimentation and uplift.

1735

Swedish botanist Carl Linnaeus first publishes *Systema Naturae* in which he spells out his system for classifying life "for the greater glory of God." Linnaeus originally held to the fixity of species but changed his mind because of his scientific observations. He viewed similarities in organisms as the mark of the Creator.

Charles Lyell

Like Hutton, Scottish geologist Charles Lyell rejected the biblical concepts of a young earth and a worldwide Flood. In 1830 he published the first volume of an extremely influential work: *Principles of Geology*. While aboard HMS *Beagle*, Captain FitzRoy loaned his personal copy of Lyell's book to Darwin. It made a profound impact on the young naturalist. Lyell argued that the observable processes on the earth, such as sedimentation, erosion, or volcanic activity, happened at the same rate in the past as they do in the present. This concept became known as **uniformitarianism**. Today we observe that most geologic processes do in fact happen very slowly. According to uniformitarian principles, this must mean that vast amounts of time were necessary to create the geological features that we can see today. The unavoidable conclusion of such thinking is that the earth must be unimaginably old. Today we sometimes refer to the concept of a very remote past in Earth's history as *deep time*.

Pierre Belon's comparative anatomy skeletons

The Concept of Homology

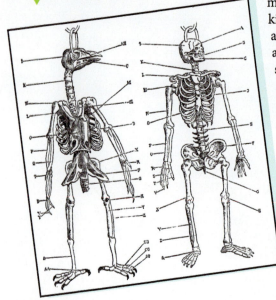

As far back as the time of ancient Greece, humans had remarked on the similarity in physical structure between different kinds of organisms. A French naturalist, Pierre Belon, had even analyzed the similarity between human and bird skeletons in a book published in 1555. Today we call this comparison of structure **comparative anatomy**. When doing comparative anatomy, it is sometimes noted that two organisms accomplish the same function with completely different anatomical structures. For instance, bats and butterflies are both capable of sustained flight, but their wings are different. Bat wings are flaps of skin, while butterfly wings are membranes covered in tiny scales. Comparative anatomists call this kind of similar function by means of different structures *analogy*. Cat legs and dog legs show a different kind of similarity. Their legs contain the same kinds of bones and muscles and are used for the same purposes. This kind of similarity is called *homology*. Cat and dog legs are examples of **homologous structures**—physical features with the same anatomical structure.

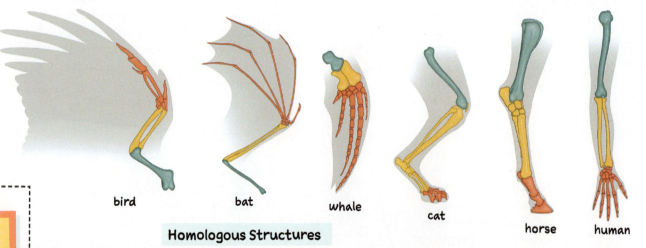

bird   bat   whale   cat   horse   human

Homologous Structures

An extension of the idea of homology is the concept of *vestigiality*. A **vestigial structure** is thought to be the remnant of a structure that has lost its original function. Early naturalists thought they saw many examples of vestigiality, such as the reduced eyes of moles and the wings of flightless birds. Although they could observe vestigial structures, these naturalists could only guess at why they existed. One such naturalist, a colleague of Lamarck named Étienne Geoffroy Saint-Hilaire, speculated that nature kept vestigial structures in some organisms as long as that structure still played an important role in any related organisms.

Early naturalists saw homology as evidence of a concept they called the *great chain of being*, a hierarchy of all matter and life. At the top of the chain was God, and beneath Him in a descending order of attributes were angelic beings, humans, animals, plants, and minerals. To them, animals were similar because they were on the same link of the great chain. Darwin would later view this similarity in a different light.

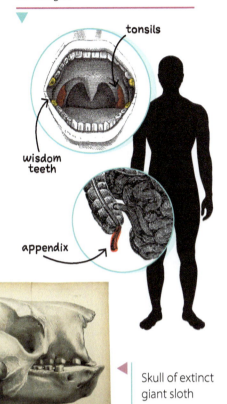

Human structures once claimed to be vestigial

tonsils

wisdom teeth

appendix

COLLECTING, OBSERVING, AND REFLECTING

Darwin showed great zeal for his work. He collected specimens of over 1500 species of plants and animals. He carefully preserved almost 500 bird skins and filled 15 journals with field notes. On the Galápagos Islands off the coast of Ecuador Darwin noted that some animals, particularly mockingbirds, were similar to those he saw on the South American mainland, yet slightly different on each of the islands he visited. Darwin not only collected examples of living organisms—he was an avid fossil hunter as well. **Fossils** are the preserved remains or traces of organisms that lived in the past. During his travels Darwin collected the fossilized remains of many large mammals near the coastal towns he visited in Argentina and Uruguay. Later, in Chile, he would see marine fossils high in the Andes Mountains.

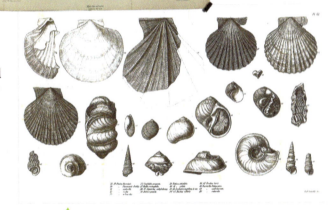

Skull of extinct giant sloth

Fossils that Darwin collected from the Andes Mountains

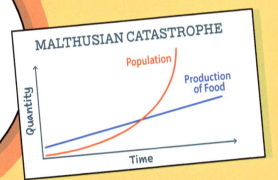

1798
Thomas Malthus, an English pastor and scholar, writes an essay that predicts that if human population growth were not checked by disease or famine, it would eventually grow beyond what the food supply could support since population grows faster than food production. This would lead to a *Malthusian catastrophe*.

MALTHUSIAN CATASTROPHE

Population

Production of Food

Quantity

Time

Galápagos mockingbird
(*Mimus parvulus*)

Hood mockingbird
(*Mimus macdonaldi*)

After returning to England, Darwin began to formulate his idea of **descent with modification**: the idea that all living things came from *common ancestors* and change a little with each passing generation. A **common ancestor** is an ancestral organism whose offspring diverge into two or more species. Darwin saw evidence of common ancestry in the Galápagos mockingbirds. Only one species of mockingbird existed on the mainland, which Darwin believed must have been the ancestral population. He believed that those mockingbirds must have colonized the islands, where the isolated populations on each island slowly changed into the species Darwin observed. Darwin supposed that over millions of years, descent with modification could have transformed simple, primitive life forms into the great variety of living things observable today.

Darwin saw further evidence for his theory in the work of comparative anatomists. It seemed logical to Darwin that homologous structures in different organisms must owe their mutual existence to common ancestors. According to Darwin's interpretation, cat and dog legs are homologous because cats and dogs diverged from a common mammalian ancestor. Darwin viewed vestigial structures similarly. Through natural selection, one organism might keep and further develop a structure that proved useful in a particular environment; a related organism might gradually lose the same structure in a different environment. Thus, for instance, land mammals kept their very useful hind legs, while those same structures became vestigial in marine mammals.

But why change at all? Darwin had earlier read the works of English economist Thomas Malthus. It was Malthus who suggested that there was a struggle for survival among humans whenever resources were scarce. Darwin extended Malthus's ideas to the rest of the natural world. He believed that plants and animals must also struggle for scarce resources. Darwin reasoned that organisms with beneficial new traits that help them survive, such as he observed in the various Galápagos mockingbird species, would produce more offspring than organisms without the new trait. Any of these heritable new traits that make an organism more fit and is passed down to the next generation is called an **adaptation**. Darwin later popularized this idea of adaptation and reproductive success in response to competition for resources as **natural selection**. Many people today think of this concept as *survival of the fittest*, a phrase coined by Herbert Spencer in 1866 after reading Darwin's theories. But take note that

1809
French biology professor Jean-Baptiste Lamarck publishes his idea that if an organism stops using a structure, it will disappear from the species (the *theory of use and disuse*). If an organism acquires a useful characteristic, it can pass this along to its offspring (the *theory of inheritance of acquired characteristics*). Experts call Lamarck's ideas the first systematic theory of evolution. An example is a giraffe that stretches its neck more and more to reach leaves. Over time, giraffe necks get longer and longer.

being "fit" in an evolutionary sense does not mean that an organism is necessarily bigger or stronger than its competitors; it simply means that the organism is successful in producing more offspring.

natural selection

A terrifying earthquake that Darwin experienced in Chile in 1835 helped persuade him that Lyell's *Principles of Geology* was correct regarding Earth's age. After the earthquake Darwin observed uplifts—places where Earth's crust had shifted upward—of as much as eight feet. According to Lyell relatively rare events like earthquakes and the relatively small geological changes they caused meant that Earth's geological features were produced very slowly. Lyell estimated Earth's age to be several hundred million years—far older than the thousands of years seen in the Bible. On that basis Darwin concluded that the marine fossils he observed in the Andes must likewise have lived millions of years ago; the mammal fossils he found near sea level along the coast would necessarily be younger. The long ages that Darwin thought he saw in the fossil record led him to conclude that the transformation of organisms, like the uplifting of the mountains, must also be a very slow process.

1831
Darwin's voyage aboard HMS *Beagle* begins. After this voyage Darwin is willing to view God as the source of morality but not the source of living things. He observed nature through Lyell's uniformitarianism, Lamarck's naturalism, and Malthus's ideas on population to formulate his theory.

1830
British geologist Charles Lyell publishes his *Principles of Geology* in which he describes the laws of nature and the forces of geology operating slowly at the same rate over time. This idea, not original to Lyell, is called *uniformitarianism*. Lyell was influenced by fellow geologist James Hutton, who intended to remove God from science. Hutton and Lyell both thought that the earth was much older than thousands of years.

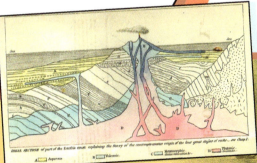

PUTTING IT ALL TOGETHER

For reasons that his biographers debate to this day, Darwin delayed publishing his theory for many years, even though he himself stated that it was clearly conceived in his mind by 1839. Darwin finally published *On the Origin of Species by Means of Natural Selection* in November of 1859. The book was an instant success, though controversial. The theory of descent with modification that Darwin laid out revolutionized both science and theology because of its philosophical implications. Darwinian evolution called into question the basic beliefs held by many people in Darwin's time. But by the late 1900s most people had come to uncritically accept the theory of evolution as scientific dogma. Today it is one of the most widely accepted and cherished theories in all of science. Some might even believe that evolution is a well-established and unassailable scientific fact. As we'll see later, there is very good reason to reject that belief. In the next section we'll look at how the theory of evolution has changed since 1859. You might say we will examine the *evolution* of evolution!

10.1 SECTION REVIEW

1. Define *biological evolution*.

2. Why is artificial selection considered a conservative process?

3. Your friend tells you that prior to Darwin, scientists believed in the fixity of species, that is, that species did not change over time. How would you respond?

4. What is deep time?

5. List at least three people who influenced Darwin's thinking about change in living things and describe the contribution of each.

6. What did Darwin mean by the phrase "descent with modification"?

7. Scientists prior to Darwin believed that organisms could change, so what set Darwin's theory apart from others?

1859
Darwin publishes *On the Origin of Species*, spurred by the publication the previous year of papers on natural selection by British biologist Alfred Russel Wallace. Wallace and Darwin collaborated on their publications, with Wallace later dedicating a book to Darwin on his travels in Indonesia collecting specimens.

THE MODERN THEORY OF EVOLUTION

The Road to a Modern Theory

Darwin spent the rest of his life after the publication of *On the Origin of Species* reviewing the responses of critics and friends and refining his theory of evolution despite recurring illness. He extended his theory to formulate a theory of human descent from animals. But the development of evolution didn't end with Darwin's death in 1882. Just as Darwin had synthesized many of the ideas of his time, people would continue to combine Darwin's ideas with further advances in science in an effort to make the theory of evolution more workable.

Most significantly, Darwin never proposed a satisfactory mechanism that could produce changes in organisms, especially one that could change one kind of organism into another. Natural selection could work only if organisms produced new traits for nature to select! Scientists could observe natural selection in action, but what could create *new* traits, as opposed to the traits that *already existed* in populations? That remained a mystery to Darwin and the early evolutionists.

REDISCOVERING MENDEL

The twentieth century saw a refinement of Darwin's ideas to form a modern evolutionary theory. The key to explaining how evolution could plausibly work turned out to be genetics. Two botanists working independently each became aware of Gregor Mendel's earlier work on patterns of inheritance. Dutchman Hugo de Vries and Carl Correns of Germany discovered that Mendelian genetics explained the variations that they were observing in plants. Both recognized a possible link between Mendelian genetics and Darwin's theory of natural selection. De Vries, in particular, noted that his primroses could produce new varieties that appeared suddenly. He coined the word *mutation* to describe these sudden appearances of new traits. De Vries suggested that these sudden inheritable characteristics introduced variations into the population that could drive evolution. Ideas such as these later led to the emerging field of population genetics (see Chapter 9). Within just a few years of De Vries's work, further advances in genetics were made. An American geneticist, Walter Sutton, and a German biologist, Theodor Boveri, each independently discovered the role of chromosomes as carriers of genetic information.

Hugo de Vries

? How do evolutionists figure out where today's living things come from?

Questions

How has the theory of evolution changed since Darwin's day?

How do modern evolutionists view the history of changing life on Earth?

How do evolutionists interpret evidence to support their theory?

Terms

modern synthesis
behavioral ecology
geologic timescale
index fossil
radiometric dating
relative dating
big bang
molecular evolution
molecular clock
adaptive radiation
speciation
reproductive barrier
hominid

With these discoveries, supporters of evolution gradually crafted a new theory. The **modern synthesis** modified classical Darwinism by adding mutations as the engine of change, coupled with population genetics, to explain how natural selection chose which organisms were most fit. In time biochemists added more details to the model, such as the structure and function of genes and how they can mutate. Evolutionists continue to hone the modern synthesis model, especially as technology makes it possible for us to learn more and more about DNA, genetics, and the cell and its processes.

To recap so far, we can summarize the modern synthesis as shown below.

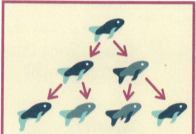

1. Genetics is the foundation for the observed variation necessary for biological evolution.

2. Natural selection is the main driver of biological evolution and depends on genetic diversity.

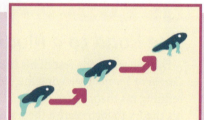

3. Biological evolution is the result of small, gradual changes over long periods of time.

BEHAVIORAL ECOLOGY

In the mid-19th century many naturalists, including Darwin, began to realize that animal behavior may play a role in natural selection. The dedicated study of animal behavior is *ethology*, and ethology as it relates specifically to evolution is the study of **behavioral ecology**. The main idea of behavioral ecology is that changes in animal behavior may provide a competitive advantage to an organism in the same manner as a change in structure. An example of this may be seen in animal threat displays. Many animals, such as frilled lizards, have physical and behavioral adaptations that enable them to ward off predators. They may make themselves appear larger or more dangerous to a predator than they actually are. Thus, the bigger and scarier the display, the more likely an organism is to survive and pass on that behavior to its offspring. Similar reasoning is used to explain mating displays. Females that choose mates with fancy feathers or large antlers, will, by this behavior, drive selection in the direction of even fancier feathers and larger antlers.

frilled lizard

moth exhibiting deimatic behavior

READING EVOLUTION INTO THE ROCKS

Dating the Geologic Column

Evolutionists see more support for their theory in the fossil record. There is a general progression of fossils in the geologic column from a relatively few, simple kinds of organisms in the lowest layers to more diverse and complex organisms in higher levels. Uniformitarian principles dictate that the lowest rock layers must be the oldest, supposedly laid down hundreds of millions of years ago. Older rocks to younger, simple organisms to complex—the conclusion seemed obvious: the fossil record must be a chronicle of the evolution of life on Earth. This conclusion seemed to be strengthened by *radiometric dating* methods that gave very old ages for fossil-bearing rocks. Paleontologists—scientists who study and classify ancient life—have used evolutionary theory to create trees of life according to the fossil record. Their goal is to trace all life back to a hypothetical common ancestor.

Just as geneticists use proteins, DNA, and RNA to construct an organism's evolutionary history, paleontologists use fossils to construct the evolutionary history of life all over the earth. Beginning with Lyell, evolutionary paleontologists have attempted to associate rock layers, called *strata*, in different parts of the world with different periods of time in life's evolutionary history. The periods of evolutionary history are called the **geologic timescale**. The rock layers that correspond to these periods of time are called the *geologic column*.

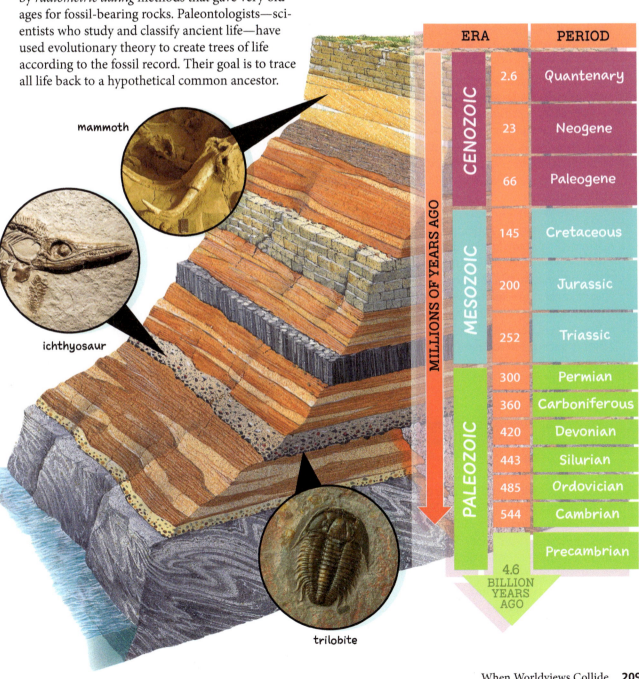

mammoth

ichthyosaur

trilobite

MILLIONS OF YEARS AGO	ERA		PERIOD
	CENOZOIC	2.6	Quantenary
		23	Neogene
		66	Paleogene
	MESOZOIC	145	Cretaceous
		200	Jurassic
		252	Triassic
	PALEOZOIC	300	Permian
		360	Carboniferous
		420	Devonian
		443	Silurian
		485	Ordovician
		544	Cambrian
			Precambrian

4.6 BILLION YEARS AGO

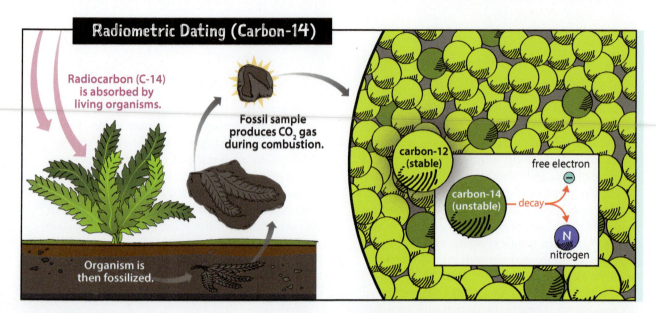

Radiometric Dating (Carbon-14)

Radiocarbon (C-14) is absorbed by living organisms.

Fossil sample produces CO_2 gas during combustion.

Organism is then fossilized.

carbon-12 (stable)

carbon-14 (unstable) — decay

free electron —

N nitrogen

Evolutionists assign ages and names from the geologic timescale to rock layers in the geologic column. Ages of rock layers are estimated on the basis of (1) the absolute or assumed ages of nearby strata and (2) the presence of certain fossils of accepted ages called **index fossils**. Sometimes the assigned ages come from **radiometric dating**, a process that analyzes radioactive elements in rocks and uses the rate at which these elements decay to calculate a sample's age. The reasoning behind radiometric dating is similar to that used for establishing molecular clocks (p. 212), but for rocks instead of organisms. Another method of rock dating depends on the relative positions of different layers. Geologists believe that if they know the ages of strata above and below a given stratum, then they can estimate the age of the strata in between. This is called **relative dating**. Interestingly, there is no one place on Earth where an entire geologic column exists.

younger

Relative Dating

older

Dating Fossils

Fossils don't come with date tags! Evolutionists generally assign ages to fossils according to the geologic ages of the rocks that contain them. They can also use radiometric dating on mineralized fossils or use the ages of index fossils found in the same strata. Just as Darwin had observed, paleontologists expect to find simple organisms in the oldest and most deeply buried rock layers and more complex organisms in the most recent rock layers. But to complicate interpreting the geologic column, some strata have moved as the result of major earth-rebuilding processes such as plate tectonics. Evolutionary geologists say that

plate tectonics has built mountains and spread sea floors over extensive periods of time, moving the continents to their present locations. Plate tectonics includes the same processes that trigger earthquakes and cause volcanoes to erupt today.

Evolutionists interpret the evidence in the geologic column as supporting biological evolution. Within the rock layers they see a progression of simple organisms, then plants and animals, and ultimately man, all by means of mutations and natural selection. This history has been punctuated with massive catastrophes, such as multiple ice ages, meteoroid impacts, and changing climate; each is thought to have caused massive extinctions and shaped the course of evolutionary history. In fact, evolutionists suggest that 99% of all the species that have ever lived on Earth are now extinct.

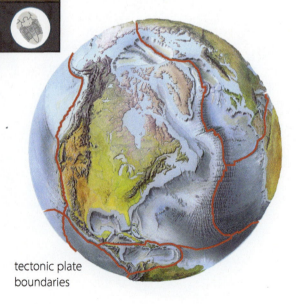

tectonic plate boundaries

MORE THAN CHANGE: THE EARLY EARTH AND ORIGIN OF LIFE

But if the simplest organisms in the oldest rocks are evidence of very early life forms, the question naturally follows: How did these organisms come to exist at all? In the more than a century and a half since the publication of *On the Origin of Species*, the significance of the theory of evolution has grown. It has become the vantage point from which much of science views our world. The grand narrative of evolution has been expanded to explain not just how living things change, but also how they got here.

Evolutionary astronomers and physicists believe that the universe began as a very hot, dense, infinitesimally small dot called a *singularity*. Then, 13.8 billion years ago, all that matter and energy began to expand. As the universe expanded, subatomic particles formed. Evolutionary scientists call this beginning of space and time that is our universe the **big bang**. After the first three minutes, the nuclei of the simple elements hydrogen and helium began to form. After 560 million years, gravity gathered gases together, and stars began to wink in the emerging dawn as fusion began. As these stars aged, they fused lighter elements into heavier ones,

including carbon, nitrogen, and oxygen. This is part of *chemical evolution*, the naturalistic formation of the elements. When old stars eventually died in violent explosions, they spewed their heavy elements out into the universe. This stardust landed in places like Earth, a ball of dust that had coalesced from a nebula in the aftermath of the big bang.

Evolutionists suggest that the necessary biochemicals for life began to form in water pooled on Earth's surface. Darwin himself referred to the possibility of such a pool as a "warm little pond." There, amino acids, proteins, and eventually RNA and DNA formed and eventually began to replicate. Many evolutionists think that lipids began to form and circle into balls—the precursors of cell membranes—enclosing biological molecules. Eventually, all the chemicals needed for life were there, and the first simple organism formed. Evolutionists posit that this ancestor of all living things was likely an anaerobic, prokaryotic bacterium that could survive in the harsh conditions on ancient Earth, similar to the way that extremophiles live today.

Years after the Big Bang

400 thousand · 0.1 billion · 1 billion · 4 billion · 8 billion · 13.8 billion

The Big Bang · Recombination · The Dark Ages · Formation of the First Astronomical Objects · Reionization · Present Day

Fully Ionized · Neutral · Fully Ionized

1000 · 100 · 10 · 1

Red Shift + 1

Evolutionists believe that over five hundred closely related species of cichlids in Lake Malawi in East Africa diverged from a common ancestor in less than a million years—a blink of an evolutionary eye.

Lake Malawi

Molecular Clock Diagram

high rate — species 1
species 2
species 3
species 4
species 5
species 6
species 7
species 8
low rate
species 9

TIME

THE HISTORY OF SPECIES

Secular scientists say that more and more complex organisms began to form as the earth's surface began to change. The DNA in these organisms mutated, yielding new traits and variations, which led to newer and more complex organisms that were better able to survive in the slowly changing environment. Evolutionists call the slow, gradual changes in genetic material and the proteins produced by them **molecular evolution**. They view genetic information and proteins as a history of an organism's evolution. Geneticists can measure the rate at which changes accumulate in a species' genome. The average rate of this accumulating change in an existing organism is called a **molecular clock**. On the basis of an organism's molecular clock, geneticists can estimate how long it took for two related species to evolutionarily diverge.

When evolutionary paleontologists look at individual fossils, they aren't thinking about how those specific fossilized organisms evolved. Rather, they are thinking about how populations evolved. Over the span of millions of years, populations develop mutations, some of which increase their fitness. If these changes are heritable (not all are), they are passed down to the next generation and increase the variation of that population. This is where evolutionists bring in population genetics (see Chapter 9) to investigate how allele frequency, genetic drift, and the bottleneck and founder effects direct the course of biological evolution. The process of change happens slowly in a population depending on whether that population is in contact with other populations or isolated, such as on an island. Species can change fairly rapidly in small, isolated populations compared with large gene pools where mixing happens more freely. Population genetics can explain how allele frequencies change in a population, but it can't predict the direction of future changes. Evolutionary biologists recognize that the process of evolution doesn't seem to have a particular destination.

When a population develops adaptations, the variation and allele frequency within its gene pool changes. The divergence of Darwin's Galápagos Island mockingbirds is an example of such adaptation. Each isolated mockingbird population changed slightly as it adapted to local conditions. This kind of change that results from a population spreading out into new environments is known as **adaptive radiation**. Over time, natural selection acted on the spreading populations to create several distinct species of mockingbirds that differed from the ancestral mainland mockingbird population. The formation of new varieties of related organisms in this manner is called **speciation**. Organisms of the same species can successfully mate to produce offspring. When a new species forms, members of that species usually don't mate with organisms from the original population to produce offspring. Anything that prevents two species from reproducing with each other is called a **reproductive barrier**. Some reproductive barriers are geographical—Darwin's mockingbirds lived on separate islands isolated by water. Other barriers may be due to an organism's behavior, physical structure, habitat, or the fitness of the offspring produced.

Changes in populations due to reproductive barriers are not unusual. We observe them happening today. But evolutionists take the narrative one step further. Evolutionists believe that the small changes that we can observe can produce very large changes over vast quantities of time. In other words, natural selection can not only produce different but related species—it can also produce entirely different kinds of organisms as well. All it takes is speciation coupled with eons of time, plus an occasional population bottleneck caused by a mass extinction event.

Evolutionists have brought together observations from diverse fields such as genetics, molecular biology, and paleontology to suggest how organisms that we observe today are evolutionary relatives of each other. Biologists diagram these relations with a chart similar to a family tree called a *phylogenetic tree*. The tips of the phylogenetic tree's branches show known organisms that are either alive today or found in the fossil record. The places where lines come together indicate hypothetical common ancestors. We'll learn more about phylogenetic trees when we study classification in Chapter 11.

Adaptive Radiation

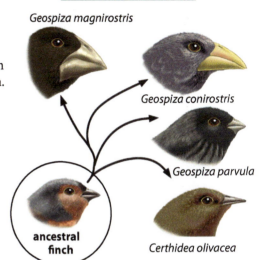

Geospiza magnirostris

Geospiza conirostris

ancestral finch

Geospiza parvula

Certhidea olivacea

Separated islands form a geographical reproductive barrier.

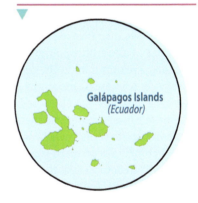

Galápagos Islands
(Ecuador)

Phylogenetic Tree

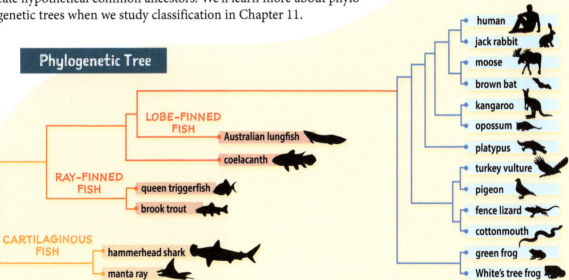

LOBE-FINNED FISH
Australian lungfish
coelacanth

RAY-FINNED FISH
queen triggerfish
brook trout

CARTILAGINOUS FISH
hammerhead shark
manta ray

TETRAPODS
human
jack rabbit
moose
brown bat
kangaroo
opossum
platypus
turkey vulture
pigeon
fence lizard
cottonmouth
green frog
White's tree frog

THE HISTORY OF MAN

Evolutionists don't end their narrative with highly evolved plants and animals. They believe that man has also evolved from earlier common ancestors. Paleoanthropology—the study of ancient humans—relates man to the evolution of mammals, especially primates. Our ability to walk on two feet, our long legs, our barrel-shaped rib cage, and the size of our brain are several factors that evolutionists compare to other members of the phylogenetic tree that includes modern-day humans. Evolutionists credit structures like these to the rise of the human species, *Homo sapiens*, and suggest that modern humans first appeared between 260,000 and 350,000 years ago. Anthropologists have found fossils of organisms that they label as predecessors of modern-day humans. They refer to this group of organisms including modern man as **hominids**. Evolutionary anthropologists believe that apes and hominids diverged from a common ancestor at least 7–8 million years ago.

AN IRONCLAD THEORY?

In today's scientific community evolution is the dominant theory used to explain how our universe came to be what it is today. We can easily find information on biological evolution, chemical evolution, and even things like stellar evolution and galactic evolution—changes in stars and galaxies. In fact, there is a field known as *cosmic evolution* that encompasses *all* change in matter, space, and time from the big bang down to the present day. Evolution is such an entrenched concept in our society that daring to challenge it can potentially cost a scientist his job. But despite this, evolution is not as blindly accepted as it might appear. Even in the secular scientific community, there are some who question the theory of evolution's explanatory power. In the next section we'll examine some serious problems with Darwin's big idea.

10.2 SECTION REVIEW

1. What was combined with Darwinian natural selection to produce the modern synthesis?

2. How did the field of behavioral ecology contribute to the modern synthesis?

3. How do evolutionists interpret the geologic column and the fossil record contained in it?

4. Does the term *evolution* these days signify only biological evolution? Explain.

5. How are molecular clocks used to support evolution?

6. How do reproductive barriers relate to the process of adaptive radiation?

7. Why do evolutionists assume that humans are merely highly evolved hominids?

Artist's impression of hypothetical early Earth

10.3 EVALUATING MODERN EVOLUTIONARY THEORY

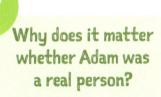

Evolutionists view their theory as the key to understanding the natural world. They endeavor to interpret evidence from the natural world without resorting to miraculous or supernatural explanations. Because of this presupposition, they reject the Bible's account of creation out of hand. Their worldview is a completely naturalistic and materialistic one in which all that matters is what is natural and material. But a naturalistic worldview has some severe limitations, and these can pose big problems for the evolutionary story. While most evolutionary scientists would say that we just need more time to figure out solutions, some of the problems are fundamental.

BEFORE LIFE

The evolutionary story's difficulties go all the way back to the very beginning—the big bang itself. The problem is a serious lack of matter. Simply put, there simply isn't enough observable matter to account for the observed effects of gravity in our universe. Worse, there doesn't appear to be enough energy to explain the gravitational interactions that we observe in our universe either. Physicists call these unobservable phenomena *dark matter* and *dark energy*. Although there is good evidence for the existence of both dark matter and dark energy, no one knows exactly what they are. Thus, their contributions to the big bang model are a matter of conjecture—astrophysicists interpret the evidence to fit their worldview. Of course, trying to explain phenomena that we *can* observe on the basis of phenomena that we *cannot* directly observe, measure, or test is hardly the way that we expect scientific inquiry to work.

Why does it matter whether Adam was a real person?

Questions

What are some weaknesses of modern evolutionary theory?

Is the modern synthesis a workable model?

How have Christians attempted to accommodate evolutionary ideas?

Terms
abiogenesis
transitional form
genetic load

Evolutionists also haven't been able to explain the evolution of information-rich molecules like DNA. In 1952 a famous experiment conducted by chemists Stanley Miller and Harold Urey seemed to demonstrate that amino acids could be formed under the conditions that existed on the early earth. But scientists don't agree on what exactly those conditions were. And other scientists have pointed out the irony of the Miller-Urey experiment demonstrating that even "simple" amino acids require precise, controlled conditions to form spontaneously. Even if simple biological compounds could form under primitive conditions, scientists have yet to demonstrate how they might further develop into the sophisticated molecules required for life. As we saw in our study of cells, proteins and DNA have an amazing amount of information packed into them. We don't know of any natural processes that can make them form effortlessly in simulated "warm little ponds."

FROM NONLIFE TO LIFE

Recall from Chapter 5 that an important aspect of the cell theory is that cells come only from preexisting cells. Of course, for evolution to be true, there had to be at least one instance where that wasn't the case. The possibility of life arising from nonlife is known as **abiogenesis**. At one time people commonly believed in the idea of *spontaneous generation*. It was a simple explanation of why, for instance, maggots might mysteriously appear on rotting meat. You may have read about how the French microbiologist Louis Pasteur disproved spontaneous generation. In a series of experiments in 1869 he showed that broth kept in sterile conditions would not spoil—no bacteria would spontaneously generate in his flasks.

Evolutionists try to argue that their modern theories of abiogenesis are different from past groundless ideas about spontaneous generation. Even so, they still have not demonstrated any mechanism by which life can arise from nonlife, regardless of what the process is called. This is such an insurmountable problem that some evolutionists have suggested that life didn't *form* on Earth at all but instead *came* to Earth from somewhere else. This hypothesis, called *panspermia*, sounds like science fiction, but some rather famous scientists have seriously considered it. Of course, the panspermia hypothesis doesn't explain how life could arise from nonlife—it just moves the problem from Earth to somewhere else.

Miller-Urey Experiment Setup

electrodes

electrical spark (lightning)

gases (primitive atmosphere)

to vacuum pump

condenser

sampling probe

water (ocean)

cooled water (containing organic compounds)

sampling probe

heat source

trap

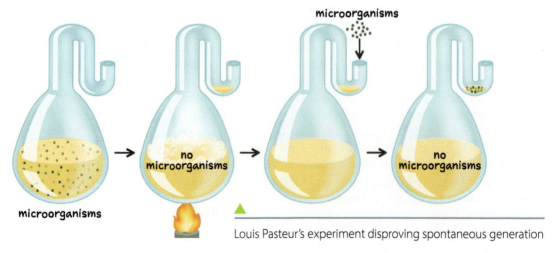

microorganisms

microorganisms

no microorganisms

no microorganisms

Louis Pasteur's experiment disproving spontaneous generation

FOSSILS

Even the fossil record does not always conform to the expectations of evolutionary paleontologists, as we have seen with dinosaur soft tissue finds. In fact, paleontologists often seem only too eager to see evidence in the fossil record that *isn't* truly there while overlooking evidence that plainly *is* there! A major potential problem with the fossil record was identified by Darwin himself. Darwin recognized that if evolution did occur gradually, as he supposed, then the fossil record should include many examples of **transitional forms**— organisms whose structures are intermediate between their ancestors and their descendants. But Darwin observed that the fossil record was lacking—there was no place in the geologic column where a complete and gradual progression from simple forms to complex could be seen. In fact, transitional forms were so conspicuously absent that some of Darwin's contemporaries saw the fossil record as evidence of fixity of species! Darwin blamed the lack of transitional forms on an incomplete fossil record. He assumed that the missing transitional forms would eventually be found. But since his time paleontologists have unearthed an unimaginable number of fossils, and still the hypothetical transitional forms are absent. Evolutionists are hard-pressed to identify even one.

Instead of gradual transition, paleontologists have discovered many instances of life forms suddenly appearing in the fossil record with no obvious ancestors. One of the most famous of these is the Cambrian explosion. Evolutionary paleontologists say that this happened about 530 million years ago. In a relatively short period of evolutionary time there suddenly appeared in the fossil record

▲
Artist's impression of Cambrian life-forms

a great diversity of animals. Nearly all the major groups of animals that we observe today are represented. Taken together, the many instances of fully formed kinds of life in the fossil record have caused some evolutionists to rethink the idea of gradual transformation. A popular theory to explain the phenomenon is known as *punctuated equilibrium*. This theory suggests that evolution can happen in relatively short bursts interspersed by long periods of little or no change.

The fossil record is fraught with other difficulties for the evolutionary story. One such difficulty is fossils that don't fit the narrative. A prime example of this is fossilized pollen found in rock layers thought to be hundreds of millions of years older than the oldest known flowering plants. Another is fossils that show evidence of extremely rapid burial and compression instead of the slow sedimentation expected by uniformitarianism. There are other significant issues as well. But the one constant in all this is that evolutionists don't view these problems as problems—they simply rewrite the evolutionary story to fit what they observe.

Perhaps the greatest problem with the fossil record is its location—in the geologic column itself. The column is formed in large measure from enormously thick layers of sedimentary rock, much thicker than deposits that we see being formed today. Recall that sedimentary rocks are formed from sand, silt, mud, and other materials transported and deposited *by water*. There is a surprisingly obvious explanation for this fact but one that evolutionists choose to ignore. We will look at that explanation in Section 10.4.

A coelacanth's fins, despite their leglike appearance, are used only for swimming.

THE LIMITS OF SPECIATION

There is a serious problem with the directions of change required by evolution and seen in natural selection. For evolution to be a valid theory, a small amount of information in a population must somehow lead to increasingly larger amounts of information. For instance, the standard evolutionary story claims that the legs of land-dwelling animals developed over time from the fins of certain kinds of fish; at one time, coelacanths (above) were a popular candidate for the transitional form. But the structure of a mammalian leg is obviously very different from that of a fish fin. Such a radical change in structure would require a *gain* of genetic information, not a loss. This is not what we see happening in our world today. Many evolutionists claim that there simply hasn't been sufficient time for us to see this kind of change happening. But on the basis of the kinds of change that we *can* see happening now, it would seem that fins could never evolve into legs, no matter how much time is allowed for the process.

As we learned at the beginning of this chapter, we can readily observe change taking place within species. Such change can actually happen rather quickly, as we see in artificial selection. It is commonly seen in fish that are bred in hatcheries. Even if wild fish are used for breeding, the first generation of offspring typically shows a measurable change in allele frequency. Typically, though, such change is seen in a *loss* of genetic diversity, not the gain needed for evolution to occur. And fish in hatcheries are far from being the only example of genetic loss—it happens in wild populations too. Whenever and wherever we see adaption taking place, it is invariably a result of changes within existing genomes, either a segregation or loss of alleles. So far as has ever been observed, natural selection begins with a large amount of information in a population that is whittled down to smaller amounts of information in the DNA of the fittest organisms. The two processes—evolution and natural selection—appear to be at cross-purposes.

MUTATION: ENGINE OF CHANGE OR DEAD END?

The theory of evolution has an even more serious problem than that of natural selection producing a loss of genetic diversity. The bigger problem is with evolution's supposed engine for creating new alleles: mutation. As we learned in Chapter 9, mutations are random changes in an organism's genetic code. At best, mutations are sometimes neutral, meaning that they don't harm affected organisms. But in the vast majority of instances mutations are harmful, even lethal. And just as random changes in a perfectly functioning machine, like a car, ultimately degrade the car's ability to do useful work, random changes—even neutral ones—ultimately degrade the information in DNA as well. Evolutionists have often been challenged to cite instances of mutations adding useful information to an organism's genome. The examples they usually give can nearly always be shown to be instances of natural selection; that is, existing alleles are sorted or lost, but no new alleles are added.

If mutations are nearly always bad, then a lot of mutations should be really bad. And that is in fact what we see in organisms today. In every generation relatively small numbers of mutations occur. With every passing generation, the total number of mutations carried by a population gets larger. Geneticists call the increasing lack of fitness caused by an increasing number of mutations a population's **genetic load**. As a population's genetic load increases, so does its genetic entropy—the degeneration of its cellular information systems. And just as a certain number of random changes made to a car will eventually render it nonfunctioning, genetic entropy has consequences too. As a population's genetic load increases, the entire population becomes increasingly less fit for survival. Eventually, the population will become extinct. Unhappily for evolutionary theory, observable rates of genetic entropy simply do not allow for organisms to be the necessary millions of years old.

Certain populations of nonmigratory Glanville fritillary butterflies, isolated on islands in the Baltic Sea, carry high genetic loads because of inbreeding, putting them at risk for extinction.

flagellum

Some scientists have challenged the idea that mutation can work as the engine to drive slow, stepwise change. These scientists have pointed out the problem of *irreducible complexity*. Simply put, there are many complex structures and chemical processes in living things that cannot function unless all the parts are in place. One example is the tiny biological "motor" that turns a flagellum (left); remove any single part from the system and the whole thing fails to work. Evolutionists try to argue that evolution can produce such a motor if all the many intermediate steps provide some kind of survival advantage. But while we can observe the advantage provided by a working flagellum, we can only guess at any advantage provided by a partial flagellum. It requires a great deal of faith to believe that an organism can gain an advantage by having any kind of partial system, much less an entire series of them! A much more natural explanation for complex systems suggests itself in our everyday experience. When humans observe a functioning, complex system, such as a car or plane, we don't assume that the system built itself. It is far more logical to assume that the system is a product of an intelligent mind—a designer. Scientists in the *intelligent design* movement defend the idea of an intelligent designer as a better way to explain the complexity that we see in nature than the accumulation of random changes.

DATING DILEMMAS

Evolutionists place great faith in the supposed reliability of radiometric dating methods. The great ages obtained by these methods seem to confirm evolution's foundational belief in an ancient Earth. But radiometric dating methods are neither foolproof nor free of bias. Radiometric dating begins with several important but untestable assumptions. These include the ideas that the relative abundance of parent and daughter isotopes can be known with certainty and that the rates of radioactive decay for the parent isotopes have remained constant. In practice, radiometric dating has yielded wildly inaccurate results for rocks of known ages. And in other instances when testing has produced a wide range of possible ages, evolutionist geologists have tended to simply discard dates that don't agree with their notions of how old the tested samples should be. How can they tell whether their ages obtained through radiometric dating are accurate or not? Often it is by comparing the radiometric dating age to the "known" age of a sample on the basis of relative dating or the presence of index fossils. Untestable assumptions and biased screening of results make the ages obtained by radiometric dating questionable at best and useless at worst.

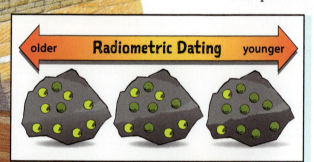

older   Radiometric Dating   younger

A LICENSE FOR EVIL

Perhaps the biggest concerns with the evolutionary, naturalist world-view are philosophical. Why is there something instead of nothing? How can consciousness be formed from unconsciousness? And how can we have standards of right and wrong behavior if the whole world is headed for a meaningless oblivion and extinction? Evolution's inability to answer basic questions about human life call into question the materialistic worldview.

The theory of evolution has very real moral consequences. While most people don't think through evolution this far, it leads to a de-humanized view of people and makes them little more than highly developed animals with no moral or spiritual anchor. In the guise of social Darwinism, the ideas of evolution have been applied to society with disastrous consequences. Commonly held racist beliefs of the nineteenth century found scientific "proof" in evolutionary theory. The idea that the white "race" (the concept of race itself is based on an evolutionary worldview) was more highly evolved than other "races" fueled the *eugenics movement*. Eugenicists like Margaret Sanger, the founder of Planned Parenthood, believed that the truth of evolutionary theory gave them license to nudge the process along by preventing those deemed unfit from reproducing. Initially this meant strategies such as forced sterilization of the poor or mentally disabled. Taken to its logical extreme, the best way of course to prevent such breeding is to simply eliminate the unfit entirely. It should come as little surprise then that the Nazi Germans who perpetrated the Holo-caust were steeped in eugenicist philosophy. While any idea can be twisted for wrong purposes, genocide is logically consistent with an evolutionary worldview.

There are scientists who believe that evolution is a theory in crisis. With each passing year, more evidence accumulates showing that even the "simplest" cells are far from simple and apparently always have been. As modern biology reveals additional insight into the cell's inner machinery, scientists struggle to construct evolutionary expla-nations for the complexity they find. The vast amounts of genetic information coded in cells strongly imply that there is intelligence at work in the cell's design. As we will see in Section 10.4, there is an-other story that we can use to understand where this superabundance of information came from.

Lebensborn birth houses were created in Nazi Germany with the intention of raising the birth rate of "Aryan" children from the extramarital relations of "racially pure and healthy" parents.

ROOM FOR COMPROMISE?

Many of the scientists who contributed to the de-velopment of evolutionary theory believed in God to one degree or another. Some were merely deists, but others believed in the authority of Scripture. But all of them were persuaded by their naturalistic worldview that the Bible's narrative of a young Earth could not be valid. This is true today as well. There are many Christians who believe that God's Word, including Genesis, is true and yet struggle to accept a literal six-day view of Creation. Many believers feel a need to reinterpret the opening chapters of Genesis in light of today's prevailing naturalistic worldview, a method of interpretation called *con-cordism*. They don't do this to discredit the Bible. In fact, they are trying to defend the Bible by showing that it is compatible with evolutionary science. But their attempts to reconcile biblical and naturalistic worldviews are dangerous. If there was no Creation, then the Fall must be redefined. If we redefine the Fall, then we must also reinterpret the Bible's teaching about Redemption through Jesus Christ. Interpreting Genesis 1 as anything other than real history shakes the very foundations of biblical Christianity.

Efforts to support concordism generally take one of three approaches: deny that Genesis 1 and 2 describe real events, assert that the days of Creation were long periods of time instead of literal days, or suggest that a gap exists in the Genesis timeline. Here we will look at some of these attempts to reconcile the Bible with billions of years of evolution.

Analogical Day and Framework Views

The *analogical day* view begins by pointing out that Genesis 2:1–3 never says that the seventh day ended. People who hold this view say that God is still resting from His creation. Since the seventh day is interpreted as an analogy that corresponds to the day on which God's people are to rest, then the other six days must simply be analogies too. The days of Genesis 1 and 2 are seen as "God days," not Earth days.

The *framework* view goes further. It says that God did not even follow the order of the days presented in Genesis 1—the order is topical rather than historical. This view proposes that Days 1–3 involved the creation of kingdoms, and Days 4–6 provided creatures to fill and rule over those kingdoms. This view suggests that the days represent a literary structure that pairs Days 1 and 4, Days 2 and 5, and Days 3 and 6.

Both views assert that Genesis 1 is not meant to be understood as a historical account. This means that Genesis does not give any information about how long Creation took or how precisely God created. These interpretations view Genesis 1 as creative literature, not history. This makes the whole passage pliable enough to "fit" evolution's millions of years of change into the narrative.

Progressive Creationism

Progressive creationists claim that the days of Genesis 1 are not twenty-four-hour days but long periods of time. This concept is sometimes referred to as the *day-age theory*. Progressive creationists conclude that Earth is billions of years old and that it began with a God-initiated big bang. Unlike the framework view, progressive creationists believe that Genesis 1 does provide the order in which God created the universe. Many progressive creationists also believe that God intervened many times in Earth's long history to directly create the different kinds of animals that exist today. Some believe that mankind is the result of one such intervention.

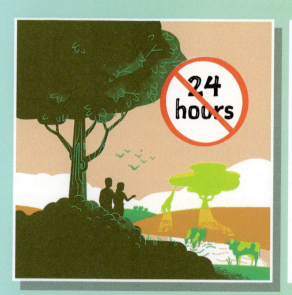

Theistic Evolution

Theistic evolutionists, or evolutionary creationists, believe that evolutionary science is correct in its narrative of Earth's origins and the development of life. They believe that God initiated everything with the big bang and that everything then proceeded in accordance with the evolutionary model. They do not believe that He specially intervened in the world at any time to create new kinds of organisms.

Theistic evolutionists offer differing views on how to harmonize Scripture and evolution. One popular theory is the *cosmic temple inauguration theory*. This theory teaches that Genesis 1 is not about the creation of the physical world at all. Instead, Genesis 1 is about the creation of the universe as a cosmic temple. Humans are given the function of serving as priests in that temple.

Gap Theory

Gap theorists contend that there is an unspecified length of time missing from the Genesis account of the Creation week, hence the name *gap theory*. Gap theory suggests that the "formless and void" language of Genesis 1:2 refers to an earlier Earth ruled by Lucifer. When Lucifer rebelled against God, God destroyed this earlier world. Classical gap theorists suggest that the fossil record consists of organisms destroyed by this first judgment, not the Flood of Genesis 6. After God's judgment of the first Earth, He then remade it in six days; thus, this theory is sometimes called the *ruin-reconstruction theory*.

Death before Sin?

A major obstacle for all the views just described is the way that the original Hebrew text is written. Just as modern English literature does, ancient Hebrew writings use language and conventions that indicate to the reader when real, historical events are being described. That style is distinctly different from those used for either poetry or allegory. Moses' ancient Jewish audience would have clearly understood that Genesis 1 and 2 are meant to describe real creative acts completed in six normal days. A proper interpretation of Scripture requires us to do likewise. In contrast, concordist alternatives are based on flawed interpretations that ignore the original intent of the author. Their theories must be forced *into* the text instead of being logical conclusions drawn *from* the text.

In addition to faulty interpretation, the alternative creation views all share one other huge problem. By accommodating evolution, each theory either makes man the direct result of an evolutionary process or inserts him somewhere in the process already in motion. None of these theories makes death the real consequence of Adam's sin. As we learned in the introduction to this section, interpreting the first chapters of Genesis as anything other than real history creates problems for the doctrines of the Fall and Redemption.

There is one verse of Scripture in particular that should put an end to attempts to force-fit evolution into the biblical narrative. Paul wrote in Romans 5:12, "Wherefore, as by one man sin entered into the world, and death by sin; and so death passed upon all men, for that all have sinned." The plain teaching of this passage and indeed of all of Scripture is that death is the result of the Fall. The Bible even tells us that death is the last enemy that Christ will destroy (1 Cor. 15:26). But if evolution is true, then we are not Adam and Eve's descendants but the result of a slow progression from microbes to mankind. Untold numbers of our supposed ancestors would have lived and died during this process. Evolution's story flips the biblical narrative on its head by making death the hero instead of the enemy.

Conflating Evolution and Natural Selection

The verb *conflate* means to merge two ideas or concepts together. Sometimes this can cause confusion, as when evolutionists conflate evolution with natural selection. As we've seen, the evolutionary narrative requires a continuous *gain* of genetic information—new traits for natural selection to act upon. By this means, every kind of living thing we see today—redwoods, bumblebees, dolphins, for example—has supposedly evolved from a hypothetical single-celled common ancestor. But many, if not most, of the examples put forth by evolutionists as evidence of evolution in action are not examples of evolution at all. In this activity we'll examine some of these examples.

Are there examples of evolution in action today?

Materials
computer with internet access

PROCEDURE

A Do an internet search for examples of evolution. Look for examples drawn from organisms that are alive today rather than examples from the fossil record.

B Choose one or two examples to discuss with a classmate. In particular, be on the lookout for information that can answer the following questions.

- Does the example actually demonstrate evolution (change from one kind of organism into a different kind of organism) or merely natural selection (change within one kind of organism)?

- Does the example demonstrate a gain of genetic information or a loss?

ANALYSIS

1. Summarize the example that you gave as an example of evolution in action.

2. Is this an example of evolution or natural selection? Explain.

3. Suggest a genetic basis for the change observed.

1. Why is the big bang an insufficient scientific explanation for the origin of our universe?

2. Compare spontaneous generation to abiogenesis.

3. Did the Miller-Urey experiment provide a workable explanation for abiogenesis? Explain.

4. The image on the right shows a dinosaur fossil in the classic "death pose," with its head thrown back and mouth wide open. Laboratory tests have shown that vertebrate animals assume this posture when drowned. Evaluate the workability of the evolutionary model with regard to this fossil.

5. What is the difference between natural selection and evolution? Which of the two can we observe happening today?

6. Evolutionist paleontologists often discard dates derived for fossils by radiometric dating if the dates are considered to be too young. Why do they do this?

7. What is irreducible complexity? How do evolutionists explain the development of irreducibly complex systems?

8. Describe at least two ways that Christians have attempted to reconcile the deep time requirements of evolutionary theory with the biblical Creation account.

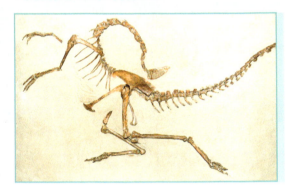

10.4 THE BIBLICAL ACCOUNT

A History of Change

The Bible's account of Creation, Fall, and Redemption offers a consistent, logical framework to properly understand the world of scientific evidence. People who accept the Bible's story work with the same limitations as evolutionists—no human witnessed Creation, and the Bible provides only a tantalizingly brief glimpse into the details. But those who believe God's Word have one significant advantage— an infallible account written by God Himself.

THE EARLY EARTH

A straightforward reading of Genesis 1 and 2 tells us that God created all things by the power of His Word in six normal, twenty-four-hour days. And God spells out a particular sequence of creative events in the early chapters of Genesis. Significantly, the order given there is not how life would have evolved, even if it were possible. For example, God created plants on Day 3 before creating the sun on Day 4. Plants could not have existed for eons of time without sunlight. Similarly, evolution teaches that birds evolved from dinosaurs, but the Creation narrative tells us that birds were created on Day 5 before dinosaurs on Day 6. God finished His creative work on Day 6. While there is a little room for debate, the best biblical scholarship suggests that Creation happened sometime between about 6000 and 4000 BC. Attempts to force the millions of years required for evolution into the biblical narrative make little sense, as we shall see later.

? How is change in nature explained by the creation model?

Questions

What does the Bible teach us about origins?

How do creationists explain how life changes?

How is the evidence for change consistent with the creation model?

Why are people special?

Term
kind

The CREATION Week

Day 1
God creates the heavens (space), Earth, and light. He separates light (day) from darkness (night). The length of a day is set (evening and morning, that is, twenty-four hours).

Day 2
God creates the firmament (atmosphere) and divides the waters between those above the firmament and those beneath.

Day 3
God gathers the waters beneath the firmament, forming seas and dry land. The land brings forth plants.

General revelation (what we can learn from the created order) and special revelation (what is revealed in Scripture) show us that Earth was designed for life (Isa. 45:18). Even many secular scientists recognize the extreme suitability of Earth as a home for sentient life, a truth some have called the *anthropic principle*. By this we mean that Earth is just the right distance from the sun, has just the right blend of gases for an atmosphere, and meets many other criteria that make life possible. Just as importantly, we *don't* see this combination of factors at work anywhere else in the known universe.

God filled His new world with an amazing variety of life. A phrase we see repeated multiple times in the Genesis account of creation is "after his kind." As we will see in Chapter 11, God didn't create different *species* of organisms. The species concept is a human one, developed many centuries after the events in Genesis. Instead, God created many kinds of organisms. A **kind** is a group of organisms that God created as distinct from other groups of organisms. In other words, according to the creation model one kind of organism cannot gradually change into another. But it makes perfect sense that God would create a great amount of diversity in the DNA within each original kind. After the Fall and over time, the original created kinds could diverge into an enormous variety of living things as they multiplied and filled the earth. Natural selection is thus completely compatible with the biblical narrative.

The Bible tells us that God's world was created "very good" (Gen. 1:31), meaning that everything was just as God had purposed it to be, free from any defect. Sadly, it didn't stay that way for long.

Day 4

God forms the sun, moon, stars, and planets.

Day 5

God forms sea creatures and flying creatures, including birds.

Day 6

God forms land animals and humans.

GEOLOGICAL CHANGE

Many people today, even non-Christians, have a sense that something isn't quite right with our world. It was Adam's sin that plunged the whole world into its current broken condition. This sin eventually accumulated to such an extent that the Bible describes the thoughts of human hearts as "only evil" (Gen. 6:5). God was very grieved to see His once-beautiful world marred in this manner. He decided to judge man's sin with a worldwide flood.

The Flood was the most significant physical event in Earth's history after Creation. From what we learn in Genesis 7 and 8, the Flood completely covered the earth's surface. Local floods today can be devastating, but the destructive forces unleashed by The Great Deluge can scarcely be imagined. The Flood marks the dividing line between the originally created world and the world that we live in today. Land experienced major reshaping as massive erosion shaped pliable Earth. Vast quantities of sediment were washed away, then sorted and deposited by the rising waters. These formed the layers of rock that we now see in the geologic column. As the waters receded, the newly laid, soft layers of sediment, not yet metamorphosed into rock, experienced additional shaping. Huge plateaus were scraped flat and deep canyons were carved by the receding flood waters. Some portions of Earth's crust were uplifted, forming high mountains, while other areas subsided to form ocean basins. All land animals not on the ark died. Many were buried and preserved in rock layers all over the earth, accounting for most of the fossils that we see today.

Day 7

God rests from His creative works.

Grand Canyon

The Flood provides a powerful and workable explanation for the different kinds of fossils seen in each layer of the geologic column. Genesis 7 tells us that first "the fountains of the great deep [were] broken up," after which "the windows of heaven were opened." In the earliest stages of the Flood, communities on the ocean floor, consisting mainly of less mobile or sedentary forms, would have been the first to be buried. More mobile creatures, such as fish, could have escaped the destruction—for a while. Eventually, they too succumbed. On land, the process was repeated. Plants and smaller, less mobile animals would be the first to be caught by the rising waters. Larger animals could escape for a time by moving to higher ground. Humans would be the last to perish as they sought refuge on the highest ground of all. Thus, the Flood provides an explanation of the basic order seen in the fossil record, with the simplest organisms in the lowest layers.

Just as important as the progression seen in the fossil record is the fact that the record exists at all. Here, too, the Flood has superior explanatory power. Organisms that die and are covered by only slowly accumulating sediment are scavenged and decay long before they can be preserved as fossils. Fossilization requires organisms to be buried and compressed rapidly, which is exactly what the Flood did. And today we see not just an occasional fossil here and there, but enormous quantities of fossils, often buried in mass fossil graveyards. This burial happened so quickly that in many instances physical structures, even soft tissues, were preserved in extra-ordinary detail. Coal is formed similarly. Today's coal seams, often hundreds of meters thick, are evidence of the enor-mous amount of plant material buried by the Flood.

Arches National Park

The effects of the Flood didn't end when the waters dried up. The changes brought about by the worldwide event greatly affected Earth's weather. Though the Bible doesn't mention an ice age, most creation geologists see clear evidence for a glacial period after the Flood. Some have created models to suggest how the events that resulted in the Ice Age would have unfolded. An ice age would lower sea level, making it easier to repopulate the world after the Flood. It would make Mesopotamia and Egypt wetter and more moderate in temperature, which would explain why people settled there.

The forces that changed Earth's features didn't stop completely with the end of the Flood. Creation geologists agree with their secular counterparts that there is a lot of evidence for the continuing earth-building and earth-shaping processes of plate tectonics. They disagree, though, on the timing and magnitude of these processes. The Genesis Flood model can explain the dramatic features that we see today in ways that the uniformitarian model cannot. For example, geologists can readily observe erosion and uplift in action today. But the mountain-building caused by uplift is not keeping pace with the rate of erosion. Earth's physical fea-tures are being worn away faster than they are being formed. A stark example of this can be seen in the rock arches of Arches National Park in Utah. These arches, formed from sedimentary rock, are collaps-ing at a rate of about one per year. At that rate they will be completely gone in about 2000 years since no new ones are being formed. This observation doesn't fit the uniformitarian story of millions of years of continuous weathering. But the rate of arch collapse does agree very nicely with the young age for Earth seen in a plain reading of Genesis.

BIOLOGICAL CHANGE

Noah took two of all unclean land animals on the ark and seven of all clean animals. We should not assume that the animals that boarded the ark looked the same as those we see today. The Bible tells us, not that two of every *species* were preserved, but rather two of every *kind*. When these kinds of animals left the ark, they stepped into a world that was very different from the one they left behind. The forces that contributed to the destructiveness of the Flood would take many centuries to wind down and reach a new equilibrium. Climates were radically altered. Much of the earth was quickly covered in the deep snows of the Ice Age. Other parts that would later become deserts were cooler and wetter than they are now. There was more volcanic and seismic activity than today—the diminishing echoes of the disaster just passed. This new world presented God's creatures with new environments. It provided a unique opportunity for the different kinds of animals to spread out and adapt. God didn't leave His creatures ill-prepared for the task of refilling the earth. He designed each animal kind with the genetic diversity it would need to survive in the post-Flood world. Through natural selection, founding populations of animals in new environments lost or conserved different sets of alleles, creating the species we see today. Today, just a few thousand years later, our world teems with plants and animals. It also provides a home for over eight billion people! God's mercy thus preserved life, even amid judgment, and allowed the earth to dramatically recover after the Flood.

Christians who believe what the Bible tells us about the Creation event can agree with Darwin on some issues. We do see that there is descent with modification among living things. We have no problem with Darwin's changing mockingbirds or de Vries's changing primroses. We observe changes over time happening even today. What we don't see is any of these modifications turning one kind of animal into another, as is required by the modern synthesis model.

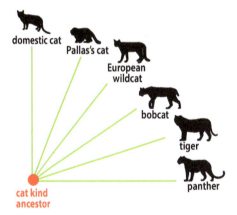

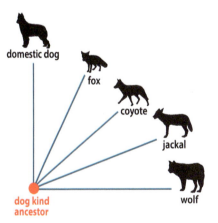

But as we observe changing life forms, we also see the effects of the Fall at work, and again the creation model provides a workable explanation. Because God originally created everything "very good," we would expect something like mutations to be generally harmful. This is in fact what we see, as we learned in Section 10.3. Indeed, our current understanding of genetics shows us that all living things, with their increasing genetic loads, are on an irreversible, downward path toward extinction. This is completely counter to the increasing complexity predicted by evolution.

THE HISTORY OF MANKIND

The Bible treats Adam and Eve as real people, created in God's image and given the task of filling the earth and wisely using it. We are the descendants of those first created humans. In the Bible's story there is a sharp distinction between humans and animals. Evolutionists can come up with only unsatisfactory attempts to explain why humans are the only creatures on our planet capable of things like complex speech, abstract thought, and the ability to create and enjoy art and music. But Genesis 1 presents man as the pinnacle of God's creative work, the only creatures made in His image. To be human is *not* to be a highly evolved, highly fit animal that has clawed its way to the top of the food chain to survive through brute force. Being God's image-bearers explains why we are like God in certain relational, emotional, and rational ways. Because of this, we have great worth. God demonstrated the value that He places on people when He became a man and suffered and died for our sins.

Even sin itself is a serious problem for evolutionists. They can't explain why humans have innate value systems. All of us are born with a sense of right and wrong. Most human cultures, for example, believe that murder is wrong. But if, as evolutionists would have us believe, humans are merely highly evolved apes, then anything that increases our reproductive success suits the evolutionary narrative—even killing off our competitors. Evolution can't explain why humans feel guilty about murder and all the other sinful behaviors that we engage in. Only the Bible presents a real, credible explanation for why sin exists. Our real first parents committed a real first sin, leading to a real fall and a real need for a savior. Happily, the Bible also presents us with a real cure for our sin problem—faith in Christ, whose death on a cross paid the price for our sin and whose resurrection sealed His promise to us of eternal life.

As we do the work of science, we must view science and the question of origins through the Bible's story. The theist and the atheist, the creationist and the evolutionist are all pressed to faith. A Christian's faith leads one to interpret the evidence in the natural world in ways that submit to the established authority for truth, God's Word. "Now faith is the substance of things hoped for, the evidence of things not seen.[…] Through faith we understand that the worlds were framed by the word of God, so that things which are seen were not made of things which do appear" (Heb. 11:1, 3).

case study

THE EVOLUTIONARY ROOTS OF PLANNED PARENTHOOD

In Section 10.3 we learned of the link between Margaret Sanger, founder of Planned Parenthood, and the eugenics movement. Like many eugenicists of her time, Sanger believed that different races were the product of evolution and that white people were the most highly evolved humans. She wrote prolifically, leaving us a detailed view of her opinions about races and people groups that she considered inferior. For instance, Sanger believed that Aboriginal Australians were only slightly more evolved than chimpanzees and less evolved than other races. She advocated for state-enforced sterilization programs for those she deemed "less-capable." Sanger founded Planned Parenthood for the purpose of providing birth control to improve the human race by preventing the birth of "defectives." She believed that such efforts would eliminate disease, poverty, and crime. Sanger's eugenic beliefs were so radical that she even opposed acts of charity to those she thought unfit. Her first abortion clinics were opened in minority and immigrant neighborhoods, places that Sanger believed were affected by "over breeding." Today Sanger is mostly remembered for her work in making birth control widely available, but her admirers tend to gloss over her racist ideas.

10.4 SECTION REVIEW

1. State what was created or formed on each day of the Creation week.

2. How does the fossil record support the Bible's Flood narrative?

3. Do creationists believe that organisms can change over time? Explain.

4. Compare the role of mutations in the modern synthesis and creation models.

5. What sets humans apart from the rest of the created order?

Use the case study above to answer Questions 6–7.

6. In what ways were Margaret Sanger's beliefs consistent with an evolutionary worldview?

7. What elements of the creation model contradict Sanger's beliefs about race and eugenics?

10 CHAPTER REVIEW
Chapter Summary

10.1 DARWIN'S THEORY OF EVOLUTION

- Darwin's theory of natural selection provided an explanation of how transformation, which was not an idea original to Darwin, might occur in living things.

- Darwin was persuaded of the concept of uniformitarianism by the geologists of his day.

- Darwin saw evidence of gradual change in organisms during his voyage aboard HMS *Beagle*, during which he made observations of the fossil record and the differences between isolated varieties of birds in the Galápagos Islands.

- Darwin expounded his ideas about descent with modification in his book *On the Origin of Species by Means of Natural Selection*, published in 1859.

Terms
biological evolution · artificial selection · uniformitarianism · comparative anatomy · homologous structures · vestigial structure · fossil · descent with modification · common ancestor · adaptation · natural selection

10.2 THE MODERN THEORY OF EVOLUTION

- The modern synthesis combines Darwinian natural selection with Mendelian genetics as the source of variation upon which natural selection can act.

- The modern synthesis also sees changes in animal behavior as a factor in natural selection.

- The evolutionary story of life on Earth is closely tied to the fossil record, which is believed to have formed over millions of years and chronicles the history of increasingly complex life forms.

- Evolutionists believe that the first life on Earth formed from biological molecules that acquired the ability to self-replicate.

- Population genetics is used to explain the process of speciation through adaptive radiation.

- Modern evolutionists believe that mankind has evolved from a line of ancestral forms called hominids.

Terms
modern synthesis · behavioral ecology · geologic timescale · index fossil · radiometric dating · relative dating · big bang · molecular evolution · molecular clock · adaptive radiation · speciation · reproductive barrier · hominid

10.3 EVALUATING MODERN EVOLUTIONARY THEORY

- Many aspects of the modern evolutionary theory face serious scientific challenges, even among scientists who identify as evolutionists.

- Significant challenges to the evolutionary story include a lack of sufficient energy and matter to explain the big bang, lack of a viable mechanism for abiogenesis, lack of transitional forms in the fossil record, and the tendency of population genetics to result in a net loss of genetic information rather than the gain required for evolution.

- Historically, evolution has been used as a pretext for racist policies such as eugenics, resulting in much human suffering.

- Some Christians have attempted to reconcile the Genesis account of Creation with evolution. Many of Christianity's foundational doctrines are undermined by these efforts.

Terms
abiogenesis · transitional form · genetic load

10.4 THE BIBLICAL ACCOUNT

- Genesis 1 and 2 are written in the language and style used for historical narrative. The clear meaning of the text is that God created the heavens and Earth, along with everything in them, in six consecutive, regular-length days.

- Creationists view the geologic column and fossil record as evidence of the Flood. Major topographical features, such as plateaus and canyons, and geologic processes, such as plate tectonics, can also be explained by the Flood model.

- Creationists understand that organisms can change, but only within the limits imposed by the biblical kind.

- The generally harmful effects of mutation are consistent with the expected effects of sin on an originally "very good" creation.

- Humans occupy a place of significance in God's creation because they alone are created in the image of God.

Term
kind

Chapter Review Questions

RECALLING FACTS

1. What did Darwin call his idea that all living things came from common ancestors and change a little with each passing generation?

2. Match the following scientists to their contribution to Darwin's theory of natural selection: Hutton, Lamarck, Lyell, Malthus.

   a. Organisms can transform over time.

   b. Earth's features are not part of the original Creation event, but the result of secondary processes.

   c. Geologic processes happened at the same rate in the past as can be observed in the present.

   d. Populations produce more offspring than can survive.

3. What did Hugo deVries contribute toward the formation of the modern synthesis?

4. According to evolutionists, where did the elements that formed the first molecules of life originate?

5. How did evolutionists in Darwin's day explain a lack of transitional forms in the fossil record? How have some modern evolutionists attempted to explain this absence?

6. Why should Genesis 1 and 2 be interpreted as real history rather than poetry or allegory?

7. Identify the concordist view associated with each of the following statements.

   a. The original creation was ruined as a consequence of Lucifer's rebellion.

   b. The days of Genesis 1 are long periods of time interspersed with multiple individual creative acts.

   c. God created the world but uses evolution as the means for creating new species.

   d. Genesis is intended to be read only as creative literature.

8. Which of the following is the correct order of created things according to Genesis 1?

   a. humans, sea creatures, land animals, plants

   b. plants, sea creatures, land animals, humans

   c. sea creatures, plants, land animals, humans

9. Which of the following would *not* be considered part of the concept of biological evolution?

   a. atoms arranging themselves into molecules in a pond on ancient Earth

   b. a population of birds producing more offspring than their environment can support

   c. a population of lizards on an isolated island developing longer tails than the same lizards on the mainland

   d. horses and zebras diverging from a common ancestor

UNDERSTANDING CONCEPTS

10. What one event do you think changed the course of Darwin's life? Explain.

11. Were Darwin's ideas about descent with modification and natural selection over long periods of time original to him? Explain.

12. How did Darwin's and deVries's ideas come together in the modern synthesis?

13. Summarize the modern synthesis in a single sentence.

14. What is one major problem with the "warm little pond" theory of evolution?

15. In what way do you think an organism is more likely to experience genetic change: through mutations or through normal reproduction? Justify your answer.

16. What do evolutionists use to establish dates for the following?

    a. rocks

    b. fossils

    c. ancestors of living organisms

17. Are evolution and natural selection the same thing? Explain.

18. Early critics of Darwinism accused its adherents of believing that humans descended from apes. Is this a valid criticism?

19. Relate the geologic timescale to the geologic column.

20. Explain why our growing understanding of cells is problematic for evolutionary biologists.

21. Which compromise view(s) of Creation would be most likely to acknowledge that Adam was a real person? Which most likely would not? Explain.

22. What are the two main differences between a straightforward interpretation of Genesis and other ideas about origins?

23. A friend states that creationists don't believe in natural selection. How would you respond?

24. The cichlids from Lake Malawi (see page 212) can easily hybridize and produce fertile offspring. How would creationists explain this?

25. Compare the early Earth in the evolutionary and Creation narratives.

26. The fossil record is usually sorted from simple organisms at the bottom to complex organisms at the top. What are the two different ways to interpret this arrangement of fossils?

27. Compare how evolutionists and creationists think that Earth has changed during its history.

28. Are there points upon which evolutionists and creationists can agree regarding how living things change? Explain.

29. A crocodile is born with a mutation that causes albinism. This condition makes it harder for the crocodile to hide and catch prey. It is eventually captured and housed at an aquarium, and it mates to produce several generations, including more albino crocs. Is this an example of evolution? Justify your answer.

30. Describe a scenario in which the interactions between individuals in a population might affect natural selection.

31. Since the fossil record is used to support evolution, should creationists oppose the search for new fossils? Defend your answer.

32. Your friend says that the vast number of species that we see in the world today are a testament to God's creative power. Is your friend correct? Justify your answer.

33. Should Christians work together with scientists in the intelligent design movement? Justify your answer.

34. In your own words, explain why the biblical Creation-Fall model alone can provide a satisfactory explanation for the presence of sin, sickness, and death in the world.

35. Are humans special? Justify your answer by comparing the place accorded to modern humans in the natural order by the evolutionary and creationist models.

case study

EXPERIMENT IN EVOLUTION

One of the major problems with origins theories is that they are hard to test. Evolutionary biologist Richard Lenski has been growing *E. coli* bacteria since 1988. He started with 12 populations. *E. coli* reproduce rapidly—a new generation is produced every 20 minutes. This allows many generations to be observed in a relatively short period of time. By 2018 Lenski's bacteria were approaching 70,000 generations. The graph on the right shows how Lenski's bacteria have changed.

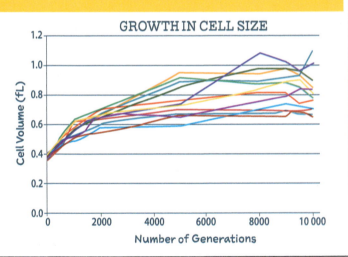

Escherichia coli

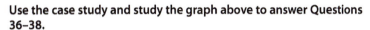

Use the case study and study the graph above to answer Questions 36–38.

36. How have Lenski's bacterial populations changed?

37. How is this diversity possible?

38. Do the results of this study provide better support for evolution or creation? Justify your answer.

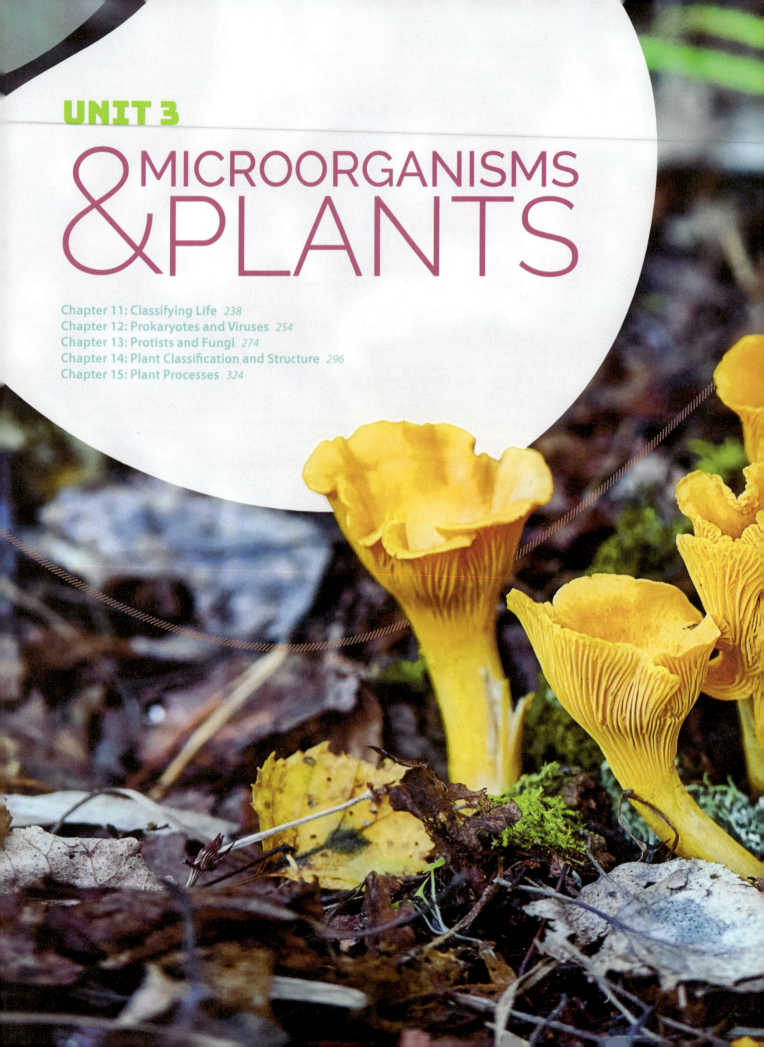

UNIT 3

& MICROORGANISMS & PLANTS

11 CLASSIFYING LIFE

11.1 TAXONOMY

What's in a Name?

From toads to hippos, alpacas, and zebras, someone had to study and give every creature a name. But why do they need a name? Without a way to identify and organize living things, managing God's creation would be almost impossible.

For people to obey God's command to manage the earth, they first need to figure out what they're managing. One of God's first commands to Adam was to name the beasts of the fields, livestock, and birds, allowing him to inspect the creatures under his care (Gen. 2:19–20). He may have studied the animals and compared them with each other, noting that a seagull acts and looks different than a robin does, and so gave them different names. He may have realized that even though seagulls and robins are two different birds, they can also be grouped together to show that they share many characteristics. The science of classifying organisms, called **taxonomy**, was as important long ago as it is today. We are still driven to name things because it helps us use and manage creation better.

The Hierarchy of Organisms

You're out in the middle of nowhere in Honduras, tagging along with a group of botanists who are looking for new species of *Aegiphila*, plants that are related to magnolias. One of the biologists lets you in on a little secret: a local in a nearby town had shown them a few *Aegiphila* flowers, and they looked like no flowers the team had seen before. You then hear a shout from a botanist farther up the trail. Everyone tramps through the brush to where she is, and you see it—a gigantic flowering plant. It looks a lot like the plants that you've seen as you hiked up the trail, but according to one of the biologists, those plants are called *Aegiphila pernambucensis*. The plant in front of you, with its towering limbs and odd flower arrangement, is dubbed *Aegiphila monstrosa*.

If another group of botanists had discovered *A. monstrosa*, they might have given it a different name. And as more information is gathered about this plant, its name may change. That's the nature of taxonomy—the names of organisms and the way they are classified change all the time! It's a model that people have created and can change to make it as useful as possible.

Why and how do scientists classify living things?

Questions

Why is classification useful?

How are organisms named and organized?

Terms

taxonomy
taxon
domain
kingdom
phylum
class
order
family
genus
species
binomial nomenclature
scientific name

▲ *Aegiphila pernambucensis* (left) and *Aegiphila monstrosa* (right) are considered two different species because people chose to classify them that way on the basis of the structure of their flowers.

TAXONOMY EXAMPLES

TAXON	Dragon Tree	Reef Octopus
DOMAIN	Eukarya	Eukarya
KINGDOM	Plantae	Animalia
PHYLUM	Tracheophyta	Mollusca
CLASS	Liliopsida	Cephalopoda
ORDER	Asparagales	Octopoda
FAMILY	Asparagaceae	Octopodidae
GENUS	*Dracaena*	*Octopus*
SPECIES	*draco*	*cyanea*

Taxonomy is in a state of flux because scientists are attempting to classify organisms on the basis of their genetic information rather than their phenotypes. The most common taxonomic system contains eight classification groups called **taxa** (s. taxon) which scientists use to classify organisms.

Actually, we think like a taxonomist every day. For example, if we were to go to the grocery store to find a specific item, such as King Arthur Baking Company bread flour, we would need to work through a hierarchy of information. First, we would need to find which aisle flour is on, then which shelf, then which kind of flour, and then which brand. We would actually be working our way through the taxonomy of the grocery store. If the items in a store weren't organized, we would never be able to find anything. The table on the left shows how a dragon tree and a reef octopus fit into this hierarchy of taxa.

DOMAINS OF LIFE

	ARCHAEA	BACTERIA	EUKARYA
Characteristics	prokaryotic cells that don't contain peptidoglycan in cell walls; contain isoprene in cell membranes, which makes them heat-resistant	prokaryotic cells that contain peptidoglycan in cell walls, which protects them and triggers other organisms' immune systems	eukaryotic cells
Kingdom	**Archaea**	**Bacteria**	**Chromista**
Defining Characteristics	• live in harsh conditions, such as extreme temperatures and acidic or salty environments • unicellular, may form colonies • heterotrophic or autotrophic	• break down materials in the soil for plants; can either cause or prevent disease in other organisms • unicellular, may form colonies • heterotrophic or autotrophic	• plantlike and fungus-like; have motile gametes (i.e., gametes that move); have cell walls made of cellulose • unicellular, colonial, or multicellular (no true organs) • photosynthetic (most) • autotrophic, a few heterotrophic
Examples	thermophiles, acidophiles, halophiles, methanogens	*E. coli*, bacteria found in yogurt, *Salmonella*	kelp and other algae, water molds, diatoms

DOMAINS AND KINGDOMS

In the fourth century BC Aristotle created the first taxonomy system, dividing living things into plants and animals. This two-kingdom system remained in place until the 1860s, when Ernst Haeckel, a German biologist and illustrator, proposed a third kingdom to include all the single-celled protists that scientists had discovered with microscopes. In the 1930s American biologist Herbert F. Copeland removed bacteria from the protist kingdom and created a fourth kingdom for them. In 1969, while the world was watching the first man walk on the moon, the American ecologist Robert Whittaker created a fifth kingdom to separate fungi from plants.

During the 1970s scientists began looking at the genetics of organisms to see which kingdom an organism belonged to. That's when Carl Woese, an American microbiologist, made an interesting discovery—organisms that looked like bacteria but had different sequences of RNA in their ribosomes. He named them *archaea* (s. archaeon), and his work led not only to a new kingdom but to a new taxon higher than a kingdom—the domain—to give genetic differences a place in the hierarchy.

Carl Woese

EUKARYA			
eukaryotic cells	eukaryotic cells	eukaryotic cells	eukaryotic cells
Protozoa	**Fungi**	**Plantae**	**Animalia**
• animal-like; motile as adults; covered by a membrane called a *pellicle* • unicellular or colonial • non-photosynthetic • heterotrophic, a few autotrophic	• cell walls made of chitin • mostly multicellular, but can be unicellular; may form colonies • heterotrophic	• cell walls made of cellulose • multicellular, with true tissues and organs • autotrophic	• no cell walls • multicellular, with true tissues and organs • heterotrophic
euglenas, amoebas, *Plasmodium* (causes malaria)	mushrooms, mildew, truffles, yeast	trees, ferns, flowering plants, mosses	sand dollars, sponges, worms, fish, reptiles, birds, mammals

The model continued to change to fit what biologists observed in the world. In the 1990s British biologist Tom Cavalier-Smith suggested that the protist kingdom, which contained both autotrophic and heterotrophic organisms, be split in two to create a seven-kingdom system.

Because of the changing nature of scientific models, there may be no system of taxonomy that is perfect. The goal is simply to develop a workable model with categories that fit the creatures that we observe in the world so that we may better understand and manage them.

SPECIES

To understand and organize creation, organisms are classified into different **species**. The most widely used definition for species is a group of organisms with similar characteristics that are capable of interbreeding to produce fertile offspring. But organizing species according to this definition can actually be quite challenging since organisms do not always neatly fit within these guidelines.

In the image on the left it's easy to see why the royal starfish (*Astropecten articulatus*) is a separate species from the two-spined sea star (*Astropecten duplicatus*) —they share many characteristics, but they look very different from each other. But the two-spined sea star and the sand sea star (*Astropecten irregularis*, not shown) are also two different species, even though they look almost exactly alike. They live in two different places—one in the Caribbean Sea and the other in the Mediterranean Sea. Also, each arm on a two-spined sea star has a pair of spines; sand sea stars don't have these. And that's enough to make these two different species according to the definition just given.

The ability to interbreed is another distinction where the rules don't always apply. Some species, such as bacteria, are asexual, thus incapable of interbreeding. There are also many species that can interbreed with other species, such as macaws. The Harlequin macaw is a product of two species, the green-winged macaw and the blue-and-gold macaw. The Catalina macaw is also a hybrid produced by a blue-and-gold macaw and a scarlet macaw. Hybrid Harlequins and Catalinas can produce Maui sunrise macaws.

The royal starfish (top) and two-spined sea star (bottom)—two species in the same genus

Many macaws have genes from different species of macaws (genus *Ara*).

Ara chloroptera (green-winged macaw)

Ara ararauna (blue-and-gold macaw)

Ara ararauna × *Ara chloroptera* (Harlequin macaw)

case study

GOPHER TORTOISE BURROWS

"I'm sorry, sir, but we can't start your project yet. You have two gopher tortoise burrows near your building site." After much research your dad learns that unless he relocates the burrows, he's going to have to wait to build the house. And it seems like relocating these current land residents may not be as easy as it sounds!

Since you've lived in Florida for quite a while, you're used to hearing about gopher tortoises. You know they are protected, and you also recall hearing that gopher tortoises generally have multiple burrows. Some male gopher tortoises are known to have as many as thirty-five burrows! If there are more burrows, that can mean more time and money spent to relocate them to another piece of land. After researching what is needed to relocate these gopher tortoises, you learn that a permit is needed only if the burrow is currently occupied by a gopher tortoise. You decide to look a little more into the signs of occupied burrows and realize that there may be a way to resolve this situation.

Do an internet search on gopher tortoises. Then think through these questions.

1. Signs of unoccupied burrows are seen in the shape and condition of the burrow. One of the burrows that you see is gently sloped down and appears to have fresh sand around the entrance. Is it occupied by a gopher tortoise? Explain your answer.

2. You check the condition of the other burrow on the land and believe that it is unoccupied by a gopher tortoise. What else should you consider?

3. After walking the land, you realize that there are more burrows. They are farther away from the building site, so they won't require relocation. In what ways can you show good stewardship of the burrows that you found?

Clearly, the current definition of species cannot reliably distinguish between living things because there are many exceptions to the definition. But scientists have found a new way to help distinguish species—by looking at their patterns of DNA. This advancement has given researchers the ability to differentiate between species and has aided in the conservation efforts of some less-known organisms.

Naming Organisms

What's the difference between plants that produce the coriander seeds found in curry and those that produce the cilantro leaves found in salsa? Nothing—they come from the same plant! This plant, popular for its strong, citrusy flavor, also goes by the names Chinese parsley, Mexican parsley, dhania, and kothmir. So how does anyone know what plant they're eating?

In 1735 Swedish botanist Carl Linnaeus came up with a solution to that problem. He invented a system, called **binomial nomenclature**, in which an organism is assigned a genus name and a species name. Linnaeus chose Latin, the language of ancient Italy, for his system because Latin is a dead language. Living languages, like the language that we speak today, change over time, and Linnaeus was looking for a language that was more stable. Latin is also a very descriptive language used by scholars. Linnaeus wanted a language that would never change, that could describe every organism discovered, and that was already familiar to people.

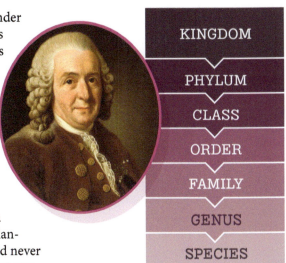

KINGDOM

PHYLUM

CLASS

ORDER

FAMILY

GENUS

SPECIES

The genus-species name of an organism is its **scientific name**. A scientific name is always printed in italics (or underlined if it is hand-written). The first part of the name is the genus name, which is always capitalized. The second part of the name, the species name, is lower-cased. For example, the scientific name of the plant that produces coriander seeds and cilantro leaves is *Coriandrum sativum*. *Coriandrum* is the plant's genus and is what coriander was called in ancient Italy, and *sativum* is its species and means "cultivated" in Latin. If there's a lot of diversity within a species, that species may be divided into *subspecies*. All dogs are part of the same species, but there are hundreds of subspecies—Siberian huskies, bulldogs, corgis, cocker spaniels, and more.

Canis familiaris
(domestic dog)

Canis lupus
(gray wolf)

Canis simensis
(Ethiopian wolf)

11.1 SECTION REVIEW

1. How does taxonomy help people fulfill the Creation Mandate?

2. *Boloceroides daphneae* was a type of sea anemone discovered in 2006. But in 2014 it was reclassified into a new family separate from sea anemones and renamed *Relicanthus daphneae*. What does this tell you about the nature of taxonomy?

3. Compare the characteristics of the three domains of life.

4. Create a table outlining the different domains and kingdoms. Include the name of each domain and kingdom, one defining characteristic that sets each kingdom apart from the others, and one example of an organism in that kingdom.

5. While testing the Green Dragon Springs in Yellowstone National Park, a scientist came across a new type of microbe that is able to survive the water's near-boiling, acidic conditions. In which domain and kingdom does this organism likely belong?

6. Which kingdom was split into two kingdoms? Why?

7. Why did Linnaeus choose Latin to create scientific names instead of his native Swedish language?

8. What is the correct way to write the scientific name for a moose?

   a. *alces Americanus*

   b. *Alces Americanus*

   c. *Alces americanus*

   d. None of these is correct.

Evolutionists look for similarities between organisms that can lead to classifications that seem forced rather than useful. Relating reptiles with frills like this frilled lizard to birds with feathers could fall into this category.

11.2 UNITY AND DIVERSITY

Changing Relationships

The reason that organisms are classified together or separately is usually obvious. Birds are grouped together because they're warm-blooded, have feathers, and lay eggs. Reptiles are grouped separately from birds because though they also lay eggs, they're cold-blooded and have scaly skin. So whenever you see a parrot or an eagle, you immediately recognize it as a bird and not a reptile.

But what if someone told you that when you think of reptiles, you should also think *bird*? This type of thought process governs the way some scientists try to make evolutionary connections between organisms. The field of **systematics** is the science of classification. Evolutionists engaging in systematics try to piece together the evolutionary relationships between both living and extinct organisms. The idea that birds evolved from reptiles came out of systematics.

Systematics has two sides: **phylogeny**, which deals with constructing the evolutionary history of organisms, and **cladistics**, which is the reclassifying of organisms according to their evolutionary history. For centuries scientists compared the anatomy and embryos of organisms to determine their evolutionary relationships (see Chapter 10, pages 202–3). Currently, genetics and biochemistry (the study of the chemical processes and products of organisms) are the foundation of evolutionary classification.

?

How is classification used to support evolutionary theory?

Questions

What classification model is used today?

How do scientists use classification to explain evolutionary relationships?

How can we describe the variety of life from a biblical worldview?

Terms
systematics
phylogeny
cladistics
phylogenetic tree
clade
divergence
derived trait
convergence

By studying the panda's genes, scientists were able to classify giant pandas as a type of bear.

Though supporting evolution is the main goal of systematics, scientists have made useful discoveries with genetics as their guide to classification. For example, one genetic difference discovered between domains Archaea and Bacteria is that they react differently to antibiotics, an important fact for physicians as they treat patients with archaeal infections. Also, genetics can help us classify organisms that have been difficult to classify in the past, such as pandas. Scientists have always debated whether pandas should be classified in the bear family or the raccoon family. But a study of the panda genome revealed that they are a type of bear.

Sadly, most scientists today use classification to try to establish evolutionary relationships—a departure from Linnaeus's original intention to organize living things in a useful way. So is it actually useful to classify birds with reptiles, or is that just a way for scientists to interpret data in a way that supports their worldview?

Classification and a Secular Worldview

Traditionally, when a biologist discovered a new creature, he compared it to other similar creatures, examining the creature's habits and structure to figure out which taxon it belonged in. Charles Darwin, in his book *On the Origin of Species*, described another way to classify living things. He felt that it wasn't enough to sort and group living things into taxa; he wanted to group them according to their evolutionary history. To do that, he drew a **phylogenetic tree**, often called a *tree of life*, which shows the supposed evolutionary relationships between groups of organisms. These trees map out relationships not only between living organisms but also between living and extinct organisms, creating a timeline of how a simple organism might have evolved into a complex one.

Charles Darwin and his tree of life

Biologists continue to use phylogenetic trees and have created thousands of them, each one unique because biologists often disagree on what they think an organism's evolutionary ancestors are. Each tree is broken up into **clades**—branches of a phylogenetic tree that include all the descendants of an evolutionary ancestor. Look at the clade below, which shows how some scientists believe certain mammals are related.

Instead of classifying this *Mesohippus* fossil as an extinct type of horse, evolutionists classify it as the horse's ancestor—the creature that evolved into a horse.

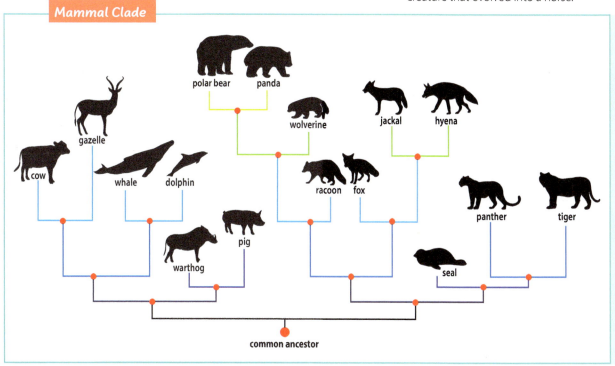

The dots on the clade above represent **divergence**, which is the process of two similar species becoming more and more different. Divergence is usually marked by the appearance of a **derived trait**, which is a trait that arises within a clade and is shared by all future members of that clade. The example clade suggests that the ability to live in water is a derived trait, leading to brand new types of animals, like whales and dolphins.

But why are whales and dolphins so far away from seals in this clade when both groups live in water? Evolutionists explain this kind of relationship as **convergence**, the process of two dissimilar species evolving the same derived trait. According to evolutionary theory, convergence is the reason that unrelated species like bats, birds, and insects can all fly—they evolved the ability to fly independently.

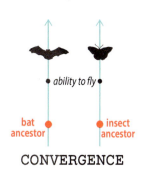

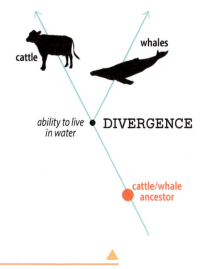

According to evolutionary theory, derived traits can lead to divergence in related organisms, or they can evolve independently in unrelated organisms through convergence.

GENETICS IN CLASSIFICATION

Classifying an organism according to its biochemistry can lead to strange connections. In 2013 researchers studying the RNA of great white sharks found that sharks produce many proteins that are more like human proteins than those of bony fish. Of course, this seems odd to us because sharks and bony fish are outwardly more similar to each other than either is to humans, regardless of what their genetics may say. Yet mind-bending genetic discoveries like this are increasingly commonplace today as more and more genomes are sequenced.

Comparing the DNA of different organisms can often lead to restructuring long-held classification schemes. The order Crocodilia, which includes alligators and crocodiles, is an example of this. Prior to 2003 only three species of African crocodiles were recognized: the Nile crocodile, the West African slender-snouted crocodile, and the African dwarf crocodile. Since then, according to genetic studies, Nile crocodiles and West African slender-snouted crocodiles were each split into two separate species. And more recent studies have suggested that African dwarf crocodiles should be separated into three distinct species as well. If the findings of those studies are accepted, the number of African crocodile species will have increased from three to seven in just a couple of decades. Such taxonomic shuffling isn't done because new species were discovered in previously unexplored areas. The frequent splitting, regrouping, and renaming is done in an attempt to show presumed evolutionary relationships that are based on genetics.

Classification and Biblical Worldview

Taxonomy and classification are manmade models designed to help us organize life on Earth. But every model is influenced by a worldview. Though genetics is being used to support a naturalistic, evolutionary view of how life came to exist, it can also be used to support a biblical perspective.

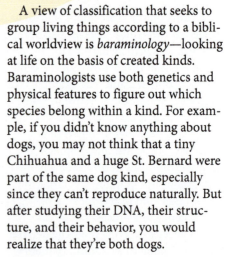

A view of classification that seeks to group living things according to a biblical worldview is *baraminology*—looking at life on the basis of created kinds. Baraminologists use both genetics and physical features to figure out which species belong within a kind. For example, if you didn't know anything about dogs, you may not think that a tiny Chihuahua and a huge St. Bernard were part of the same dog kind, especially since they can't reproduce naturally. But after studying their DNA, their structure, and their behavior, you would realize that they're both dogs.

The potential for great genetic diversity isn't evidence for evolution; it's a sign that God included a lot of information in the cells of each of His creatures. That information continues to reveal itself in the form of new species and varieties. But the kinds that God created in the beginning remain the same—the dog kind didn't evolve from another kind. Baraminologists represent this idea with an *orchard of life*, which shows the special creation of each kind and the species that arise within each kind. The diagram below shows how the mammals from an evolutionary clade might be grouped according to their kind.

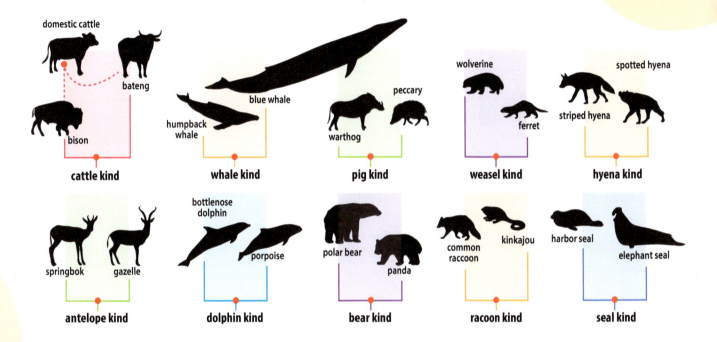

From a straightforward reading of Genesis 1, we can conclude that each group of common ancestors—the first of each kind—was created by God just a few thousand years ago. God records in Genesis 1:11–12 that He created several kinds of plants, specifying that they would continue to reproduce after their kind. In Genesis 1:21, 24–25 the Bible says the same about the different kinds of animals. That means that the first pigs that God created produced a variety of pigs in a short amount of time; pigs and other very dissimilar mammals—like whales—didn't evolve from a common ancestor over millions of years as evolutionists suggest.

According to a biblical worldview, divergence occurs only within a kind. Consider the example of biblical kinds shown on the right. The original cats that were created diverged into a variety of cat species. God not only created each kind with an amazing capacity for genetic diversity; He also made speciation a speedy process. We see today that new species within a kind can appear very quickly. But we don't see evidence of a species, such as a butterfly, diverging into a completely different kind of creature, such as a bat.

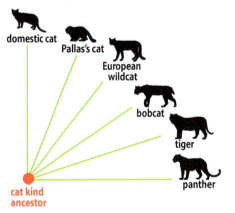

What about convergence? The same trait appearing in very different creatures is evidence of design—a Designer chose to give certain traits, such as the ability to fly, to a wide array of organisms so that they could thrive in His world.

Because someone with a biblical worldview knows where life on Earth came from, he can view organisms in a straightforward, organized way. Using the taxonomy that scientists around the world use, he can have a biblical understanding of the many varieties of organisms observed in nature. Someone with a naturalistic worldview believes that the origin of life is an evolutionary mystery, so he uses classification to try to figure it out. Using taxonomy in this way distracts people from one of the tasks that God gave to us all—naming, cataloging, and wisely managing His creatures.

◀ In just a few decades scientists have observed the Ecuadorian butterfly splitting into two species—one white and one yellow.

MINI LAB

Inquiring into Baraminology

Baraminology is a system of classification used to relate different forms of life from a biblical worldview. Unlike cladistics, baraminology takes into account different kinds of species and subspecies and classifies them into different groups. In this inquiry mini lab activity we will use baraminology to create a way to organize organisms according to their kind.

Is there another way to classify life?

Materials
none

PROCEDURE

A Along with a lab partner, make a list of thirty different animals (including insects) and plants.

B From this list write procedures for classifying each organism. Decide how you plan to organize the different organisms.

C Following the procedures, classify each animal according to its kind. Make any necessary changes to your procedures.

D Share your list and procedures with one other group. Each group will follow the procedures and classification steps to organize another group's organisms.

E Share your classification results with the other group. Analyze similarities and differences between the two different classifications.

QUESTIONS

1. When organizing the organisms according to their kinds, what characteristics were you looking for?

2. What was the most difficult aspect of organizing the different species of organisms?

3. Did any of the organisms in your list seem to fit into more than one kind? If so, how did you determine what kind they should belong to?

4. Identify any limitations present in your procedures or method of classifying.

5. Identify other characteristics that should have been considered when organizing the organisms.

GOING FURTHER

6. Do an internet search on baraminology and illustrate the various ways that organisms can be classified.

case study

ANALYZING A CLADOGRAM

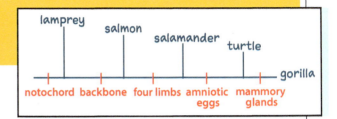

Cladograms are often used to show evolutionary relationships between different living and extinct groups of organisms. Can a cladogram accurately connect different species to a common ancestor? While cladograms are helpful in comparing different types of organisms, they struggle to take into account complex structures found within organisms. Let's examine a modern cladogram to see whether it's an effective model for explaining similarities between species.

Questions to Consider

1. Examine the cladogram above right. What relationships do you observe between the different organisms?

2. What differences do you observe between different organisms?

3. The goal in studying cladistics is to demonstrate evolution by analyzing possible links between derived traits (traits from a recent ancestor). After analyzing the cladogram, in what order does it appear that the derived traits evolved?

4. Evaluate this evolutionary model. Identify any weaknesses in showing evolutionary lineages.

5. Perform a quick internet search on cladograms or cladistics to view several other examples of cladograms. Describe the limitations of these models.

11.2 SECTION REVIEW

1. How is systematics used to classify organisms?

2. What two areas of study are the foundation for today's evolutionary classification?

3. How is traditional classification different from today's classification? How are they the same?

4. What is the difference between divergence and convergence?

5. *Pliohippus*, an extinct genus that evolutionists say lived fifteen million years ago, was once thought to be the common ancestor of horses, zebras, and donkeys. But after studying the fossils of several extinct genera, researchers now say that horses, zebras, and donkeys arose from a different common ancestor, *Dinohippus*, which lived seven million years ago. How did worldview play a part in the researchers' decision after they studied the fossils?

Use the clade diagram on the right to answer Questions 6–8.

6. According to a naturalistic worldview, which animal is the evolutionary ancestor to all the other animals? Which three animals diverged and didn't continue to evolve into modern-day elephants?

7. According to a biblical worldview, how should the three divergent animals be classified?

8. According to a straightforward interpretation of Genesis 1, how long did it take for the first created kind of elephant to produce the variety of elephants that we observe today? How does this timeline compare to an evolutionary one?

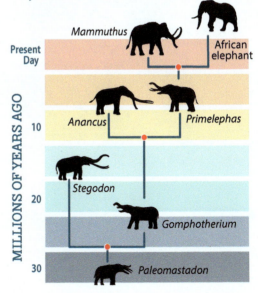

11.1 TAXONOMY

- Classification is used to organize, identify, and compare organisms.

- Scientists classify organisms into eight classification levels called taxa.

- The three domains of life are Archaea, Bacteria, and Eukarya.

- The seven kingdoms include Archaea, Bacteria, Chromista, Protozoa, Fungi, Plantae, and Animalia.

- Carl Linnaeus invented binomial nomenclature to assign genus and species names to organisms.

Terms
taxonomy • taxon • domain • kingdom • phylum • class • order • family • genus • species • binomial nomenclature • scientific name

11.2 UNITY AND DIVERSITY

- Evolutionists use systematics to classify organisms on the basis of evolutionary relationships and common ancestry.

- Traditionally, scientists compared the anatomy and embryos of organisms to classify organisms.

- Scientists today classify organisms on the basis of genetics and biochemistry.

- Those with a biblical worldview accept that genetic diversity stems from God's capacity to create and design many different kinds of organisms.

Terms
systematics • phylogeny • cladistics • phylogenetic tree • clade • divergence • derived trait • convergence

Chapter Review Questions

RECALLING FACTS

1. What is the purpose of classification? How is this purpose in line with fulfilling the Creation Mandate?

2. What evidence led scientists to create a third domain for the classification system?

3. Which taxon includes both plants and animals?

4. Using the Domains of Life table on pages 240–41, describe the main difference between kingdom Chromista and kingdom Plantae.

5. What are two possible definitions for the term *species*? Why do these definitions have limited usefulness?

6. Why would the English language not be a good choice for scientific names?

7. A newly discovered mollusk has the species name "harlingae" and the genus name "ittibitium." Write the scientific name for this mollusk, using the correct form.

8. Mention two ways that genetics can be helpful for classification.

9. A seal sharing an ancestor with a panda is part of a(n) _____ worldview, while a seal sharing an ancestor with other seals *only* is part of a(n) _____ worldview.

UNDERSTANDING CONCEPTS

10. Do some research to find out why *Papaver somniferum*, a species of poppy, is illegal to grow in Canada. Why would it be important to be able to identify this plant?

11. How does classification relate to species diversity?

12. Look at the Catalogue of Life online to find each item: the scientific name for the Ohio buckeye, the common name for *Macropus rufus*, the family that jewel beetles belong to, and the common name for organisms in the genus *Daucus*.

13. A paleontologist finds a fossil that has the body of a bird but the head of a lizard. He decides that this creature must have evolved from dinosaurs and is the common ancestor of birds. He then gives the creature the genus name *Protoavis*, which means "first bird," and includes it in class Dinosauria to show its evolutionary relationship to dinosaurs. Which part of this scenario is phylogeny? Which part is cladistics?

14. Why is it difficult to classify organisms using their DNA?

15. Consider the phylogenetic tree on page 247. Which part of the tree do you think creationists could agree with? Which part could they not agree with? Explain your answers.

16. Frogs and mice look nothing alike, are genetically very different, and live in completely different environments. Yet evolutionists believe that they both descended from a common ancestor. What would a creationist believe about the ancestry of frogs and mice?

CRITICAL THINKING

17. Explain how it is possible for two different biologists to construct two different phylogenic trees using the same organisms.

18. Consider the concept of convergence seen in whales and seals. Give the secular and biblical worldview explanations for this occurrence.

19. How would you respond to the statement, "Speciation is evidence for evolution"?

20. Bats, whales, cats, and humans are different species, but they have similar bone structure in the forelimbs. Evolutionists believe that this is due to divergent evolution. How would someone with a naturalistic worldview and someone with a biblical worldview describe the origin of these organisms and divide the species?

21. If you were told that long ago cattle and whales were similar and then slowly became more and more different, how would you respond?

12 PROKARYOTES & VIRUSES &

12.1 Prokaryotes | 12.2 Viruses

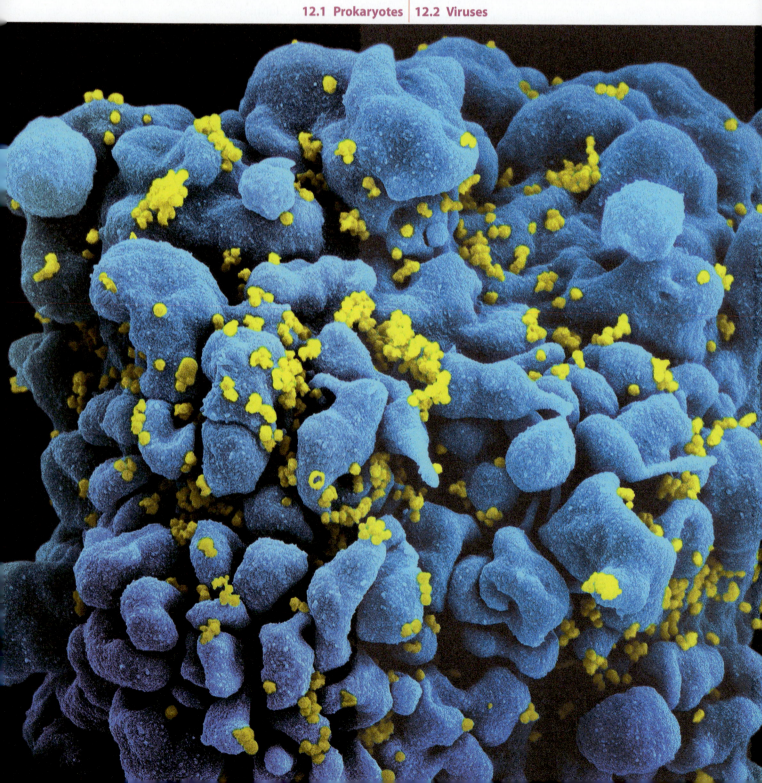

12.1 PROKARYOTES

Domain Archaea

The world under the microscope greatly affects the rest of our world. Diseases like polio are caused by *microbes*, microscopic organisms. Some people that develop a poliovirus infection escape with a mere fever and headache. Others are left with useless limbs or, worse yet, paralyzed lungs that once required a machine to keep them alive. In the 1950s and '60s Americans feared polio as much as an atomic bomb.

Poliovirus is one member of the microscopic world that affects our health and immunity. Belonging to the world of the prokaryotes, these tiny viruses, as well as organisms in the domains Archaea and Bacteria, have some interesting characteristics. **Archaea** are prokaryotes that can act as decomposers in extreme environments. They also share a lot of characteristics with eukaryotes, such as their methods of replicating RNA and producing energy, as well as the complexity of their polymerase molecules. They were originally classified with bacteria until scientists discovered several key differences between bacteria and archaea.

- Many archaea are found in extreme environments, such as deep-sea hydrothermal vents, acidic springs, volcanoes, and salt mines. Except for cyanobacteria, many bacteria are not designed to handle extreme environments.

- The cell walls of bacteria contain the protein **peptidoglycan**, which protects them and triggers other organisms' immune systems. The cell walls of archaea do not have peptidoglycan. Their cell membranes contain different proteins and rubbery *isoprene*, which make them more stable and heat resistant.

- Transcription and translation cell processes in archaea are different from those in bacteria.

Are all germs bad?

Questions

How are prokaryotes classified?

What makes archaea and bacteria different?

How do bacteria grow and reproduce?

How do bacteria interact with people and the environment?

Terms

archaea
peptidoglycan
methanogen
thermophile
halophile
acidophile
bacterium
binary fission
conjugation
transformation
transduction
bacteriophage
microbiome
pathogen

ARCHAEA

Methanosarcina mazei

Pyrococcus furiosus

Methanogens are the largest group of archaea and, like the *Methanosarcina mazei* above, are found in anoxic, or very low-oxygen, environments like swamps and the intestines of animals and humans. They produce the gas methane.

Thermophiles live in hot environments. This *Pyrococcus furiosus* lives in deep sea hydro-thermal vents that reach temperatures well above water's boiling point. They die if the sur-rounding temperature falls too far below boiling.

unidentified
halophilic archaea

Sulfolobus sp.

Halophiles are found in water that is saltier than the ocean, such as the Great Salt Lake in Utah or the Dead Sea between Israel and Jordan. The halophile *Halobacterium salinarum* is often found in salt-cured ham.

Acidophiles prosper in acidic conditions and were the first archaea discovered. *Sulfolobus acidocaldarius* thrives in the hot, acidic springs of Yellowstone National Park.

Domain Bacteria

Unless you plan to visit extreme environments, you're more likely to encounter bacteria than archaea. **Bacteria** (s. bacterium) are also prokaryotes and often act as decomposers like archaea, breaking down dead materials, such as fallen trees, and returning them to the environment. They also help break down old cells, proteins, and lipids on your skin and in your intestines.

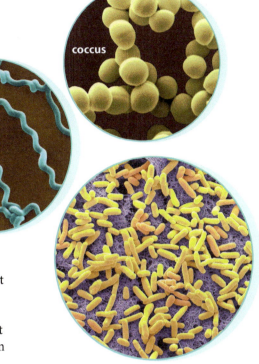

coccus

spirillum

bacillus

SHAPE AND STRUCTURE OF BACTERIA

Whatever their shape, bacteria have peptidoglycan in their cell walls. In a technique called *Gram staining*, scientists use a violet dye to classify bacteria into two groups according to the amount of peptidoglycan in their cell walls. Gram-positive organisms have greater amounts of peptidoglycan and will absorb so much dye that they turn purple. Gram-negative organisms have less peptidoglycan and will absorb only a little dye and turn pink. Certain antibiotics such as *penicillin* are medications that either kill bacteria or slow down their growth. In both cases, antibiotics weaken the bacteria enough for a person's immune system to clear up the infection. They affect gram-positive organisms by stopping peptidoglycan formation.

Many bacteria also have *capsules* outside their cell walls, which protect bacteria from drying out. If you've ever had a bacterial respiratory infection like pneumonia, then you may have experienced the protective power of a bacterium's capsule. The capsule makes the bacterial cells slippery and difficult for immune cells to engulf. Antibiotics help the body eliminate bacteria by disrupting the bacteria's metabolism, interfering with its ability to build cell walls. Antibiotics also keep bacterial cells from synthesizing the proteins and enzymes that they need to function. This allows the immune system to jump in and form specific antibodies to remove the bacteria.

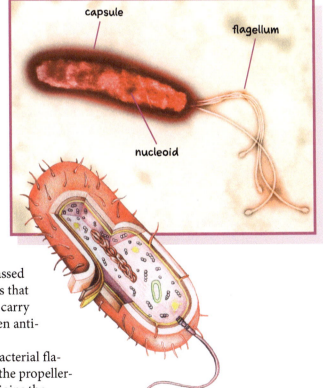

capsule

flagellum

nucleoid

Most bacteria have two forms of DNA, both in circular shapes. The *nucleoid*, which is the DNA region in most bacteria, is not bound by a membrane and is made of one circular chromosome. Bacteria often have smaller circular DNA molecules called *plasmids* (see page 178) in addition to their chromosomal DNA. Plasmids are passed between bacteria during reproduction and contain genes that help bacteria during stressful situations. Some plasmids carry antibiotic-resistant genes that allow them to survive when antibiotics are present.

Many bacteria are able to move using a flagellum. A bacterial flagellum is typically made of a protein motor, a filament (the propeller-like portion of the flagellum), and a hook (the part that joins the motor to the filament).

Reproduction & Genetic Transfer *in* BACTERIA

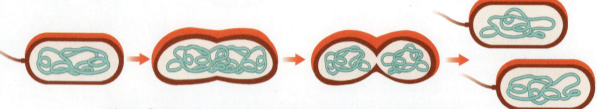

Binary fission, the most common form of reproduction in bacteria, is asexual. The nucleoid replicates and then the cell divides, producing two identical bacterial cells.

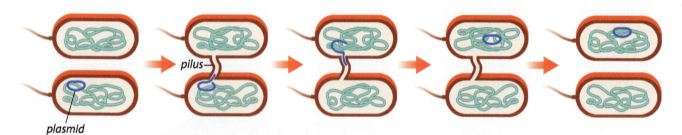

pilus

plasmid

In **conjugation**, a plasmid replicates and is transferred from one bacterium to another by use of a *pilus*. The bacteria don't exchange genetic information—the transfer is one-way.

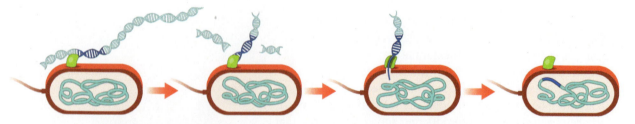

In **transformation**, a bacterium takes in a piece of bacterial DNA that is floating free in the environment. The bacterium then expresses the traits that are contained in the new piece of DNA if it is incorporated into the bacterium's DNA.

For example, a bacterium that couldn't form a capsule can do so if it takes in a stray capsule-forming gene. Transformation is known to occur in only a few types of bacteria.

Finally, **transduction** is the transfer of genetic information from one bacterium to another by way of a **bacteriophage** (a virus that infects bacteria).

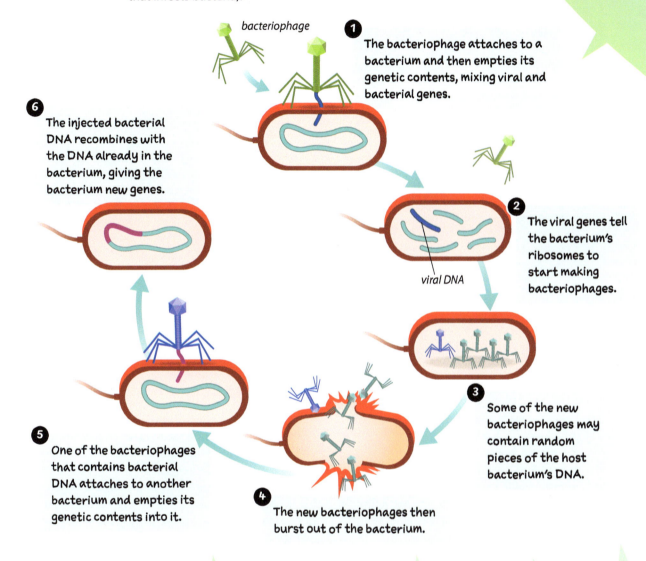

bacteriophage

1 The bacteriophage attaches to a bacterium and then empties its genetic contents, mixing viral and bacterial genes.

6 The injected bacterial DNA recombines with the DNA already in the bacterium, giving the bacterium new genes.

viral DNA

2 The viral genes tell the bacterium's ribosomes to start making bacteriophages.

3 Some of the new bacteriophages may contain random pieces of the host bacterium's DNA.

5 One of the bacteriophages that contains bacterial DNA attaches to another bacterium and empties its genetic contents into it.

4 The new bacteriophages then burst out of the bacterium.

A little bacterium can make a significant impact only when it experiences environmental factors that are favorable to its growth and reproduction. All bacteria require water to grow, though they can survive for short periods without it. Certain bacteria operate best within a specific range of temperatures, usually about 30 °C. For example, the bacteria in the gut do best around normal body temperature (37 °C). Though many bacteria can grow in nearly neutral pH conditions, raising or lowering the pH of their environment can kill them. Bacteria also need a food source if they are heterotrophic. Some bacteria grow best without oxygen, some grow best with it, and some can operate in either condition. Others are very picky, operating under a specific set of environmental conditions.

Bacteria, when combined with other microorganisms and water, can turn a pile of leaves into nutrient-rich compost that will improve the soil for plants.

How Bacteria Function in the Environment

Bacteria are some of the most versatile organisms on the earth. They're in our food, water, air, soil, and even our bodies. Bacteria break down wastes, help plants grow, and keep our digestion balanced. Bacteria are responsible for specific links in the biogeochemical cycles for carbon, nitrogen, and sulfur. Without them the cycles would not continue. Without them there would be no life on Earth!

GOOD BACTERIA

One cutting-edge area of scientific study today is the **microbiome**—the collective DNA from the microorganisms that live on and in an organism. *Microbiota*—the bacteria of our microbiome—are heterotrophic, obtaining their energy by digesting our food. They remove toxins, prevent disease, synthesize vitamins B12 and K, and absorb many other vitamins and minerals for us.

What we eat determines which bacteria we'll have in our digestive systems. Scientists theorize that eating lots of fat and sugar throws off the balance of the gut's bacteria and creates an environment for them to produce toxins and damage proteins, leading to health problems. Eating a diet high in fiber along with non-starchy fruits and vegetables keeps the bacteria in our digestive systems on track.

Some antibiotics can destroy good microbes in the digestive system, leaving the body open to harmful microbes called *opportunistic pathogens* that might not cause disease under otherwise normal conditions. Sometimes health-care professionals recommend that people take *probiotics*—microbial cultures that they can ingest—to help prevent the growth of these pathogens and contribute to gut health. Because antibiotics kill normal microbiota, they should be taken only when the body can't fight off a bacterial infection by itself.

Other kinds of helpful bacteria are *decomposers*— bacteria that get their energy from dead and decaying matter. Even if you never raked the leaves out of your backyard, the combination of water and bacteria would break down the leaves, and the nutrients trapped in the leaves would be added back to the soil. But bacteria's usefulness doesn't stop there. Bacteria in the soil can also defend plants against certain fungal infections.

Cyanobacteria are *autotrophic*; that is, they get their energy through photosynthesis. Like decomposers, cyanobacteria help cycle materials such as nitrogen and carbon through the environment. They produce oxygen and are a source of food for many organisms.

BAD BACTERIA

Ever since Adam and Eve fell into sin and brought God's curse on the whole earth, bacteria have been hurting us. Our microbiome and immune systems can't always defend us against every **pathogen**—a substance that causes disease—that comes our way. Pathogens sometimes rely on other organisms for transportation. Organisms that are carriers of pathogens are called *vectors*. There are many more good bacteria in the world than pathogenic ones, but pathogens get more attention.

EXAMPLES OF PATHOGENIC BACTERIA

SPECIES		DISEASE	ALWAYS A PATHOGEN?
	*Streptococcus pyogenes*	scarlet fever, pharyngitis (strep throat), necrotizing fasciitis (flesh-eating infection)	yes
	Streptococcus pneumoniae	pneumonia	no; can live harmlessly in the nose and throat
	Staphylococcus aureus	boils and other skin infections, food poisoning, sinusitis, MRSA	no; can live harmlessly in the respiratory system and on the skin
	Listeria monocytogenes	food poisoning, meningitis in newborns	yes
	Rickettsia spp.	trench fever, typhus, Rocky Mountain spotted fever	no; found in many lice, ticks, or mosquitos that act as vectors of bacteria to people
	Borrelia burgdorferi	Lyme disease	yes; transmitted to humans by a vector, usually a tick
	Mycoplasma spp.	pneumonia	no; can live harmlessly in the respiratory system and urinary tract

Green film on a body of water caused by a bloom of cyanobacteria

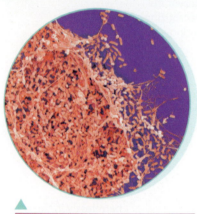

Serratia marcescens. This is what your shower curtain looks like under a microscope.

Some bacteria can act as both helpful bacteria and opportunistic pathogens, such as *Escherichia coli* (*E. coli*). They don't cause disease in their normal habitat, but they may in a different environment. Many strains of *E. coli* are helpful, producing vitamin K and fighting off pathogens in the gut. But some strains can cause severe diarrhea and dehydration. And in some cases, a strain of *E. coli* that is helpful in one person can be a pathogen in another. Even a single strain of *E. coli* that is helpful in one organ, such as the large intestine, can be a dangerous pathogen in another part of the body, such as the abdominal cavity.

Cyanobacteria can be harmful as well. Because they typically grow in bodies of water, they can quickly take over a lake or even a section of an ocean. They cut off nutrients to other organisms and produce toxins that affect the look and quality of the water. Some bacteria can simply be a nuisance, such as certain strains of *Serratia marcescens*. These airborne bacteria grow and reproduce whenever they enter a moist environment. *S. marcescens* appears as a pink smear in your shower and bathroom sink. Since it is airborne, the only way to prevent its pink colonies is to either filter the air or eliminate moisture.

12.1 SECTION REVIEW

1. Which two types of archaea thrive in hot environments?

2. How are archaea and bacteria similar? How are they different?

3. The prefix *staphylo-* means "cluster." What would *Staphylococcus* bacteria look like?

4. Draw a bacterium and label the following parts: capsule, cell membrane, cell wall, cytoplasm, flagellum, nucleoid, plasmid, and ribosome.

5. Which type of bacterial genetic replication involves asexual reproduction and is most common in bacteria?

6. How could a harmless bacterium suddenly become capable of producing a dangerous toxin?

7. What are two functions of good bacteria in the human body?

Use the table on page 261 to answer Questions 8–10.

8. Which two pathogenic bacteria are generally transmitted by a vector?

9. Folliculitis is an inflammation of the hair follicles in skin. One type occurs when skin is irritated by shaving. Which species of bacterium is responsible for this condition?

10. Thoroughly cooking meat and washing fruits and vegetables can prevent food poisoning from which two pathogens?

12.2 VIRUSES

Structure of Viruses

Polio, once a devastating childhood disease, is now close to eradication thanks to the vaccines developed by Jonas Salk and Albert Sabin. Polio is caused by a **virus**, a carrier of genetic information. The poliovirus (*Enterovirus*) causes disease when it spreads from person to person. Sometimes it causes only minor inconvenience, but it can be a serious disease that attacks the central nervous system. It infects healthy cells in the intestines. Its RNA core is enclosed by a protein coat called a *capsid*.

Like bacteria, viruses can infect living things. But viruses aren't made of cells as bacteria are; they infect a host cell to continue activity. Scientists are not even sure that these microscopic structures are alive. Though they aren't cellular, they contain a few parts that resemble structures found in cells.

Are viruses even alive?

Questions

How are viruses similar to but different from bacteria?

How do viruses grow and reproduce?

How do viruses interact with humans and the environment?

How did COVID-19 affect the world?

Terms
virus
nucleic acid
envelope
capsid
spike
prion
lytic cycle
lysogenic cycle
retrovirus

nucleic acid
either DNA or RNA, but never both

capsid
protein coat that surrounds the nucleic acid molecule

envelope
lipid bilayer that surrounds the capsid; studded with proteins; made from the cell membrane of the cell that it infects; not found in all viruses

spike
protein that attaches virus to specific cell surfaces

PRIONS

A **prion** is an infectious particle made only of abnormal proteins; it contains no genetic information. Prions tend to infect the brains of animals and people. They act as a template for brain proteins, causing healthy proteins to turn into abnormal ones. Prions cause Creutzfeldt-Jakob disease in humans and bovine spongiform encephalopathy (mad cow disease) in cattle.

Prion protein (stained in red)
▼

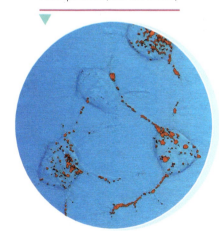

Reproduction in Viruses

A virus is like a flash drive, full of information but unable to do anything unless connected, in the case of a virus, to a living organism. A virus cannot replicate its DNA, transcribe it into RNA, or translate it into a protein on its own. So how does it pass on its genetic information in order to make more viruses?

Because a virus has no ribosomes and mitochondria for reproduction, it uses the few proteins that its genes code for to enter a cell, called a *host*, and hijack its machinery. The virus triggers the host cell to begin producing more virus particles, causing the cell to lyse, or burst open, to flood the surrounding area with new viruses that can infect new host cells.

LYTIC CYCLE

A virus has a couple of ways that it can enter a cell and use it to produce more viruses. One of those ways is the **lytic cycle**, which is the rapid infection and destruction of a host cell, resulting in more virus particles.

We saw the lytic cycle from a bacterium's point of view during transduction (p. 259)—a bacteriophage attached to a bacterium, emptied its genetic contents into the bacterium, and forced the bacterium's organelles to produce virus particles until the bacterium burst open. The lytic cycle also occurs in organisms besides bacteria. In humans, cold and flu viruses attack cells and cause *lysis*, which is the process of releasing viral particles from a cell. The main reason we feel tired when sick is that our bodies are using more energy to fight off these viral invaders during periods of infection.

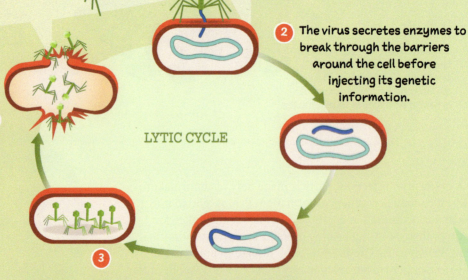

LYTIC CYCLE

1 The virus attaches to a cell.

2 The virus secretes enzymes to break through the barriers around the cell before injecting its genetic information.

3 The virus's genetic material incorporates itself into the cell's genetic material to produce new viruses.

4 The cell is then forced to assemble the virus parts into fully formed viruses. When the cell can no longer make viruses, the virus produces more enzymes to cause lysis.

LYSOGENIC
CYCLE

toddler with chicken pox

Other viruses will wait inside a cell, sometimes for years, until conditions are right for them to activate. While they're waiting, they go through the **lysogenic cycle**, a process in which a virus infects a cell but does not destroy it, allowing the cell to naturally copy the virus's genetic material and pass it to daughter cells during mitosis.

The lysogenic cycle is related to the second part of bacterial transduction—a bacteriophage empties its genetic contents into a bacterium, and the bacterium simply incorporates the virus genes into its own. Before the bacterial cell divides, virus genes replicate and are passed along through cell division.

Two common viruses in humans that go through the lysogenic cycle are herpes simplex and varicella-zoster. Herpes simplex is a virus that can be transmitted by drinking from someone else's cup or using someone else's lip balm. It then lies dormant in the cells around a person's mouth. Something that upsets the body's homeostasis, such as a lot of stress or a cold, can trigger the virus, which then leaves the lysogenic cycle and

begins the lytic cycle. Once this happens, the cells are destroyed, resulting in cold sores. The virus soon returns to the lysogenic cycle in the new cell that it has infected, and the cold sores clear up. But the viral genes in the host DNA wait for another chance to activate. The virus continues to spread its genes to other cells by host-cell division.

Varicella-zoster is even more patient than herpes simplex and can lie dormant in nerve cells for decades. Often, the virus is never activated. The infection begins as a lytic one in the form of chickenpox. Once the chickenpox outbreak is gone, the virus enters the lysogenic cycle. Sometimes varicella-zoster reenters the lytic cycle many years later, causing shingles.

3 The cell then goes through the normal process of cell division. The reproducing virus continues to remain dormant until an event triggers the lytic cycle.

LYSOGENIC CYCLE

1 The virus's genetic material incorporates itself into the cell's genetic material as in the lytic cycle.

2 The virus's genes replicate as the cell's genes replicate.

Some lysogenic viruses are known as **retroviruses**—viruses that can force a cell to transcribe the virus's RNA into the cell's DNA. Cells usually transcribe DNA into RNA, but retroviruses contain an enzyme called *reverse transcriptase* that reverses the process. The most well-known and well-studied retrovirus is the human immunodeficiency virus, or HIV, which can stay in the lysogenic cycle for many years. If it enters the lytic cycle, it triggers the disease known as acquired immunodeficiency syndrome (AIDS).

How Viruses Function in the Environment

USEFUL VIRUSES

Until about 150 years ago, people didn't know anything about viruses. One of the important contributions of learning about viruses is how it has expanded our knowledge of genetics. Viruses were the first to have their genomes sequenced, unlocking the door to the genomes of living things.

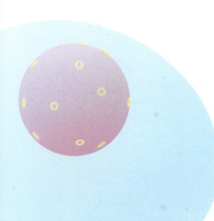

Viruses are also quickly becoming a vital part of medical research and treatment. A virus can have healthy genes inserted into it and then be used to deliver the genes to genetically damaged cells. For example, one virus currently in clinical trials is being used to cure several different types of hereditary blindness. The gene-carrying virus is injected directly into a person's eyes, and the virus inserts genetic material into the DNA of host cells to start making the proteins that are needed to give the person sight. This kind of gene therapy has the potential to cure diseases that were once considered incurable, though there are challenges to overcome to make the benefits of gene therapy long-lasting.

Bacteriophages (p. 259) are increasingly being researched as a new kind of antibiotic. As bacteria become resistant to our current antibiotics, researchers have been working to invent one that contains a bacteria-killing virus, which may help fight against super bacteria like MRSA and drug-resistant tuberculosis. Research has also shown that our microbiota may contain bacteriophages that defend us against pathogenic bacteria.

Viruses are also useful in agriculture. Certain viruses can be sprayed on plants to ward off common plant pests, such as caterpillars, beetles, and even rabbits. Researchers are working on ways to insert genes in a virus that will allow it to build long carbon molecules. These molecules act like the electrodes in a normal battery, only much smaller and with a much better capacity to hold a charge.

bacteriophage

Mapping Outbreaks

In 2020 our world was hit with a major pandemic. The virus responsible was COVID-19. Originating in China, COVID-19 quickly spread across the globe and essentially put the world on pause. The data collected (see table below) shows that COVID-19 affected many countries. But what did the spread of this disease actually look like?

How much of the world did COVID-19 actually affect?

Materials
red, green, blue, and purple markers (5)

PROCEDURE

A Using the data from the table, create a bar graph showing the number of cases per time interval for each country. Place these on a single graph. Use a red marker for the first time-interval bar, a green marker for the second time-interval bar, a blue marker for the third time-interval bar, and a purple marker for the fourth time-interval bar.

QUESTIONS

1. Every time interval on the data table is a four-month period. According to the data table, which time interval showed the most significant increase of COVID-19 cases in the United States? in India?

2. What do the results from Question 1 indicate about the differences in the spread of COVID-19 between the United States and India?

3. Dividing the number of COVID-19 cases at the end of April 2021 by a country's population will give the percentage of a population that was affected by the disease. Use the table to find the percentage of COVID-19 cases for each country.

4. According to your calculations above, which country had the highest percentage of cases of COVID-19? Which had the least?

5. Use your bar graph to make predictions on the current number of COVID-19 cases for each country. Perform an internet search for the current number of COVID-19 cases in each country. Were your predictions accurate? Suggest an explanation for any differences you observe between your predictions and the actual values.

CUMULATIVE CASES OF COVID-19

Country	Population in 2021	Jan–Apr 2020	May–Aug 2020	Sep–Dec 2020	Jan–Apr 2021
Brazil	213.4 million	85,380	3,910,901	7,675,973	14,665,962
England	56 million	118,343	290,138	2,190,702	3,856,836
Germany	79.9 million	163,009	243,295	1,745,518	3,392,232
India	1.3 billion	34,863	3,687,939	10,267,283	19,157,094
United States	331.4 million	1,124,568	6,339,212	20,628,348	33,181,681

HARMFUL VIRUSES

Though viruses have many uses, many more are pathogenic, like the poliovirus. The polio vaccine developed by Jonas Salk in 1951 is still administered today by injection. It uses an inactivated virus, which after injection triggers a child's body to build an immunity, protecting him from future exposure to poliovirus without actually causing the disease. Albert Sabin's vaccine is also still used today to protect people from the poliovirus.

Until the 1980s pathogenic viruses could not be directly treated. In 1957 two scientists described a protein called *interferon*. They discovered that cells produce this protein after they have been infected with a virus. When a cell is lysed and the viral particles spill out, the interferon protein gets released and signals surrounding uninfected cells. Interferon then signals proteins that block viral replication in newly infected cells. Interferon has created a surge in modern medicine. Today there are drugs that use this handy protein to treat diseases caused by herpes viruses, HIV, hepatitis viruses, multiple sclerosis, flu viruses, and even cancers. Compared with antibiotics, however, there are very few antiviral drugs. It is difficult to find antivirals that are selective for viruses, flexible enough to be effective when viruses mutate, and harmless to their host cells.

poliovirus

Ebola virus

Our medical technology can't always keep up with the viruses that are out there. Some viruses seem to appear out of nowhere, decimating a group of people in just a few weeks. Other viruses that were thought to be eradicated through vaccines have made a comeback. These viruses are part of a larger group called *emerging infectious diseases*. One example is the Ebola virus, which surfaced in 1976, infecting people in Africa and claiming the lives of many people before it subsided to sporadic cases in the following years. Then in 2014 and again in 2018 the virus reemerged in cities where it was harder to contain. But just as in the H1N1 flu virus outbreaks in 1918 and 2009, scientists were able to create a vaccine to fight against the deadly virus.

12.2 SECTION REVIEW

1. What characteristics of viruses make them seem like living things? What characteristics make them seem nonliving?

2. Draw and label a virus with the following parts: nucleic acid, envelope, spike, and capsid.

3. Particles that are similar to viruses but are made of proteins only are called _____.

4. Describe how a virus can duplicate its genetic information and spread.

5. Hepatitis C is a viral infection of the liver that can lie dormant for years before a person ever shows symptoms. While the person has no symptoms, what cycle is the virus in? Once symptoms begin to manifest themselves, what cycle has the virus entered?

6. List three ways that viruses can be useful.

7. You have gone on a mission trip to Ghana in West Africa to help build a new wing on a village clinic. While you're working, you notice that the physicians at the clinic are treating a lot of people with a mysterious illness that no one has seen before. One physician mentioned that he thought it might be Buruli ulcer disease, which is caused by a bacterium and was last seen in Ghana in the 1980s. What kind of disease might this be?

Use the case study on the facing page to answer Questions 8–11.

8. Why is it important to study dangerous bacteria and viruses?

9. Explain why some laboratories study deadly bacteria and viruses.

10. Identify two possible ways that harmful bacteria and viruses can be mishandled, causing them to be released from laboratories.

11. Identify two deadly bacteria or viruses, other than those listed in the case study, that have vaccines.

VIALS OF TERROR

Some of the most well-protected items in our country are not priceless artifacts, jewels, or even people. They're strains of deadly viruses and bacteria, all housed at the Centers for Disease Control and Prevention (CDC) headquarters in Atlanta, Georgia. Laboratory technicians at the CDC Laboratories wear protective suits with a filtered air source when handling biohazardous materials. Safety and security are foremost on the minds of everyone who works there, yet occasionally something goes wrong—a microbe is mishandled and released, and people die. Why do we not destroy these microbes instead of keeping them to study? Why take the risk that these lethal vials might fall into the wrong hands?

The main reason that dangerous bacteria, like anthrax and tuberculosis, and viruses, like smallpox, avian flu, SARS, and COVID-19, are kept alive and accessible is to study them and find defenses against them. Even though it's hazardous, researchers have greatly reduced the impact of diseases like Ebola, malaria, polio, measles, rabies, tetanus, and influenza because of the tests, vaccines, and treatments that have been developed.

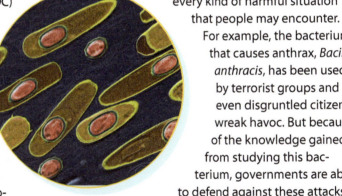

Bacillus anthracis

Another reason for keeping these bacteria and viruses around is to combat those who want to turn microbes into weapons. Part of wise stewardship in a fallen world is to prepare for every kind of harmful situation that people may encounter. For example, the bacterium that causes anthrax, *Bacillus anthracis*, has been used by terrorist groups and even disgruntled citizens to wreak havoc. But because of the knowledge gained from studying this bacterium, governments are able to defend against these attacks and treat people who are exposed to anthrax toxins. Some laboratories, such as the medical center at Wake Forest University, are going a step further. They are using deadly microbes to create antidotes against them.

Laboratories like the CDC need to keep their bacteria and viruses secure to minimize the damage that they could cause. But researchers also need to learn about them because these microbes aren't just found in laboratories; they can occur naturally in the environment, and we need to be able to care for God's image-bearers when they are affected by the lethal efforts of fallen men.

12.1 PROKARYOTES

- Archaea and bacteria are both prokaryotes that can act as decomposers. Archaea can be found in extreme environments. Bacteria have cell walls containing peptidoglycan.

- Archaea are categorized according to their habitats: thermophiles (hot environments), methanogens (anoxic environments), halophiles (salty environments), and acidophiles (hot, acidic environments).

- Bacteria have three different shapes: coccus, spirillum, and bacillus. The amount of color taken in by the cell wall can determine whether it is a gram-positive (purple) or gram-negative (pink) bacterium.

- Bacteria have capsules and nucleoids; some have plasmids and move using flagella.

- Bacteria most often undergo binary fission to reproduce. Some bacteria also reproduce asexually by exchanging of information with other bacteria through either transformation, transduction, or conjugation.

- Bacteria can either be beneficial or harmful. Beneficial bacteria, such as the human microbiota, help to digest food, remove toxins, and absorb vitamins. Harmful bacteria are often pathogens that cause diseases.

Terms
archaea · peptidoglycan · methanogen · thermophile · halophile · acidophile · bacterium · binary fission · conjugation · transformation · transduction · bacteriophage · microbiome · pathogen

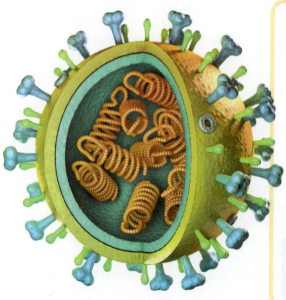

12.2 VIRUSES

- Viruses disrupt cell functioning and may cause illness. Scientists debate whether viruses are alive since they cannot reproduce on their own.

- To reproduce, viruses enter cells to produce new viruses in two ways: the lytic cycle (rapid cell infection and destruction of cells) or the lysogenic cycle (cell infection and replication without destruction of cells).

- Although viruses can cause disease and may even be deadly, some can be useful in medical research, treatment of disease, and agriculture.

Terms
virus · nucleic acid · envelope · capsid · spike · prion · lytic cycle · lysogenic cycle · retrovirus

Chapter Review Questions

RECALLING FACTS

1. What shape are the bacteria in the image shown on the right?

2. How does an antibiotic clear up a bacterial infection?

3. What are the three parts of a bacterium's flagellum?

4. What are the two forms that DNA takes in a bacterial cell?

5. How do bacteria typically reproduce?

6. What is a bacteriophage? What kind of bacterial genetic transfer involves a bacteriophage?

7. List three things that bacteria do for the environment.

8. List three things that the bacteria of our microbiota do for us.

9. What structures do bacteria and viruses have in common?

10. Where does a virus get the envelope that surrounds it?

11. How do viruses translate their genetic information into proteins?

12. What is the most well-known retrovirus?

13. What protein produced by cells can be used in antiviral drugs?

14. Name a virus that goes through both the lytic and lysogenic cycles.

UNDERSTANDING CONCEPTS

15. Explain how organisms are classified in domain Archaea.

16. Create a hierarchy chart relating the following terms: *acidophile, archaea, bacillus, bacteria, coccus, halophile, methanogen, prokaryote, spirillum,* and *thermophile.*

17. How are archaea and bacteria similar? How are they different?

18. What determines whether a bacterium is gram-positive or gram-negative? Explain.

19. Explain how the process of transformation causes the addition of a capsule to a bacterium.

20. Describe how a bacteriophage is used in the process of transduction.

21. What is the difference between the lytic and lysogenic cycles?

22. How might viruses be helpful in the fight against antibiotic-resistant bacteria?

23. How are viruses used in medical research?

24. How does a vaccine work?

CRITICAL THINKING

25. If you were visiting a lake and observed a film of residue on the surface of the water, what could you assume was the cause of this?

26. One morning you wake up feeling more tired than usual and suddenly begin observing the signs of an oncoming illness. Explain what is happening to your body on a cellular level.

27. The apples on the tree in your yard are being destroyed by codling moth caterpillars, while the apples on your neighbor's tree look perfectly healthy. You notice that your neighbor sprays his tree once a week, but he says he doesn't use chemical pesticides. What might he be using to keep the codling moths away?

28. How are deadly microbes and our earthly stewardship related?

29. Suppose you were a scientist who had the goal of developing a new vaccine. From a biblical perspective, why would you do this? What would you need to consider?

30. Considering the attributes of life on pages 8–9, write out an argument for why viruses should be considered either living or nonliving.

Use the table below to answer Questions 31–33.

31. Create a graph that plots the ratio of tuberculosis (TB) patients per 100,000 people every 5 years, starting with 1953.

32. What does the trend in your graph signify?

33. What happened to the number of TB cases from 1988 to 1992?

YEAR	Number of TB Cases	Ratio per 100,000 People	YEAR	Number of TB Cases	Ratio per 100,000 People
1953	84,304	52.6	1988	22,436	9.2
1958	63,534	36.3	1989	23,495	9.5
1963	54,042	28.6	1990	25,701	10.3
1968	42,623	21.1	1991	26,283	10.4
1973	30,998	14.6	1992	26,673	10.4
1974	30,122	14.1	1993	25,103	9.7
1975	33,989	15.7	1994	24,205	9.2
1976	32,105	14.7	1998	18,287	6.6
1977	30,145	13.7	2003	14,835	5.1
1978	28,521	12.8	2008	12,895	4.2
1983	23,846	10.2	2013	9,582	3.0

COMPARING COVID-19 CASES

When COVID-19 disrupted the world in 2020, the news was filled with statistics, scientific data, and new possible treatments. To isolate those most susceptible to the disease, many states required isolation, masks, social distancing, virtual learning, and even quarantine after traveling to other states. In the beginning of the outbreak not much was known about the disease. Scientists scrambled to understand more about the virus so that they could create a vaccine. As time went on, scientists learned about its transmission, structure, and method of attacking the body. Eventually, this new research led scientists to create a much-anticipated vaccine.

Despite the increasing number of COVID-19 cases, many states differed in how they managed the virus. Two such states were Florida and California. Having a substantially smaller population than California, Florida took a less restrictive approach to preventing the spread of COVID-19. Florida initially had lockdown orders in place but eventually withdrew them and gradually reduced mask mandates and reopened businesses. California chose more rigid measures and had multiple lockdowns as well as a mask mandate for an entire year. Despite the different approaches, both states recorded similar numbers of cases of COVID-19 per capita: Florida had 8700 cases per 100,000 people and California had 8900 cases per 100,000 people. How could these two states, with two different approaches, have very similar outcomes?

Use the case study above and the graph at right to answer Questions 34–38.

34. Compare the line graph of the two states and describe the similarities and differences you observe.

35. The case study states that the number of cases per capita for both Florida and California were similar. So why is the highest peak of the graph for California significantly higher than that for Florida? In what ways can this difference in peaks lead people to misinterpret the data?

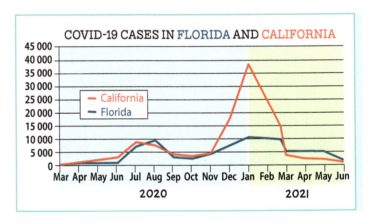

36. California had more restrictions in place that were intended to prevent the spread of COVID-19. What factors do you think caused these different approaches to produce similar results?

37. COVID-19 proved to be harmless for some people but debilitating and deadly for others. At the time of the pandemic there was an overall estimated death rate due to COVID-19 of 2.1%. In Europe the bubonic plague of the 1300s had a death rate of nearly 60%. Why do you think the bubonic plague had a higher death rate? What have we learned about how to prevent the spread of viruses and bacteria since the Middle Ages?

38. Do you think that Florida or California had a more appropriate response to COVID-19? Do some additional research on this topic and explain your answer.

13 PROTISTS & FUNGI

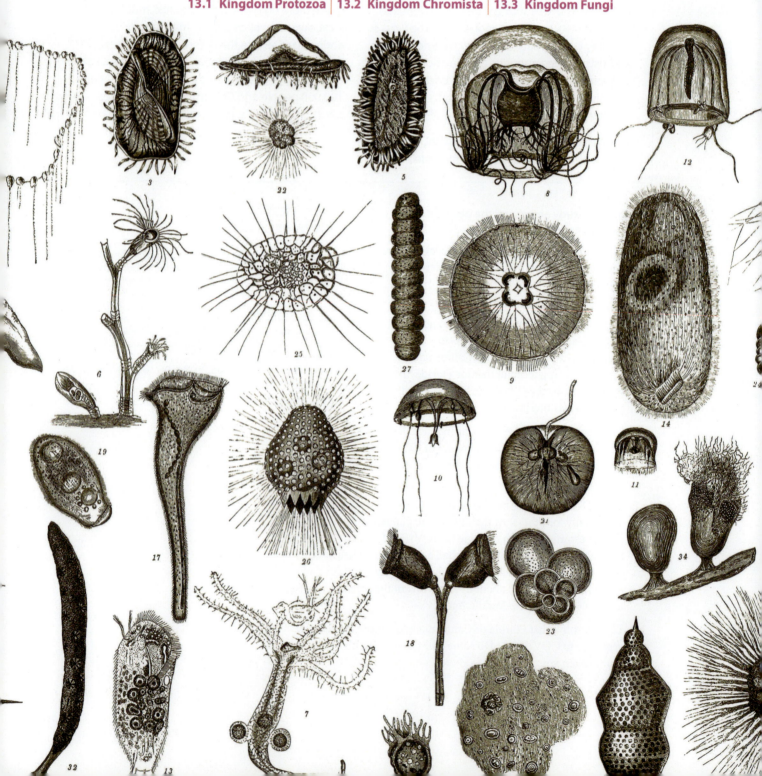

Blood Monsters

Plasmodium falciparum is a blood monster. It invades the bloodstream, taking over a person's red blood cells. This malaria-causing parasite starts out as a single-celled organism that enters the blood through a mosquito bite. The cell multiplies quickly, and within two weeks an infected person can die if not treated early enough. About half the world's population lives in places where malarial parasites thrive. Those most affected are sub-Saharan African children. Though there are other species, almost all deaths from the disease are caused by *P. falciparum*.

In 1880 Charles Laveran, a French doctor in Algeria, found the key to solving the age-old problem of malaria. He observed *P. falciparum* under a microscope in a blood sample from a malaria patient. That beginning led scientists to learn about the parasite's life cycle. How can we use what we know about *P. falciparum* to eradicate malaria and protect God's image-bearers?

13.1 KINGDOM PROTOZOA
Classifying Protozoans

Biologists have probed the microscopic world for centuries to understand, model, and classify microorganisms. *P. falciparum* is one example of a **protist**, a single-celled or multicellular eukaryotic microorganism. When Dutch biologist Antonie Van Leeuwenhoek observed protists through a microscope he said, "No more pleasant sight has met my eye than this, of so many thousands of living creatures in one small drop of water." Protists ooze, they propel, they float through their watery environments in the hidden world revealed by the microscope.

? Why are protists so hard to classify?

Questions

How is kingdom Protozoa distinct from the other kingdoms?

How do protozoans grow and reproduce?

In what ways are protozoans both useful and harmful to humans and the environment?

Why are protists important to evolutionary theory?

Terms
protist
protozoan
pseudopodium
eyespot
cyst
spore

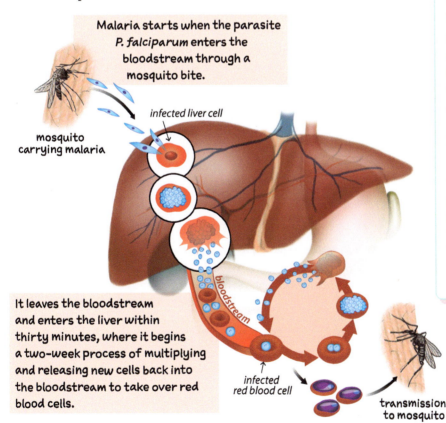

Malaria starts when the parasite *P. falciparum* enters the bloodstream through a mosquito bite.

mosquito carrying malaria

infected liver cell

It leaves the bloodstream and enters the liver within thirty minutes, where it begins a two-week process of multiplying and releasing new cells back into the bloodstream to take over red blood cells.

bloodstream

infected red blood cell

transmission to mosquito

The name "protist" comes from an outdated biological kingdom, the kingdom Protista. This kingdom previously served as a kind of catchall kingdom for organisms that were difficult to classify. Because of genetics and the drive of evolutionists to link classification to the phylogenetic tree, this kingdom has been restructured into several new kingdoms. Also driving the change is the fact that classification is constantly changing as scientists learn new information about organisms.

Protists are amazingly diverse. Most protists are made of only one cell, but some work together to form *colonies*, free-living but related cells of the same kind that interact with each other. Some protist colonies can be large enough to see without a microscope.

Protists are broken up into two kingdoms. Those that are more like animals are grouped in kingdom Protozoa, while those that are more like plants fall into kingdom Chromista (you'll learn more about them in the next section). Both kingdoms include protists that are more like fungi than either plants or animals. In general, organisms in kingdom Protozoa are heterotrophs and those in kingdom Chromista are autotrophs, though there are many exceptions to this general rule. All protists need water in which to live and function.

Plasmodium falciparum in the form of rings inside a human erythrocyte (red blood cell)

P. falciparum is a protist that is grouped with other animal-like protists in kingdom Protozoa. This kingdom is the first of five included in the domain Eukarya. This domain also includes some protists that are more fungus-like. Protists in kingdom Protozoa are referred to as **protozoans**. As adults, they are usually *motile*, that is, able to move—they slide around and propel themselves. They are often found wherever water abounds—oceans, freshwater ponds, streams, lakes, and even moist soil. They can appear as unicellular organisms or can group in multicellular colonies. Most protozoans are heterotrophs, meaning that they get their energy from other organisms, though a few can carry out photosynthesis. Protozoans that are autotrophs are green, while other heterotrophic protozoans come in other colors or are colorless. *Euglena* are autotrophic and possess chloroplasts that make them green. Many heterotrophic protozoans live inside other organisms, carrying out a symbiotic relationship with their host. Some protozoans are parasitic, while others benefit their host organisms. *P. falciparum*, for example, is a sporozoan, a protozoan that is parasitic with no means of locomotion. It relies on its host for movement and nourishment.

Structure of Protozoans

We've seen that many protozoans have the ability to propel themselves. Some protozoans have cilia, a fringe of hairlike projections that ripple along in rhythmic waves as a protozoan moves. Cilia work in a way that is like that of a rowing team on a boat. Some protozoans such as *dinoflagellates* propel themselves with a whip-like structure called a *flagellum* (pl. flagella).

Some protozoans, such as *Amoeba proteus*, have no fixed body shape. They ooze along in what is called *amoeboid movement*. They develop bulges called **pseudopods** (s. pseudopodium), filled with cytoplasm, that they extend in front to pull them along. Amoebas also use pseudopods to engulf their food, forming an enclosure in the cytoplasm called a *food vacuole*. When food is digested, its nutrients are distributed throughout the cell.

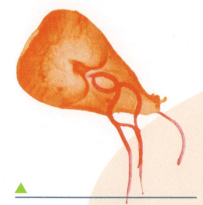

Giardia (*Giardia lamblia*), a protozoan with flagella, lives in contaminated water and can cause diarrhea in humans.

PROTOZOANS

Radiolarians are protozoans and have silicon enclosures called *tests*. They contribute to a type of sediment on the ocean floor called *ooze*. Ooze may be several miles thick in places, with one gram of ooze being made of 50,000 radiolarian tests!

Ciliates are protozoans that are surrounded with cilia. A paramecium is a classic example of a ciliate. A stentor is an example of a trumpet-shaped ciliate. Both are free-swimming inhabitants of stagnant lakes and ponds.

Dog vomit slime mold is classed in kingdom Protozoa and consists of one large cell with multiple nuclei that form when individual cells merge. They can grow as large as a few square meters.

Euglena are very common protozoans with a flagellum that live in fresh and salt water. If the water is calm enough, they can make the surface of a pond appear green. Some euglenas are heterotrophs but also have chloroplasts that enable them to perform photosynthesis. An organism like a euglena that is both a heterotroph and an autotroph is called a *mixotroph*.

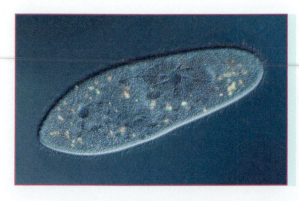

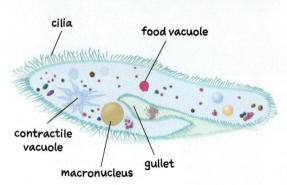

cilia

food vacuole

contractile vacuole

macronucleus

gullet

Since most protozoans have only one cell, they need a few more organelles to fill their needs. Many protozoans, such as *Paramecia*, have a *contractile vacuole*, a storage container that rids the cell of excess water to both maintain a healthy pressure and control osmosis in the cell. Contractile vacuoles regulate homeostasis between a protozoan and its environment. *Paramecia* also have something like a mouth called a *mouth pore*. The cilia sweep the food into the mouth pore, and the food travels down the gullet into a food vacuole. If a paramecium's food vacuole is like a human stomach, then its gullet can be likened to a human esophagus. *Paramecia* also have both a *micronucleus* and *macronucleus* for storing genetic information.

Many species of protozoans have specialized structures that can respond to changes in their environments. Some have areas of pigment called **eyespots** that can detect light so that the protozoan can move either toward or away from light. Other protozoans can detect and respond to chemical changes in their environments. During times of harsh environmental conditions, many protozoans survive by forming a hard, protective coating called a **cyst**. While in the cyst form, the protozoan's metabolism slows down to allow it to survive without food.

Reproduction in Protozoans

Protozoans can form colonies, resulting in a stronger defense against other organisms. Some reproduce asexually, others sexually, while many reproduce both ways.

ASEXUAL REPRODUCTION IN PROTOZOANS

Protozoans such as *Paramecia* can reproduce asexually by mitosis just as bacteria do (see page 258). During this process the micronucleus divides by mitosis and the macronucleus enlarges and divides in half. Organelles such as the gullet are duplicated, and the cell divides. Other protozoans such as *Plasmodium* produce several new nuclei through mitosis before separating into daughter cells. This is called *multiple fission*. This process is what allows malaria parasites to multiply quickly in the human body.

Some protozoans use mitosis to form **spores**, reproductive cells that are protected by a thick cell wall. These spores have genetic material that is identical to the parent cell.

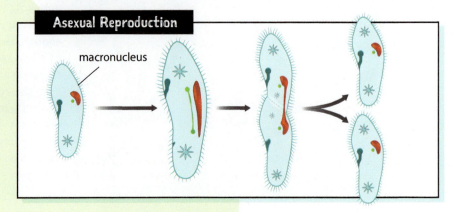

Asexual Reproduction

macronucleus

SEXUAL REPRODUCTION IN PROTOZOANS

Protozoans can also reproduce by a type of sexual reproduction called *conjugation*. Note that this is different from the bacterial conjugation that we studied in Chapter 12. For example, when paramecia undergo conjugation, two of them attach at their oral grooves. Each paramecium undergoes a variety of nuclear changes. At some point, they exchange genetic material through the cytoplasm bridge that connects them, which is similar to the pilus used in bacterial conjugation. The cells separate and undergo more nuclear changes. Finally, each paramecium divides to form a total of four new individuals. Conjugation allows protozoans to mix genetic material in a way that is impossible in asexual reproduction.

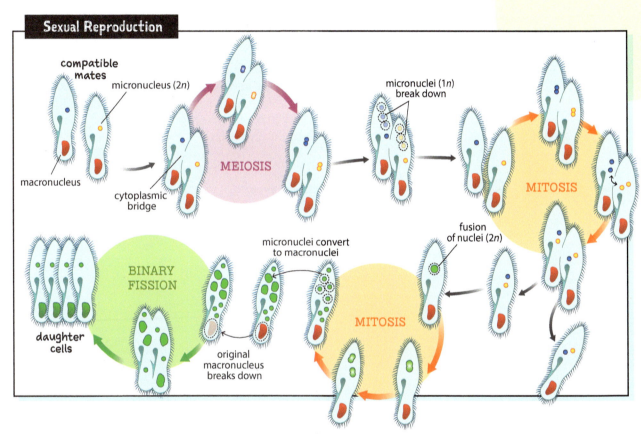

Sexual Reproduction

compatible mates

micronucleus (2n)

macronucleus

cytoplasmic bridge

MEIOSIS

micronuclei (1n) break down

MITOSIS

fusion of nuclei (2n)

BINARY FISSION

micronuclei convert to macronuclei

MITOSIS

daughter cells

original macronucleus breaks down

How Protozoans Function in the Environment

Protozoans are especially important to an aquatic environment because they serve as decomposers and form a key part of the marine food chain as they transport nutrients in the different zones of the ocean. They also keep bacteria populations in check.

But protozoans also often act as parasites, weakening their hosts. The malaria parasite *P. falciparum*, mentioned at the beginning of the chapter, and the protozoan *Trypsanosoma*, which causes African sleeping sickness, are examples of parasitic protozoans. Plants,

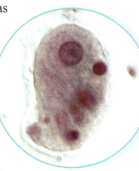

Entamoeba histolytica

insects, fish, and even humans can suffer from the parasitic effects of protozoans.

Many protozoans are pathogens—disease-causing organisms. For example, *Giardia* and *Entamoeba*, which live in water or in water-dwelling organisms, cause diarrhea and dysentery in humans. So clean, fresh water is a high priority to protect God's image-bearers. Managing protozoan populations helps protect the plants and animals that people rely on for their lives and livelihood.

Protists and Biological Evolution

Because protists are so difficult to classify, biologists have frequently changed the way that they classify them. Earlier in this chapter we saw that the word *protist* comes from the outdated kingdom Protista, which has been replaced by kingdoms Protozoa and Chromista. But what has driven all these changes?

As we learned in Chapter 11 and mentioned earlier in this section, some biologists have strayed from the original intent of classification, which was to organize living things in a useful way. Instead, they use classification to try to establish evolutionary relationships between organisms. This has led to some confusing findings and bizarre conclusions, especially since scientists began using genetics to create kingdoms within domain Eukarya. These scientists are trying to deduce how simple, unicellular eukaryotic protists may have evolved into much larger and more complex eukaryotes such as plants, animals, and even humans.

Evolutionists theorize that the earliest lifeforms were prokaryotic. These early cells grew larger, and their cell membranes folded as they developed. Eventually these folds of membranes enclosed the nucleus to form the first eukaryotic cell. These organisms engulfed other organisms, which became organelles.

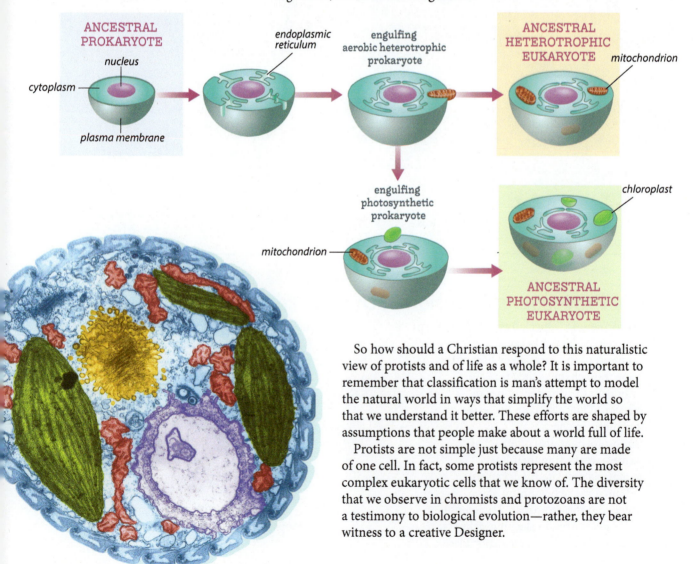

So how should a Christian respond to this naturalistic view of protists and of life as a whole? It is important to remember that classification is man's attempt to model the natural world in ways that simplify the world so that we understand it better. These efforts are shaped by assumptions that people make about a world full of life.

Protists are not simple just because many are made of one cell. In fact, some protists represent the most complex eukaryotic cells that we know of. The diversity that we observe in chromists and protozoans are not a testimony to biological evolution—rather, they bear witness to a creative Designer.

1. Using the illustration on page 275 as a guide, how do you think malaria spreads from person to person?

2. How do protists, kingdom Protozoa, kingdom Chromista, and domain Eukarya relate?

3. Draw a series of pictures that shows how an amoeba eats. Label each diagram appropriately to communicate the amoeba's motion and eating.

4. *Euglena* have both a gullet and chloroplasts. What does this information tell you about what they eat?

5. Trace or sketch a euglena in the margin on your paper and label the following parts: chloroplast, contractile vacuole, eyespot, flagellum, gullet, and nucleus. Compare this diagram to the paramecium on page 278 to correctly label the parts.

6 What are the different ways that protozoans reproduce?

7. List two examples of a protozoan that is harmful to humans or the environment. Describe the danger posed by each.

8. What is the difference between a parasite and a pathogen?

9. The pathogenic protozoan *Cryptosporidium* can resist chemicals like chlorine that are often used to sanitize public drinking water. This protozoan is introduced to water supplies through contact with animal or human waste. Suggest one way to limit *Cryptosporidium* in water supplies to provide people around the world with fresh drinking water.

10. Why do evolutionists believe that all life came from prokaryotic cells?

11. As a result of their genetic testing to develop the phylogenetic tree for eukaryotic organisms, scientists have found that the malaria parasite *Plasmodium* contains a chloroplast-like plastid. This is an unexpected explanation for the mystery of why medications that attacked chloroplasts were effective in treating malaria. Many scientists are pointing to this finding as a relic of evolution, proof that *Plasmodium* is an evolutionary relative of photosynthetic organisms. How would you respond to this claim?

13.2 KINGDOM CHROMISTA
Classifying Chromists

How do chromists affect my life?

Protists that are plantlike are grouped in kingdom Chromista, making it a kingdom that contains more photosynthetic protists than kingdom Protozoa does. This kingdom is the second of five in domain Eukarya. Protists in kingdom Chromista are referred to as **chromists**. Chromists can be unicellular, multicellular, or colonial. They have motile gametes, and many have cell walls made of cellulose as plants do. Most chromists are autotrophic, though a few are heterotrophic. Many chromists are green, indicating that they are photosynthetic because of the presence of chlorophyll.

Chromists include all the **algae** (s. alga)—kelp, seaweed, green algae, and red algae—which are protists that have chloroplasts, making them photosynthetic organisms. Algae are unicellular and can form colonies. Chromists also include *diatoms*, found in the oceans and fresh water, and *foraminifera*, which are chromists with shells that are found as fossils all over the world. Some foraminifera can be as large as 20 cm long. Chromists also include one type of slime mold, called a *slime net*, and *water molds*, fungus-like protists that are heterotrophic. The latter have cell walls made of cellulose and so are categorized in Chromista with autotrophic protists.

Questions

How are the kingdoms Chromista and Protozoa distinct from each other?

How do chromists grow and reproduce?

How do chromists contribute to the environment?

Terms
chromist
alga
plankton
algal bloom

CHROMISTS

Water molds like the potato blight are microscopic, fungus-like chromists that prefer high-humidity environments. Many are parasitic.

Algae form a major part of kingdom Chromista. Shown on the right is a colony of volvox, a green alga. Other algae include brown algae (kelp and seaweed) and red algae. Notice that many of the volvox in this photo hold daughter colonies.

Foraminifera like the ones shown above are chromists with tests. Almost all are aquatic. Scientists have found fossilized foraminifera buried in rock strata all over the world.

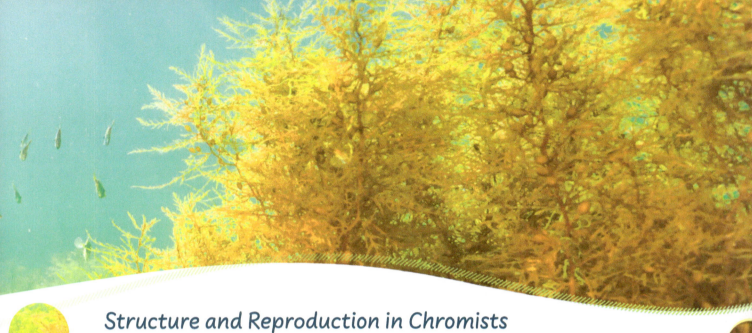

Structure and Reproduction in Chromists

Most chromists lack the specialized structures of protozoans since they are usually autotrophic. Their chloroplasts allow them to respond to their environments to manufacture food for their sustenance. Giant kelp, a type of brown algae, is one of the few chromists that form complex structures. These structures allow them to grow very large, but most lack cilia or flagella to propel themselves.

Many chromists form colonies and can undergo both asexual and sexual reproduction. Like protozoans, they can reproduce asexually using multiple fission and mitosis. They can also sexually reproduce through conjugation. During conjugation each chromist, such as an alga, experiences a variety of nuclear changes. At some point two algae cells exchange genetic material through a cytoplasmic bridge—called a *conjugation tube*—that connects them during conjugation. The cells separate and undergo more nuclear changes. Finally, each cell divides to form a total of four algal cells. Conjugation allows chromists to mix genetic material in a way that is impossible in asexual reproduction.

Some chromists such as algae alternate between sexual and asexual reproduction as shown below. The adult chromist is a haploid organism. During asexual reproduction, the cell undergoes mitosis, producing daughter cells that are haploid. In the next generation haploid cells can fuse genetic information through fertilization to form diploid zygotes that multiply through meiosis.

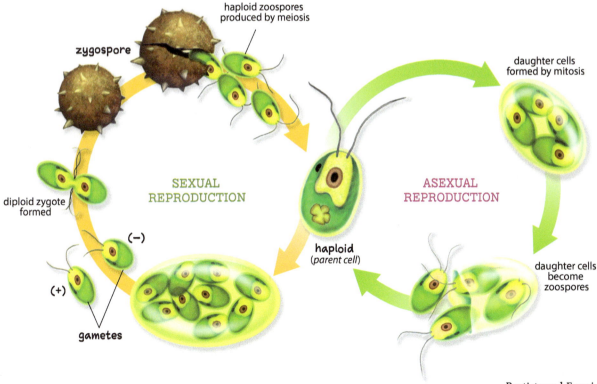

zygospore

haploid zoospores produced by meiosis

daughter cells formed by mitosis

diploid zygote formed

SEXUAL REPRODUCTION

ASEXUAL REPRODUCTION

daughter cells become zoospores

(−)

(+)

haploid (*parent cell*)

gametes

How Chromists Function in the Environment

Microorganisms like protists are very small inhabitants of our world, but they can have a big impact on living things around the world. Even a unicellular alga can be part of a bigger colony that helps produce the oxygen that makes Earth a haven for life. Recall from Chapter 7 that 70%–80% of the oxygen that we breathe comes from algae in the oceans.

Marine chromists like algae and diatoms form the base of the ocean's food chain. Many larger animals such as fish, whales, and some seals eat **plankton**, organisms that float in the ocean and don't swim against the current. Plankton include two groups: *phytoplankton*, which are autotrophic organisms, and *zooplankton*, which are heterotrophic organisms, including protozoans. There are protists in both groups. Algae, diatoms, and dinoflagellates are all phytoplankton that feed jellyfish, krill, and other zooplankton. Zooplankton include many species of protists such as foraminifera and radiolarians, as well as eggs, larvae, and hordes of other multicellular organisms. Some chromists provide shelter for animals, such as sea otters that wrap themselves in kelp to keep from drifting while they eat fish and shellfish. Certain algae such as seaweed are edible and are a healthy diet choice for many people.

Under the right conditions—warm water and the presence of oxygen, sufficient nutrients, and sunlight—protists in the ocean such as dinoflagellates can multiply rapidly. This can produce a surge in the population called an **algal bloom**. Though these protists can fill a helpful role in consuming wastes in ponds, lakes, and oceans, algal blooms generally deplete the water of nutrients and can release toxins that are harmful to living things.

Some chromists have symbiotic relationships with other animals as, for example, coral reefs with algae: the algae provide nutrients for the coral and the coral's hard covering provides support and protection for the algae.

Shown above is a photo of an algal bloom, called a *red tide*, washing ashore at Mount Maunganui, New Zealand. A red tide is caused by a booming population of phytoplankton and can lead to a serious problem for people in this dry area: the red tide clogs pipes used in desalination plants that provide clean, fresh water.

13.2 SECTION REVIEW

1. Identify each of the following protists as protozoans (P) or chromists (C).

   a. water mold

   b. algae

   c. foraminifera

   d. ciliates

2. Create a table that compares protozoans and chromists. Include the following in your comparison: motility, energy source, and the ability to form colonies.

3. Why don't chromists usually operate as parasites in their environments as other protists do?

4. How do chromists reproduce?

5. Why are chromists in ocean waters important for life on Earth?

Managing Algae Growth

Imagine walking out to the pool one hot summer day and noticing that the pool is green! How did this happen? Just yesterday the pool was a clear blue color! The answer is algae. While various types of algae play an important role in our ecosystem, we don't want them growing in our pools. So how can we prevent algae from growing? And how can we deal with this situation it if it happens?

How do we prevent and treat algae growth in swimming pools?

Materials
test tubes (4) • organic chlorella powder, 1 tsp • distilled water, ½ cup • chlorine, 1 cup • disposable pipette, 5 mL (2)

PROCEDURE

A Label four test tubes 1–4. Mix 1 tsp chlorella powder with 1/2 cup distilled water.

B To each of the test tubes add 10 mL of the chlorella solution. Tube 1 will be your control.

C To Tube 2 add 5 mL of chlorine into the chlorella solution and swirl gently. Record your observations.

D To Tube 3 add 7 mL of chlorine and swirl gently. Record your observations.

E To Tube 4 add 10 mL of chlorine and swirl gently. Record your observations.

F If your solution has not turned white at this point, add 1 mL of chlorine at a time to Tube 4 until the solution has turned white. Record how many milliliters were required to make this happen.

QUESTIONS

1. What happened to the chlorella solution when you added each amount of chlorine?

2. How many milliliters of chlorine were needed to turn your chlorella solution white and kill the algae?

3. Explain what could happen if more chlorella solution were added to the test tubes.

CONCLUSION

4. Under the right conditions, algae are able to reproduce quickly. How can you relate algal blooms to the reproduction of algae?

GOING FURTHER

5. Perform an internet search on factors that affect the growth of algae. List the ideal conditions for algae growth.

6. If you wanted to predict when an algal bloom might occur, what should be considered?

13.3 KINGDOM FUNGI
Classifying Fungi

Questions

How are fungi classified?

How do fungi grow and reproduce?

How can fungi be useful or harmful to humans and the environment?

Terms

chitin
hypha
mycelium
fruiting body
stipe
cap
gill
budding
fragmentation
mycorrhiza
lichen

Do you know what the largest living thing is? It's not an African elephant, the blue whale, or even the giant sequoia named General Sherman. It's a small, humble fungus called the honey mushroom. This huge organism doesn't look so big—it isn't even visible except in the fall when little brown mushrooms push up through the moss and leaf debris on the forest floor.

The part of any mushroom that we can see above the ground isn't all there is to a mushroom. The honey mushroom creates an underground network that can cover hundreds of acres, killing stands of trees. In Oregon there is a network of honey mushrooms over two thousand acres in size thought to be possibly over two thousand years old. That's one large fungus!

But what is a fungus anyway? A *fungus* is a member of the third kingdom in the domain Eukarya—kingdom Fungi. Fungi often grow from the ground like plants, but they lack chlorophyll and are heterotrophs. Some are parasitic, but many serve as decomposers, filling an important role in the environment by breaking down organic matter in the soil. Fungi have cell walls as plants do, but instead of cellulose, their cell walls are made of a substance called **chitin**, a large sugar molecule also found in the outer coating of insects. The majority of fungi are multicellular, but some, such as yeasts, are unicellular. Like many protists, they can form colonies. Fungi include the abovementioned honey mushroom, molds, mildews, lichens, and yeast. Fungi are grouped by the kinds of spores that they produce during sexual reproduction.

Structure of Fungi

Most fungal colonies are made of slender filaments called **hyphae** (s. hypha) that look like very fine plant roots. Some hyphae have partitions called *septae* (s. septum) that create segments made of individual cells connected by pores to allow the exchange of cytoplasm, organelles, and nuclei. Other hyphae are not segmented like this.

Even large fungi like mushrooms are made of interwoven hyphae called **mycelia** (s. mycelium). Mycelia are visible without the aid of a microscope and can be as simple as clumps of bread mold or as highly organized and specialized as a morel mushroom.

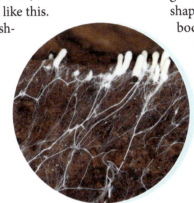

Note the fine, root-like mycelia that have worked through a piece of rotting wood.

Though hyphae may be present in the ground year-round, you might never know a fungus is there until it sends up a **fruiting body**, a special fungal structure made of many hyphae that is responsible for reproduction. In club fungi, which are the only group of fungi that form traditional mushroom-shaped fruiting bodies, the stalk of the fruiting body is known as the **stipe**, and the top of the fruiting body is called the **cap**. The underside of the cap often contains ribs called **gills** that radiate outward from the stipe. The gills can produce thousands of spores, reproductive cells that have thick walls and are able to survive a variety of hostile conditions and pass on genetic information from the parent organism. Spores are the reproductive units of fungi, just as seeds are the reproductive units of plants. Fungi are just one of many types of organisms that use spores to reproduce. Most fungi don't have gills—they let their spores out through pores in the fruiting body or from knobs that hang down from the cap.

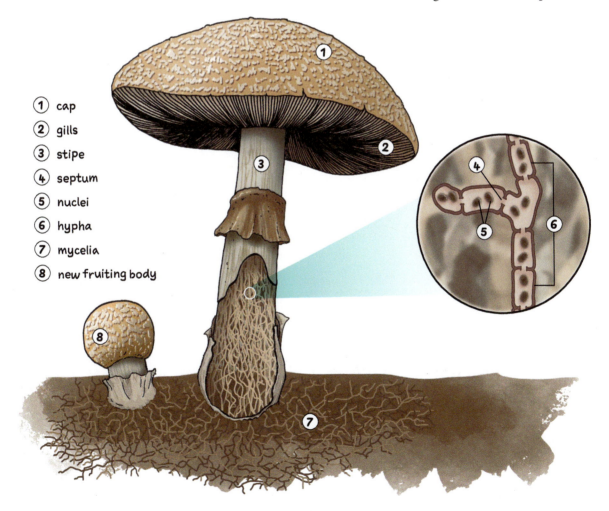

1. cap
2. gills
3. stipe
4. septum
5. nuclei
6. hypha
7. mycelia
8. new fruiting body

FUNGI

The largest phylum in kingdom Fungi is Ascomycota, or the *sac fungi*. Sac fungi have a tiny pouch called the *ascus* (pl. asci), where spores form. These fungi are cup shaped. *Penicillium* is a fungus in this phylum and produces the antibiotic *penicillin*. Shown at right is the orange peel fungus, an edible fungus that looks great in a salad!

Sometimes mushrooms in wooded areas and grasslands form an arc called a *fairy ring* as their underground network spreads out in a circle to feed on decaying organic matter in the soil.

Some fungi produce odors, like this stinky devil's fingers fungi (below). This fungus is not a sac fungus—it's in the phylum Basidiomycota, which includes the most familiar *club fungi*. This phylum is the only group of fungi that can correctly be called "mushrooms." It includes puffballs, stinkhorns, and shelf fungi. Basidiomycota is the next largest phylum in the kingdom Fungi after Ascomycota. Devil's fingers feeds on decaying wood and smells like rotting meat, attracting insects, which spread its spores. It is edible, but who would want to eat it with a smell like that!

Cordyceps is a family of brutal fungal parasites in phylum Ascomycota. Some parasitic fungi have tree or plant hosts; others like the one shown above left have insect hosts. Cordyceps turn insects into zombies, absorbing nutrients from their hosts' internal organs while they are still alive. Eventually, each insect's body is filled with the fungus's reproductive spores. When the insect's body bursts, the spores are released.

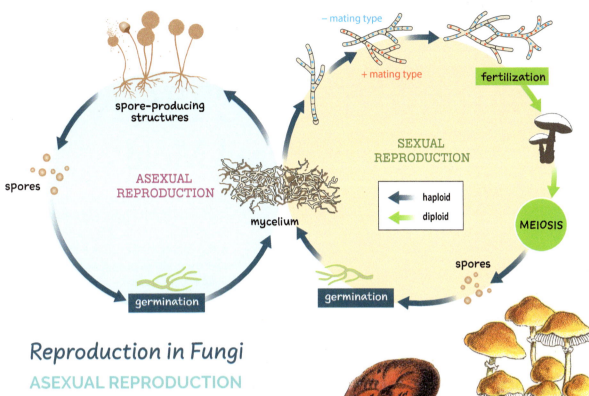

Reproduction in Fungi

ASEXUAL REPRODUCTION

Like protists, fungi can reproduce through sexual reproduction, asexual reproduction, or a combination of the two, though asexual reproduction is most common. Some unicellular fungi use mitosis for asexual reproduction, as some bacteria and protists do.

Yeast cells reproduce by **budding**, a process that involves part of the cell pinching off to produce a new yeast cell (see diagram below). The nucleus first divides by mitosis. Then a small pouch forms on the side of the yeast cell and one of the new nuclei moves into it, forming the bud. After a period of growth, the bud separates from the parent.

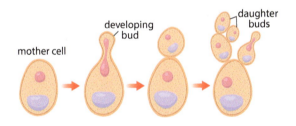

Fungi can also reproduce asexually through a process called **fragmentation**. In this process part of a fungus that has broken off can form a new fungus as cells multiply through mitosis. Normally a hypha or mycelium forms the fragment. Other organisms, such as some lichens, algae, worms, starfish, and bacteria, can also reproduce asexually through fragmentation.

SEXUAL REPRODUCTION

The most common form of sexual reproduction in fungi is the formation of spores. Spores can form sexually through meiosis inside an enclosure or sac, such as in the asci of sac fungi. In some types of fungi they can also form sexually outside of a sac. Some fungi can form spores through sexual reproduction when environmental conditions are unfavorable for asexual reproduction. There are no male and female fungi; instead, they have mating types that are symbolized by + and − signs. Sexual reproduction happens when the hyphae of different mating types contact each other. The fused hyphae produce a fruiting body that forms and releases spores through sexual reproduction.

Fungi such as sac fungi, club fungi, and yeasts that can produce both sexually and asexually are called *perfect fungi*. Fungi that have not been observed to reproduce sexually are called *imperfect fungi*. Some examples of the latter are *Penicillium* and *Candida albicans*, which can live in a person's mouth, and some fungi that are used to produce cheeses.

REPRODUCTION in
Black Bread Mold

Black bread mold is an example of the third largest phylum of kingdom Fungi: Zygomycota. This mold uses a combination of asexual and sexual reproduction to produce spores called *zygospores*.

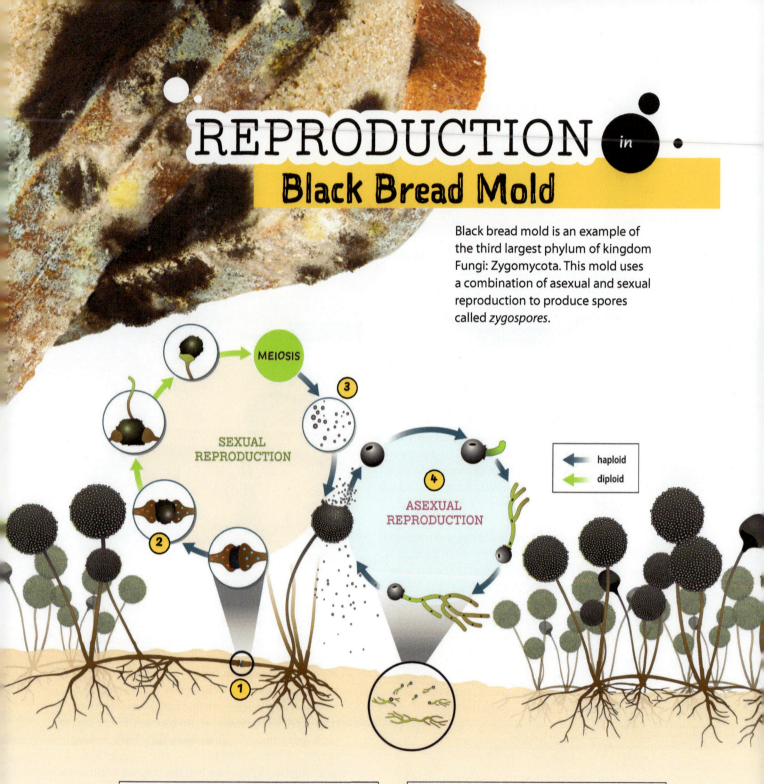

MEIOSIS

SEXUAL REPRODUCTION

ASEXUAL REPRODUCTION

→ haploid
→ diploid

1. Two hyphae from different mating types grow until they touch each other. Nuclei in the tip of each hypha divide several times.

2. The hyphae fuse where they touch to form a zygosporangium, a diploid zygote (2*n*), through sexual reproduction. It forms a thick outer covering and enters dormancy.

3. By means of meiosis, zygosporangia form spores, which germinate to form hyphae through asexual reproduction.

4. As hyphae grow, asexually formed spores called *sporangiospores* form and are released.

Lichen is formed from interwoven algae and hyphae.

- fungal hyphae
- interwoven algae and hyphae
- fungal hyphae

How Fungi Function in the Environment

Humans aren't the only farmers. There's an unlikely farmer less than an inch long that also builds skyscrapers on the African savanna. It's the African termite. Termites have a symbiotic relationship with a fungus, the *Termitomyces* mushroom. The termites cultivate this mushroom, gathering spores to deposit in their colonies. It's hard work for termites to decompose the cellulose in wood, so they get these fungi to do it for them! The fungi turn the cellulose into sugars, which the termites eat. When things get especially wet in Africa, some of the *Termitomyces* mushrooms become quite large (caps 1 m wide!) and start growing outside the termite mound. People gather and enjoy these mushrooms as a culinary delicacy.

Many fungi live with plant roots in a symbiotic relationship called **mycorrhiza**. Over 90% of plants have fungi associated with their roots. The fungi effectively increase the surface area of a plant's roots to enhance its ability to absorb minerals from the soil. In return, the fungi receive carbohydrates from the plant. This relationship allows plants and fungi to live in environments that would otherwise be unable to support them.

A **lichen** consists of a fungus and algae living in symbiosis. Much like the chromists in a coral reef do, the alga in a lichen captures energy from sunlight and manufactures sugars for itself and the fungus. The fungus provides support and protection for the alga. Lichens are commonly found on tree bark, rock formations, fence posts, exposed rock, and brick walls.

These two examples demonstrate that fungi serve a very important role as decomposers in the environment. They take nutrients and waste products and convert them into a form that other living things can use. It is fungi that are responsible for that distinctive flavor in the mushroom pizza that many enjoy!

But like protists, fungi can have a more sinister role in the environment. They can function as pathogens. Plants get fungal infections from fungi (rusts and smuts) that drastically reduce harvests. Animals and insects get fungal infections that can eventually kill them. Even humans can get thrush, ringworm, and athlete's foot, which are all caused by fungi.

With both protists and fungi, we must manage and use their populations in ways that maximize their benefits and minimize their dangers. We can develop treatments for malaria and athlete's foot, grow portabella mushrooms for food, and harness the photosynthetic power of algae to power our homes and cars. When we control and use protists and fungi to wisely manage the resources that God has given us to help people, God's image-bearers, we bring Him glory.

The fungus *Ustilago maydis* causes corn smut. But that's not a bad thing in Mexico since the locals prize it as a delicacy! The infected corn is used in quesadillas because it contains more protein than uninfected corn does.

FIGHTING MALARIA WITH A FUNGUS

Malaria, the disease caused by the protist *Plasmodium*, has affected people's lives for many years. Even with modern medicine, hundreds of millions of people will be stricken with malaria this year. But there's a light at the end of this very dark tunnel. In 1950 a protein known as the *Duffy antigen* was discovered on the surface of blood cells. Once *Plasmodium* has entered a person's body, it latches on to the Duffy antigen, infecting the person. In 1996 the source of the antigen was discovered—it is produced by the DARC gene (*D*uffy *a*ntigen *r*eceptor for *c*hemokines) of human chromosome 1. Researchers have found that if the antigen is turned off in the DARC gene, *Plasmodium* has nothing to latch on to, making a person resistant to malaria. Turning off this tiny protein may become the greatest weapon we have against this devastating disease.

Would it also be possible to defeat the malaria-causing parasites in a mosquito vector by infecting the mosquito with a fungus? This odd idea for killing a parasite is born out of centuries of frustrating attempts to control *Plasmodium*, most of which have resulted in either the mosquito or *Plasmodium* developing a resistance to the treatment used. Could fighting a protist with a fungus really reduce malaria?

Questions to Consider

1. After *Plasmodium* enters the body, how can it make someone ill?

2. What did scientists find must be done for humans to become resistant to malaria?

3. What should we consider when turning off this protein to prevent malaria?

13.3 SECTION REVIEW

1. How are protists different from fungi?

2. How are mycelia and hyphae related?

3. Draw a picture of the *Amanita muscaria* shown below, and label the cap, gills, mycelia, and stipe on the mushroom. Indicate where you think the mycelia would be.

4. (True or False) Mycelia appear only in the ground below club mushrooms.

5. How do fungi reproduce asexually?

6. How do fungi reproduce sexually?

7. Identify the following as perfect or imperfect fungi.

   a. *Candida*

   b. fruiting bodies of sac fungi

   c. *Penicillium camemberti*, a fungus used to produce Camembert cheese

   d. budding yeast

8. List in descending order and describe the three largest phyla in kingdom Fungi.

9. Why are fungi so important?

10. What harmful effects do fungi have on the environment?

11. What kind of symbiotic relationship (mutualism, commensalism, neutralism, amensalism, parasitism) does each of the following pairs of organisms exhibit?

    a. humans and ringworm

    b. termites and the *Termitomyces* mushroom

    c. algae and a fungus in lichen

    d. mycorrhizae between fungi and trees

13 CHAPTER REVIEW
Chapter Summary

13.1 KINGDOM PROTOZOA

- Protists are single-celled or multicellular eukaryotic microorganisms that are broken down into two different kingdoms, kingdom Protozoa and kingdom Chromista.

- Organisms in kingdom Protozoa are animal-like and are often found in water.

- Most protozoans are heterotrophic and move by using cilia, flagella, or pseudopods. Some protozoans have specialized structures that respond to changes in the environment.

- Protozoans may form colonies or reproduce sexually, asexually, or both. Some protozoans form spores, which are reproductive cells, through mitosis.

- Protozoans are heterotrophs and function to provide nutrients for aquatic food chains. They may also be pathogenic.

- Knowing that unicellular protists can be just as complex as eukaryotic cells weakens the evolutionary claim that unicellular life gave rise to multicellular life.

Terms
protist • protozoan • pseudopodium • eyespot • cyst • spore

13.2 KINGDOM CHROMISTA

- Chromists are plantlike protists that are usually unicellular and may form colonies.

- Chromists are usually autotrophic, lack specialized structures, and reproduce either sexually or asexually.

- Chromists play an important role in providing nutrients and shelter to aquatic life and oxygen in the air that we breathe.

Terms
chromist • alga • plankton • algal bloom

13.3 KINGDOM FUNGI

- Fungi are heterotrophic organisms that may serve as decomposers or parasites.

- Fungi have cell walls made of chitin and may be unicellular or multicellular.

- The fruiting body of the mushroom is only a small portion of the mushroom; much of the organism is underground.

- Fungi are neither male nor female. Sexual reproduction occurs when the hyphae of different mating types fuse together to form a fruiting body.

- Lichens are organisms that consist of a fungus and algae living in symbiosis. They can be found on tree bark, fences, rocks, and brick walls.

Terms
chitin • hypha • mycelium • fruiting body • stipe • cap • gill • budding • fragmentation • mycorrhiza • lichen

RECALLING FACTS

1. What structures allow different protists to propel themselves?

2. Draw an amoeba, a paramecium, and a euglena, and use arrows or descriptions to show how they move.

3. When do protozoans need contractile vacuoles?

4. Identify the two ways that protists exchange genetic information.

5. Summarize the two main contributions of chromists to their environments.

6. What kind of fungus has fruiting bodies with caps, stipes, and gills?

7. What part of a fungal colony is responsible for creating fairy rings?

8. Create a hierarchy chart that relates domain Eukarya, protists, kingdom Fungi, kingdom Chromista, and kingdom Protozoa. Include ciliates, slime molds, foraminifera, water molds, algae, sac fungi, and club fungi in your chart.

9. What is the difference between perfect and imperfect fungi?

10. Name the three ways that fungi can reproduce asexually.

11. How can both protists and fungi be harmful to their environments?

12. Compare the ecosystems that are home to protists with those that are home to fungi.

UNDERSTANDING CONCEPTS

13. Why are chromists and protozoans classified into two separate kingdoms even though they are both protists?

14. How do some species of protozoans respond to changes in their environments?

15. Why do you think that some biologists believe that trees and algae are related to a common evolutionary ancestor?

16. Explain how conjugation in chromists mixes genetic information more than asexual reproduction does.

17. Explain why the term *plankton* includes more than just protists.

18. Why do you think that both unicellular protists and fungi form colonies?

19. Explain why it may not be apparent that fungi are present.

20. Describe how protists are different from fungi.

21. Summarize how reproduction is similar in protists and fungi.

22. How do the algae and the fungus in a lichen help each other?

23. Summarize the main contribution of fungi to the environment.

CRITICAL THINKING

24. *A. proteus*, though a very commonly studied protist, is a protozoan without an assigned phylum. We've learned how this protozoan uses pseudopods to move and capture food. It once had a different Latin name: *Chaos diffluens*. Why do you think this protist is difficult for biologists to classify?

25. What does the classification of *A. proteus* tell you about the nature of classification?

26. If a new protist is discovered, what characteristics would help scientists determine the kingdom it belongs in?

Use the case study on the right to answer Questions 27–32.

27. Using a map of the United States, determine where you think algal blooms most commonly occur.

28. What do you think would cause algae populations to grow so quickly in these areas? (*Hint:* What makes plants grow quickly?)

29. Algal blooms in the oceans can cause large dead zones that contain very little oxygen. Why does this happen?

30. Why are dead zones so dangerous to marine organisms?

31. Suggest one way that scientists can use technology and scientific models to monitor algal blooms in the Gulf of Mexico.

32. How could scientists warn people of a dangerous algal bloom in an area?

Use the case study on the right to answer Questions 33–36.

33. On the basis of what you know about malaria, suggest why there is no vaccine for African sleeping sickness.

34. Suggest one way to prevent infection in people in Africa without an effective vaccine.

35. Scientists have found that *T. b. gambiense* uses specific proteins to build its flagellum. Without this flagellum, the protist can't survive in the human bloodstream. How do you think doctors could use this information to treat African sleeping sickness?

36. Doctors do have treatments for African sleeping sickness, but they have bad side effects, and the protozoan that causes the disease seems to be developing the ability to resist the drugs. How could doctors avoid developing resistance in this pathogen?

case study

HAB ALERT!

In the Gulf of Mexico harmful algal blooms (HABs) are becoming more frequent. These blooms can have a devastating effect as they produce toxins that poison shellfish, birds, and mammals living in these areas. They also have serious local effects on the oxygen content of the water. The National Oceanic and Atmospheric Administration (NOAA) monitors the Gulf of Mexico to warn about any development of a dangerous algal bloom.

case study

AFRICAN SLEEPING SICKNESS

African sleeping sickness is caused by the flagellated protozoan *Trypanosoma brucei gambiense*. The disease begins with joint pain, then over a series of weeks produces confusion and trouble sleeping. Eventually, it crosses the brain barrier and wreaks havoc. If untreated, a victim will almost always die.

The parasitic protist *T. b. gambiense* is spread through the bite of a large fly called the tsetse fly. *T. b. gambiense* has over 9000 genes that manufacture a confusing array of proteins that help it continually change its outer coating to evade the human immune system. There are currently no vaccines available to prevent this disease.

Pain-Relieving Plants

Think back to the most painful experience you have ever had. People who have been in extreme pain because of an injury, sickness, or surgery know the relief that medicines like morphine or oxycodone can provide. These two drugs come from the opium poppy, a plant that has been used medicinally for centuries. But for all the relief that the opium poppy has brought to people, it has also brought another pain—addiction.

Researchers continually try to make new opium-based pain relievers that are *not* addicting, while government regulators and even militaries try to control the growth of opium poppies—but neither has had much success. What is the right way to respond to a part of creation that can be both incredibly useful and detrimental to people at the same time?

14.1 KINGDOM PLANTAE
Characteristics of Plants

Plants sustain life. God created them to feed every living thing on the earth (Gen. 1:29–30) and to be admired for their beauty (Matt. 6:28–29). But ever since Creation, people have been doing more with plants than just eating them and appreciating their beauty. Your favorite cotton T-shirt, the grounds for your coffee, the aloe vera gel that you put on a sunburn, the paper that you doodle on, the biodegradable containers that your Chinese takeout comes in—all are made from plants.

Plants are a lot like other living things. They are multicellular, and their cells are eukaryotic. But plants also have a group of characteristics that set them apart from other created kinds. Though other organisms may share some of these characteristics with plants, no other group has all of them.

What makes something a plant?

Questions

What makes a plant different from other living things?

What are the different kinds of plants?

Terms
cuticle
cellulose
alternation of generations
vascular plant
nonvascular plant
gymnosperm
bryophyte
angiosperm

PLANT ATTRIBUTES

Cuticle. The outer surfaces of plants are covered by a **cuticle**, a waxy substance that protects the leaves, green stems, and fruits. The cuticle helps plants keep moisture in and pests and diseases out.

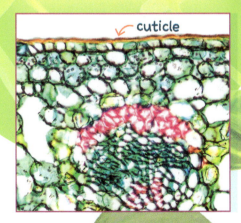
cuticle

Cellulose. In their cell walls, plants have **cellulose**, the tough fiber that gives plants their rigidity and strength. Cotton and linen fibers are almost 100% cellulose.

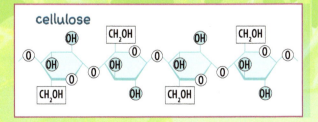
cellulose

PLANT ATTRIBUTES

Multicellular Embryo. Plant gametes fuse to form a zygote that develops into a multi-cellular embryo.

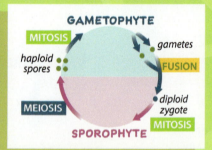

Reproduction. Plants reproduce sexually by **alternation of generations**, which is a cycle that includes a spore-forming stage and a gamete-forming stage.

Photosynthesis. Almost all plants are autotrophs—they produce their own food through photosynthesis. A few plants are parasitic, getting their nutrition from other living things.

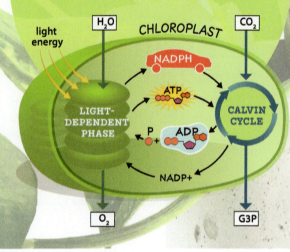

Classifying Plants

It takes an army of botanists and groundskeepers to maintain the Royal Botanic Gardens, Kew, England, which are home to almost 14,000 different species of plants that come from around the world. But that's only a fraction of the hundreds of thousands of plant species in the world, so the Royal Botanic Gardens are always adding to their greenhouses. The caretakers there have collected some that look so strange that it's hard to think of them as plants, such as the giant water lilies that are native to the Amazon. With so many species of plants and so much diversity, what's the best way to group them?

A plant is classified in one of several ways. It can be classified as a **vascular plant**, which is a plant that has tissues capable of transporting water and other materials. Without vascular tissues, tall plants wouldn't be able to carry nutrients from their roots to their leaves. Some very small plants are classified as **nonvascular plants**; that is, they have no vascular tissues. Nonvascular plants must be small enough to absorb their water and nutrients directly through their leaflike structures, and they're usually found in moist environments.

Plants are also divided according to whether they produce *seeds*—the diploid, multicellular structures that nurture a plant's developing embryos. All plants produce *spores*, which are unicellular and haploid, and *gametes*, unicellular cells that fuse to form multicellular embryos. But nonvascular plants and a few vascular plants don't protect their embryos inside seeds. Vascular plants that do produce seeds can be classified according to whether the seeds are formed in a flower and enclosed in a fruit or are borne in a cone and exposed.

A gardener carrying a giant water lily in a Berlin greenhouse

PLANTS

Plants can be classified into four groups, primarily on the basis of the way that they reproduce.

Seedless vascular plants like this club moss have tissues that circulate nutrients but don't produce seeds. Plants in this category also include horsetails, ferns, and whisk ferns.

Some vascular plants like this cycad produce seeds but no fruits or flowers. These kinds of plants, called **gymnosperms**, bear their seeds out in the open on a cone. Gymnosperms also include pines and ginkgo trees.

Seed-bearing vascular plants that produce flowers and fruits are called **angiosperms**, the largest group of plants. These saguaros are a species of cactus with large, fleshy trunks and arms that store water.

Nonvascular plants like this moss are commonly called **bryophytes**, though this is an informal taxon. This group of plants includes mosses, hornworts, and liverworts.

14.1 SECTION REVIEW

1. What characteristics do plants share with many other living things?

2. Describe at least two characteristics that make plants distinct from animals.

3. How is alternation of generations different from the sexual reproduction in many other organisms?

4. What characteristics are used to classify plants into four main groups?

5. Do seed-bearing vascular plants produce spores? Explain your answer.

6. How are vascular plants able to grow so large?

7. A pine tree produces seeds in cones rather than in flowers and fruit. Which of the four types of plants is it?

THE STRUCTURE OF PLANTS

Plant Tissues

How is the exotic jade vine, found deep in the tropics of the Philippines, like your mom's little basil plant that sits in the kitchen window? They are made of the same four kinds of tissue: dermal, vascular, ground, and meristematic.

DERMAL TISSUE

The **dermal tissue** of a plant does for the plant the same thing that your skin does for you. It protects the outside of the plant and helps it either retain or release water. In a plant's roots dermal tissue also aids in water and nutrient absorption.

The outer layer of dermal cells forms the **epidermis**. The epidermis is more than just a passive barrier that insects and disease can easily bypass; it secretes the cuticle to seal in moisture and keep out bacteria and viruses, and it produces different kinds of extensions, such as the prickles on roses or the stinging hairs of nettles, to discourage creatures from venturing too close.

In a plant with woody stems and roots, such as a tree, the epidermis provides another service. The cells of the epidermis die to create thick, tough layers of dead cells called **cork**, which protects and supports the weight of large plants.

What is a plant made of?

cork → epidermis

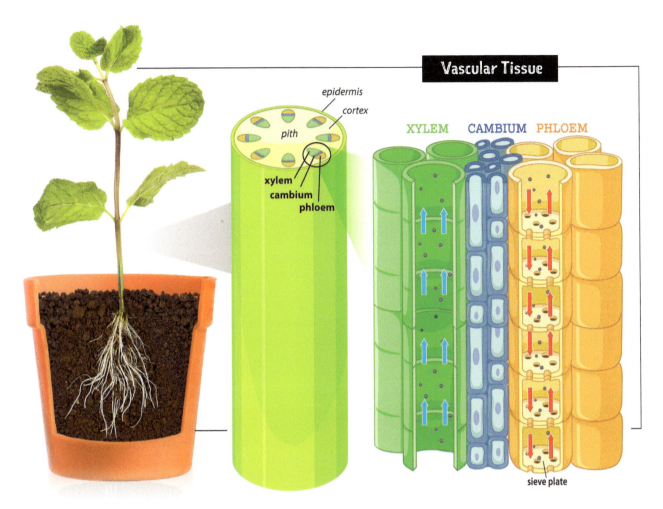

epidermis
cortex
pith
xylem
cambium
phloem

XYLEM   CAMBIUM   PHLOEM

sieve plate

VASCULAR TISSUE

Vascular tissue acts as the transport system for a plant, shuttling water (shown with blue arrows above) and nutrients from the roots through the stems to the leaves, and then carrying sugar (produced by photosynthesis) from the leaves to other parts of the plant for food or storage.

One kind of vascular tissue, **xylem**, transports water and dissolved minerals from the roots to the rest of the plant. Xylem is made of two types of cells. *Vessel cells* are wide and hollow with small pits along their sides; grid-like perforation plates can be found between each vessel cell. *Tracheid cells* are slender and hollow; large pits can be found on the sides of and in between each tracheid cell.

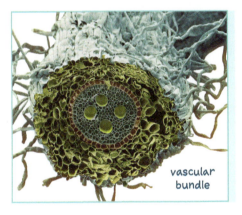

vascular bundle

Xylem cells die at maturity, and their hollow cell walls are connected end to end, forming a pipeline for fluids to travel through. The xylem's pits allow fluids to flow between the xylem pipelines. In woody plants, xylem cell walls harden and fill with a polymer called *lignin* to form wood.

Another vascular tissue, **phloem**, transports food (shown with red arrows above) made in the leaves to the rest of the plant. Like xylem cells, phloem cells are connected end to end, but they contain *sieve plates* at each connection. A sieve plate is like the wire mesh that's found at the end of a sink's faucet that helps control the flow of water from the faucet. Unlike xylem cells, phloem cells are still alive at maturity.

While xylem carries materials in only one direction—from the bottom of the plant to the top—phloem can carry materials in either direction, depending on the season. While the leaves are actively producing sugar, phloem cells shuttle it to other places in the plant. But after a season of dormancy, such as winter, phloem cells carry sugar and water back to the leaves as the plant gets ready to produce new stems, leaves, and flowers for spring. In some plants, phloem cells form flexible fibers, such as linen from flax and natural rope fibers from the hemp plant.

Plant Classification and Structure   **301**

parenchyma
cells

collenchyma
cells

GROUND TISSUE

Ground tissue is both the workhorse and the packing material of plant tissues, responsible for producing sugar, storing materials, and supporting the plant. The majority of a plant is made of ground tissue.

Ground tissue includes three types of cells. *Parenchyma* cells contain chloroplasts for photosynthesis and also store the products of photosynthesis. The fleshy parts of fruits and vegetables are generally made of parenchyma cells. Parenchyma cells also surround the tubes of xylem and phloem to keep them packed together tightly. Parenchyma cells can divide quickly, making them important for healing wounds that a plant may get. *Collenchyma cells* are very thick and provide flexible support for a plant. They often form long fibers in plants—the "strings" in vegetables like celery. *Sclerenchyma* cells are thick support cells like collenchyma, but they are much more rigid. When sclerenchyma cells die, their hardened, lignin-filled walls contribute to the woody structures within a plant.

MERISTEMATIC TISSUE

Why do the bushes in your front yard keep growing back, no matter how much you prune them? Plants have the ability to regenerate because of meristematic tissue. **Meristematic tissue** consists of undifferentiated cells (stem cells) that can become any type of tissue. They are often found in areas of fast growth, such as new leaves and shoots, flower buds, and root tips. Meristematic tissue is crucial during asexual reproduction, helping a new plant generate leaves, stems, and roots before it breaks away from its parent plant.

meristematic
tissue

petiole

stipule

blade

Leaves

Leaves are a plant's primary organs for producing energy through photosynthesis and exchanging gases with the environment. Some plants are called *evergreens* because they retain their leaves throughout the year. *Deciduous plants* lose their leaves each fall and grow them again the following spring. (Leaves are also used to classify plants as either monocots or dicots; see page 304.)

The broad part of a leaf is the **blade**. Attaching the blade to the stem of a plant is a tiny stalk called the **petiole**. Many monocots lack a petiole, so their blades are attached directly to the stem. Some leaves have *stipules*, which are tiny, leaflike structures at the base of the petiole. They can protect a newly forming leaf, harden into spines for further protection, produce nectar, or lengthen into tendrils.

TYPES OF LEAVES

simple—not divided into leaflets

simple, lobed

simple, palmate

American elm

English oak

maple

Compound

pinnately compound—leaf divided into multiple leaflets connected along one main stem called the *rachis*

bipinnately compound— leaf divided into leaflets that further divide into smaller leaflets

trifoliate—leaf divided into three leaflets connected in one central location

black walnut

mimosa

white clover

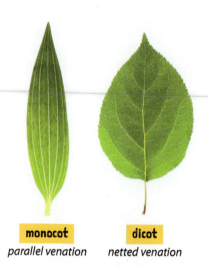

monocot
parallel venation

dicot
netted venation

VEINS IN LEAVES

Bundles of phloem and xylem form the veins that all leaves have. The arrangement of veins depends on the kind of plant. In monocot leaves, all the veins run parallel to each other; we call this *parallel venation*. In dicot leaves, the veins branch and form a network of smaller veins, a pattern known as *netted venation*.

GAS EXCHANGE IN LEAVES

Though plants don't technically breathe, they do have a way to bring in oxygen for cellular respiration and carbon dioxide for photosynthesis and then release oxygen and water vapor as byproducts. This gas exchange occurs on the underside of each leaf through little openings called **stomata** (s. stoma). Around each stoma is a set of **guard cells**, which are modified epidermal cells that open and close a stoma. As a plant actively pumps solutes like potassium ions into the guard cells, water passively follows the solutes through osmosis. This pulls the guard cells tight and opens the stoma. When ions are pumped out from the guard cells, the water goes with them, causing the guard cells to shrink and relax, closing the stoma.

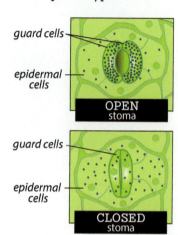

guard cells
epidermal cells
OPEN stoma

guard cells
epidermal cells
CLOSED stoma

STRUCTURE OF A LEAF

The topside and underside of a leaf are covered by the epidermis, which secretes a waxy cuticle to seal in moisture.

Like other plant parts, leaves are made of a mix of dermal, vascular, and ground tissue, with some areas of meristematic tissue.

Below the topside layer of epidermis is the palisade mesophyll, an orderly row of tightly packed, column-shaped cells that contain chloroplasts.

Chloroplasts circulate from top to bottom within the cells so that each chloroplast can be exposed to the sun.

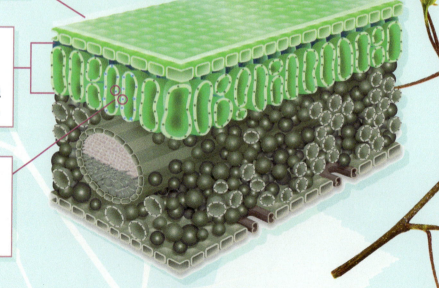

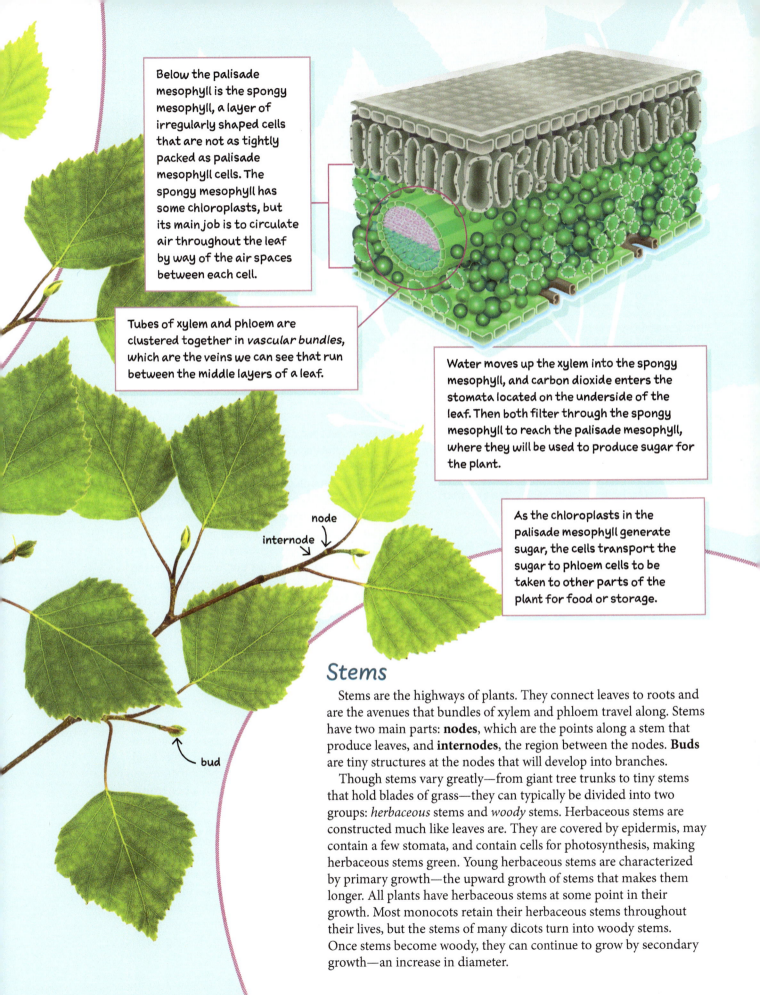

Below the palisade mesophyll is the spongy mesophyll, a layer of irregularly shaped cells that are not as tightly packed as palisade mesophyll cells. The spongy mesophyll has some chloroplasts, but its main job is to circulate air throughout the leaf by way of the air spaces between each cell.

Tubes of xylem and phloem are clustered together in *vascular bundles*, which are the veins we can see that run between the middle layers of a leaf.

Water moves up the xylem into the spongy mesophyll, and carbon dioxide enters the stomata located on the underside of the leaf. Then both filter through the spongy mesophyll to reach the palisade mesophyll, where they will be used to produce sugar for the plant.

As the chloroplasts in the palisade mesophyll generate sugar, the cells transport the sugar to phloem cells to be taken to other parts of the plant for food or storage.

node

internode

bud

Stems

Stems are the highways of plants. They connect leaves to roots and are the avenues that bundles of xylem and phloem travel along. Stems have two main parts: **nodes**, which are the points along a stem that produce leaves, and **internodes**, the region between the nodes. **Buds** are tiny structures at the nodes that will develop into branches.

Though stems vary greatly—from giant tree trunks to tiny stems that hold blades of grass—they can typically be divided into two groups: *herbaceous* stems and *woody* stems. Herbaceous stems are constructed much like leaves are. They are covered by epidermis, may contain a few stomata, and contain cells for photosynthesis, making herbaceous stems green. Young herbaceous stems are characterized by primary growth—the upward growth of stems that makes them longer. All plants have herbaceous stems at some point in their growth. Most monocots retain their herbaceous stems throughout their lives, but the stems of many dicots turn into woody stems. Once stems become woody, they can continue to grow by secondary growth—an increase in diameter.

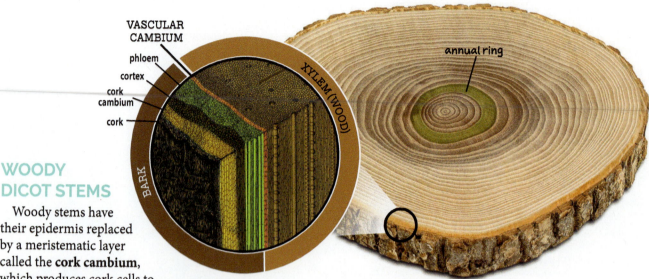

WOODY DICOT STEMS

Woody stems have their epidermis replaced by a meristematic layer called the **cork cambium**, which produces cork cells to protect and support the stem. Behind the cork cambium is the cortex, a layer of ground tissue that can fill several roles. In some plants the cortex acts as storage. In others the cortex cells have chloroplasts, making them capable of photosynthesis. But the cortex's main function in a stem is to allow water to flow between inner and outer layers of tissue and to exchange air with the environment through tiny pores (called **lenticels**) in the surface of a woody stem.

The next layer in a woody stem is the **vascular cambium**, which is a thin layer of meristematic tissue that produces a plant's xylem and phloem. Despite its thinness, vascular cambium is the most active layer of living cells in a woody stem, producing new cells as long as the plant lives. It also separates a woody plant's *bark* (cork, cork cambium, cortex, and phloem) from the plant's *wood* (hardened, lignin-filled xylem cells).

YEARLY GROWTH OF A WOODY STEM

Plants not only put out new primary growth but also increase in size through secondary growth. Each spring, a woody stem produces lots of new xylem to get ready for the flurry of growth that's about to take place, decreasing that production as summer comes on. This difference in seasonal xylem production creates **annual rings** that can give us a good estimate of how old a tree is. Annual rings can also tell the story of a tree. If a tree lives through a very rainy spring, the spring ring for that year will be wide and light colored; we call it *springwood*. As temperatures get hotter and rain lessens, the summer ring will be thinner and darker, forming *summerwood*. One summerwood ring and one springwood ring make up one annual ring.

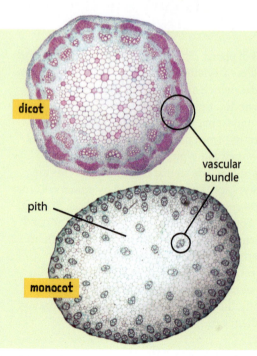

HERBACEOUS DICOT AND MONOCOT STEMS

The spongy ground tissue in the center of herbaceous stems is known as **pith**. It serves primarily to store food and water. As stems age, the pith is crushed by the expanding xylem layers. In a mature stem, such as a tree trunk, most of the stem is xylem with either no pith or only a small core of pith remaining.

As in leaves, xylem and phloem are arranged in vascular bundles in stems. In herbaceous dicots, vascular bundles are arranged in a ring; in monocots, vascular bundles are scattered throughout the pith. The main difference between dicot and monocot stems is that monocots typically have no secondary growth because they lack a cork cambium, cortex, and vascular cambium. See page 315 for a discussion of monocots and dicots.

Roots

How does a tree that's 20 m tall stay upright? How do vines of ivy climb walls? They use their roots. Roots are the primary organs for anchoring a plant and normally grow from the stem. They also absorb water and nutrients and store food that was made in the leaves. Roots are also important in asexual reproduction. If you can coax a part of a plant to grow roots, then that part can become a separate plant.

Plants have one of two types of root systems. A **taproot system** consists of one main root—the taproot—and lots of small roots branching from it. Many dicots have taproot systems. Carrots and beets are actually enormous taproots. Monocots typically have **fibrous root systems**, which are many small roots that come straight from the stem and not a taproot.

STRUCTURE OF A ROOT

Like stems, roots are capable of both primary and secondary growth. But a root doesn't have nodes or buds, so its primary growth—elongating the root—is limited to the meristematic tissue at its tip. The root tip is covered by a root cap, which is made of dead cells for protecting the delicate, fast-growing tissue.

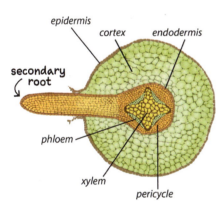

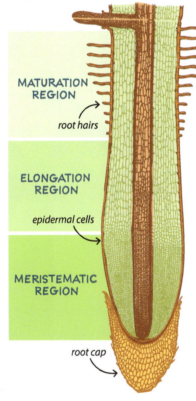

Roots are covered by an epidermis. Along certain portions of a root, the epidermis will extend to form root hairs, which are used to increase a root system's surface area, allowing a plant to find and absorb water that the main roots can't reach. Under the epidermis is the *cortex*, which stores materials, helps a plant take in water and minerals, and is made of ground tissue. Just inside the cortex is the *endodermis*, a thin layer that is considered part of the cortex and is made of tightly packed cells that carefully regulate the flow of materials between the cortex and the vascular tissues. In between the endodermis and vascular tissues is the *pericycle*, a meristematic tissue used to form secondary roots that branch off the main root. The pericycle produces roots that are different from the root hairs of the epidermis: roots that grow from the pericycle are covered in dead cork cells instead of an epidermis, so they are able to absorb water only through their tips.

MINI LAB

Using Plant Parts

Humans have learned how to use a bewildering variety of plants for foods, medicines, materials, and other applications. You probably know that a carrot is a root and an apple is a fruit, but what about other plant-based foods? Let's see how well you know your roots, stems, and leaves!

Materials
examples of plant-based foods

PROCEDURE

Examine the specimens provided by your teacher. Discuss with your partners whether each specimen is derived from the leaves, stem, or roots of a plant. If possible, you may be even more specific; for example, a specimen might be derived from seeds rather than whole fruits. Have a member of your group record your conclusions and be prepared to discuss them with the rest of the class.

Which parts of plants do we eat?

14.2 SECTION REVIEW

1. Determine the tissue type for each plant part in the table.

Plant Part	Structure	Tissue Type
vein	leaf	
cortex	stem	
root tip	root	
spongy mesophyll	leaf	
root hair	root	

2. What would happen if the palisade mesophyll were below the spongy mesophyll?

3. You are asked to identify a plant. Its leaves exhibit netted venation, and it does not appear to experience any secondary growth. Is this plant a monocot or a dicot? Explain.

4. Explain the difference between primary and secondary growth.

5. Why does a woody stem require a cork cambium but a herbaceous stem does not?

Use the case study on page 307 to answer Questions 6–8.

6. Why is a long taproot normally beneficial for a large woody dicot such as a tree?

7. How do the roots of a coast redwood support the tree?

8. Coast redwoods are common ornamental trees, but arborists (experts in the care and cultivation of trees) do not recommend planting them as individual specimens. Why do you think this is so?

14.3 THE LIFE CYCLES OF PLANTS

Alternation of Generations

Unlike animals, which are diploid and produce haploid gametes through meiosis, plants produce an entire haploid structure called a **gametophyte**. The gametophyte then produces haploid *gametes* (e.g., sperm and eggs). These gametes combine to form a diploid zygote (e.g., a seed). Once the zygote germinates, the diploid plant that emerges is known as a **sporophyte**. A few cells of the sporophyte undergo meiosis to produce haploid spores. The spores germinate to form a gametophyte, beginning the process again. This is the process of alternation of generations, the sexual reproductive cycle of all plants.

In ferns, cone-producing plants, and flowering plants, the sporophyte stage is dominant and is therefore the most visible. In non-vascular plants like mosses, the gametophyte stage is dominant.

Why don't all plants have flowers?

Questions

What are the various ways that plants reproduce?

How does a plant grow from a seed and produce seeds of its own?

How can the parts of a plant be used for both beneficial and harmful purposes?

Terms

gametophyte
sporophyte
sorus
frond
petal
sepal
pedicel
receptacle
carpel
stamen
ovule
double fertilization
endosperm
cotyledon
monocot
dicot
germination
fruit

LIFE CYCLE of MOSSES

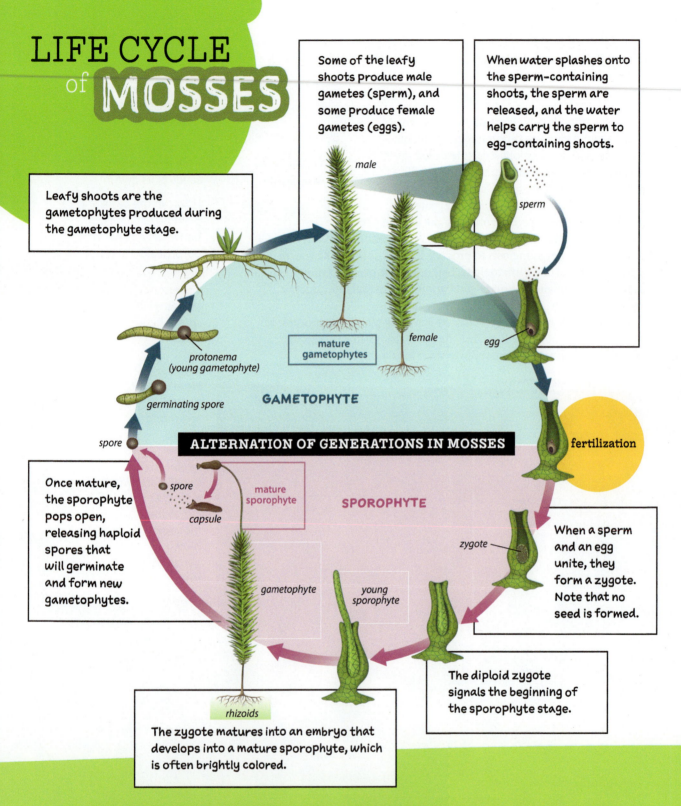

Leafy shoots are the gametophytes produced during the gametophyte stage.

Some of the leafy shoots produce male gametes (sperm), and some produce female gametes (eggs).

When water splashes onto the sperm-containing shoots, the sperm are released, and the water helps carry the sperm to egg-containing shoots.

male

sperm

protonema (young gametophyte)

mature gametophytes

female

egg

germinating spore

GAMETOPHYTE

spore

ALTERNATION OF GENERATIONS IN MOSSES

fertilization

Once mature, the sporophyte pops open, releasing haploid spores that will germinate and form new gametophytes.

spore

mature sporophyte

capsule

SPOROPHYTE

zygote

When a sperm and an egg unite, they form a zygote. Note that no seed is formed.

gametophyte

young sporophyte

rhizoids

The diploid zygote signals the beginning of the sporophyte stage.

The zygote matures into an embryo that develops into a mature sporophyte, which is often brightly colored.

Mosses, shown above with both sporophyte and gametophyte stages visible, are the largest group of bryophytes and are incredibly adaptable. They can be found anywhere from the harsh cold of the tundra to the hot humidity of the tropical rainforest. You've likely seen moss growing on a rock beside a stream. The velvet green that's visible is the *leafy shoot* of the moss plant. Leafy shoots are responsible for collecting water and nutrients for the plant, much like the way that a paper towel absorbs water. *Rhizoids* are the "roots" that anchor a moss plant in place. They can't absorb water and nutrients like true roots can, so water from the leafy shoots keeps rhizoids moist and nourished.

On the underside of fern fronds are rows of tiny bumps called **sori** (s. sorus). Sori are structures that house and protect a fern's *sporangia* (s. sporangium), which produce spores.

After release, the spores germinate to begin the gametophyte stage of the fern's sexual reproduction.

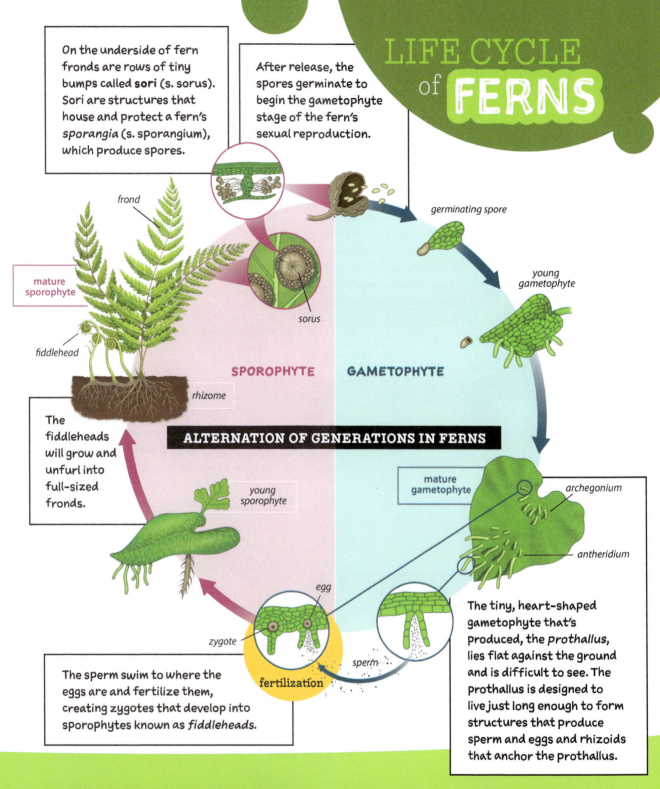

frond

mature sporophyte

fiddlehead

rhizome

sorus

SPOROPHYTE

GAMETOPHYTE

germinating spore

young gametophyte

ALTERNATION OF GENERATIONS IN FERNS

The fiddleheads will grow and unfurl into full-sized fronds.

young sporophyte

mature gametophyte

archegonium

antheridium

egg

zygote

fertilization

sperm

The sperm swim to where the eggs are and fertilize them, creating zygotes that develop into sporophytes known as *fiddleheads*.

The tiny, heart-shaped gametophyte that's produced, the *prothallus*, lies flat against the ground and is difficult to see. The prothallus is designed to live just long enough to form structures that produce sperm and eggs and rhizoids that anchor the prothallus.

Under the shade of tall oaks in Louisiana grows the southern wood fern. The oaks do more than just shade wood ferns; they also play host to resurrection ferns, which grow on the trees' branches. Resurrection ferns go dormant during dry periods and can remain that way for up to a hundred years. Though they often look delicate, ferns and their allies (club mosses, horsetails, and whisk ferns) are an amazingly tenacious and diverse group.

The leaves of a fern are usually called **fronds**. Fronds are vascular, so they can grow to be quite large in some species. A fern's underground stems are known as *rhizomes*, which produce roots, store nutrients for the rest of the fern, and help the fern multiply and spread through asexual reproduction.

LIFE CYCLE of GYMNOSPERMS

Every Christmas, gymnosperms get plenty of attention as their cones and boughs are used to celebrate the season. Gymnosperms can be found in every climate in the world, but the largest group—the conifers—are famous for enduring bitterly cold, snowy weather. Part of the reason for their durability is that, instead of using fragile flowers and soft fruits to produce and protect their seeds, they make their seeds in hardy cones.

Conifers and most other gymnosperms produce two types of cones. Pollen cones are small male cones that house structures producing sperm-containing *pollen*.

MEIOSIS

microspore mother cells

FORMATION OF POLLEN

microspore

pollen (gametophyte)

pollen-bearing cone

seed-bearing cone

Seed cones are large female cones that are made of scales, with each scale containing two eggs. Before fertilization, seed cones are green.

As pollen makes its way into seed cones and fertilizes the eggs inside, the seed cone closes to protect the developing zygotes and seeds.

LIFE CYCLE OF A PINE

mature sporophyte

stages of seedling

pine seeds

mature cone

megaspore mother cell

MEIOSIS

FORMATION OF EGGS AND FERTILIZATION

megaspore

embryonic sporophyte

egg (gametophyte)

pollen

fertilization

When conditions are right, the seed cone, now brown and woody, opens up to release its seeds.

Instead of the zygotes turning directly into sporophytes, they remain inside a seed.

LIFE CYCLE of ANGIOSPERMS

American beautyberries, mustard plants, magnolias, and grape vines are just a few of the hundreds of thousands of angiosperms that give us flowers and fruits to enjoy. Because angiosperms make up over 80% of all plant species on the earth, we'll focus our study of plants on them. Angiosperms have three basic reproductive parts—flowers, seeds, and fruits.

Flower Structure and Function
Though flowers come in all shapes and sizes, they have the same general parts.

Petals are the often brightly colored, scented part of plants and are useful in attracting pollinators.

Sepals are usually green, leaflike structures that protect flower buds as they form and support the flower's petals after it blooms.

The **pedicel** is the stalk that connects the flower to the rest of the plant.

The **receptacle** is the thickened top part of the pedicel that supports the actual flower parts.

LIFE CYCLE of ANGIOSPERMS

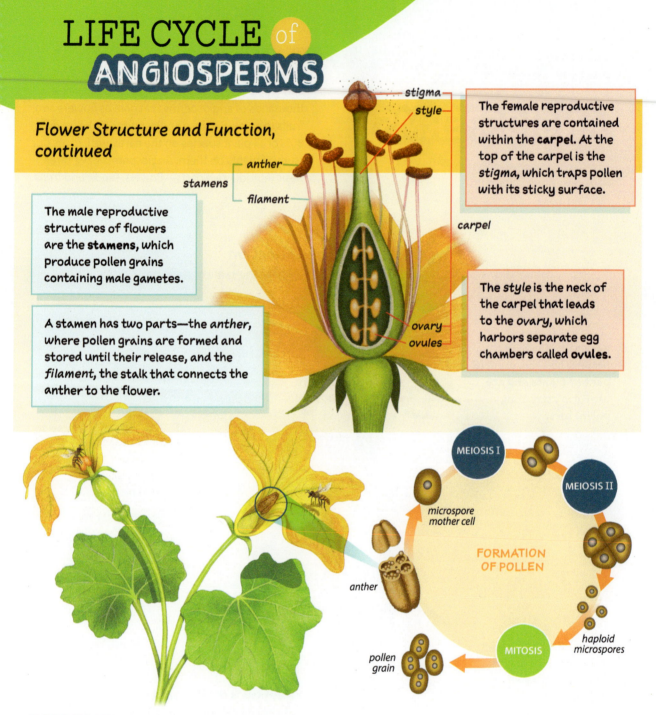

Flower Structure and Function, continued

The male reproductive structures of flowers are the **stamens**, which produce pollen grains containing male gametes.

A stamen has two parts—the *anther*, where pollen grains are formed and stored until their release, and the *filament*, the stalk that connects the anther to the flower.

stigma
style

anther
stamens
filament

The female reproductive structures are contained within the **carpel**. At the top of the carpel is the *stigma*, which traps pollen with its sticky surface.

carpel

The *style* is the neck of the carpel that leads to the *ovary*, which harbors separate egg chambers called **ovules**.

ovary
ovules

MEIOSIS I

MEIOSIS II

microspore mother cell

FORMATION OF POLLEN

anther

haploid microspores

MITOSIS

pollen grain

POLLINATION

The parts of a flower are designed to produce gametes and seeds so that a flowering plant can generate offspring. But how do these parts work together?

Pollen grains leave the anther when carried by the wind, water, or an animal. To attract birds, insects, and other potential pollinators, flowers produce *nectar*, a sweet-tasting liquid secreted at the base of the petals. As pollinators flit from flower to flower drinking nectar, they brush against the anthers and take pollen with them. Plants depend on pollinators because most plants cannot *self-pollinate* (produce

seeds from the pollen and eggs of the same flower). They must be *cross-pollinated*, exchanging their pollen with other plants of their species.

During pollination, pollen grains land on the sticky stigma of a flower and produce a *pollen tube*, which begins to tunnel into the style. The pollen tube allows the male gametes to reach the female gametes inside the ovary's ovules. In gymnosperms, pollination is similar. The difference is that pollen forms a pollen tube directly into the ovules of a gymnosperm—there are no flower parts to tunnel through.

FERTILIZATION

Once the pollen tube reaches an ovule, one sperm fertilizes the egg cell (forming the zygote), and other sperm fertilize other cells in the ovule, called *polar bodies*. The fertilization of both the egg and the polar bodies is known as **double fertilization** and is unique to plant reproduction. While the zygote develops into an embryo, the fertilized polar bodies become the **endosperm**, which is the stored food inside a seed that nourishes the embryo.

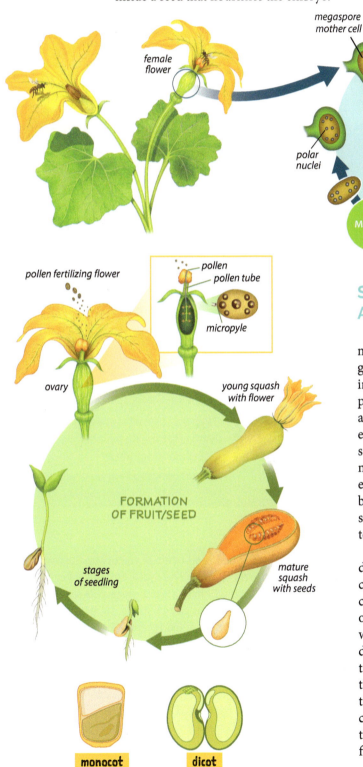

female flower

megaspore mother cell

MEIOSIS I

MEIOSIS II

haploid cells (4)

ovule

megaspore

degenerated cells

polar nuclei

FORMATION OF OVA

MITOSIS

MITOSIS

MITOSIS

MITOSIS

pollen fertilizing flower

pollen

pollen tube

micropyle

ovary

young squash with flower

FORMATION OF FRUIT/SEED

stages of seedling

mature squash with seeds

monocot

dicot

SEED FORMATION AND GERMINATION

The flower has done its job and is beginning to fade. As the embryo continues to grow inside the ovule, the tissue surrounding the ovule begins to harden, forming a protective seed coat around the embryo and endosperm. As the seed matures, the embryo becomes surrounded by leaflike structures called **cotyledons**, which provide nutrition for the embryo in addition to the endosperm. Once the seed is mature, it will break away from the ovary wall, leaving a scar on the seed called a *hilum* that is similar to a person's navel.

Some differences between monocots and dicots were mentioned in Section 14.2. That classification is based on the number of cotyledons that a plant has. Plants that have one cotyledon are **monocots**, while plants with two cotyledons are **dicots**. In most dicot seeds, the embryo uses up all the nutrients in the endosperm and then relies on the cotyledons for its food; a few species use these resources in reverse order. In monocots the single cotyledon remains small, and the embryo continues to get its nutrients from the endosperm.

From the embryo grows a *plumule*, the structure that will eventually produce the leaves and stems of the plant. The embryo also forms a *radicle*, which will act as the seedling's first root. But before these structures can emerge from the seed, something has to crack open the seed coat. The right combination of three different factors will cause **germination**: a seed forms a new plant, as explained below.

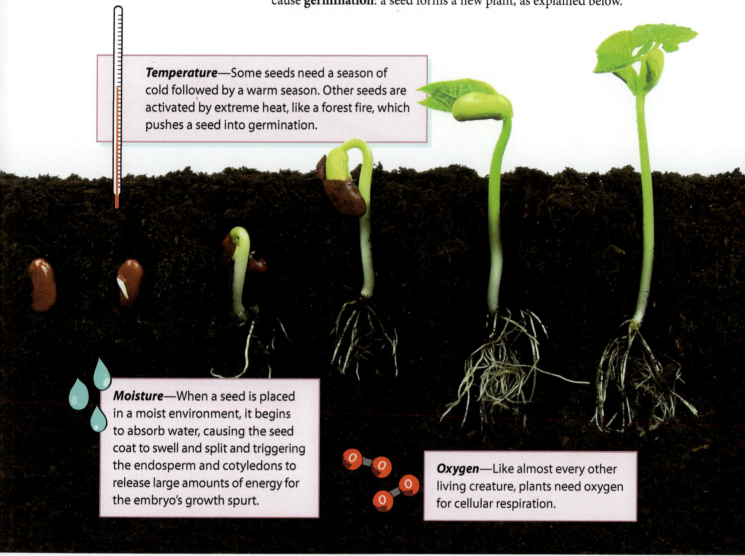

Temperature—Some seeds need a season of cold followed by a warm season. Other seeds are activated by extreme heat, like a forest fire, which pushes a seed into germination.

Moisture—When a seed is placed in a moist environment, it begins to absorb water, causing the seed coat to swell and split and triggering the endosperm and cotyledons to release large amounts of energy for the embryo's growth spurt.

Oxygen—Like almost every other living creature, plants need oxygen for cellular respiration.

FRUIT FORMATION

While the seed is forming inside an ovule in the ovary, the ovary is forming something else on its own. The ovary grows thicker and thicker to form a **fruit**, which will help a plant protect its seeds and disperse them when the time is right. Some plants, such as apples and pears, form their fruits inside an enlarged receptacle. So when we bite into an apple, we're eating what's known as an *accessory fruit*; the true fruit lies in the apple core where a thin ovary wall protects the seeds.

There are several ways to classify plants according to their fruits. *Simple fruits* form from a single ovary, while *complex fruits* grow from more than one ovary. Simple fruits can be further classified as either *dry*, such as peanuts and sunflowers, or *fleshy*, such as tomatoes and peaches.

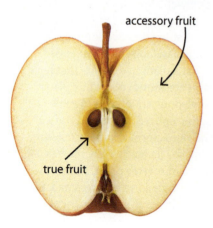

accessory fruit

true fruit

CLASSIFYING FRUITS

Simple Fruit — comes from a single ovary

Dry Fruit
hardened or leathery ovary wall at maturity

TYPE	EXAMPLE
legume—ovary wall splits in half	bean, pea, peanut
capsule—ovary wall splits open in many directions	poppy, mustard
grain—thin seed coat fused to the ovary wall	corn, wheat
nut—thick, tough ovary wall	acorn, chestnut
achene—thin ovary wall	sunflower, dandelion
samara—winged achene	maple, elm

Fleshy Fruit
thickened, fleshy ovary wall at maturity

TYPE	EXAMPLE
drupe—fleshy fruit with one seed	olive, peach, coconut
pome—thickened receptacle with a thin ovary wall around the seeds	apple, pear
true berry—very thin ovary wall	tomato, grape, blueberry
pepo—modified berry with hard rind	squash, cantaloupe
hesperidium—modified berry with leathery partitions and skin	orange, lemon

Compound Fruit — comes from multiple ovaries

TYPE	EXAMPLE
aggregate fruit—forms from several ovaries within one flower	raspberry, blackberry
multiple fruit—forms from the ovaries of several flowers that fuse together	pineapple, fig

Using Plants Wisely

A single plant can be a valuable medicine, an addictive drug, or even a deadly poison. The common foxglove is very poisonous, but its toxin, *digoxin*, is an important heart medicine. Digoxin must be used very carefully since there is only a small difference between an effective dose and a toxic one. Only a trained medical professional should ever prescribe digoxin or any other prescription drug to a patient.

A plant that epitomizes this dichotomy is the opium poppy. Opium and its derivatives, like oxycodone, are powerful painkillers that have brought relief to many patients suffering with pain resulting from accidents or surgery. But many people who are not dealing with pain take these painkillers simply for the thrill of it or to escape their problems, bringing other and often much more terrible problems into their lives.

Since ancient times, people have used and abused the medicinal power of opium. Researchers have combed through the plant kingdom for centuries to find something to combat the addictive power of opium products, collectively known as *opiates*. During the American Civil War soldiers were given the opiate *morphine* to treat pain and diarrhea. But so many soldiers became addicted to it that the addiction became known as "soldier's disease." So pharmacists created another opium-based medicine that they thought wouldn't be as addictive as morphine. They named it *heroin*. Within just a few years, physicians realized that heroin was even worse for patients than morphine. In 1924 it was declared illegal in the United States.

During World War II Polish scientist Jack Fishman began experimenting with opiates to see whether he could find a way to keep their medicinal properties and do away with their addictiveness. Fishman came to the United States after the war and soon developed naloxone. Naloxone quickly became the most popular drug on the market to fight opiate addictions and overdoses and is still commonly used today because of its relatively low cost and lack of side effects.

Poppy seeds grow in a pod at the base of the poppy flower. Opium is extracted from a white liquid that leaks out of the pod when it is slit with a knife.

Another plant that can both powerfully help and hurt people is the coca plant, native to South America and used medicinally for centuries by people who chew its leaves or brew it to make tea. Chemicals in coca leaves can relieve altitude sickness, which is caused by breathing thin air in mountainous regions. The leaves also provide vitamins and minerals that help regulate blood sugar and give colas their signature flavor. But those same chemicals can be extracted and concentrated to form the highly addictive drug *cocaine*.

God created things like opium poppies and coca plants for our good, but in our fallen state we have defiled their good uses. Countries must deal with both the drug cartels that grow these plants to make illegal drugs and the people who fall prey to them. How should we manage something that God created but people corrupted?

Many countries regulate or prohibit opium poppies and coca plants in an attempt to keep the illegal drug trade under control. But the plants then often appear elsewhere. The ultimate solution to living with easily misused plants is Christ. When people, transformed by the Spirit, rely on Christ's strength, they can resist the temptation to be controlled by drugs and allow Him to lead them out of their addiction. God also uses people to help others, giving scientists the ability to create medicines like naloxone that have helped thousands of people break away from the drugs that held them captive. Naloxone can also be used in combination with addictive pain killers like oxycodone so that addiction never happens in the first place. We should study and wisely use these potent plants so that we can gain the greatest benefit from them and help people fight the corruption that often comes with fallen man's attempts to manage the earth.

Most countries have declared coca plants illegal, but they're still legal to grow in Peru and other South American countries.

14.3 SECTION REVIEW

1. Identify the dominant generation, the gametophyte or the sporophyte, in each of the four types of plants.

2. Place the following structures of a fern in the correct order of its life cycle, following the fiddlehead: frond, gametes, prothallus, spore, zygote.

3. Are gymnosperm cones part of the gametophyte or the sporophyte structure?

4. List the two types of haploid cells that are fertilized in a plant's double fertilization. What do they form in the seed?

5. Create a flow chart to show reproduction in an angiosperm. Use the following terms: *angiosperm, anther, female flower, fruit, male flower, ovary, ovules, pollen, seeds*.

6. How could a forest fire be beneficial to a seed?

7. Pick a fruit not listed on the table on page 317. What type of fruit is it?

8. Some states allow naloxone to be sold without a prescription to drug users in case they overdose. But it's possible that drug users who buy naloxone may be less likely to seek help with their addictions. From a biblical viewpoint, is naloxone beneficial to humans? Explain.

14 CHAPTER REVIEW
Chapter Summary

14.1 KINGDOM PLANTAE

- Plants are multicellular, eukaryotic organisms that perform photosynthesis to make their own food. Plant cells have cell walls made of cellulose.

- Plant reproduction alternates between a spore-forming stage and a gamete-forming stage.

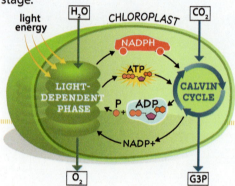

- Plant gametes fuse to form a multicellular embryo that develops into a sporophyte.

- Plants can be classified into four main groups on the basis of their means of reproduction and the presence or absence of vascular tissue: nonvascular plants (bryophytes), seedless vascular plants, flowering plants (angiosperms), and cone-bearing plants (gymnosperms).

Terms
cuticle · cellulose · alternation of generations · vascular plant · nonvascular plant · gymnosperm · bryophyte · angiosperm

14.2 THE STRUCTURE OF PLANTS

- Plants have four types of tissue: dermal, vascular, ground, and meristematic.

- Dermal tissue protects the outside of the plant, including secreting the waxy cuticle that helps seal in moisture.

- Vascular tissue consists of two types: xylem carries water and dissolved minerals from the roots to the rest of the plant, and phloem carries sugars from the leaves to the other parts of the plant.

- Most of a plant is made of ground tissue. Parenchyma cells contain chloroplasts for photosynthesis, while collenchyma and sclerenchyma cells provide support and structure.

- Meristematic tissue consists of stem cells located in the fast-growing parts of a plant, including new leaves, shoots, buds, and root tips.

- Leaves produce a plant's energy through photosynthesis. The arrangement of cells, tissues, and structures within a leaf maximizes food production.

- Stems support a plant's leaves and transport water, minerals, and sugars between the plant's roots and leaves. Stems may be either herbaceous or woody, and the arrangement of their vascular tissues differs between monocots and dicots.

- Roots anchor a plant in place and absorb water and nutrients from the surrounding soil.

Terms
dermal tissue · epidermis · cork · vascular tissue · xylem · phloem · ground tissue · meristematic tissue · blade · petiole · stoma · guard cell · node · internode · bud · cork cambium · lenticel · vascular cambium · annual ring · pith · taproot system · fibrous root system

14.3 THE LIFE CYCLES OF PLANTS

- All plants exhibit an alternation of generations between a haploid gametophyte and a diploid sporophyte. Gametophytes are the most readily seen generation in bryophytes; in other plants, the sporophyte generation is dominant.

- Bryophytes and seedless vascular plants such as ferns do not produce seeds; fertilized eggs on gametophytes mature directly into sporophytes that release spores, which germinate to form new gametophytes.

- Gymnosperm and angiosperm zygotes are protected within seeds. Gymnosperm seeds are borne by cones. Angiosperm seeds mature within fruits.

Terms
gametophyte · sporophyte · sorus · frond · petal · sepal · pedicel · receptacle · carpel · stamen · ovule · double fertilization · endosperm · cotyledon · monocot · dicot · germination · fruit

Chapter Review Questions

RECALLING FACTS

1. Why did God create plants?

2. What five characteristics distinguish plants from other living things?

3. What are two differences between a seed and a spore?

4. Which cell type provides a plant with structural support even after the cells die?

5. In the stem of a plant experiencing secondary growth, where is the vascular cambium located?

6. Identify the function of each structure listed below as either producing energy (E), providing support (S), or transporting water, food, or nutrients (T). More than one answer is possible for each.

   a. leaves

   b. stems

   c. roots

7. In the alternation of generations, is the structure that produces spores haploid or diploid?

8. What is the main structure of the gametophyte in the fern life cycle?

9. Do plants produce spores by mitosis or meiosis?

10. Trace the image above on your paper. On your drawing label the flower's carpel, pedicel, petals, sepals, and stamens.

Use the diagram on the right to answer Questions 11–12.

11. Draw the germinating seed in the figure on your paper. Label the plumule, the radicle, and the one or two cotyledons.

12. Is this plant a monocot or a dicot?

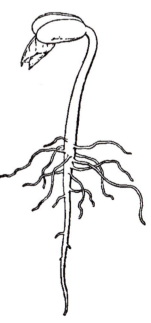

UNDERSTANDING CONCEPTS

13. Do liverworts produce seeds? Explain.

14. Which of the four basic plant tissues would liverworts not have? Explain.

15. Explain why a redwood tree growing by a stream is large, but the moss that grows on the stream bank remains small, even though it has access to the same water and nutrients.

16. Copy the table below on your own paper. Identify each plant type as nonvascular, seedless vascular, gymnosperm, or angiosperm.

Characteristics	Type
seeds produced without flowers or fruit	
always small; gametophyte the dominant structure	
largest group of plants	
spores produced in sori	

17. Would a tree be able to grow tall without cell walls? Explain.

18. What type of leaf is illustrated on the right?

19. Does the palisade mesophyll or the spongy mesophyll have a greater role to play in transpiration? Explain.

20. How are a tree's annual rings formed?

Use the diagram below to answer Questions 21–22.

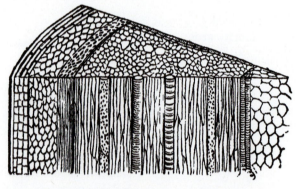

21. Draw the diagram on your paper and label the cork, cork cambium, vascular cambium, phloem, and xylem.

22. Is this stem a monocot or a dicot?

23. In the diagram on the right, correctly identify structures A, B, and C.

24. Of what type of tissue is a root's endodermis made? Explain.

25. How are pedicels and the petioles related?

26. Name one similarity and one difference between gymnosperm and angiosperm reproduction.

CRITICAL THINKING

Use the graph on the right to answer Questions 27–29.

27. Lines A and B represent two different types of leaves and how fast they perform photosynthesis in different levels of light. Sun leaves and shade leaves can be on the same plant. Which is a shade leaf, A or B? How can you tell?

28. Why can the same plant have sun and shade leaves on it?

29. Shade-loving plants have more chlorophyll in them than sun-loving plants do. Why do you think this is?

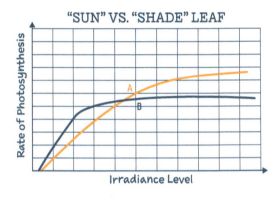

"SUN" VS. "SHADE" LEAF

30. Draw and label a process map of alternation of generations in a moss, including the words *gamete*, *gametophyte*, *spore*, and *sporophyte*. (See Appendix B.)

31. Draw a process map of the life cycle of ferns, using the terms *fiddlehead*, *frond*, *prothallus*, *rhizoid*, and *spore*.

32. Do you think that a wind-pollinated flower is likely to produce nectar? Explain.

33. Why do you think that seed cones close only after the eggs have been pollinated?

34. Classify a lime as either a simple fruit (specify dry or fleshy) or a compound fruit.

Use the case study below to answer Question 35.

35. A friend tells you that since God created plants, including *Cannabis*, we are free to use them in any manner we choose. How would you respond to your friend's claim?

case study

CANNABIS—THE GOOD AND THE BAD

Just as opium poppies and coca, *Cannabis sativum* has a long history of cultivation and can be used for many purposes, some more noble than others. Cannabis plants contain a psychoactive drug called *tetrahydrocannabinol*, or *THC*. Cannabis with very low levels of THC is known as *hemp*. Hemp fiber can be used to make paper, textiles, and biodegradable plastic, among many other useful things. But the THC in cannabis has also been used as a recreational drug. Some cannabis strains, usually referred to in this context as *marijuana*, have been selectively bred to have higher amounts of THC. The amount is often many times more than that found in industrial hemp.

Most Christians will probably agree that honoring God with our bodies and lifestyle choices means avoiding the use of recreational drugs. But in more recent years a new cannabis product has become widely available—*cannabidiol*, or *CBD*. CBD is not a psychoactive drug like THC, and many lofty claims have been made regarding its effectiveness in treating a variety of health issues. Most of these claims have not been thoroughly tested. In fact, as of this writing, prescription-strength CBD has been approved by the FDA only for the prevention of seizures. But despite a lack of supporting scientific evidence, many people continue to use products containing CBD, including dietary supplements and cosmetics.

15 PLANT PROCESSES

Weeding Out the Supervillains

Sam Walter is a soybean farmer who needs a good crop to have the money to support his family. But in years past it seemed as though each time his plants sprouted, insects ate them before they got a chance to produce beans. And if his plants survived the insects, they were quickly choked by weeds. So for the past few years he has bought the latest in genetically modified (GM) soy plants, which are resistant to pests and to the weed killer glyphosate. But the weeds have begun to fight back, and several species are now resistant to glyphosate. Though GM plants seem like the perfect solution to food-shortage problems, many people are fearful of their effects on humans and the environment. Many others balk at the price of GM plants, which can cost twice as much as traditional plants. Is there a way to create safe and inexpensive food crops that will save the day?

15.1 TRANSPORTING NUTRIENTS

Rivers of Life

?

What happens when I overfertilize a plant?

Questions

How do plants get nutrients from the soil?

How are water and nutrients transported throughout a plant?

How does sap move throughout a plant?

Terms

sap
transpiration
capillary action
cohesion-tension theory
turgor pressure
pressure-flow hypothesis

We rely on plants in many ways, so it's important for us to understand them. A plant doesn't have a heart or blood to move nutrients throughout its body. Instead, it relies on columns of water in its vascular tissues to transport necessary nutrients. God has designed some marvelous mechanisms for plants to deliver minerals and sugars where they are needed. As we learned in Chapter 14, plants have two types of vascular structures: xylem and phloem. Xylem carries primarily water and dissolved minerals from the soil. Phloem carries sugars from leaves to storage organs or areas of growth. While the fluids in xylem and phloem are different from one another, they are both often called **sap**.

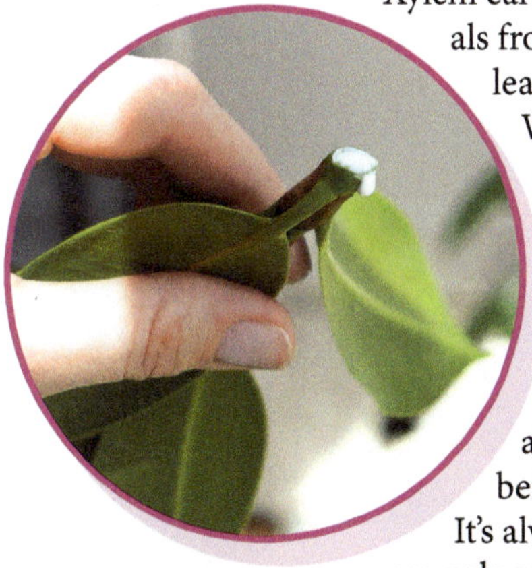

Scientists have created models to explain how xylem and phloem move sap around the plant. While these models are accepted by scientists, they do not always explain all the evidence, despite being the best models currently available. It's always important to remember that models are only representations of reality and so are always subject to revision or even replacement. For this reason, scientists must be careful not to cling to an old model when a better one comes along. Constantly testing and revising models is one of the strengths of science, but it also makes science a poor source for absolute truth. A model that is accepted today may be replaced by another model tomorrow. If we are looking for absolute, unchanging truth, we must look to the Word of God.

leaf cells

stomata

xylem

root cells

Rivers Flowing Upward

In California's Redwood National Park stands Hyperion. Scientists believe that this coast redwood, named for one of the gigantic Titans of Greek mythology, is the tallest tree in the world, standing just over 116 m tall. To survive, this tree's xylem must transport hundreds of liters of water from the ground through its leaves into the atmosphere each day in a process called **transpiration**.

The roots provide a small portion of the force necessary to move water all the way to Hyperion's highest needles. The roots' epidermis, especially the root hairs, absorbs water and dissolved minerals from the soil. Root cells use active transport to increase the amount of dissolved minerals inside the root so that osmotic pressure moves water from the soil into the root cortex. The incoming water pushes water already in the root deeper into the cortex. The water is then filtered through the endodermis as it moves into the xylem. The root pressure caused by the incoming water pushes the water and dissolved minerals up the xylem.

Tubes of various widths displaying capillary action

Water's properties also help it rise in the xylem through a process called **capillary action**. Water molecules often stick to glass due to the property of adhesion, an attraction between molecules of different substances. Adhesion is important to plants because it causes water to rise, against gravity, when inside a thin tube. Capillary action, which is greater in thinner tubes, is also responsible for some of water's movement from roots to leaves.

But root pressure and capillary action can move water only a short distance. If they were the only forces moving the *xylem sap*, the tallest trees would never be more than ten or fifteen feet tall. Any tree taller than that would die because water would not be able to make it all the way to its leaves. Root pressure and capillary action operate by pushing water up the xylem from the bottom, but most of the force that moves the water up the xylem operates *by pulling the water up from the top*. Water molecules are attracted to each other because of their strong hydrogen bonds. In Chapter 2 we learned about cohesion—the tendency of substances to stick together. Cohesion is responsible for the surface tension that allows water striders to glide across the surface of water. According to the **cohesion-tension theory**, as transpiration occurs, water molecules in the leaves' spongy mesophyll exit the leaf as water vapor through the stomata. More water moves from the xylem into the spongy mesophyll. As these molecules move out of the xylem, cohesion pulls on the other water molecules in the xylem, causing all the water to move up the entire length of the xylem.

As water moves up the xylem, it accomplishes several purposes. Some water enters into cells' central vacuoles. Full central vacuoles press cytoplasm and cell membranes against cell walls, maintaining **turgor pressure**, which gives a plant rigidity and explains why a plant that lacks water looks wilted. Some water is used along with carbon dioxide in photosynthesis to make sugar for the plant. The moving water also carries necessary minerals such as nitrates and phosphates from the soil throughout the plant. But as much as 99% of the water is simply released into the atmosphere as water vapor.

cohesion-tension theory

vessel cells

tracheid cells

low turgor pressure

H_2O H_2O

high turgor pressure

H_2O H_2O

Sweet Rivers

Xylem sap flows only upward, but the sugar that the leaves make needs to reach every part of the plant in order to provide all the plant cells with the nourishment they need. Sugars move in the phloem rather than in the xylem since *phloem sap* can move upward or downward in the plant. Cells requiring large amounts of sugars are called *sinks* and are typically either rapidly dividing meristematic tissue or storage structures such as roots and fruits. Cells that make the sugars are called *sources* and are found in the leaves and sometimes in herbaceous stems.

When we connect a hose to a spigot and turn on the water, the incoming water creates an area of high pressure at one end of the hose. Meanwhile, the water flowing out the other end of the hose creates an area of low pressure there. The water flows from the high-pressure area to the low-pressure area. This is similar to what happens in a plant. According to the **pressure-flow hypothesis**, plants move sugars through the phloem in much the same way. Source cells use active transport to move sugars into the water-filled phloem. The higher concentration of sugar in the phloem causes water to move osmotically into the phloem from nearby xylem cells. This influx of water creates an area of high pressure near the source. At the same time, the sink cells remove sugar molecules from the phloem by active transport, and the water in that part of the phloem is directed back into xylem cells, creating an area of low pressure near the sink. This pressure gradient causes the water to flow from the source to the sink, carrying the sugar molecules along with it. In an ongoing process, the sink's cells remove sugar molecules from the phloem, the source's cells export sugar molecules into the phloem, and water enters and exits the phloem, causing the movement of sugars in the phloem to continue.

Since the flow of phloem sap seems to be driven by a difference in pressures, phloem sap is able to move in any direction toward a sink. So sugars made in the cells of leaves on one branch can be sent to an area of growth at the top of the tree in the late spring, to nearby fruit in the summer, and to storage regions in the roots in the early fall.

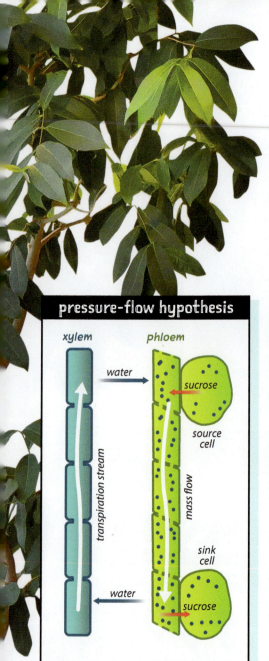

pressure-flow hypothesis

xylem   phloem

water

sucrose

source cell

transpiration stream

mass flow

sink cell

water

sucrose

15.1 SECTION REVIEW

1. Is the difference between xylem sap and phloem sap due more to their solutes or their solvents? Explain.

2. Place these structures in the order that water passes through them during transpiration: endodermis, root cortex, root hair, spongy mesophyll, stomata, and xylem.

3. What would happen to water in the stems of plants if water molecules did not exhibit hydrogen bonding?

4. What is the main difference between the direction of flow in xylem and phloem?

5. If a farmer were to apply too much fertilizer to the ground, the solute concentration would be higher in the soil than in the roots. Explain how this would affect plants.

6. Does a plant use energy when moving sap in the xylem and phloem? Explain.

7. According to the pressure-flow hypothesis, where does the water in the phloem go after the sugar has been removed by the sink cells?

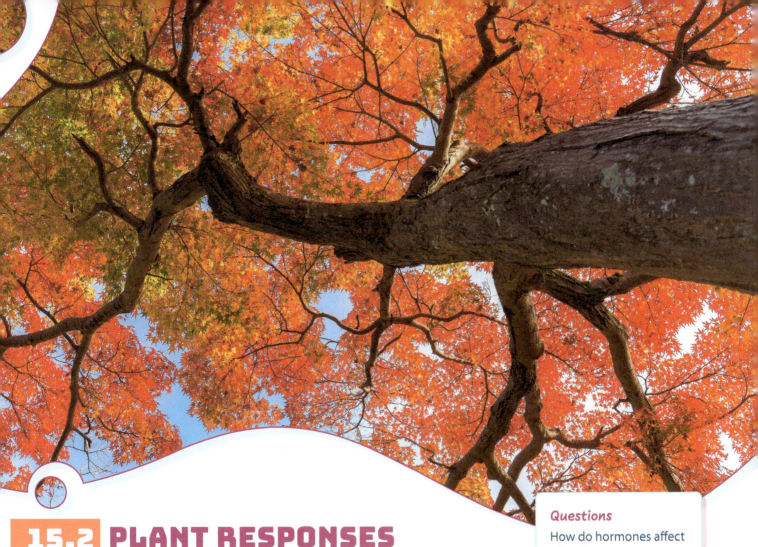

15.2 PLANT RESPONSES

Hormones

All through the cold months of winter, a black walnut sapling has stood dormant, half buried in snow. But as the warmth of the spring sun melts the snow, the sapling begins to leaf out and grow taller. How does a sapling tell its meristematic tissue to start growing in the springtime? How does a red maple lose its leaves in the fall without losing large amounts of sap from its vascular tissues? Plants are able to undergo these kinds of changes through the use of plant **hormones**—messenger molecules that are typically produced by one tissue to produce a response in other tissues. While there are numerous plant hormones, many of them can be grouped into one of five major categories: abscisic acid, auxins, cytokinins, ethylene, and gibberellins.

Scientists have learned that plant hormones very rarely work alone. Auxins may cause ethylene production, auxins and cytokinins may work together and against each other in different parts of the plant, and abscisic acid may counteract several hormones all at once. The ways that the different plant hormones affect each other and influence plant cells are still areas of intense research as scientists seek to unravel the intricacies of plants, showing to all the creativity of our Heavenly Father.

?

Can plants move?

Questions

How do hormones affect plants?

How do plants respond to their environments?

What are the effects of light on a plant?

Terms

hormone
auxin
cytokinin
gibberellin
ethylene
abscisic acid
tropism
phototropism
gravitropism
thigmotropism
hydrotropism
photoperiodism
short-day plant
long-day plant
phytochrome
day-neutral plant

PLANT HORMONES

Auxins are produced in the apical meristem of a plant shoot. They promote cell elongation and suppress cell division in the shoot. Auxins from the apical meristem suppress development of the lateral buds, so if the top of the main shoot of a plant is removed, the plant often becomes bushier. Auxins also promote lateral root development.

Cytokinins promote cell division and work with auxins to control the growth and division of cells. They also promote lateral bud development. While scientists have traditionally viewed cytokinins and auxins as working against each other, some evidence suggests that their relationship is more complex.

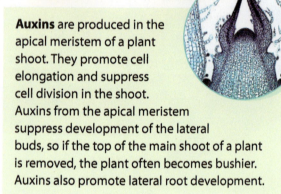

Gibberellins are a large group of plant hormones that are produced in a plant's roots. Their primary purpose is to encourage stem elongation and cell division in the shoots and leaves. They are also involved in fruit formation and seed germination. Various families of plants produce slightly different gibberellins.

Abscisic acid is an inhibitor of several of the other plant hormones, slowing both cell growth and cell division. It also causes dormancy in buds and seeds. Abscisic acid is the primary reason that seeds do not germinate when they are first released from a parent plant. A variety of factors such as heavy rains or a period of cold remove the abscisic acid from the seed, allowing gibberellins to cause the seed to germinate.

Ethylene, unlike the other hormones, is a gas. It causes fruit to ripen and trees to form abscission layers between twigs and leaf petioles, allowing leaves to drop in the fall. It also is responsible for many of a plant's responses to stress, such as injury or dehydration.

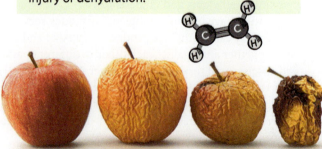

Response to Stimuli

Sometimes a stimulus causes a plant to respond without affecting the direction of that response. For example, the leaves of a Venus flytrap snap shut when triggered by an insect, no matter which direction the insect pressed the trigger hair on the leaf. This type of directionally independent response is called a *nastic movement* and is based on turgor pressure changes. Other responses, however, do respond differently to the same stimulus depending on the direction of the stimulus. These directionally dependent responses are called **tropisms**. While auxins are involved in some way with most tropisms, tropisms are complex processes involving many different molecules.

Tropisms

Since plants need light to conduct photosynthesis, plant stems exhibit positive **phototropism** by growing toward the light. For many years scientists believed that light caused auxin to concentrate in the dark side of the plant, causing that side to experience greater cell elongation. But with further experimentation scientists have observed no difference in the amount of auxin on either side of the plant. Instead, light appears to cause growth inhibitors to concentrate on the light side of the plant. Auxin, therefore, has a greater effect on the dark side of the stem. The lengthening of the dark side of the stem causes the plant to bend toward the light. Roots, on the other hand, bend away from the light, exhibiting negative phototropism.

phototropism

Light isn't the only stimulus that plants respond to. Remember from Chapter 14 that a germinating seed produces a radicle and a plumule, the plant's future root and stem, respectively. When a seed germinates, its radicle must grow downward and its plumule must grow upward, or the young plant will die. However, the seed is underground, so phototropism cannot guide the radicle or the plumule. Instead, the radicle grows toward the pull of gravity, exhibiting positive **gravitropism**. The plumule exhibits negative gravitropism, growing against the pull of gravity. Adult roots and stems continue to exhibit gravitropism throughout a plant's life. Auxins are involved in gravitropism, but scientists are still investigating the mechanisms that God created to ensure the survival of plant seedlings.

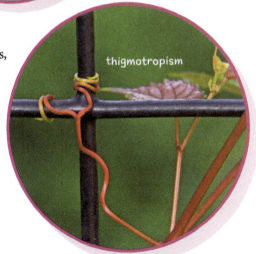
gravitropism

All plants exhibit phototropism and gravitropism, but some plants, primarily climbing plants such as vines, exhibit **thigmotropism**. Plants such as grapevines have specialized stems called *tendrils*. When a tendril touches an object, the cells on the side opposite the object elongate so that the tendril begins to bend around the object. Eventually, the tendril twines itself around the object, providing support for the entire plant. Some vines, such as morning glories, don't have tendrils, but their entire stems demonstrate thigmotropism by wrapping around other objects.

thigmotropism

Demonstrating a Plant Response

The Venus flytrap is known for its ability to trap prey within the clutches of its leaves. In response to external stimuli, the plant quickly closes its leaves and releases enzymes to dissolve its meal. Let's take a look at what it takes to stimulate the Venus flytrap's trapping abilities!

?

How does a Venus flytrap know when to snap shut?

Materials
Venus flytrap • thin pin (2)

PROCEDURE

A Analyze the open leaves of the Venus flytrap and record what you see.

B Using the pin, touch one of the hairs on the inside of the plant leaves. Record the plant response.

C Perform the following scenarios. Record the plant response.

- Touch one hair and then one minute later touch the same hair again.
- Touch one hair and then ten seconds later touch the same hair again.
- Touch two hairs at the same time.

D If there are any available leaves to continue testing the plant's response, create a scenario and record whether it triggered a response.

QUESTIONS

1. What causes a plant to close its leaves when triggered?

2. What did you observe when you touched one hair of the plant's leaves just one time?

3. What stimulus (or stimuli) triggered the Venus flytrap's leaves to close?

4. If a Venus flytrap has not caught any prey, its leaves will remain closed for twelve to forty-eight hours. If it has caught an insect, it may not reopen for about five to twelve days. Why do you think the plant takes longer to reopen after it has caught an insect?

GOING FURTHER

5. Perform an internet search on carnivorous plants. Research one other kind of carnivorous plant and create a Venn diagram comparing it with a Venus flytrap.

Of all the tropisms, **hydrotropism** is the least understood. While several scientists studied it in the nineteenth century, very little further research was conducted on the ability of plant roots to grow toward water until the twenty-first century. A root's positive gravitropism is much stronger than its positive hydrotropism, making hydrotropism hard to study. But with the ability to create mutants lacking gravitropism as well as the opportunity to grow plants in microgravity environments such as aboard the International Space Station, scientists have begun to take another look at hydrotropism. Like the other root tropisms, much of the ability to sense water appears to be concentrated in the root cap. To find water, a root grows and encounters moisture in the soil. When moisture is found, the root forms additional branches that supply the plant with more water.

hydrotropism

Photoperiodism

Plants can't live without light. Light provides the energy that drives photosynthesis, and phototropism causes stems to grow upward and roots to grow downward. But light has an even greater effect on plants in controlling the time of year they flower. You could plant chrysanthemums early in the spring, hoping to have some nice blossoms for July 4th. But your mums are far more likely to bloom for the autumnal equinox, since they flower according to the length of daylight and darkness, a phenomenon called **photoperiodism**.

Some plants won't flower until nighttime is longer than a certain critical length of time. The exact length of time the plant needs to be in darkness varies from one species to another. Early researchers thought that the plants required less light in a specific amount of time, so they called them **short-day plants**. We now know that it is the length of the nights and *not* the days that controls photoperiodism. If a bright light shines briefly on a short-day plant during the night, it will not flower. It perceives the interrupted long night as two short nights separated by a short day. In the world's temperate regions, some of the most familiar fall flowers—such as chrysanthemums, goldenrod, and aster—are short-day plants.

Long-day plants won't flower until the length of darkness is less than a certain amount. Just as with short-day plants, the critical length of time varies from one species of plant to another. If a bright light is shined on a long-day plant at night, it may induce it to flower by fooling it into thinking that it has experienced two short nights separated by a short day. Long-day plants, such as irises, sunflowers, and clovers, typically blossom in early summer when the nights are shortest.

short-day plants

day period <12 hrs | dark period >12 hrs
flowering

light
day period <12 hrs | dark | dark
no flowering

long-day plants

day period >12 hrs | dark period <12 hrs
flowering

day period <12 hrs | dark period >12 hrs
no flowering

light
day period <12 hrs | dark | dark
flowering

A tomato is a day-neutral plant.

The way that a plant responds to photoperiodism is a complex process, but much of it is controlled by pigments called **phytochromes**. The phytochromes active in photoperiodism exist in two different forms, P_r and P_{fr}. Plants produce P_r, and light converts it into P_{fr} during the day. At night, the P_r becomes far more abundant, as P_{fr} either reverts back to the first form or is broken down by the plant's enzymes. The exact ratio of these two forms of phytochromes present in the plant controls when it flowers. This system is very precise, sometimes preventing flowering if the dark period is as little as a minute too short or too long. A single night of the correct length is enough for some plant species to begin blossoming, while several nights of the correct length are required by other species.

While many plants are affected by photoperiodism, other plants will grow until they reach maturity and then will blossom regardless of the length of day or night if other conditions, such as moisture and temperature, are right. These plants are called **day-neutral plants** and include tomatoes, beans, and dandelions.

15.2 SECTION REVIEW

1. Considering what you have learned about the effects of ethylene gas on plants, what do you think would be the most obvious characteristic of a yellow birch that did not produce ethylene? Explain.

2. Would an acorn without abscisic acid be likely to produce an oak tree? Explain.

3. Would a mutant plant without growth inhibitors experience positive phototropism? Explain.

4. What two plant structures tend to exhibit opposite tropisms? Explain.

5. Why is thigmotropism not a nastic movement?

6. Your friend tells you that last year a plant in his yard began to flower after a night that lasted eleven hours. Can you assume that it is a long-day plant? Explain.

7. Is a mutant variety of a long-day plant without P_{fr} likely to survive in the wild? Explain.

USING PLANTS WISELY

Asexual Plant Reproduction

In Chapter 14 we discussed sexual reproduction through the uniting of pollen and egg in a plant's flowers, but some plants also reproduce asexually. Asexual reproduction in plants, also called **vegetative propagation**, produces a **clone**, an organism that is genetically identical to the parent plant. Many plants reproduce asexually using specialized underground storage structures such as *bulbs*, *corms*, or *tubers*.

? **Will we ever run out of food?**

Questions

How can plants be produced without seeds?

Why are plants important to people and the environment?

Is it right to use genetic enhancements in food crops?

Terms
vegetative propagation
clone

Bulbs, underground buds like those of an onion, produce bulblets that become new plants.

Corms, underground stems such as those on a crocus plant, often produce small cormlets that become new plants. Corms are solid, while bulbs have scales or layers.

Tubers produce new plants by sending out buds, such as the eyes on a potato. Tubers are underground stems that swell to store nutrients.

Other plants, including the strawberry, produce specialized stem structures called *stolons* that form new plants that take root some distance from the parent plant. Other plants produce new plantlets that drop off from specialized stems or leaves and take root right next to the parent plant. While many plants use stem structures to reproduce asexually, many trees like aspens use their roots to send up new shoots. As a result, whole groves of trees may be clones of the original tree connected to each other through a vast underground root system. Sometimes adjacent groves may have slight genetic differences that cause them to undergo seasonal changes at slightly different times, making the boundaries of the clonal groups obvious.

New plants rarely grow from leaves. However, the leaves of a few plants, such as the African violet, will produce new plants when they are placed in soil. Dipping the leaf in rooting hormone greatly increases the leaf's chances of forming a new plant.

strawberry plant stolon

GOING BANANAS

The popularity of bananas owes much to the fact that domesticated varieties have no seeds, unlike wild varieties. This means that they must be vegetatively propagated. In the 1950s the primary variety of banana, the Gros Michel, and the banana industry that depended on it in South America, Africa, and Asia, were all but wiped out by Panama disease, which is caused by a fungus. It was discovered that another variety, the Cavendish, was resistant to the fungal disease, so most of the banana industry began growing the Cavendish variety.

But in the latter part of the twentieth century a new strain of *F. oxysporum* appeared that attacked even the Cavendish banana. This variety of Panama disease has already spread throughout Asia, Australia, and Latin America. While the destruction of bananas would be only an inconvenience in Europe and the United States, it would be a disaster in Africa and Latin America. Bananas are a major part of the diets of those regions, and the banana industry employs millions of workers throughout the world.

Introduction
Imagine that you are a researcher working for a major banana producer. To keep your company's banana business going, you and your team of researchers must find a solution and create a plan to produce disease-resistant bananas.

Task
Your mission is to research possible ways to save the banana plantations. You will present your findings in a four- to six-minute speech.

Procedure
1. Research Panama disease, especially how it affects banana plants and how it spreads.

2. Research some possible solutions to the issue and some of the research being done in affected areas.

3. Plan your presentation. Describe what you think is the most promising solution and what additional work needs to be done to implement it.

4. Create your presentation and show it to another person.

5. Deliver your presentation to your class or family.

Conclusion
This example highlights the long-term effects of modifying and vegetatively propagating plants, especially those used for food. While there is currently no cure for Panama disease, much research is being done to find a solution to the problem. Given that bananas provide livelihoods for millions of people and are a major source of nutrition for millions more, investing in this research is a good example of using science to show love to our fellow human beings.

People and Plants

When God gave the Creation Mandate to Adam and Eve and their descendants, His command included using plants. Plants fulfill many human needs. For example, biofuels made from wheat, corn, soybean, and sugarcane can be used in transportation. Houses are made from wood, long grasses, or refined materials made from other plant products. Some clothing is made from cotton fibers and linen from the flax plant. In much of the world even today, people keep their houses warm by burning wood or other plant products. But plants' most significant contribution to our well-being is in the production of food.

Most plants that we eat typically reproduce through sexual reproduction. Sexual reproduction can create a problem for farmers since it almost always results in a mixing of alleles. As a result, a farmer who plants seeds hoping to grow a whole crop like the parent plant will probably end up with plants that are very different from each other and from the parent plant, unlike the trees in the orchard shown at right, which are probably genetically identical to each other and to an original parent plant.

While it is possible to completely isolate a crop from other varieties of the same species so that the seeds will produce offspring just like the parents, doing so presents several major problems. To isolate a group of plants, they would need to be placed inside a greenhouse. But since the greenhouse will also keep out both the wind and pollinating insects, the plants would need to be hand pollinated. And if the plant in question were a tree like an apple or pear, then the greenhouse wouldn't be feasible even if hand pollination were used because the tree would grow too large to be enclosed. Furthermore, some plants, including many fruit trees, will not form fruit or seeds if pollinated by a plant of the same variety. Because of these difficulties, some crops, such as potatoes, are raised entirely by asexual reproduction. When a farmer wants to grow a crop of potatoes, he just plants other potatoes. Or if he is concerned about plant diseases ruining his crop, he can buy seed potatoes—ordinary potatoes that have been certified to be free of disease.

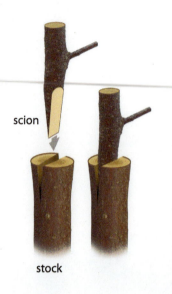

scion

stock

Asexual reproduction is natural to potatoes, but some other crops present more of a challenge. Since most fruit trees must be cross-pollinated by a tree of another variety, a fruit tree of the parent's variety cannot be grown from a seed. Instead, an orchardist who wants to plant a new apple tree uses a technique called *grafting*. A branch called a *scion* is cut from a tree and inserted into a rooted sapling, called the *stock*, during its dormant period. Different types of trees require different grafting techniques, but in all of them the goal is to connect the xylem and phloem of the scion and the stock. The scion then grows to be a tree using the stock's root system. Typically, only the scion grows, and any branches and leaves of the stock are pruned, but some orchard companies sell apple trees that have branches of several varieties of apple trees grafted together. At least one company even sells a fruit tree that bears apricots, peaches, and plums on different branches!

Some plants are produced by *layering*, a process in which a branch of the plant, usually a bushy plant such as a holly or azalea, is bent to the ground and a portion of it is buried. The remaining part of the branch is staked to force it to grow upward. The branch continues to receive nourishment from the parent plant until sufficient roots develop. At that time, the new plant is cut from the parent plant.

simple layering

Some of the plants that people use would not be able to survive on their own without human intervention. (See the worldview investigation on page 336.) As we learned in Chapter 9, bananas are triploid, which causes them to have no seeds. While this is good for banana lovers, banana growers must grow new banana plants from small daughter plants that the parent plant produces. Seedless watermelons are also triploid plants and cannot reproduce at all.

Vegetative propagation may cause other problems too. It produces a large group of plants with identical genotypes. Because of the limited genetic diversity, these plants are very susceptible to diseases. A disease that kills one plant may very well kill most of them. In a species propagated with seeds, the larger amount of genetic diversity makes it much more likely that at least a few plants will be resistant to a particular disease or pest.

Plants in the Environment

Besides providing us with food, clothing, and shelter, plants also serve an important role in ecosystems. Transpiration releases water from the ground into the atmosphere where it joins water evaporated from oceans, lakes, and rivers to form clouds. In addition to being a major part of the water cycle, plants also help to filter water, especially in watersheds. Often fertilizer and other nutrients enter runoff water from fields. When plants absorb the water, they also use these nutrient molecules. While these nutrients help plants, unabsorbed fertilizer pollutes *aquifers*, underground water reservoirs that are the source of water for most wells. Plants filter water before it enters aquifers, streams, and lakes, helping to ensure that people have clean drinking water.

Plants can also have a huge effect on local climate. In some cities the absence of plants and the large areas covered by buildings, concrete, and asphalt cause the temperature in urban areas to be significantly higher than the temperature in the surrounding rural areas. To counter this, urban planners in recent years have increased the number of green spaces in cities. Some are even growing plants on top of buildings.

Plants are also very useful for controlling erosion. Their roots hold soil particles in place, lessen the force of moving water, and act as windbreaks. For example, during the 1920s many farmers on the Great Plains plowed up the grasslands to plant wheat. But a series of droughts during the 1930s killed the wheat, leaving only dry, bare ground. Then strong winds picked up the dry soil and blew it away, sometimes as far as New York City and Washington DC. Many farmers in the Great Plains were forced to abandon their farms. The Dust Bowl, the name given to these events, is an example of massive wind erosion. Erosion occurs when loose soil is moved from one place to another, typically washed away by water or blown away by the wind.

Areas that lose soil often cannot support plant life for some time afterward. In addition, eroded soil causes other problems ranging from landslides to large amounts of fertilizer in runoff that cause algal blooms in lakes and rivers. Farmers use plants' ability to stop erosion by planting cover crops in the winter to hold soil in place.

Virtually all organisms on Earth depend on the sun to keep a supply of energy entering the ecosystem. But many organisms can't use sunlight directly. Plants use the sun's energy to power photosynthesis, producing sugars that they can store in tissues to feed animals and people. Plants also use carbon dioxide and release oxygen during the photosynthetic process to replenish the atmosphere for people and animals to breathe. Without plants, life on Earth would be difficult, if not impossible. God has given humans the gift of plants to wisely use and manage for His glory and for the good of all.

15.3 SECTION REVIEW

1. What is the primary distinction between a plant produced by vegetative propagation and one produced by sexual reproduction?

2. Corms, tubers, and stolons are technically examples of what plant structure?

3. If you were a gardener who wanted to collect seeds from a zucchini, what steps could you take to prevent the zucchini flowers from being pollinated by another variety of squash?

4. Why are most apple trees produced by grafting rather than by planting apple seeds?

5. List three ways that people use plants other than for food.

6. How do plants help keep fertilizer and other pollutants out of aquifers?

7. How do plants help prevent erosion?

15.1 TRANSPORTING NUTRIENTS

- Transpiration occurs when water is released through the stomata of the leaves of a plant.

- Root pressure and capillary action are responsible for pushing water in the xylem upward a short distance.

- The cohesion-tension theory describes how tension pulls water through the xylem from the roots to the leaves.

- Xylem moves sap upward, and phloem moves sugars around to different locations in a plant.

- Plants move sugar through the phloem from a high-pressure area to a low-pressure area. Sap can flow in any direction throughout a plant.

Terms
sap • transpiration • capillary action • cohesion-tension theory • turgor pressure • pressure-flow hypothesis

15.2 PLANT RESPONSES

- Abscisic acid, auxins, cytokinins, ethylene, and gibberellins are categories of plant hormones that act as messenger molecules to produce responses in plants.

- Plants respond to various stimuli. Nastic movements are directionally independent responses, and tropisms are directionally dependent responses.

- Tropisms include phototropism (response to light), gravitropism (response to gravity), thigmotropism (response to touch), and hydrotropism (response to water).

- Photoperiodism is a response of a plant to changes in the duration and intensity of light exposure.

Terms
hormone • auxin • cytokinin • gibberellin • ethylene • abscisic acid • tropism • phototropism • gravitropism • thigmotropism • hydrotropism • photoperiodism • short-day plant • long-day plant • phytochrome • day-neutral plant

15.3 USING PLANTS WISELY

- Vegetative propagation is a form of asexual reproduction in which a plant produces a clone that is genetically identical to itself.

- Plants can be used as food, fuel, building materials, and clothing fibers.

- In addition to providing food and oxygen, plant roles in the environment include transpiring water vapor into the atmosphere as part of the water cycle, stabilizing temperatures in cities, and controlling erosion.

Terms
vegetative propagation • clone

Chapter Review Questions

RECALLING FACTS

1. Which of the two types of vascular tissue is more likely to carry phosphates from fertilizer? Explain.

2. If xylem and phloem are analogous to the human circulatory system, what in a plant is analogous to human blood?

3. What property of water molecules causes sap to rise in xylem? Explain.

4. What type of water movement is essential for phloem to move sugars between sources and sinks? Explain.

5. If a plant's apical meristem produces a large amount of auxin, what general shape will the plant have?

6. The potato is a specialized form of which plant structure?

7. Plants keep runoff water from reaching aquifers. What happens to most of this water?

UNDERSTANDING CONCEPTS

8. Would the xylem or the phloem be more directly affected if a plant's stomata permanently closed? Explain.

9. Is a fruit a source or a sink? Explain.

10. Create a concept map of the pressure-flow hypothesis using the following terms: *high pressure, low pressure, sink cell, source cell, sugar, xylem*, and *water*.

11. What two plant hormones work in concert with each other?

12. What would be most obvious about a mutant grapevine lacking thigmotropism?

13. Explain how roots experience hydrotropism.

14. Would a plant's stem and roots grow in opposite directions in a microgravity environment such as that onboard the International Space Station? Explain.

15. A mutation to what substance is most likely to change a plant's photoperiodism? Explain.

16. Which type of plants tend to bloom during the late spring and early summer?

17. Why does vegetative propagation limit genetic diversity?

18. List three structures involved in vegetative propagation.

19. Does an apple tree produced by grafting most resemble the stock or the scion? Explain.

20. Could an apple tree that bears two different kinds of apples on different branches be produced by sexual reproduction? Explain.

21. How could a farmer use plants to prevent excess fertilizer from washing out of his field into a nearby stream?

22. Explain the importance of plants to biogeochemical cycles.

CRITICAL THINKING

23. Suppose you went on vacation for two weeks, and when you came home, you discovered that the plants you left near the windowsill have wilted. Explain what caused the plants to lose their rigidity. What should be done to get them to return to their natural state?

24. Which plant hormone is most likely to be used to determine whether two plants are in different families? Explain.

25. Several plants of one species are provided with plenty of nutrients and have reached maturity without blooming in a greenhouse where lights are turned off to allow for eight hours of darkness a day. When the period of darkness is increased to nine hours, they flower. What type of photoperiodism do these plants exhibit? Explain.

26. Two strawberry plants growing side by side are attacked by a fungal disease. One is genetically resistant to the disease and recovers, while the other dies. Were these two plants produced from stolons from the same plant? Explain.

case study

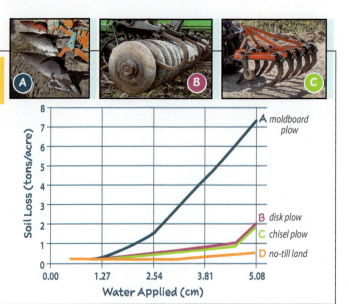

SOIL EROSION

Soil erosion is a major problem for farmers. Erosion removes nutrients from soil and often washes excess fertilizer and pesticides into nearby streams and rivers where they act as pollutants. Various methods of tilling, including not plowing at all, have been developed. The graph at right shows the amount of soil erosion associated with the different tilling methods.

Use the case study above to answer Questions 27–32.

27. How much water is applied when erosion rates suddenly increase for disk and chisel plowing?

28. Why does this spike occur?

29. Why does a similar spike not occur with no-till agriculture?

30. If the graph were extended to the right, would such a spike occur for no-till agriculture?

31. A similar increase in the rate of erosion occurs for agriculture using moldboard plows. How much water is applied when this increase occurs?

32. Is erosion affected only by the plowing method? Explain.

GENETICALLY MODIFIED FOODS

Researchers who wear personal protection equipment are usually required by federal regulations to do so when working with certain chemicals that are used with both GM and non-GM crops, including some organic crops.

ISSUE

Genetically modified foods, otherwise known as "GM foods," have been the center of ethical debate for quite some time. Those who support modifying the genes of crops see value in altering foods to fit the needs of people and support a growing population. Others question the long-term effects on people and animals who consume GM foods. Many GM foods have been created with the intent of increasing a healthy food supply by reducing insects, pests, and diseases that are harmful to crops. They also have the benefits of adding nutritional value to foods and making produce more attractive.

When the first GM tomato became available in 1994, the market for GM foods began to widen. Research companies began expanding their testing and new GM crops began popping up. Since that time, teams of researchers have studied the health effects of GM foods. Even with research suggesting that GM foods are safe to consume, people often wonder whether there has been enough testing and research done to truly understand the long-term effects. With all we know and don't know about GM foods, should we eat them?

Work through this issue using the guiding questions from the biblical ethics triad.

1. What information can I get about this issue?
2. What does the Bible say about this issue?
3. What are the acceptable and unacceptable options of consuming GM foods?
4. What are the motivations of the acceptable options?
5. What action should I take?

Use the ethics box to answer Question 33.

33. Using the biblical ethics triad questions above, formulate an essay on the Christian position of consuming genetically modified foods. Be sure to address each leg of the triad: biblical principles, biblical outcomes, and biblical motivations.

UNIT 4
ANIMALS

16 INVERTEBRATES

16.1 KINGDOM ANIMALIA

Animal Attributes

God has filled our world with animals, and He has commanded us to wisely manage them for His glory. God values animal life (Jon. 4:11), and biology can be a wonderful tool for conserving animal populations, especially ones that provide valuable resources for people. People in California have been trying for many years to reverse the decline of abalone populations (see chapter opener). Historically, this was done by limiting how abalone could be caught. Instead of using scuba gear to catch large numbers of abalone, people were permitted to fish for abalone using only free-diving techniques, such as snorkeling. Coupled with laws on the number and size of abalone that could be taken, this limited the number of individuals harvested. It was hoped that this would allow populations of abalone to recover as smaller individuals were left behind to mature. But these efforts haven't been completely successful, and even more drastic measures have been implemented. For now, California has closed the abalone fishery until at least 2026. Scientists still have work to do to ensure the survival and recovery of California's abalone.

Biologists use kingdom Animalia in domain Eukarya to classify all animals. Animals are heterotrophic organisms that are made of many eukaryotic cells. Because their cells have no cell walls, animals must have another means of support. Since they move, they require a structure for their muscles to pull against. For many animals God has met these two needs by providing them with some kind of skeleton. Some animals, such as mammals and reptiles, have an internal skeleton made of lightweight bones called an **endoskeleton**. Others, such as lobsters and insects, get their support from a stiff, plated outer covering called an **exoskeleton**. Still others, such as jellyfish and starfish, use their watery environment to fill a central cavity inside their bodies to build up pressure. This structure is called a **hydroskeleton**, or a *hydrostatic skeleton*. While the carbohydrate *cellulose* is the primary chemical providing support in plants, the protein *collagen* is an essential part of the skin, muscles, bones, and connective tissues of animals.

?

What makes an animal an animal?

Questions

How are animal bodies different yet similar?

In general, how are animals classified?

How do animals reproduce?

How do animals behave in their environments?

Terms

endoskeleton
exoskeleton
hydroskeleton
body plan
bilateral symmetry
radial symmetry
sessile
vertebrate
chordate
invertebrate
endotherm
ectotherm
blastula
germ layer
endoderm
mesoderm
ectoderm
cephalization
anus
external fertilization
internal fertilization
oviparous species
viviparous species
ovoviviparous species
behavior
instinct

bilateral symmetry

radial symmetry

external asymmetry

All animals have a **body plan**, the arrangement of physical features that contribute to structure and form. Animals with similar body plans are often grouped and classified together. Different body plans demonstrate various kinds of symmetry that physically balance body parts. Animals demonstrating **bilateral symmetry** can be divided by a plane through the center in one place to form mirror images. Animals with **radial symmetry** have no left and right sides and can be divided by a plane through their center in several places to form mirror images. Some animals have no symmetry—they're said to be *asymmetrical*.

ANIMAL ATTRIBUTES

What sets an animal apart from other types of life? All animal cells are eukaryotic, which places animals in the domain Eukarya and distinguishes them from bacteria. Animals are also heterotrophs, which obtain the nutrients they need from other organisms, unlike autotrophic plants and algae. Animal cells don't have the cell walls that plants and fungi have. Finally, all animals are multicellular, unlike most prokaryotes, which are primarily unicellular or colonial.

Respiration. Animals need a way to exchange gases with their environment to continue life-sustaining cell processes. They have organs that help them get the oxygen they need for cellular respiration from the environment. Some get oxygen from air, and some from water. Animals excrete carbon dioxide as a waste product.

take in O_2

release CO_2

Nutrition. All animals are heterotrophs since they obtain their food for energy and growth from other organisms. Getting nutrition requires taking in food through ingestion, breaking down food through digestion, and getting rid of waste products through egestion. As cells digest proteins, they produce nitrogenous wastes that are expelled through excretion. Without these processes, an animal would not be able to get needed nutrients and would build up harmful waste products. Such a disruption in homeostasis would eventually kill an animal.

Circulation. Animals transport materials through their bodies to get nutrients to where they're needed. Some animals have blood or a similar fluid that circulates nutrients in vessels or bathes internal organs. Some organisms, like sponges, circulate water through their bodies to transport needed substances.

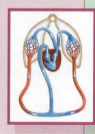

Reproduction. All animals reproduce animals after their own kind. Most animals reproduce sexually, requiring haploid gametes from two individuals to form one unique diploid zygote. Many animals also reproduce asexually, and quite a few of them reproduce both ways.

Movement. Animals have some kind of locomotion—the ability to move through their environments. Many animals move using appendages, limbs that protrude from their bodies. But not all animals move at the same speed. Some animals, such as sponges, are quite capable of motion as larvae but become **sessile** (fixed in one place) as adults, attaching themselves to objects.

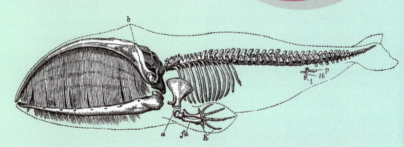

Support. Most animals have some form of support to hold their weight, with larger animals generally needing more support than smaller ones. This support can be inside or outside the body, or even separate from the body. Blue whales, for example, can grow to nearly 200 tons because ocean water provides additional support for their huge bulk.

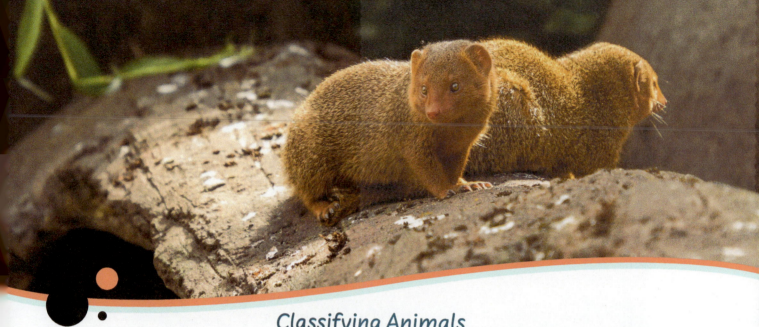

Classifying Animals

Kingdom Animalia includes many different phyla, but the animals in these phyla are grouped in different ways. Animals with backbones are called **vertebrates**. All vertebrates are classified in one phylum, phylum Chordata, which includes a few **chordates** without backbones. When we think of animals, we usually think of vertebrates such as mammals, fish, snakes, and birds. But animals that do not have backbones, called **invertebrates**, make up about 97% of all animals and cover about thirty phyla in kingdom Animalia.

Another model of classification that biologists use to group animals is based on the way that they regulate their body temperature to stay at the optimum temperature for homeostasis. An animal's temperature can affect how fast or slow the life-sustaining chemical reactions in its cells take place or even whether they take place at all. Animals must keep their body temperatures sufficiently high for their cellular processes to work fast enough to sustain life but also cool enough so that those processes don't stop working.

Animals that derive much of their body heat from their metabolism are **endotherms**, and they often possess insulating body coverings such as blubber, feathers, fur, and hair. They also have complex internal mechanisms that regulate their body temperatures to the right level for life. This is not usually the same temperature as their surrounding environment. When an endotherm becomes too hot or too cold, its body activates regulatory mechanisms that involuntarily cause sweating, panting, or shivering to restore its body to the proper temperature.

But most animals are **ectotherms**. They often have body temperatures much closer to that of their environment. Instead of using internal mechanisms, they use behaviors to establish and maintain their optimum body temperature. A green iguana may sun itself on a rock to increase its body temperature or move into the shade to cool off. A honeybee vibrates its body to generate heat before flight.

Aquatic animals live in environments that don't have big swings in temperature, but land animals in some environments must deal with temperatures during certain times of the year that are hostile to life. Some animals, both endothermic and ectothermic, escape extreme temperatures by entering a state of torpor during which their body temperatures drop and their life processes slow to a fraction of their normal levels.

Cells make up body systems, like the respiratory, nervous, and skeletal systems. Systems work together to maintain homeostasis in an organism.

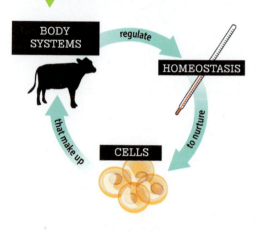

Structure in Animals

Animals are basically tubes, usually exhibiting radial or bilateral symmetry. This body plan begins when an animal is still just a fertilized egg. The zygote undergoes division to form a hollow ball of cells called a **blastula**. The cells in the wall of the blastula begin to move into the inside of the ball through a hole called the *blastopore*. As a result of this inward movement, layers of cells called **germ layers** develop in the embryo. Most animals have germ layers, typically three. The inner layer, the **endoderm**, forms the digestive tract. The **mesoderm**, the middle layer, forms the muscle and circulatory systems along with others. The **ectoderm** forms the outer layer, including the outer covering and related systems.

Germ layers influence how an animal's digestive system, circulatory system, and body cavity take shape to enable it to get nourishment, rid itself of wastes, and respond to changes in its environment so that it can maintain homeostasis. Some animals with radial symmetry, such as jellyfish, have two germ layers. Animals with three germ layers often exhibit bilateral symmetry, but some radial animals, such as sea urchins, also have three germ layers. Bilaterally symmetrical animals also tend to have their sensory organs and mouth concentrated at one end. This concentration of sensory and nerve cells is called **cephalization**. This is usually something like an animal's head. In many cases, an **anus**, where the animal expels waste from its food, is located at the other end of the body, away from the head.

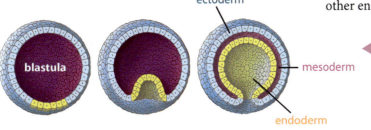

◄ The blastula folds in on itself to form three germ layers in animals that exhibit bilateral symmetry. The way that the blastula folds determines which end forms the mouth and which end forms the anus. This process is called *gastrulation*.

GERM LAYERS IN ANIMALS

	SPONGES	CNIDARIANS	SEGMENTED WORMS	MOLLUSKS
■ ectoderm ■ mesoderm ■ endoderm				
Example	sea sponge	jellyfish, sea anemone	earthworm	clam, snail, octopus, squid
Germ Layers	none	two	three	three
Body Symmetry	asymmetrical	radial	bilateral	bilateral
Cephalization	no	no	yes	yes

	ECHINODERMS	ARTHROPODS	CHORDATES
■ ectoderm ■ mesoderm ■ endoderm			
Example	starfish, sand dollar, sea urchin	butterfly, lobster, spider	deer, lizard, eagle, salmon
Germ Layers	three	three	three
Body Symmetry	radial	bilateral	bilateral
Cephalization	no	yes	yes

Reproduction in Animals

Animals, like all living things, reproduce after their kind, a clear reflection of God's created order (Gen. 1:21–22). Most animals reproduce sexually when the sperm from a male fertilizes an egg from the female to produce a zygote. This process can happen either outside or inside the female's body. In a process called **external fertilization**, many fish and other animals that live in water release eggs and sperm into the water, allowing the eggs to be fertilized outside the female's body. Land animals and some aquatic animals fertilize their eggs when sperm travels into the female's body to fertilize the eggs there; this is called **internal fertilization**. The male inserts these sperm into the female's body, or the male deposits the sperm into the environment, and the female later takes them in.

When an animal egg is fertilized, it begins to divide and develop into a multicellular embryo. This process happens in different places in different animals. The embryo of an **oviparous species** develops outside its mother's body but within the protective shelter of an egg, eventually hatching. Birds and most fish, amphibians, and insects are oviparous. The embryo often uses the yolk of the egg for nutrients needed for growth.

A developing animal that draws its nutrients directly from its mother's body is a **viviparous species**. Most mammals, some fish, some amphibians, and a few insects develop this way.

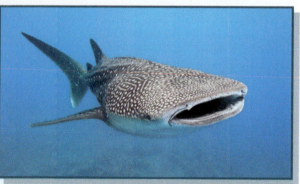

The embryo of an **ovoviviparous species** develops inside its mother's body, without being directly connected to her. Some fish and amphibians develop this way, and there are many variations on this method of reproduction.

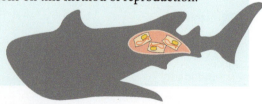

If you've ever known a mom expecting a baby, you know that reproduction places a burden on her body. The demand to support another life upsets homeostasis. For all animals, reproduction works against homeostasis in an individual yet is a necessary and vital part of being alive and sustaining a population.

MINI LAB

Identifying Animals

As we saw in this section, not all animals fit our preconceptions of what animals should look like. In this activity we'll test your ability to tell whether something is an animal.

How can I tell whether something is an animal?

Materials
preserved specimens • hand lens

PROCEDURE

A Obtain two specimens from your teacher. One is an animal, and the other is not.

B Examine each of the specimens. You may use a hand lens to get a closer look. As you do so, consider whether each specimen exhibits any of the characteristics of animals as described in Section 16.1.

QUESTIONS

1. Which characteristics of animals does each of your specimens exhibit, if any?

2. Which specimen do you think is an animal? Defend your choice.

3. Why do you think the other specimen is *not* an animal? Explain.

GOING FURTHER

4. Do some research and see whether you can identify your specimens. What is the identity of each?

Animals in the Environment

Fall is in the air in New England. Brightly colored maple leaves drift toward the forest floor while squirrels scurry among them, gathering acorns for the winter. A big black bear grazes on a few late blueberries to fatten up before turning into its cave to sleep off the winter. The "Honk! Honk!" of Canada geese migrating south passes overhead, and a cardinal on a pine branch fluffs up its feathers to prepare to weather the winter.

God's creative order and care for living things is evident in the way that He has created animals. Geese, squirrels, and black bears respond to their environment, such as by detecting the changing of seasons, according to God's design. Their bodies receive signals from the environment that activate processes and changes to deal with the changing conditions. That's why bears and cardinals increase their insulation, and snakes and woodchucks prepare to hibernate to survive the bitter cold of winter. The seasonal cycles are an example of how an animal changes what is called its **behavior**—the way that an organism responds to certain environmental conditions. Animals don't just change with the seasons; they change their behaviors every day as the sun begins to set and night comes to the forest. This is called a *circadian rhythm*.

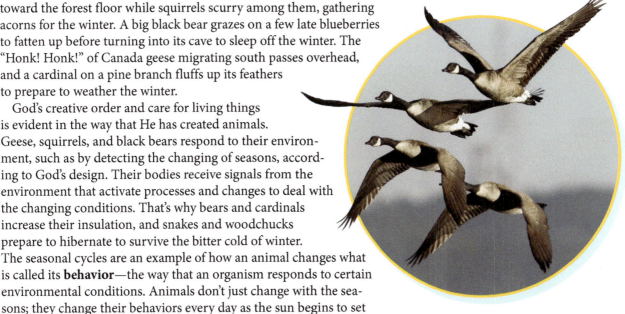

Some animal behaviors are *learned behaviors*; some are not. How do Canada geese even know which way is south? Spiders spin egg sacs in the fall without ever being taught, using a built-in response to the environment called an **instinct**. But animals must learn certain behaviors. The Canada goose must learn to fly. The black bear and squirrel must learn to forage. Where do they learn these things? Many animals learn these behaviors from their parents.

God's care and love for animals is evident in the way that He provides for their needs and gives them the knowledge and skill to respond to their environments. We can use biology to learn about animals and their behaviors so that we can use them to meet people's needs and take care of God's world.

16.1 SECTION REVIEW

1. Give an example of an animal with bilateral symmetry not mentioned in your textbook.

2. Use the characteristics of animals to show that a Holstein cow used for milk production on a dairy farm is an animal.

3. Rotifers, like the one shown below, are actually tiny animals that are only a fraction of a millimeter long. They are usually found in fresh water. Why do you think rotifers are classified as animals and not protozoans?

4. Are most animals vertebrates or invertebrates?

5. Create a T-Chart graphic organizer (see Appendix B) that compares endothermic and ectothermic animals. Include how their body temperatures compare with their environments and how they regulate their body temperatures.

6. What is the connection between the number of germ layers and the presence of cephalization in the animals in the table on page 351?

7. What is the general relationship between animal symmetry and the number of germ layers?

8. What is distinctive about how animals reproduce compared with other organisms?

9. *Echidna* is a genus of mammals from Australia and New Guinea that lays eggs. Identify this group as oviparous, ovoviviparous, or viviparous. Why is this unusual?

10. Identify each of the following as a learned behavior or an instinct.

   a. A rattlesnake moves to the sun to warm itself.

   b. A monarch butterfly navigates over Lake Superior toward its winter mating grounds in Mexico.

   c. A duckling follows its mother and siblings to swim in the local pond.

   d. A wolf cub joins the pack on its first hunt for food.

16.2 SPONGES AND CNIDARIANS

Classifying Sponges

Now that we know about the characteristics of animals, let's consider the largest category of animals—invertebrates. Every phylum in kingdom Animalia contains invertebrates except one. We'll begin by studying the different invertebrates in kingdom Animalia on the basis of their germ layers.

Sponges are odd creatures. They are invertebrates with no germ layers and no nervous, digestive, or circulatory systems. They are sessile, attaching themselves to rocks and basically staying in one place most of their lives. They are asymmetric and were actually thought by some of the ancient Greeks to be plants! So why do biologists still consider a sponge to be an animal?

When microscopes came into use in the mid-1800s, we learned that sponges have eukaryotic cells and are multicellular. Though sponges are sessile, they *move their environment through them*. They develop from zygotes to form mature organisms. Along with these factors, sponges are heterotrophic, leading biologists to classify them as animals. Sponges are classified in kingdom Animalia in the phylum Porifera and are sometimes called *poriferans*, meaning "pore bearers." If you've ever used a natural sponge for bathing, you know that sponges have lots of holes in them. Sponges can be as tiny as 1 cm across to as big as 2 m across. There are about 10,000 species of sponges around the world, some marine and some freshwater.

Why are sponges and jellyfish considered animals?

Questions

What is a sponge?

How do sponges feed and reproduce?

What is a cnidarian?

How do cnidarians feed and reproduce?

How do sponges and cnidarians contribute to the environment?

Terms

intracellular digestion
hermaphrodite
cnidarian
nerve net
polyp
medusa

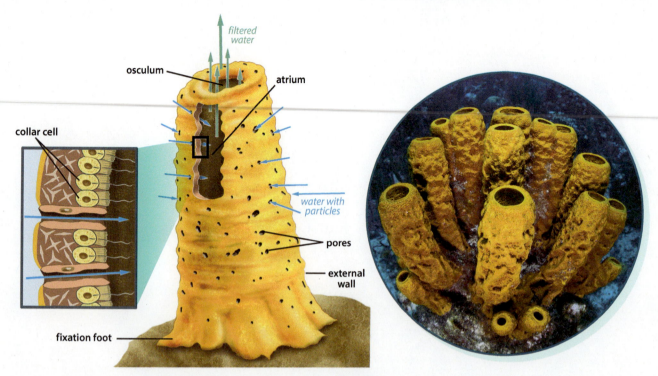

Structure and Reproduction in Sponges

SPONGE STRUCTURE

Sponges, like other animals, are basically tubes. But they are a little different in that they have only one open end. This structure is similar to a test tube. The walls of this tube are supported by spiky, stiff, needlelike structures called *spicules*. Cells in the sponge that transport nutrients also create these spicules. Some sponges are softer and contain both spicules and a spongier material called *spongin*.

Flagellated cells called *collar cells* that line the inside of the sponge beat vigorously to create a current through the sponge. The water flow through a sponge is similar to the flow of air through a chimney—in the bottom and out the top. Little pores in the sponge take in water from the environment, bringing food and nutrients with it. Collar cells engulf food particles through phagocytosis. These cells partially digest food, sending it to other cells to finish the digestion process. This allows nutrients to spread in a sponge. Waste products are ejected into the current through the sponge's *osculum*. When digestion happens inside cells instead of inside a digestive system, it is called **intracellular digestion**.

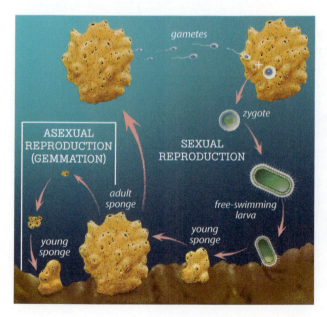

SPONGE REPRODUCTION

Sponges reproduce both sexually and asexually. They produce offspring asexually through budding, similar to a new branch growing off a tree. They also reproduce asexually by creating structures called *gemmules* that are released to grow into a clone of the parent sponge. However, sponges reproduce most often through sexual reproduction. Sponges are **hermaphrodites**; that is, they have both male and female reproductive organs. They produce both sperm and eggs. They typically release their sperm into the ocean to fertilize the eggs of another sponge. Eggs stay attached to the "mother" sponge until they hatch to release larvae capable of motion. The larvae find places to call home and attach there to mature and live the rest of their lives.

Classifying Cnidarians

While sponges are stiff, immobile animals, cnidarians are the exact opposite. Their soft, radially symmetric bodies ripple in their watery environment. **Cnidarians** are members of kingdom Animalia classified in phylum Cnidaria. They possess two germ layers. Familiar cnidarians include jellyfish, or sea jellies, with their bell shapes and painful stings; corals with their hard skeletons; and sea anemones that look like flowers of the sea.

Exploring

CNIDARIANS

Corals and Sea Anemones

Many of the different colors in this picture represent different kinds of corals. Together with the anemones, corals are classified as anthozoans. They are sessile, soft animals that build a stonelike coating for protection on top of the coats of past coral polyps. Types of coral include lobe, brain, antler, firecracker, and tongue corals.

Like corals, sea anemones are sessile and rely on a hydrostatic skeleton for support. Unlike coral, they do not produce a hard coating for protection. Instead, their ability to sting predators gives them some protection.

Jellyfish and Hydrozoans

Sea jellies like this Nomura's jellyfish are classified as medusozoans. They are free swimmers, pulsating their bell for locomotion. Nomura's jellyfish is one of the largest sea jellies, reaching a width of almost 2 m. Like other jellies, Nomura's jellyfish has tentacles that it uses to sting and capture its food. Believe it or not, people in Asia capture and eat it for food!

Hydrozoans are another group of medusozoans. Most are very small, like the familiar freshwater hydra. A few hydrozoans form large colonies, like the Portuguese man o' war shown on the right.

Myxozoans

All myxozoans are parasites. This subphylum includes the smallest animals known to science, with some as small as 10 μm in length. Most myxozoan life cycles are poorly understood—only about one hundred have so far been fully described.

Structure and Reproduction in Cnidarians

CNIDARIAN STRUCTURE

Cnidarians show much variety in their structures. They have a nervous system that forms a network of nerves called a **nerve net**, but beyond this they have no true brain. Excretion and respiration are carried out primarily on the cellular level. Each cell extracts its own oxygen from the surrounding water and releases its own nitrogenous wastes and carbon dioxide. All rely on water pressure for support, while the corals excrete calcium carbonate (limestone) to produce exoskeletons for extra protection.

Cnidarians exist in two forms: a polyp form and a medusa form. A **polyp**, as seen in the hydra on the right, is a cup-shaped or tubular form with a mouth and tentacles at one end and a sticky *basal disk* for attaching to hard surfaces at the other end. It looks similar to a plant with a suction cup at the bottom instead of roots. The tentacles pull food into its mouth where it enters the *gastrovascular cavity*, a hollow place in the polyp where digestion and food circulation take place. In both cnidarian forms, the gastrovascular cavity has only one open end; a cnidarian does not have an anus.

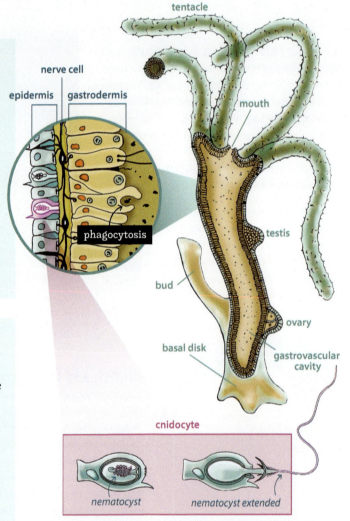

Other cnidarians have an umbrella-shaped body and swim freely—the **medusa** form. By pumping its body in a way similar to the opening and closing of an umbrella, the medusa glides through the water in a jerky upward motion. Many cnidarians spend part of their lives in each form, but some cnidarians, such as corals and sea anemones, exist only in the polyp form. Others, including some sea jellies, have only a medusa stage.

Cnidarians are named for their stinging cells called *cnidocytes*. Both forms of cnidarians have thousands of them on their tentacles, and those of some species are dangerous to humans. The cnidocyte holds a coiled tube that acts like a spring attached to a barb. When prey or an unwary swimmer touches the cnidocyte's trigger, the barb launches and discharges a toxin that paralyzes the cnidarian's prey.

CNIDARIAN REPRODUCTION

Cnidarians, like sponges, reproduce both asexually and sexually. Some reproduce asexually by dividing themselves in half, while many cnidarians, especially in the polyp form, also reproduce asexually through budding. In fact, in many coral species, all the polyps that live on a single exoskeleton are genetically identical. They are the result of asexual reproduction.

However, cnidarians reproduce most often through sexual reproduction. Unlike the hermaphroditic sponges, an individual cnidarian is typically either male or female. The male releases sperm into the ocean, which can fertilize ova that a female has also released. The cells of a fertilized zygote multiply until they form a hollow blastula that develops into a larval form. This larval form matures to take on a polyp or medusa form.

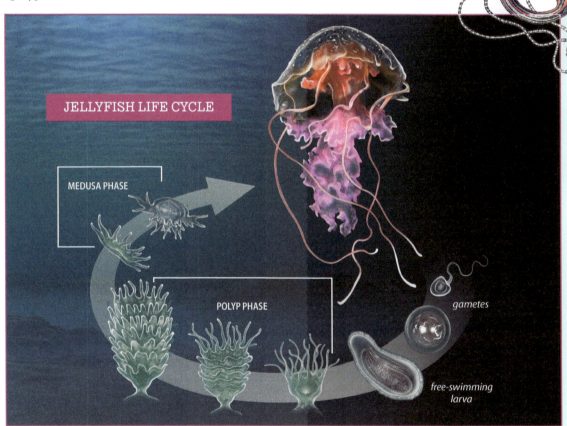

JELLYFISH LIFE CYCLE

MEDUSA PHASE

POLYP PHASE

gametes

free-swimming larva

How Sponges and Cnidarians Function in the Environment

Sponges and cnidarians have an effect on their environment as part of the food web. The larval forms of many sponges and cnidarians contribute to plankton, which is the foundation of the marine food chain. Most sponges are filter feeders, eating detritus, though some sponges in the deep ocean are carnivores. These sponges have hooks for grabbing and entangling live prey, such as crustaceans. Cnidarians are usually carnivorous, eating plankton, invertebrates, and sometimes larger prey. They use the cnidocytes on their tentacles to stun and capture larger prey, such as fish. However, they can also eat detritus.

coral bleaching

Some sponges and cnidarians are parasitic. Sponges are usually parasites to moving animals, such as mollusks, breaking down and boring into the shells of clams and barnacles. Cnidarians can be parasites to other moving organisms such as sea jellies and fish. In general, parasitic sponges and cnidarians are sessile and choose free-swimming hosts.

But sponges and cnidarians don't just take from other organisms; sometimes they give too. Sponges form mutualistic symbiotic relationships, such as they do with algae, harboring them in the material that makes up their spongy walls. The algae produce food through photosynthesis, passing on some of this energy to the sponge. The sponge provides protection and some of the nutrients that the algae need for photosynthesis.

The best example of symbiosis is a coral reef, a whole ecosystem that relies on cnidarians. Corals also have a mutualistic relationship with algae. In fact, algae somehow seem to be essential to a coral's ability to form reefs. Those without the symbiotic algae do not form reefs, and corals on reefs that lose their algae die. Sponges, fish, sea anemones, other corals, sea jellies, and larger animals find food and shelter on the reefs. They are like underwater tropical rainforests that are home to many types of organisms. Many islands are formed by former coral reefs. Coral reefs also protect many beaches from wave erosion.

It should be noted that about 60% of the coral reefs around the world are endangered, and 10% have already been killed. Scientists are still trying to figure out why this is. They theorize that there are a variety of factors at work in each area, including rising sea temperatures and too much fertilizer runoff. These factors cause the corals' symbiotic algae to die. This leads to the corals' deaths, turning them white in a process called *coral bleaching*. Though some of these effects may be natural rather than artificial, we must do our part to keep the water free from pollution. We should avoid dumping wastes in the ocean. We should also protect watersheds, which direct runoff water into the ocean and other bodies of water. By monitoring how we use the earth's resources, we can honor God with their use and protect them for people in the future.

16.2 SECTION REVIEW

1. How are sponges different from other animals?

2. Why are sponges considered animals even though they are so different from other animals?

3. Do sponges have an exoskeleton, endoskeleton, or hydroskeleton? Explain.

4. How do sponges reproduce?

5. Compare the diets of sponges and cnidarians.

6. Regardless of whether they are parasitic or free-living, what is one structure that all cnidarians possess?

7. Sea jellies are about 95% water. Suggest what kind of support they use—an exoskeleton, endoskeleton, or hydroskeleton.

8. Create a concept definition map for cnidarians. (See Appendix B.)

9. Male and female sea anemones release eggs and sperm for fertilization. Do they practice internal or external fertilization? Explain.

10. Give an example of mutualism involving either sponges or cnidarians.

16.3 WORMS
Classifying Worms

Three of the phyla in kingdom Animalia contain the three types of worms—roundworms, flatworms, and segmented worms. Worms are animals with soft, long bodies, three germ layers, and few to no appendages. Explore the various types of worms shown in the images below.

? How do worms breathe?

Questions

What are worms and how are they classified?

What are the differences between flatworms, roundworms, and segmented worms?

How do worms feed and reproduce?

How can we manage and use worms in the environment?

Terms

extracellular digestion
ganglion
closed circulatory system

Exploring
WORMS

Segmented Worms
The most familiar segmented, or *annelid*, worm is the earthworm. The segmented worms of phylum Annelida have a bumpy body that looks like a vacuum cleaner hose. Most segments of a worm's body contain the same organs; this is called a *segmented* body plan. Annelids have three germ layers. The giant Gippsland earthworm, one of the largest, can grow to be a meter long. Other segmented worms include leeches, clam worms, and feather duster worms.

feather duster worm

sand worm

Flatworms
Phylum Platyhelminthes includes the flatworms. They look just like what their name suggests. They include tapeworms, flukes, planarians, and the colorful marine flatworms known as turbellarians (above). Most flatworms are parasites, but some, like the turbellarians, are free-living inhabitants of oceans, lakes, streams, and moist terrestrial habitats. Some flatworms, like the planarians, are known for their amazing ability to regenerate tissue after an injury. A planarian that is cut in half will even grow into two new worms!

Roundworms
The roundworms of phylum Nematoda, also known as *nematodes*, are mostly tiny, round worms less than 2.5 cm long, though a few are larger. And they're everywhere—water, soil, even in our bodies. *Ascaris*, shown here, is a parasitic roundworm that can reach 30 cm long!

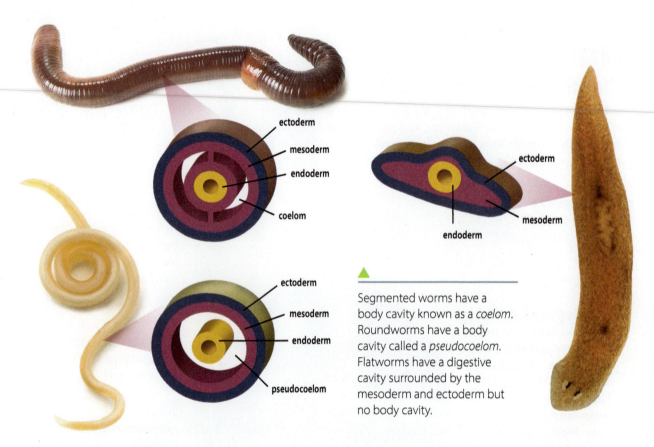

ectoderm
mesoderm
endoderm
coelom

ectoderm
mesoderm
endoderm

ectoderm
mesoderm
endoderm
pseudocoelom

Segmented worms have a body cavity known as a *coelom*. Roundworms have a body cavity called a *pseudocoelom*. Flatworms have a digestive cavity surrounded by the mesoderm and ectoderm but no body cavity.

Structure of Worms

All worms exhibit bilateral symmetry, but beyond this it's hard to make general statements about them since their body plans and lifestyles are so different. As a rule, free-living worms tend to be more complex than parasitic ones, especially internal parasites. Segmented worms and roundworms have a mouth at one end and an anus at the other. Free-living flatworms and some parasitic flatworms have a mouth that leads to a digestive system, while other parasitic worms absorb nutrition directly from their hosts' intestines. Because some worms are parasites and some are free-living, their diets vary. Free-living worms eat things like detritus, fungi, protozoans, and algae. A worm typically uses its digestive system to break down its food before it gets into the cells that need it, a strategy called **extracellular digestion**. Its digestive system forms from the endoderm during embryonic development. But a tapeworm lacks a digestive system and instead absorbs already-digested nutrients from the intestines of its host through its outer layer. All these digestive structures keep a worm alive and maintain homeostasis by allowing it to take in nutrition and expel waste products to its environment.

The ectoderm layer forms the outer layer of a worm, which typically has sensory organs that feed information from the environment to the nervous system made up of two bundles of nerve cells called **ganglia** (s. ganglion). This outer layer has several different forms and names depending on the phylum. In a flatworm the tegument is part of the living cells of the outer layer of the worm, while most segmented worms and roundworms have a non-living layer called the *cuticle*.

All worms possess a mesoderm in their embryonic form, but only segmented worms have a **closed circulatory system**, one that transports nutrients from food through vessels. Flatworms and roundworms entirely lack a circulatory system because they have no body cavity or *coelom*. Instead, flatworms use diffusion to spread nutrients to the various parts of their bodies, and roundworms transport nutrients with the fluid in their body cavities. Most worms respire through their epidermis, lacking any specialized respiratory organs, although a few aquatic worms do have gills. All worms have structures to prevent osmosis from disrupting their homeostasis. Some flatworms and segmented worms also remove nitrogenous wastes with such structures, called *flame cells*. Roundworms rid themselves of these same wastes through structures called *excretory cells* that empty outside the worm's body through excretory pores.

Reproduction in Worms

Like other animals, worms reproduce sexually, although some worms, such as the group of flatworms called *flukes*, also reproduce asexually. Flatworms and segmented worms are generally hermaphroditic. But this doesn't mean that they don't mate; actually, most hermaphroditic animals will cross-fertilize with other individuals. Self-fertilization, while sometimes possible and found in some tapeworms, is almost always detrimental to the offspring since it allows genetic problems to accumulate.

After fertilization, some worms (e.g., earthworms) lay eggs in a cocoon that they attach to a plant or rock in their environment. Worms eventually hatch inside the cocoon and then move from it into the outside world. Some species of worms that live in a watery environment have eggs that hatch to release a plankton-like larva called a *trochophore*. Parasitic worms typically release their eggs in the feces of their hosts.

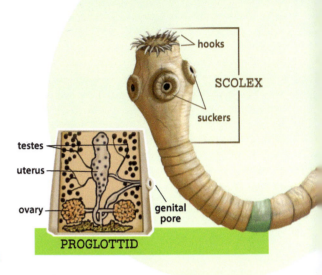

How Worms Function in the Environment

Many worms are parasites, drawing nutrition from their hosts, which could be humans or animals, such as snails. All three types of worms can be parasitic, though there are fewer parasitic segmented worms than roundworms and flatworms. Parasitic worms have fewer sensory organs and are not as capable of locomotion since they don't have to travel to find food. These parasites have the ability to avoid a host's immune system in order to survive. Many of them have complex life cycles involving at least two hosts. An adult parasite matures and sexually reproduces in a *definitive host*. After the parasite's eggs pass out of the definitive host, the larvae enter an *intermediate host* and undergo one or more stages of development. The intermediate host is then eaten by the definitive host, completing the parasite's life cycle.

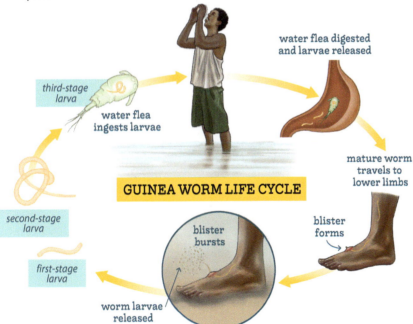

GUINEA WORM LIFE CYCLE

third-stage larva

water flea ingests larvae

second-stage larva

first-stage larva

worm larvae released

blister bursts

blister forms

mature worm travels to lower limbs

water flea digested and larvae released

case study

GUINEA WORM

The guinea worm is a round-worm that can infect a human, mating inside its host's body and emerging, usually near the foot, in an extremely painful blister. The problem starts with a drink of water. People might not know that they have a guinea worm until a parasitic worm, up to 1 m long, begins to emerge in a blister that usually forms near the foot a year after infection. The only way to eliminate the worm is to wrap the end of it on a stick and turn it gradually until the worm is extracted, a painful process that may take weeks. As the infected person bathes his foot in a local water supply to relieve the excruciating pain, the guinea worm releases more eggs, potentially beginning a new infection. The good news is that, because of the efforts of health officials, only a few cases of dracunculiasis, the guinea worm disease, are reported worldwide each year.

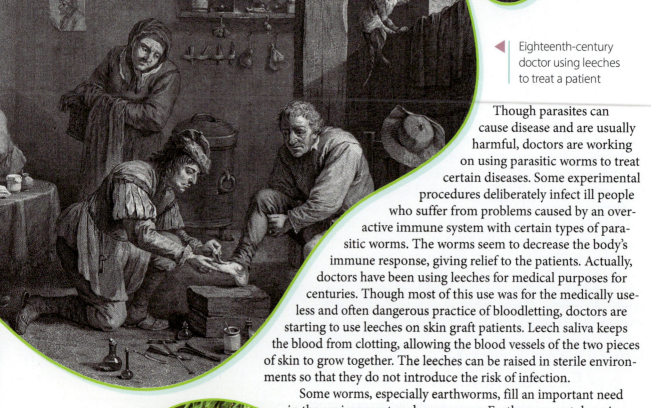

Eighteenth-century doctor using leeches to treat a patient

Though parasites can cause disease and are usually harmful, doctors are working on using parasitic worms to treat certain diseases. Some experimental procedures deliberately infect ill people who suffer from problems caused by an overactive immune system with certain types of parasitic worms. The worms seem to decrease the body's immune response, giving relief to the patients. Actually, doctors have been using leeches for medical purposes for centuries. Though most of this use was for the medically useless and often dangerous practice of bloodletting, doctors are starting to use leeches on skin graft patients. Leech saliva keeps the blood from clotting, allowing the blood vessels of the two pieces of skin to grow together. The leeches can be raised in sterile environments so that they do not introduce the risk of infection.

Some worms, especially earthworms, fill an important need in the environment as decomposers. Earthworms eat decaying organic matter in the soil, taking in some soil along with the decayed material to help grind it up. This not only sustains the worms but also enriches the soil as the earthworm expels its wastes, called *castings*. Earthworm burrows also allow oxygen and water to filter down into the ground, enabling plants to send their roots down deeper. This burrowing action also mixes organic matter into the soil. A single earthworm can produce one half pound of rich, fertile soil in one year. Tens of thousands of earthworms can inhabit one acre of land. What a significant impact one little invertebrate can have on an area's ecology!

16.3 SECTION REVIEW

1. How can you tell the difference between flatworms, roundworms, and segmented worms?

2. Describe at least two ways that worms can feed.

3. List five of the general characteristics of worms. (*Hint:* Use the table on page 351 to get started.)

4. How do worms reproduce?

5. Are worms like earthworms oviparous, viviparous, or ovoviviparous? Explain.

6. Suggest how we can use worms to help people in ways other than those mentioned in the textbook.

Use the case study on page 363 to answer Questions 7 and 8.

7. If you were a health official in Africa, how would you help people to avoid and treat dracunculiasis?

8. If you were a government official in Africa, how would you deal with an outbreak of dracunculiasis?

9. Kids typically think that earthworms eat dirt. Is this correct? Explain.

16.4 MOLLUSKS
Classifying Mollusks

? What's the smartest mollusk?

The submarine called *Nautilus* sinks to the inky depths of the ocean, attracting the attention of a giant squid, which wraps its tentacles around the sub. Captain Nemo sends an electric shock through the ship but fails to dislodge the creature. He decides to rise to the surface to fight the beast in a thunderstorm. Such is the imagined tussle of a submarine with a giant squid in the movie *Twenty Thousand Leagues Under the Sea*, based on a novel by Jules Verne.

Squids are members of phylum Mollusca in kingdom Animalia, along with slugs, oysters, octopuses, clams, cuttlefishes, nautiluses, and snails. These animals are referred to as **mollusks**. They are the second largest phylum in kingdom Animalia.

Questions

What are mollusks?

How do mollusks reproduce?

How do mollusks affect their environment?

Terms
mollusk
visceral mass
mantle
radula
siphon

Exploring

MOLLUSKS

Gastropods. Though most mollusks live in the ocean, like this sea slug, some live on land. Slugs are placed in a category of mollusks called *gastropods*, which also includes snails. A slug has a foot that is found right below its **visceral mass**, the part of a mollusk's body that contains its heart and digestive organs.

Cephalopods. This flamboyant cuttlefish is a member of the *cephalopod* category of mollusks. Cephalopods have feet that are divided into sucker-bearing arms, which are used to catch food. Both cuttlefishes and octopuses have pigment-bearing cells called *chromatophores* in their skin that allow them to change color for camouflage and communication.

Bivalves. Scallops are part of a category of mollusks—called *bivalves*—that includes oysters, mussels, and clams. The fleshy part that you can see in the image on the right is the **mantle**, a sheath of tissue that wraps around the organs of mollusks. You can also see the two shells—hard, protective coverings created by the mantle. These shells serve as support for many mollusks. A fleshy, muscular foot helps many mollusks move around.

Structure and Reproduction in Mollusks

MOLLUSK STRUCTURE

As we saw in the previous section, mollusks are diverse, though they have some common characteristics. Mollusks have three germ layers, most exhibit cephalization and bilateral symmetry, and all have digestive, circulatory, nervous, and respiratory systems.

But the three categories of mollusks—bivalves, cephalopods, and gastropods—are also very different. They can have one shell, two shells, or none at all. They can have one foot or eight arms filled with suckers. Some mollusks, including gastropods and most cephalopods, have both a mouth and a **radula**, an organ near the mouth with many tiny teeth, similar to a cheese grater, for scraping up food and pulling it into the mouth. Some mollusks have a nervous system that consists of ganglia. Cephalopods stand out among the invertebrates because they have a cluster of ganglia that are organized to form a brain similar to those of vertebrates. In fact, cephalopods have the highest brain-to-body mass ratio of any invertebrate—higher even than many vertebrate animals. Octopuses in particular have demonstrated remarkable feats of intelligence, such as being able to navigate mazes or unscrew jar lids to get at food items. Most of an octopus's nerve tissue is in its arms, allowing each arm to taste, touch, and move independently.

Mollusks that live underwater breathe using *gills*, thin-walled structures laced with blood vessels that bring oxygen to the blood. Cephalopods have closed circulatory systems, while bivalves and gastropods have open circulatory systems where blood and other fluids flow through the body cavity to bathe internal organs rather than circulate through vessels.

Mollusks that live in the water also often circulate water from their environment through two tubes, or **siphons**. One tube brings water in, while the other allows water to leave. If the mollusk wants to move quickly to avoid a predator, it jets water out of its siphon. Many cephalopods have the ability to squirt ink to confuse and distract predators.

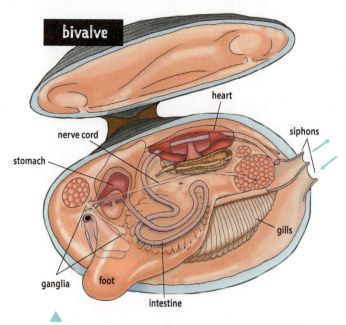

Clams have an anatomy that is representative of mollusks. The anatomy of cephalopods is different, including a brain and a closed circulatory system.

MOLLUSK REPRODUCTION

Most mollusks reproduce sexually, though they vary in the method. Simpler mollusks reproduce through external fertilization, with some species such as snails being hermaphrodites. More complex mollusks such as cephalopods mate and reproduce through internal fertilization.

In mollusks, fertilization produces eggs that hatch to release larvae. For water-dwelling mollusks, these larvae float as plankton in their watery environment. For simpler mollusks, larvae are trochophores similar to those of worms. They change and mature through a process called *metamorphosis*. In cephalopods, the young that hatch from eggs are just smaller versions of their adult form.

How Mollusks Function in the Environment

Though some mollusks live on land, most are aquatic. Mollusks can be carnivores, herbivores, and scavengers in their habitats. They eat algae, detritus, and even each other. Mollusks play a role in a variety of ecosystems, such as the coral reef community. They find shelter and food in the coral reef.

People use mollusks to enrich and improve their lives. Many mollusks are edible, such as the abalone mentioned at the beginning of this chapter, and are important fisheries resources. Scientists are also studying the regenerative power of mollusks to help create therapies and medications that can help people with specific medical problems. Cuttlefish ink has shown an ability to suppress the growth of tumors; scientists continue to explore this as a potential treatment for cancer. We can use our knowledge of biology to manage the creatures that God has put under our care to help people and bring Him glory.

16.4 SECTION REVIEW

1. Identify each animal in the images below as a bivalve, gastropod, or cephalopod. Explain your choice.

   a. nudibranch

   b. giant clam

   c. chambered nautilus

2. Give four of the general characteristics of mollusks.

3. What does a cephalopod have that every other invertebrate we have studied does *not* have? How would this affect its lifestyle?

4. What is different about the type of reproduction of mollusks compared with worms, sponges, and cnidarians?

5. The New Zealand mud snail can reproduce by a form of cloning. Why is that unusual?

6. In what kind of ecosystems do mollusks play a key role?

16.5 ECHINODERMS

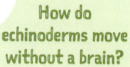

Classifying Echinoderms

Questions

What is an echinoderm?

How are echinoderms classified?

How do echinoderms reproduce?

How do echinoderms affect their environment?

Terms

echinoderm
water-vascular system
tube foot

Another phylum of invertebrates in kingdom Animalia is phylum Echinodermata. It includes marine animals ranging from prickly sea urchins to the starfish (also known as sea stars). These invertebrates can be found in all ocean zones, from the light-filled intertidal zone to the depths of the abyssal zone, though most live in the deep oceans. Animals in this phylum are called **echinoderms**. They have three germ layers and exhibit radial symmetry. Most also have an endoskeleton made of plates called *ossicles*. Explore the five different classes of echinoderms shown on the facing page.

How do echinoderms move without a brain?

ECHINODERMS

Brittle Stars

Class Ophiuroidea contains the brittle stars. They look like starfish but have long, thin arms. They're called "brittle" because they readily discard their arms when attacked or disturbed and then easily regenerate them.

Sea Stars

Sea stars of the class Asteroidea are probably the most familiar echinoderms. They can be found in a variety of colors and have various numbers of arms, though most have five.

Sea Lilies

The sea lilies of class Crinoidea are echinoderms that are sessile and look more like plants. Many sea lilies, also called *crinoids*, are found as fossils. Crinoids without stalks are called *feather stars*.

Sea Cucumbers

The oddly named curryfish is a sea cucumber from class Holothuroidea. The curryfish lives in shallow tropical seas, but sea cucumbers are widely distributed throughout the world's oceans. Some species even thrive on the deepest ocean floors. Sea cucumbers are the only class of echinoderms that lack ossicles.

Sea Urchins

Class Echinoidea, the sea urchins, are like marine porcupines or hedgehogs. In fact, they were once called *sea hedgehogs*. They have sharp spines that can create painful puncture wounds in predators.

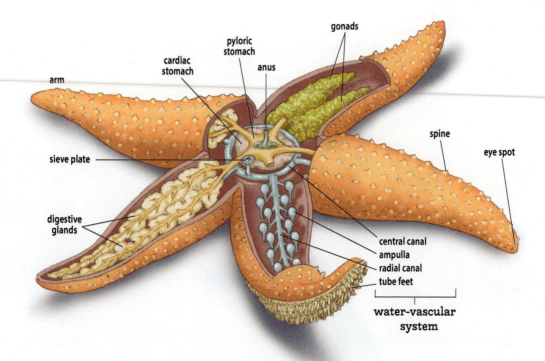

Structure and Reproduction in Echinoderms

ECHINODERM STRUCTURE

One of the identifying features of echinoderms is a **water-vascular system**—a series of canals and tubes used for circulating nutrients, moving, and capturing food. The most obvious feature of this system is hundreds of **tube feet**, knobs or leglike structures on the bodies of echinoderms. All echinoderms have tube feet. These hollow feet are joined by a water canal that extends to the central part of the echinoderm's body. This water canal system opens to an echinoderm's watery environment through a sieve-like structure called a *madreporite*.

Echinoderms have all the body systems associated with animals. They breathe through their tube feet, or in some cases through gills, allowing them to take in oxygen from their aquatic environment to maintain homeostasis in their bodies. They have a closed digestive system with a mouth on their lower surface. Their mouth scoops up detritus, algae, or prey. An anus on their upper surface expels waste, again maintaining homeostasis by ridding the body of substances that it can't use. An echinoderm has a nerve net like a cnidarian, and it can sense and respond to its environment even without a head or brain. It also has an open circulatory system.

Some echinoderms have amazing powers of regeneration and can regrow a lost limb. They can even completely regenerate from a partial organism made of part of the central body and a limb or two.

ECHINODERM REPRODUCTION

Echinoderms reproduce both asexually and sexually. Most often echinoderms reproduce sexually when male and female echinoderms mate, though there are some hermaphroditic species. Eggs are fertilized through external fertilization, with some echinoderms guarding or *brooding* their fertilized eggs. These eggs hatch to release larvae that exhibit bilateral symmetry and are capable of motion and asexual reproduction. As they mature, they develop the radial symmetry typical of echinoderms.

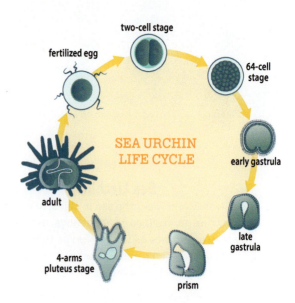

Crown-of-thorns starfish preying on coral polyps

How Echinoderms Function in the Environment

All echinoderms are marine, having no representatives on land or in fresh water. They serve a variety of roles in the environment, acting as algae grazers, scooping up detritus, and even hunting other animals in their ocean environments.

Echinoderms can be found in any of the ocean zones. They have a large presence in the dark ocean depths of the hadal and abyssal zones. This area is like an underwater desert—nothing grows because there is no light. Echinoderms serve a useful purpose as scavengers, eating organic matter and carcasses that settle to the ocean depths.

Echinoderms also have a large presence in the other extreme zone of the ocean—the intertidal zone. Sea stars and sea urchins flourish amid the bashing waves and fluctuating conditions of tide pools. Some echinoderms find shelter and food in coral reef communities. For example, the crown-of-thorns starfish is one of the few predators that corals have.

Echinoderms are useful to humans too. Sea urchins and sea cucumbers are viewed as a culinary treat in some areas of Asia, and there is a danger of them being overharvested. People are investigating ways to raise them in captivity. The shells of echinoderms are also used by the ton each year as a source of limestone to fertilize farm fields. The stunning and useful diversity and beauty in the world of invertebrates point to a God who is an intelligent, artistic, and caring Creator.

16.5 SECTION REVIEW

1. What two things about the body of an echinoderm make it different from a mollusk?

2. Create a table that compares the five classes of echinoderms, including the following characteristics: locomotion, the presence of ossicles, and the presence of tube feet.

3. Compare the number of germ layers and the symmetry of echinoderms and cnidarians.

4. Some echinoderms respond to injuries in a way similar to that of planarians. Explain.

5. What is the most common way that echinoderms reproduce?

6. Mollusks and echinoderms both play a key role in what kind of ecosystems?

16 CHAPTER REVIEW
Chapter Summary

16.1 KINGDOM ANIMALIA

- All animals are heterotrophic organisms with a variety of ways to move, respire, obtain energy, circulate materials, support their bodies, and reproduce.

- Animals can be classified by the presence (vertebrates) or absence (invertebrates) of backbones.

- Endothermic animals can regulate their internal body temperatures; ectothermic animals cannot.

- The internal structure of an animal is influenced by the number of germ layers present in its early embryonic development.

- Most animals reproduce sexually. Fertilization may be internal or external, and young may develop in eggs outside the mother's body (oviparous), in eggs carried within the mother's body (ovoviviparous), or by being connected to the mother's body (viviparous).

- Animals exhibit a wide variety of instincts, complex behaviors, and interactions with their environments.

Terms
endoskeleton · exoskeleton · hydroskeleton · body plan · bilateral symmetry · radial symmetry · sessile · vertebrate · chordate · invertebrate · endotherm · ectotherm · blastula · germ layer · endoderm · mesoderm · ectoderm · cephalization · anus · external fertilization · internal fertilization · oviparous species · viviparous species · ovoviviparous species · behavior · instinct

16.2 SPONGES AND CNIDARIANS

- Sponges are sessile, aquatic animals that filter food particles from currents generated by their collar cells.

- Sponges are hermaphroditic and can reproduce both sexually and asexually (budding). Fertilized eggs hatch into larvae that later settle from the water.

- Cnidarians possess stinging cells. They exist in one of two forms, either a polyp or medusa, and many have a life cycle that includes both forms. Polyps are sessile, while medusas are free-swimming.

- Cnidarians can reproduce both sexually and asexually.

- Larval sponges and cnidarians are an important part of marine plankton. Adult sea jellies are important marine predators, while corals and sponges provide structure in reef communities.

Terms
intracellular digestion · hermaphrodite · cnidarian · nerve net · polyp · medusa

16.3 WORMS

- Worms can be classified as either flatworms, roundworms, or segmented worms.

- All worms have a mesoderm, but only in segmented worms does this develop into a true body cavity with a closed circulatory system.

- Worms include both free-living and parasitic species.

- Roundworms are either male or female, while many flatworms and segmented worms are hermaphroditic. Some worms are capable of asexual reproduction.

- Although many worms are parasitic, others are important members of both terrestrial and aquatic ecosystems.

Terms
extracellular digestion · ganglion · closed circulatory system

16.4 MOLLUSKS

- The three main groups of mollusks are bivalves (having two shells, such as clams), gastropods (slugs and snails), and cephalopods (those having arms, such as squids). They may be either terrestrial or aquatic.

- Mollusks are diverse, but all have three germ layers and organ systems. Most exhibit cephalization and bilateral symmetry.

- Most mollusks reproduce sexually. Most practice external fertilization, while many cephalopods use internal fertilization. Many slugs and snails are hermaphroditic.

- Most mollusks are aquatic. They exhibit a wide variety of feeding strategies and include species that are herbivores, carnivores, or scavengers.

Terms
mollusk • visceral mass • mantle • radula • siphon

16.5 ECHINODERMS

- Echinoderms are marine animals that exhibit radial symmetry and have water-vascular systems that often include tube feet for movement. They can be classified as sea stars, brittle stars, sea lilies, sea urchins, or sea cucumbers.

- Most echinoderms reproduce sexually utilizing external fertilization. Some species are hermaphroditic.

- Echinoderms feed in many ways. Various species graze on algae, feed on detritus, or prey on other animals.

Terms
echinoderm • water-vascular system • tube foot

Chapter Review Questions

RECALLING FACTS

1. List the six characteristics of animals.

2. Do all animals have germ layers? Explain.

3. Give the number of germ layers in each of the following organisms.
   a. blue dragon sea slug
   b. crown-of-thorns starfish
   c. lion's mane jellyfish
   d. marine flatworm
   e. row pore rope sponge

4. Name one thing that both the polyp and medusa forms of cnidarians have in common.

5. Why are sponges and cnidarians such an important part of the marine food web?

6. Why do segmented worms have digestive, circulatory, and nervous systems while sponges do not?

7. Why do many worms not need other worms to reproduce?

8. Which two classes of worms contain most of the parasitic worms?

9. How can learning about worms help us protect people from disease?

10. What is one of the most significant features that make cephalopods different from all the other invertebrates?

11. What ability do some cephalopods have to defend themselves against predators that bivalves and gastropods lack?

12. What do all bivalves have that some gastropods and most cephalopods lack?

13. What feature do sea cucumbers lack that other echinoderms have?

14. Name two things that all echinoderms have.

15. What is different about the body plans of echinoderm larvae and adults?

16. List all the groups of invertebrates that you learned about in this chapter that have hermaphroditic species.

17. What is the most important role for benthic (bottom-dwelling) echinoderms?

UNDERSTANDING CONCEPTS

18. Use the characteristics of animals to show that a domestic cat is an animal.

19. A peacock butterfly suns itself on a rock to raise its body temperature. Do you think it is an ectothermic or endothermic animal? Explain your choice.

20. Which kind of symmetry does the peacock butterfly demonstrate?

21. Explain the relationship between bilateral symmetry and germ layers.

22. Why do you think a sponge has no nervous, circulatory, or digestive system?

23. A red kangaroo reproduces when the male's sperm fertilizes an egg inside the female's body. Eventually, a young joey the size of a lima bean emerges and migrates up to the pouch where it will nurse and continue to grow. What kind of fertilization (internal or external) and development (oviparous, ovoviviparous, or viviparous) does the red kangaroo exhibit?

24. Why must ovoviviparous and viviparous species always reproduce through internal fertilization?

25. Give an example of how an animal could respond to the environment and change how it interacts with other animals.

26. Identify each of the following worms as a flatworm, roundworm, or segmented worm. Explain your choice.

   a. *Pseudoceros ferrugineus*   **b.** *Toxascaris leonina*   **c.** *Spirobranchus giganteus*

27. Invertebrates that have digestive systems use extracellular digestion. Create a T-Chart (see Appendix B) with the headings "Intracellular Digestion" and "Extracellular Digestion." Place sponges, cnidarians, worms, mollusks, and echinoderms in each column according to their method of digestion.

28. Compare the nervous systems of sponges, cnidarians, worms, mollusks, and echinoderms.

29. Compare the circulatory systems of worms with that of sponges, mollusks, and echinoderms.

30. Create a hierarchy chart (see Appendix B) using domain Eukarya and kingdoms Protozoa, Fungi, Chromista, Plantae, and Animalia. Include vertebrates, invertebrates, sponges, cnidarians, mollusks, worms, and echinoderms. Include the different groups of cnidarians, worms, mollusks, and echinoderms in your chart.

31. Use a T-Chart to compare vertebrates and invertebrates. Include the six characteristics of animal life in your comparison.

32. In which ocean zone are echinoderms more likely to be scavengers, the abyssal zone or the intertidal zone? Explain.

CRITICAL THINKING

33. Why is it difficult to make generalizations about most animal phyla?

34. A friend says that since sea lilies are sessile, the assertion that all animals can move is false. How would you respond?

Use the cladogram below to answer Questions 35–38. Think about the germ layers as you read these questions. Some of the assumed evolutionary relationships on this chart may surprise you!

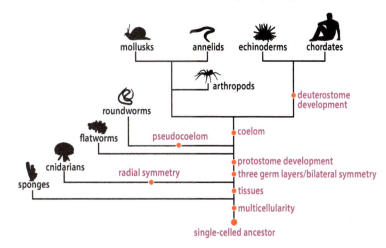

35. Why do you think sponges are the lowest on the cladogram?

36. Annelids, arthropods, and mollusks all are on the same branch of the cladogram. Why is this?

37. Why are humans grouped with chordates on this cladogram?

38. Why do evolutionists say that humans, vertebrates, and echinoderms all come from a common ancestor? You may need to do some research.

Use the graph below to answer Questions 39–43.

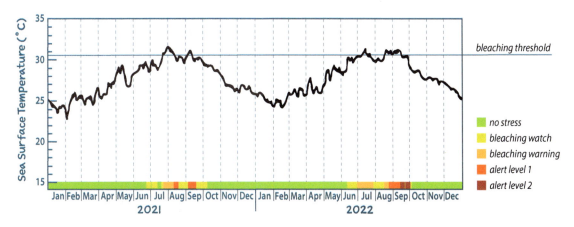

The graph shows how the conditions of the ocean off the Florida Keys change during the year. This data was collected from January 2021 to December 2022.

39. What time of year do corals experience the least stress?

40. When did the coral reefs near the Florida Keys begin to struggle?

41. What do you think is the main concern with coral bleaching (see page 360)?

42. Why is it a problem if corals die?

43. Suggest a course of action to exercise good and wise dominion of coral reefs.

17 ARTHROPODS

Insect Forensics

The small, annoying insect buzzing around the room right now may just be able to resolve the next murder case. Insects like the green bottle fly are giving forensic scientists the data they need to solve crimes. When an organism dies, insects take over to feed and lay eggs. The green bottle fly is one of the first insects to arrive at a corpse, within minutes after death. Forensic scientists know how much time these processes take, so they can use these clues to unravel mysteries and help bring perpetrators of crime to justice.

In fourteenth-century China, Sung Tzu, one of the first forensic entomologists, used the green bottle fly to solve a case in which a worker was murdered in a rice field with a sickle. Many people had sickles—so who was the culprit? The sickles of all the workers were lined up, and green bottle flies were attracted only to the sickle of the murderer, which had a small amount of blood that only the flies could detect. Case solved!

17.1 ARTHROPOD INTRODUCTION AND CHELICERATES

Arthropod Attributes

Arthropods cover our planet. They are members of the largest and most significant phylum in kingdom Animalia—phylum Arthropoda. Insects are the largest group in this phylum. Because insects are so common in our world, they are always available to police and crime scene investigators and are useful in providing the information necessary to solves crimes.

Like the other animals you learned about in the last chapter—sponges, cnidarians, worms, mollusks, and echinoderms—all arthropods are invertebrates. They include crustaceans, spiders, scorpions, insects, centipedes, and millipedes. They also include the now-extinct trilobite, which is found fossilized in rock all over the world. Let's look at the characteristics and the three major categories of arthropods—chelicerates, crustaceans, and insects.

How are spiders different from insects?

Questions

What are the characteristics of arthropods?

What is a chelicerate?

How do chelicerates feed and reproduce?

What is the role of chelicerates in their environment?

Terms

arthropod
thorax
abdomen
cephalothorax
compound eye
molting
antenna
chelicera
pedipalp
book lung
spinneret

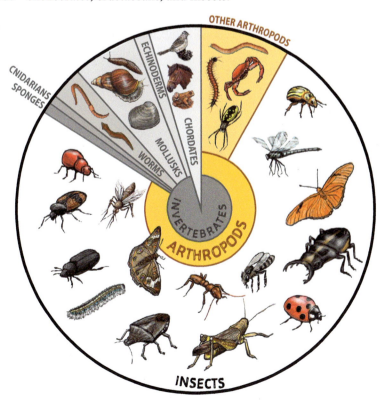

ARTHROPOD ATTRIBUTES

SEGMENTED BODY

The bodies of typical arthropods are divided into three sections: the head, the **thorax**, and the **abdomen**. The thorax and the head together are called the **cephalothorax**. Each body section has its own segments. Some arthropods have other features along with these three parts that give them special abilities to move, get food, or escape predators—for example, the double wings of a dragonfly.

cephalothorax

head

thorax

abdomen

JOINTED APPENDAGES

Arthropods have appendages that are jointed to enable movement. To fend off predators, some arthropods have swiveling stinging tails, as in scorpions; delicate, nimble legs, as in grasshoppers; or ferocious-looking claws, as in lobsters.

COMPOUND EYES

Some arthropods, such as crustaceans and insects, rely on **compound eyes** for visual information about their environments. These special eyes contain thousands of lenses set at different angles to create a mosaic image. The eyes on this fly give it an almost 360° field of view. That's like having eyes in the back of your head!

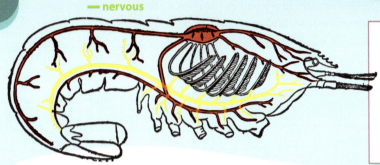

— circulatory
— nervous

OPEN CIRCULATORY SYSTEM

Arthropods have dorsal vessels that pump blood to the front of their bodies. Blood then bathes organs in these cavities.

cicada molt (chitin)

EXOSKELETON

Arthropods have stiff exoskeletons that give them support. This structure is made of layers of substances, including *chitin*. Chitin gives arthropod bodies strength and flexibility, like a good suit of armor. Because this outer coating doesn't grow with the arthropod, it must be periodically shed in a process called **molting**. After molting, an arthropod excretes a new exoskeleton that quickly hardens.

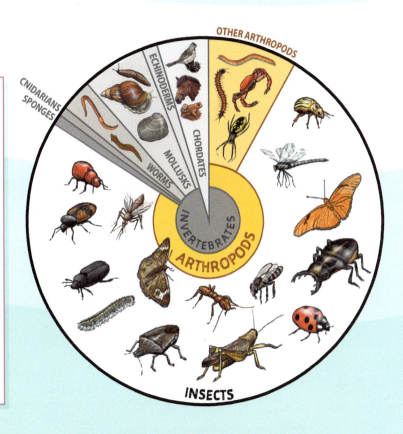

NERVOUS SYSTEM

The nervous system of an arthropod includes a pair of ganglia joined by two major nerves that together function like a brain. This "brain" gets sensory information from the appendages, including two called **antennae** (s. antenna), located in the head region. Antennae give information about taste, smell, and touch, but they can't relay visual information. Insect feet actually have sensory structures that smell and taste—think about that fly on your hamburger!

Classifying Chelicerates

Chelicerates are arthropods having one or two body sections and clawlike appendages called **chelicerae** (s. chelicera) for grabbing and immobilizing prey. Chelicerae work like a pair of tweezers or tongs. Chelicerates include horseshoe crabs and *arachnids*, a class of arthropods that includes spiders, ticks, mites, and scorpions.

CHELICERATES

The horseshoe crab belongs to a chelicerate class that is aquatic, but it's not really a crab! It has a hard carapace and clawlike chelicerae. The horseshoe crab has five pairs of legs for walking and swimming. It reproduces through external fertilization as the female lays her eggs. It has two body sections.

The arachnid that we're all probably most familiar with is the spider. The colorful peacock spider shown above displays his vibrant colors for mating. He is very tiny, only 5 mm long; his mating dance can be seen only with the help of a microscope. If he doesn't succeed in attracting a mate, he scampers off quickly—unless he wants to be eaten by disapproving, observing females!

The red bug shown below, right, sometimes called a chigger or mite, is also an arachnid. Its body sections are fused, so it looks like it has only one body section. This is characteristic of ticks and mites. Red bugs are very tiny, less than half a millimeter, but they can attach to animals and humans just like their tick relatives.

This yellow scorpion, also called by the sinister name of deathstalker, has eight legs and two body sections. Its sting makes it the most dangerous scorpion, being very painful and life-threatening to children and sickly adults.

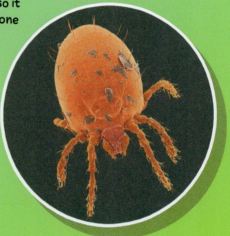

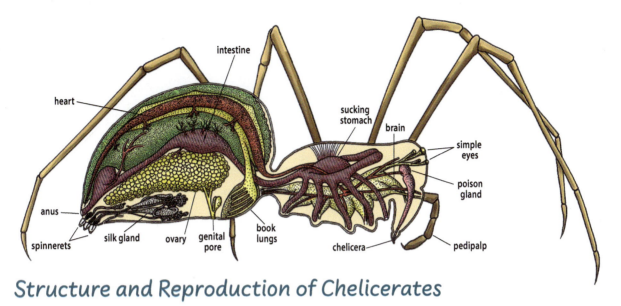

Labels on diagram: intestine, heart, sucking stomach, brain, simple eyes, poison gland, anus, spinnerets, silk gland, ovary, genital pore, book lungs, chelicera, pedipalp

Structure and Reproduction of Chelicerates

CHELICERATE STRUCTURE

Chelicerates share some common features in addition to their pincerlike chelicerae. Most have four pairs of walking legs and one or two body sections, including a cephalothorax and abdomen, much like a spider. In front of the walking legs are **pedipalps**, appendages that chelicerates use to eat, mate, and sense their environment. Chelicerates don't have antennae, unlike other arthropods. They also typically have two to four eyes, which are usually *simple eyes* with only one lens. How, then, do they get information about their environment? Sensory information is received from their feet and simple eyes.

Since chelicerates are a mix of aquatic animals, such as horseshoe crabs and sea spiders, and land animals, such as scorpions, ticks, and spiders, they have unique features for maintaining homeostasis in their own environments. Just think how different their breathing needs to be. For example, spiders,

scorpions, and mites have organs called **book lungs** that allow air to enter their bodies. This air goes straight to tissues without passing into the circulatory system. Since aquatic chelicerates live in the water, they have *book gills* that help them get oxygen from the water to transport to their tissues.

Spiders also have some unique structures. A system called the *Malpighian tubule system* allows the spider to get rid of nitrogen-containing wastes without losing necessary water. Another organ called a **spinneret** spins spider silk from a protein manufactured in the tip of the abdomen. Spiders use this silk for lots of things, including capturing prey for nutrition. Laboratories full of scientists have not been able to create a material that is as naturally strong as spider silk, and it is manufactured right in the abdomen of spiders.

CHELICERATE REPRODUCTION

Aquatic and terrestrial chelicerates also reproduce differently. Generally, aquatic chelicerates such as sea spiders and horseshoe crabs reproduce sexually through external fertilization, while terrestrial chelicerates undergo internal fertilization. Some chelicerates, such as scorpions, spiders, and sea spiders, brood their eggs until they hatch to release young.

Horseshoe crabs spawning and their eggs (inset)

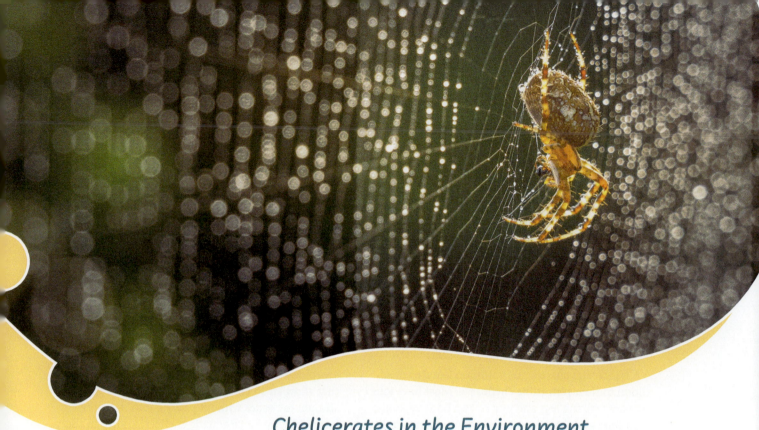

Chelicerates in the Environment

Chelicerates play various roles in the environment—they are parasites, predators, and prey. Ticks and mites are parasitic, with some being vectors for diseases such as Lyme disease and Rocky Mountain spotted fever.

But not all chelicerates are bad news. Since many are insect predators, they can also help keep insect populations under control. For example, some smaller scorpions eat dust mites and moth larvae, which eat clothes. Spiders spin webs and orb webs to catch insects.

Chelicerates are part of the food web in their roles as both predator and prey. For example, horseshoe crab eggs are an easy food source for shorebirds. Some insects even eat spiders! The tarantula hawk, a type of wasp, captures a tarantula, stings and paralyzes it, and lays a single egg, which then hatches, releasing a larva that slowly eats the spider alive.

Chelicerates could also save your life. Scientists have been "milking" horseshoe crabs for their blue blood, which has amazing antibacterial properties, forming clots around even very small amounts of bacteria that could be lethal for humans. After horseshoe crabs are "milked," they can be released back into the wild to regenerate their blood supply. Scientists are using this blood to test for the presence of bacteria in medicines and on medical equipment. Scientific work is being done to use horseshoe crab blood to detect viral and fungal infections, develop vaccines, and even treat cancers.

Horseshoe crab blood is blue because it contains a chemical called *hemocyanin* instead of the hemoglobin that makes most blood red. It is used in the LAL test, which indicates the presence of even extremely small amounts of bacteria in about forty-five minutes. Horseshoe crab blood costs over $10,000 per quart!

MINI LAB

Discovering Arthropods

There are many different types of arthropods. But what really sets them all apart? Are they really so different that scientists needed to create different classes to organize them? Let's do a field activity to see whether we can spot the differences between arthropods.

How can we observe the differences between arthropods?

Materials
notebook • pencil • insect collection containers • natural area

PROCEDURE

A Go to a natural area and capture two or three different arthropods.

B Using the information on pages 378–79, observe the physical characteristics and behaviors of the arthropods that you collected. Compare your arthropods with the arthropods of other students. Record your observations.

C Using your recorded observations, group the arthropods according to their characteristics. Create a Venn diagram to describe the similarities and differences between the different arthropods.

QUESTIONS

1. What physical characteristics and behaviors did you observe in the arthropods you collected?

2. In comparing the arthropods that you collected with those of other students, what similarities and differences did you observe?

3. On the basis of your observations, do you believe that the arthropods you collected should be in the same or a different group of arthropods?

1. Demonstrate that the cicada shown below is an arthropod using at least three characteristics of arthropods that you can observe in the photo.

2. Does the photo indicate that the cicada lives in water or on land? Explain.

3. Draw a hierarchy chart using the following terms: *arachnid, arthropod, horseshoe crab, invertebrate, mite, spider, scorpion,* and *tick.*

4. What one characteristic does every chelicerate have?

5. Name one attribute of arthropods that spiders lack.

6. Name one attribute of spiders that other arthropods lack.

7. The sea spider shown at right is an aquatic chelicerate that is not actually a spider. Would it be more likely to reproduce through internal or external fertilization? Explain.

8. The organism that permanently lives at the highest elevation on Earth is a spider—the Himalayan jumping spider. It eats insects that are occasionally carried to these elevations by mountain breezes. What role does this spider play in the food chain?

9. Before scientists used horseshoe crab blood for the LAL test, they used rabbits to test vaccines for contamination by bacteria. Live rabbits were injected with the vaccine and observed for days. Much like humans, they have a sensitivity to these bacteria. Horseshoe crab blood is about one hundred times more sensitive. Research horseshoe crab blood and give reasons why using it to test vaccines is a better way to help people and wisely use animals.

10. Suggest a way that we could improve the LAL test that would reduce our reliance on horseshoe crabs, especially since they are so useful for other things, including their role in marine ecology.

17.2 CRUSTACEANS
Classifying Crustaceans

Questions

How is a crustacean different from other arthropods?

How do crustaceans feed and reproduce?

How do crustaceans affect their environment?

Terms
carapace
swimmeret
walking leg
cheliped
mandible

Crustaceans are a large group of arthropods, second in number of species only to insects. Most crustaceans spend the majority of their time in water, although some spend a good bit of time on land. Crustaceans are as abundant in the oceans as insects are on land. Because they are usually aquatic, many share some key features, including a cephalothorax covered by a single exoskeletal plate called a **carapace**. Crustaceans have two or three body sections. The abdomens of many are covered by a series of plates. Pairs of flipper-like appendages called **swimmerets** extend from the abdomen.

Are there crustaceans that don't live in the ocean?

CRUSTACEANS

Not all crustaceans spend their days galloping across the ocean floor! The barnacles shown above are mostly sessile, though they have free-swimming larvae. They have one body section. They inhabit places with lots of tidal action, feeding on food particles that float in ocean water.

This Japanese spider crab is the largest known arthropod at a whopping fifteen feet wide! The spider crab and most other crustaceans are aquatic. Large arthropods tend to be aquatic since they lack the muscle power to move their heavy exoskeletons on land.

Krill are shrimp-like crustaceans with two body sections. They make up a huge percentage of the biomass in the oceans. Krill form a key link in the marine food chain between phytoplankton and larger forms of life. Marine life near Antarctica couldn't survive without Antarctic krill, which appear in pink, opaque clouds when millions hatch.

This may surprise you, but pillbugs (above) are also crustaceans. Also known as a roly-poly or wood louse, the pillbug is a nocturnal crustacean with seven pairs of legs. It is one of the few terrestrial crustaceans on Earth. Although pillbugs are not aquatic, they have gill-like structures used for respiration.

Some crustaceans, like this crayfish, live in fresh water along with some species of shrimp. Crayfish are very similar to lobsters in their structure and function.

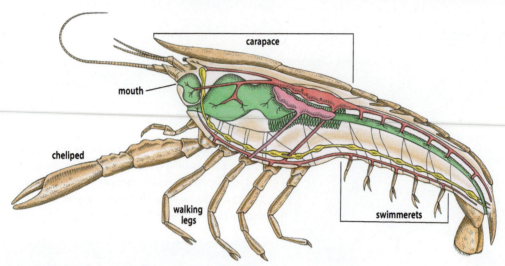

carapace

mouth

cheliped

walking legs

swimmerets

Structure and Reproduction of Crustaceans

CRUSTACEAN STRUCTURE

Crustaceans have body parts that help them function in their environment, whether it be salt or fresh water. Many crustaceans like crabs can venture out of the water onto land for shelter or to hunt for food. Their appendages help them do this. Most crustaceans have appendages that branch into two parts, unlike the other arthropods that have non-branching appendages. Five pairs of **walking legs** allow them to maneuver on land and in the water. The first three pairs often exhibit claws called **chelipeds**. Crustaceans use their chelipeds to fend off predators, spar with other crustaceans, capture prey, and groom themselves.

When hunting for prey, a crustacean collects sensory information from its environment with its compound eyes. It also uses its two sets of antennae to feed sensory information to its nervous system. Many crustaceans are also equipped with chromatophores like some mollusks, enabling them to change color in order to blend into their surroundings or to communicate. Once they capture prey, some crustaceans use **mandibles**, appendages near their

mandibles

mouths, to crush their food. They have a digestive system with both a mouth and an anus. Crustaceans have an open circulatory system to spread nutrients and gases to their bodies for maintaining life processes. Larger crustaceans use gills to breathe in their aquatic environment. Even crabs that venture onto land must often return to the water to wet their gills. Air can diffuse into the water on the gills so that they can extract oxygen from the water. Smaller crustaceans breathe through their skin.

CRUSTACEAN REPRODUCTION

Crustaceans can't reproduce asexually, but they can regenerate their limbs when one is lost in a battle with a predator. Crustaceans usually reproduce sexually when males mate with females, who store sperm in their bodies until they lay eggs. Fertilization happens externally as the eggs pass out of the female's body past the place where the sperm is stored. Some aquatic crustaceans can also exhibit internal fertilization, though that is more common with terrestrial crustaceans such as crabs.

This porcelian crab mother broods her fertilized eggs on the underside of her body until they hatch later to release larvae.

Crustaceans in the Environment

Since crustaceans are so common in the marine environment, they have their greatest effect on this biome. Crustaceans are ecosystem engineers, organisms that shape their environment in ways that often make it a better place for other organisms to live. The changes that ecosystem engineers make usually result in increasing the biodiversity of a community. We've seen other ecosystem engineers, such as the African termite, giant kelp, and corals. So what do crustaceans do to make the marine environment better for other organisms?

Many crustaceans are filter feeders and dine on detritus, returning nutrients to their environments and keeping the water free of waste products. Crayfish serve this purpose in fresh water. Other crustaceans keep the ocean floor clean, scavenging dead carcasses like fallen whales that sink slowly into the abyssal zones. Scavenging crustaceans include lobsters, prawns, shrimp, and isopods (a marine version of pillbugs).

Krill are ecosystem engineers, not because of what they eat but because they get eaten! This tiny crustacean is a giant in the marine community. Krill make up the largest biomass of any animal in the ocean. They eat plankton, sometimes scraping algae off sea ice in polar regions. Larger organisms such as whales, squid, fish, and seabirds eat krill when they gather in large groups called *swarms*. In fact, animals that eat krill migrate with their babies to the Antarctic to feast when krill swarms form as millions of krill hatch from their eggs. Because of their strategic place on the food chain, krill are called a *keystone species*.

17.2 SECTION REVIEW

1. Why can aquatic arthropods grow larger than terrestrial arthropods?

2. Suggest one way that you could tell a crustacean apart from other arthropods.

3. It's dinnertime for a common lobster, and it's not being eaten—it's eating! It climbs out of its cave, dragging along its large claws as the waters grow dark, so that it can feast on sea urchins, mollusks, and sea stars. Suggest how it captures and eats its prey.

4. Terrestrial crabs such as the blue land crab don't use external fertilization. Why not?

5. The blue mud shrimp lives in the intertidal region in Oregon. It feeds on detritus, altering the chemistry of its habitat. Because of its influence on its habitat, what is this kind of species called?

6. Suggest what would happen in the Antarctic if Antarctic krill experienced a drastic decline in population.

17.3 INSECTS
Classifying Insects

Questions

What are the characteristics of an insect?

How do insects feed and reproduce?

How can insects be controlled but still be useful to people and the environment?

Terms

Malpighian tubule
trachea
spiracle
pheromone
metamorphosis
incomplete
   metamorphosis
nymph
complete
   metamorphosis
pupa

?

Why are insects essential to life?

Insects are all over the place! They are the most abundant animals in the world. They are a part of the largest and most significant class in phylum Arthropoda—class Insecta.

So how could you tell an insect from a chelicerate or a crustacean? One easy way would be to look at the number of body sections. Insects have three body parts, while chelicerates and crustaceans have two or even only one body part. You could also look at the number of legs. Insects have six legs, while chelicerates often have eight and many crustaceans have ten. Insect legs don't branch like crustacean appendages do. Also, insects have one pair of antennae, while chelicerates have none and crustaceans have two pairs. Insects have one pair of compound eyes; chelicerates and crustaceans may or may not have compound eyes.

There's another thing that almost all insects have—wings. In fact, many of the orders of insects include the suffix *-ptera*, meaning "wings." Insects use their wings to fly and find prey and mates. If you've ever chased around a fly or mosquito, you also know that they use their wings to avoid danger! Notice the different types of insects with varying numbers of wings.

WINGS AND THINGS

No Wings

You're most likely to run into a silverfish when digging through boxes in your attic since they like to eat the cellulose in paper and clothing. They are one of the few wingless insects in the order Thysanura.

One Pair of Wings

Flies, mosquitoes, and gnats all have one pair of membranous wings, similar to wax paper or plastic wrap. They are in the order Diptera. Instead of a pair of hind wings, they have organs that keep them balanced.

Two Pairs of Wings

All termites are in the order Isoptera. Most of the termites in a colony are the wingless youngsters, though adults are winged. One identifier for the queen in a termite colony is the presence of wings, which enable her to move around for reproduction.

Insects that are scientifically called "true bugs" have two pairs of wings. Their front wings have a solid part that transitions to a membranous wing. The second pair of wings is completely membranous. These insects are in the order Hemiptera. This group includes stink bugs.

The order of insects that has the most species (300,000+) is Coleoptera. This group includes weevils and beetles like the grapevine beetle shown below. One pair of stiff, sheath wings, which are not used during flight, protects a pair of membranous wings. The membranous wings fold up and are sheathed beneath the horny wings when the beetle isn't flying.

Some scientists estimate that 10% of the organisms on Earth are in the second-largest order of insects, Lepidoptera. This group contains butterflies and moths, organisms that have bodies and wings covered with scales. Two pairs of stunning wings make these insects some of the showiest in the insect world.

Bees, ants, and wasps have two pairs of membranous wings, much like dragonflies do. Uniquely, though, the first pair of wings is hooked to the second pair. These three insects are grouped together in the third largest order of insects, Hymenoptera.

Damselflies have two pairs of membranous wings. The second pair is wider than the first. Damselflies are in the order Odonata.

Grasshoppers, crickets, and cockroaches all have two pairs of straight wings, like the *Titanacris albipes* shown on the left. These straight wings accompany legs that give these insects super jumping ability. The front pair of wings is more leathery, and the back pair is membranous. The leathery wings cover the membranous wings when the insect is resting, similar to the beetle's horny wings. These insects can use their legs and wings to produce sounds in their own form of an insect orchestra, a process called *stridulation*. And grasshoppers can hear what they play with a chamber in their abdomens called a *tympanum*. Grasshoppers, crickets, and cockroaches are in the order Orthoptera.

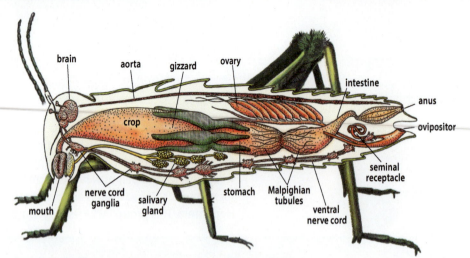

Structure and Reproduction of Insects

STRUCTURE OF INSECTS

To maintain homeostasis and carry out their life functions, insects need certain body parts. The first order of business is getting nutrition. We've seen that wings help insects move to capture prey, find a mate, gather nectar, and avoid predators. Some insects come with stingers to help them capture prey. Other insects like butterflies and bees have special appendages to help them gather nectar. Grasshoppers have mouthparts that help them gather vegetation. Mosquitoes have mouthparts that pierce through a host's skin to feed on its fluids.

Once an insect has obtained food, it needs to break it down. An insect has various kinds of mouthparts appropriate for its diet. It has something like lips and a mandible similar to that of a crustacean for grinding up food. It generates saliva just like humans do to help break down food. Ground-up food passes through its digestive system, including the gizzard in the foregut, the stomach in the midgut, and the intestine in the hindgut. This system allows the insect to absorb nutrients from its food and eliminate

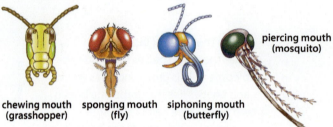

chewing mouth (grasshopper)   sponging mouth (fly)   siphoning mouth (butterfly)   piercing mouth (mosquito)

solid waste products through the anus. An open circulatory system allows the insect to remove nitrogen-containing wastes in the blood through a bundle of thin tubes between the midgut and hindgut called **Malpighian tubules**.

Insects also need to breathe. The respiratory system for an insect is made of tubes called **tracheae** (s. trachea), which branch throughout the animal. These tracheae open to the outside through a series of small pores on the sides of the insect called **spiracles**. Insects take oxygen from the air through the spiracles, distributing it throughout the body. The combination of structures in an insect's body helps it sustain life, get nutrition, eliminate wastes, and breathe.

REPRODUCTION OF INSECTS

Insects get information about their environment from their antennae and their compound eyes. When insects are ready to mate, females produce and release chemicals called **pheromones** to attract and find a mate. Insects can also use pheromones to communicate with other members of a population about danger or to give directions to a food source.

Sexual reproduction in insects takes place through internal fertilization. The male deposits sperm, which are stored by the female. As the female lays eggs, they are fertilized. The last body section of the female's abdomen forms a pointed extension called the *ovipositor*, which is used to deposit the fertilized eggs.

▲

Insects can also use pheromones to signal other individuals to form a cluster called an *aggregation*. Clustering can also be used to muster a mass attack against a predator.

METAMORPHOSIS

Each insect begins as a tiny, fertilized egg. It goes through a series of stages called **metamorphosis** in which it changes and grows. Other animals, such as sponges, crustaceans, and amphibians, also go through this process. A few insect species do not go through metamorphosis, but most follow one of two major processes—incomplete metamorphosis or complete metamorphosis. Grasshoppers, cicadas, and true bugs go through incomplete metamorphosis. Almost 90% of all insects undergo complete metamorphosis.

Incomplete Metamorphosis

Incomplete metamorphosis is the process in which an insect hatches from an egg to form a nymph, which matures into an adult.

1. The **nymph**, an immature form, hatches from the egg and looks like a miniature form of the adult. It eats and lives in the same places that adults of its species do. However, it lacks wings and external reproductive structures.

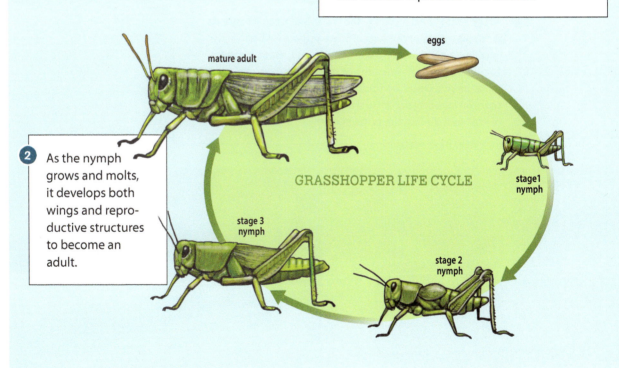

2. As the nymph grows and molts, it develops both wings and reproductive structures to become an adult.

eggs

mature adult

stage 1 nymph

GRASSHOPPER LIFE CYCLE

stage 3 nymph

stage 2 nymph

Complete Metamorphosis

Complete metamorphosis is the process in which an insect hatches from an egg to form a larva, then a pupa, and finally an adult.

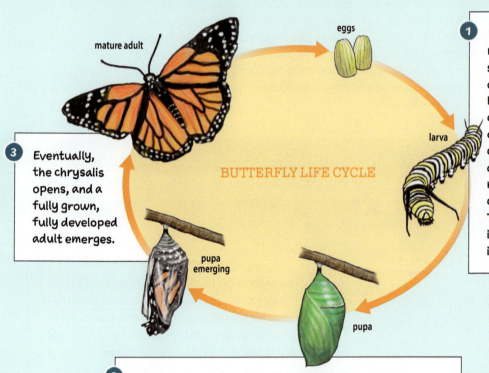

eggs

mature adult

larva

BUTTERFLY LIFE CYCLE

pupa emerging

pupa

1 The larva is a segmented, wormlike stage. Fly larvae are called *maggots*, beetle larvae are called *grubs*, mosquito larvae are called *wrigglers*, and butterfly and moth larvae are called *caterpillars*. The larva's most important activity is to eat a lot!

3 Eventually, the chrysalis opens, and a fully grown, fully developed adult emerges.

2 After eating and molting, the larva becomes a **pupa** as it forms a case around itself. The pupa of a butterfly is called a *chrysalis*. On the other hand, most moths form *cocoons*. The cocoons of certain moths are used to make silk. Don't be fooled—the larva is busy inside its cocoon! Its body structure and organs are completely re-formed during the pupa stage.

Insects in the Environment

Insects can cause great harm yet also bring great help to their environments. For example, mosquitoes and fleas can be parasitic and become vectors for diseases, such as dengue fever and malaria. Some insects, such as locusts, can greatly affect agriculture, reducing a field of crops to bare twigs, but are also prey to other insects, and so we see a kind of balance in the ecosystem. Pesticides have been created to help farmers overcome the negative side effects of insects, but caution must be taken to avoid creating an even bigger problem.

Insects serve a vital role to plants, animals, and humans as both pollinators and as a source of food. Without them, farmers would have to pollinate all their crops by hand in order to get a harvest. Pollinators like bees, moths, and butterflies help plants reproduce and spread. Insects are also a source of nutrition for other insects, animals, and humans. Insects have an important role in the food web of almost every biome, providing food for a wide variety of animals.

Insects interact with each other in complex ways. Some ants, bees, and all termites form populations that are almost like a society. There are queens in these populations and roles for different insects in a colony, such as nursery workers, foragers, guards, and those that discard the colony's trash. Older generations of individuals train younger generations to do specific jobs, such as taking care of larvae. Scouts mark food sources with pheromones so that other workers in the colony can

▲ Locusts swarm, especially in desert areas. When they destroy the crops, the locals do the only reasonable thing: they eat the locusts!

find them. Some ants even capture ants from other colonies and force them to work as slaves. As we learned in Chapter 13, some termites have underground farms, raising fungi for food.

Because of the incredible number of arthropods, especially insects, on Earth, their impact on the biosphere is almost immeasurable. Insects account for a large percentage of the biomass on land. They have shaped the course of history, led to the development of culture and art, given us medicines, and inspired us with ideas for technological design. As we saw early in the chapter, insects have even given us clues for fighting crime and bringing justice. As we work to understand, manage, and use insects and their fellow arthropods, we can make the world a better place for people. God meets our needs through His provision for all His creatures, even the tiny insects.

SAILOR BUG

Crustaceans are the arthropods of the sea, and insects are the arthropods of the land, right? Don't tell that to the ocean strider, a member of the true bug family and one of the very few marine insects. Though it lives on the ocean, it doesn't swim—it walks on water! Ocean striders eat plankton off the surface of the ocean. They have little air-trapping hairs on their legs that act as life jackets to keep them afloat. They had better look out though! Seabirds in the air and crustaceans below the surface are constantly trying to scoop them up.

Ocean striders are getting an unexpected lift from the Great Pacific Garbage Patch, an area of tiny artificial particles suspended in the water column trapped in a circular current in the North Pacific Ocean. Some experts estimate the patch

to be the size of Texas. Much of the garbage has been broken down into very small pieces after being continually bashed by wave action. Even though the particles are creating a problem for aquatic animals, they are providing a handy place for ocean striders to lay their eggs on the open sea. Thanks to the garbage patch, ocean striders are thriving.

Ecologists are concerned that the biodiversity of the ocean could be affected by all this artificial material in the ocean. And in another twist, scientists are finding that the ocean strider acts as a storage vehicle for heavy metals present in the ocean, such as cadmium. This could give scientists some clues about how pollution and toxic metals spread in the oceans and what we can do to keep our waters and insects pollutant-free.

17.3 SECTION REVIEW

1. Give three ways that you can identify the walking stick shown below as an insect.

2. How does an insect use its wings to maintain homeostasis?

3. How does an insect breathe?

4. How does the shape of an insect's mouthparts relate to its function?

5. How does an insect find a mate?

6. Does an insect reproduce sexually or asexually? Explain.

7. How does an insect change over its life?

8. What are some ways that we can control insects without using pesticides that could be harmful to the environment?

Use the case study above to answer Questions 9–11.

9. How are ocean striders able to survive on the ocean?

10. How does floating debris on the water affect marine life?

11. Explain how studying ocean striders may help scientists find new ways to keep our water clean.

17 CHAPTER REVIEW
Chapter Summary

17.1 ARTHROPOD INTRODUCTION AND CHELICERATES

- Arthropods have open circulatory systems, compound or simple eyes, segmented bodies, nervous systems, jointed appendages, and exoskeletons.

- Chelicerates are arthropods with clawlike appendages called chelicerae. Most have one or two body sections and four pairs of walking legs.

- Chelicerates use pedipalps to eat, mate, and sense their environment.

- Aquatic chelicerates reproduce sexually through external fertilization; terrestrial chelicerates reproduce through internal fertilization.

- Some chelicerates may be parasites or predators. Some chelicerates serve as food for other animals, while some are used for the treatment of diseases and infections.

Terms
arthropod • thorax • abdomen • cephalothorax • compound eye • molting • antenna • chelicera • pedipalp • book lung • spinneret

17.2 CRUSTACEANS

- Most crustaceans are aquatic arthropods that have two or three body sections, a carapace, and swimmerets. They have five pairs of walking legs and chelipeds used for defense, eating, and grooming.

- Crustaceans hunt for food using their antennae and grind food using mandibles.

- Most crustaceans reproduce sexually through external fertilization, though some undergo internal fertilization.

- Crustaceans impact their biome by feeding on detritus and serving as keystone species for aquatic food chains.

Terms
carapace • swimmeret • walking leg • cheliped • mandible

17.3 INSECTS

- Insects are arthropods that have three body segments, six legs, one pair of antennae, and one pair of compound eyes. Most insects have wings.

- Insects have mandibles used for grinding food. Ground-up food gets broken down further by saliva and travels through the digestive system.

- Females release pheromones to attract a mate. These pheromones are also used to communicate danger or directions to a food source for other insects in a colony.

- Insects begin as a fertilized egg and undergo incomplete or complete metamorphosis to reach adulthood.

- Insects are useful in their roles as pollinators; however, some are known to destroy crops and land. They are also used in medicine and forensic investigations.

Terms
Malpighian tubule • trachea • spiracle • pheromone • metamorphosis • incomplete metamorphosis • nymph • complete metamorphosis • pupa

RECALLING FACTS

1. Since an arthropod doesn't have a backbone, what provides its support?

2. List the attributes shared between the different types of arthropods.

3. What do all chelicerates have in common?

4. What three main roles do chelicerates play in the environment?

5. What group of arthropods can regenerate lost appendages?

6. Why are crustaceans on the ocean floor so crucial to their environment?

7. Create a table with one column each for chelicerates, crustaceans, and insects. Indicate the number of body sections, legs, wings, and antennae for each of these groups of arthropods.

8. Give one example of how chelicerates, crustaceans, or insects communicate with each other.

9. Identify each of the following arthropods as a chelicerate, crustacean, or insect and justify your answer.

   a. bedbug   **b.** amphipod   **c.** cellar spider

10. How does an insect eliminate wastes and keep its body free of toxins?

11. What is the difference between the nymph and adult stages of a cicada?

12. Name three different kinds of organisms that go through metamorphosis.

13. What stage is present in complete metamorphosis that is lacking in incomplete metamorphosis? What happens during this stage?

UNDERSTANDING CONCEPTS

14. Compare the ways that chelicerates, crustaceans, and insects breathe.

15. Write a sentence that relates how arthropods breathe to maintain homeostasis.

16. What is similar about the way that chelicerates, crustaceans, and insects eat?

17. What is similar about the way chelicerates, crustaceans, and insects reproduce?

18. Compare the environments in which chelicerates, crustaceans, and insects live.

CRITICAL THINKING

19. Write a general rule that describes how marine and terrestrial animals fertilize their eggs.

20. When used in a biology context, the suffix *-pod* means "foot." Crustaceans are referred to as decapods, chelicerates as octopods, and insects as hexapods. Centipedes and millipedes are in a group of arthropods called *myriapods*. Why are they referred to this way?

21. A tardigrade, or water bear (shown at right), is a microscopic animal that has body sections and eight legs and lives in the water. It is one of the hardiest organisms on Earth and is able to survive in deep space, in extreme temperatures, and under high doses of ionizing radiation. Why isn't it classified as an arachnid?

22. Give an example of an ecosystem engineer not mentioned in your textbook.

23. Why do most people use pesticides to control insect populations?

24. Why don't people in underdeveloped countries use many pesticides?

25. Why should we be careful with pesticides?

case study

TRENDS IN HONEYBEE COLONIES

In 2006 a downward trend in honeybee colonies became evident. Since honeybee pollination is one of the primary drivers in crop production, a decrease in colonies could cause significant problems with many important food sources. After investigation, it was found that the decrease in honeybee colonies was due to colony collapse disorder, which is a phenomenon that occurs when worker bees disappear and the queen, a few nurse bees, and food are left behind. This is likely due to various toxins, disease, or climate variability.

To focus attention on increasing numbers of honeybee hives, the National Institute of Food and Agriculture (NIFA) began programs to promote pollinator health. The programs worked. Over time, more and more honeybee colonies began popping up around the country. While this is good news, we still must be diligent to preserve the health of honeybee colonies around the world.

After the collapse, the numbers of honeybee colonies went through periods of growth and decline. Most of these fluctuations were due to the natural cycles of colony growth. What kinds of trends can we find throughout a typical year?

Use the case study above together with the graph on the right to answer Questions 26–33.

26. Research what could occur if honeybee populations declined and when declines are most likely to occur. What did you find?

27. What is suspected of causing colony collapse disorder?

28. Suggest three factors that could affect the health of bee colonies in America beyond what the case study mentions.

29. What kinds of changes were implemented to increase the number of honeybee colonies?

30. The number of bee colonies tends to change throughout the year. According to your research, what month tends to have the least number of bee colonies? What month tends to have the most bee colonies?

31. What can you conclude about the trends in growth and decline of bee colonies?

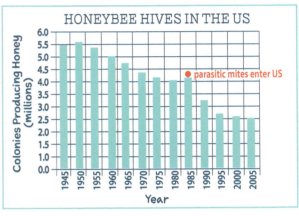

32. Why do you believe there are more bee colonies in the summer?

33. Do you think Christians should be concerned with the number of honeybee colonies in the United States? Why or why not?

18 ECTOTHERMIC VERTEBRATES

A Dragon Dilemma

As evening falls on Komodo Island in Indonesia, a long, low shape moves in the shadow of the stilt houses—a 3 m, 70 kg Komodo dragon. These giant lizards, the world's largest, have little fear of humans and perhaps even view them as just another menu option. Although the overall number of attacks is low and fatalities are rare, they do nevertheless occur. Despite being large and dangerous, Komodo dragons are also vulnerable to poaching. To protect the lizards, the Indonesian government declared the islands where they live a national park. Not surprisingly, creating a national park has impacted the traditional ways of life on the islands. In addition, the growing human population of the islands, coupled with larger numbers of tourists, has increased the number of potentially dangerous encounters with the huge lizards. To cut down on the number of such encounters, the Indonesian government has proposed raising the park's entry fee from $15 to $1000. Balancing the needs of a vulnerable species like the Komodo dragon with the needs and desires of people is like solving a math problem with constantly changing variables.

18.1 CHORDATE INTRODUCTION AND FISH

Attributes of Chordates

Can a fish drown?

Do you have a pet? Chances are it's a vertebrate! Dogs, cats, gerbils, fish, snakes—all of these are vertebrates. Perhaps being more relatable than sea jellies, worms, or bugs explains why these kinds of animals are the ones most likely to be kept as pets. It might also partly explain why vertebrates are often at the center of fierce environmental debates as in the Komodo dragon situation in Indonesia.

Just as in the rest of science, decisions about when and how to set aside resources to preserve a species are guided by models. A biblical model of ecology operates under two constraints, both outlined in the Creation Mandate. Man is created in the image of God and is to exercise dominion over the other creatures, so humans are distinct from the other creatures and have a greater inherent worth. However, man's dominion is over *God's* creation, and as custodians, humans do not have the right to abuse the creation. These two ideas—mankind's right to rule over nature and the responsibility to rule well—should guide all ecological decisions.

Part of wisely managing vertebrates is understanding them, so scientists group them by looking at the characteristics that they have in common. Vertebrates form a subphylum in phylum Chordata. Although the three subphyla of chordates have little in common, they all share four characteristics during at least some stage of their lives. Many chordates possess all these characteristics only as embryos.

Questions

What is a fish?

How are groups of fish different from each other?

How do fish breathe?

How do fish function and reproduce?

Terms

pharyngeal pouch
nerve cord
notochord
post-anal tail
cartilaginous fish
bony fish
jawless fish
scale
operculum
swim bladder
spawning

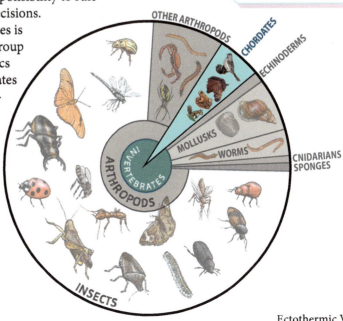

CHORDATE ATTRIBUTES

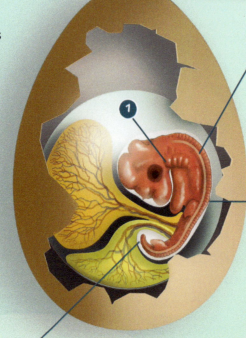

1 Chordate embryos have folds of skin along the neck called **pharyngeal pouches**. In most aquatic chordates, openings called pharyngeal slits, or gill slits, develop at these pouches. These openings permit water to flow over the gills, which develop inside the pouches. In nonaquatic chordates, the pouches never open but instead develop into structures of the lower face, neck, and upper chest.

2 Chordates have a dorsal tubular **nerve cord**. In vertebrates, this is the familiar spinal cord that forms the central nervous system with the brain.

3 Chordates have a dorsal **notochord** at some time during their life. This tough, flexible rod provides protection and support. In most vertebrates, vertebrae of the backbone grow to replace the notochord.

4 Chordates have a **post-anal tail** that extends past the anus.

Vertebrates are divided into five groups: fish, amphibians, reptiles, birds, and mammals. In this chapter we will study the ectothermic vertebrates—fish, amphibians, and reptiles—so grouped because they generally must regulate their body temperature by their behavior rather than by their internal metabolism. While it might seem that ectothermic vertebrates would be limited to living where the temperature is perfect for life, God designed them to live in all kinds of environments. Ectothermic vertebrates live in a variety of habitats—from the dark waters of ocean trenches to the sunbaked sands of the desert and the frozen tundra of the Arctic Circle. We'll begin our exploration of ectothermic vertebrates by looking at fish.

Classifying Fish

Fish are the largest group of vertebrates—over half of all known vertebrate species! As a rule, fish are strictly aquatic. Their gills dry out in air and they quickly suffocate if removed from the water. Other generalizations are difficult partly because fish are a *negative group*, meaning that if a vertebrate is *not* a bird, mammal, reptile, or amphibian, it is considered a fish. As a result, fish classification is complicated, with some taxonomists classifying fish into three classes and some spreading them over as many as six. It's generally accepted that fish can be divided into three groups.

FISH

The majority of **cartilaginous fishes**, those having skeletons entirely made of cartilage, are placed in the class Chondrichthyes. Cartilaginous fishes include some of the most graceful and some of the most notorious fish of the sea, including the stingray and the great white shark. Like all fish, they possess fins, but their fins are covered with the same type of skin as the rest of their bodies.

Most **bony fishes** are placed in the class Actinopterygii, the ray-finned fishes. The spines and soft rays that support their membranous fins give them both their common and scientific names. Since they make up over 95% of all fish species, most of our discussion of fish will concern them. The other class of bony fishes is the lobe-finned fishes in the class Sarcopterygii, including the lungfishes.

lamprey

Jawless fishes lack many of the characteristics of vertebrates; hagfish (class Myxini) even lack vertebrae. While hagfish generally feed on dead and dying fish and are known for the copious amounts of slime they produce as a defensive measure, the adults of some lampreys (class Cephalaspidomorphi) are notorious for their parasitic lifestyles. Sea lampreys attach to fish and use their rasping tongue to bore into the fish and feed on blood and other fluids.

lungfish

Fish Structure

All animals face the challenge of maintaining homeostasis in order to survive. For fish, this process is complicated by their extracting dissolved oxygen from the water and living in water that is more salty or less salty than their bodies. Their organ systems work together to maintain their homeostasis and allow them to thrive in their watery world.

The most distinguishing external feature of a fish is its fins. Fins are mainly used to steer, maintain stability, and move in an aquatic environment. Modified fins can serve other functions as well.

Most fish are covered with **scales** that give them some protection and help them move through the water. There are several types of scales, and sometimes the type of scale can help identify a fish. Fish scales are produced by the dermis layer of their skin and are made of several substances that vary somewhat with the type of fish. In most fish species the scales overlap each other, reducing drag from the water. Some species are born with roughly the same number of scales that they will have as adults, so their scales must grow as the fish grows. Others grow additional scales as their bodies grow. Bony fishes also secrete a layer of mucus over their scales that waterproofs them, further reduces drag, and helps protect them from parasites.

Along the center of each side of most fishes is a thin, dark line called the *lateral line* composed of sensory organs that detect vibrations in the water. The upper sides of many fish species are darker than their lower sides. This *countershading* camouflages the fish by preventing its being silhouetted by the sunlight filtering down through the water.

▲

The dark upper side and white underside of this spotted eagle ray are an example of countershading. The fish blends in with its surroundings whether it is being viewed from above or below.

BONY FISH
RESPIRATORY SYSTEM

gill arch

gill filaments

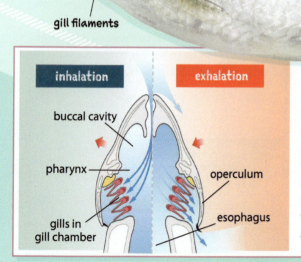

inhalation   exhalation

buccal cavity

pharynx

operculum

esophagus

gills in gill chamber

At the back of its head, a bony fish has a structure called an **operculum**, which covers its gills. With its operculum closed, a fish first fills its mouth with water. It then simultaneously raises the floor of its mouth while opening its operculum, pumping the water over its gills and out the opening. This entire process is known as *buccal pumping*.

Like all vertebrates, fish have a closed circulatory system. The two-chambered heart has one *atrium* and one *ventricle*. It pumps the blood through the gills where gas exchange takes place, and then the blood moves throughout the rest of the body before returning to the heart. Wastes are removed as the blood passes through the kidneys.

Most cartilaginous fishes have a series of gill slits instead of an operculum. Many also cannot pump water over their gills. Instead, they must constantly move to keep water flowing over their gills, a process called *ram ventilation*. Some large, fast-swimming bony fish such as tuna

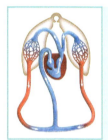

two-chambered heart (fish)

and marlin also ram ventilate. And some fish, like the sand tiger shark, can use either method, depending on whether and how fast they are swimming. Though technically fish can't "drown," since they have no lungs, ram ventilating fish that can't switch to buccal pumping will suffocate if they cannot move. Fish may also suffocate if there isn't enough dissolved oxygen in the water. Some fish, like the fancy bettas that are commonly kept as pets, have special organs that enable them to obtain oxygen by gulping air at the water's surface.

The gills consist of many gill filaments supported by bony gill arches. The delicate gills are protected by gill rakers that strain particles out of the water (and sometimes food as well).

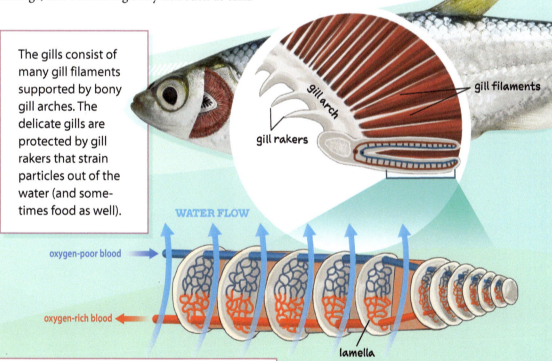

WATER FLOW

oxygen-poor blood

oxygen-rich blood

lamella

The blood vessels in each gill filament are arranged in a way that maximizes the amount of dissolved gases that can diffuse to or from the blood. Oxygen diffuses into the blood as carbon dioxide diffuses out of it and into the passing water. The water then passes out of the fish's body through the gap created as the operculum opens, taking the carbon dioxide with it.

Once oxygenated, a fish's blood does not return to the heart but circulates directly through arteries to the fish's body. Deoxygenated blood returns via veins to the atrium, the first of a fish heart's two chambers. The heart's muscular ventricle then pumps the blood back to the fish's gills.

Because fish breathe through gills, the biggest challenge to their maintaining homeostasis is regulating the proper balance of water and salt inside their bodies, a process known as *osmoregulation*. Gills must be permeable to allow gases to diffuse into and out of the blood, but that same permeability also allows water to diffuse into or out of the bloodstream. Saltwater fishes live in a hypertonic environment, so ironically, they are in danger of dehydration as water tends to leave their bodies through their gills by osmosis. To replace this lost water, saltwater bony fishes have to drink, but the only water available is salty. Salt enters the bloodstream and travels to the gills, where specialized cells pump the excess salt into the water flowing over the gills. Cartilaginous fishes, on the other hand, turn their nitrogenous wastes into *urea* and allow large amounts to build up in their tissues. This makes them effectively isotonic to the salt water around them, so they avoid the danger of dehydration altogether.

Freshwater fish live in a hypotonic environment, so water moves by osmosis into their bodies through their gills. Their kidneys remove excess water from their bloodstreams and release very dilute urine. Furthermore, like saltwater fish, freshwater fish also have gills with specialized salt-pump cells. But in freshwater fish, these cells pump salt *into* the body, helping to maintain a proper level of salt in the fish's system.

Fish, like all vertebrates, face the issue of staying afloat in water. Many bony fishes have a **swim bladder** that fills with gases from the bloodstream or digestive tract. This organ increases or decreases the amount of gas that it holds as the fish swims higher or deeper in the water, changing the fish's overall density to maintain *neutral buoyancy*. Because it is located on the dorsal side, it also helps stabilize the fish. But not all fish species possess a swim bladder. Many bottom-dwelling fishes lack one, and cartilaginous fishes rely on their large, oily livers to maintain buoyancy. Some cartilaginous fishes even have fins that provide lift like an airplane's wings as they swim through the water.

swim bladder

Both cartilaginous and bony fishes have jaws, and many have teeth used only to capture food. Most bony fish can process their food using a second set of jaws located in the fish's throat called *pharyngeal jaws*; cartilaginous fishes lack this second set of jaws. Food then moves down the esophagus into the stomach to be digested. From the stomach, digested food passes to the intestine where nutrients are absorbed. Finally, waste is released through the anus.

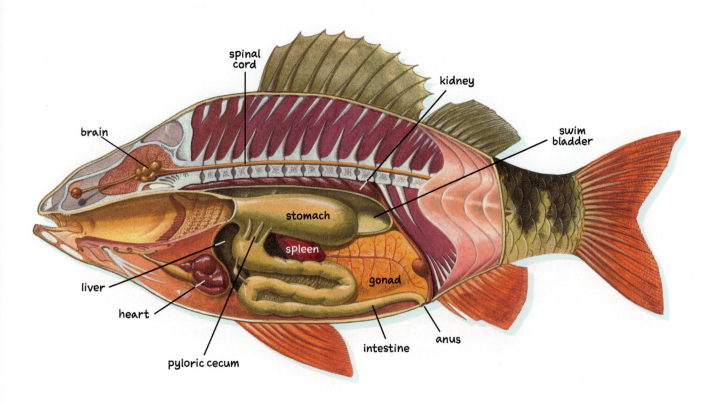

spawning salmon

Fish Reproduction

In a group as large and varied as fish, reproduction strategies are likewise varied. Many bony fishes reproduce by **spawning**. Females lay eggs that the males then fertilize with sperm. Fish can be further grouped according to how much parental care is given to eggs and young. Fish that don't guard their eggs typically produce large numbers of eggs, sometimes millions. Depending on the species, eggs may simply be released into the water or placed in a prepared nest, but once spawning is complete, no further care is given. The survival rates for the eggs and young of such fish are unsurprisingly quite low. Other fish are model parents, defending their eggs and young until they are free-swimming and able to forage for themselves. These fish generally produce much smaller numbers of eggs. Some fish species, such as the various Pacific salmons, will travel many miles from the ocean into rivers and streams to spawn in the same waters where their parents did. Many oviparous cartilaginous fishes exhibit internal fertilization, laying their eggs in an egg case commonly called a *mermaid's purse*. Still other fishes are ovoviviparous, and a few are even viviparous. These latter two groups are often called *livebearers*; they include many popular aquarium fish such as guppies and mollies. A few species of fish can reproduce asexually. The most common method is *parthenogenesis*, in which a female produces young from an unfertilized egg. Most parthenogenetic species, including the bonnethead shark, typically reproduce sexually, and females reproduce asexually only when unable to mate with a male.

Male Banggai cardinalfish are *mouthbrooders*, fish that rear developing eggs within their mouths.

Fish in the Environment

Fish are the most common vertebrates in aquatic biomes and fill many different niches. Some, such as the Achilles surgeonfish, are peaceful herbivores, grazing on plants and algae in the water. Many of these fishes help keep a check on algae populations that would otherwise overwhelm coral reefs. Others, like the rainbow trout, feed primarily on aquatic invertebrates. Still others, such as the bull shark, are on the top of the food chain, have few to no enemies, and have even attacked humans. Ironically, the very largest fish, the whale shark, eats only plankton and is so peaceful that divers have been known to ride them without provoking them.

Approximately 71% of the earth is covered by water, so a great part of the Creation Mandate involves learning to properly manage aquatic resources. Many people fish for a living, and many others for pleasure, yet an estimated 80% of the world's oceans have not yet been explored. At the same time, many fish populations, including the formerly abundant cod of the Grand Banks of Newfoundland, have drastically declined due to overfishing.

Part of being a good steward is finding ways to manage fish populations to provide for human needs without destroying the populations for future generations. This may include strategies such as limiting catches, seeking new areas to fish, growing fish in ponds (fish farming, or *aquaculture*), and eliminating some practices altogether. For instance, many shark species have greatly declined, at least partly, because of a practice called *finning*—a shark is caught, and its fins are removed to make an Asian delicacy known as shark fin soup. The shark is then thrown back into the water where, unable to swim, it soon suffocates or is killed by predators. Taking dominion of the earth does not excuse this type of waste. Biblical dominion requires maintaining sustainable levels of shark fishing and using as many parts of the fish as possible, including the fins.

18.1 SECTION REVIEW

1. What are the four characteristic physical features of all chordates?

2. At what stage of life do most chordates have all four chordate characteristics?

3. Hagfish and lampreys are grouped together because they lack what head structure found in other fish?

4. Describe ram ventilation. With which group of fishes is ram ventilation most closely associated?

5. Place the following structures in the order that fish blood passes through them after leaving the fish's ventricle: artery, atrium, gill filament, vein.

6. You dissect a fish with jaws and a large oily liver but no swim bladder. What type of fish is it? Explain.

7. What fish organ systems are involved in osmoregulation?

8. Salmon live in the ocean but return to fresh water to spawn. What process has to change its normal function to allow the salmon to make this journey and maintain homeostasis? Explain.

9. Why is it difficult to make generalizations about the roles of fish in their environments?

10. Is it wrong to make shark fin soup? Explain.

11. Coelacanths, deepwater fish of the Indian Ocean, were once known to exist only in the fossil record. Evolutionists viewed them as part of a lineage that gave rise to land animals. Then in 1938 the scientific world was shocked to learn that a coelacanth had been caught off the coast of South Africa, alive and well! What does the coelacanth show us about the nature of science?

MINI LAB

New Tank Syndrome

Imagine this scenario: You've purchased a new aquarium and have just finished setting it up. Everything is in place—water, filter, gravel, some rocks, and a few plants. You add some brightly colored fish to the tank and look forward to watching them enjoy their new home. But within just a few days, your fish start showing obvious signs of stress. Some fish appear to be gasping for air at the surface, while others are lying on the bottom. What is going on?

Your fish are suffering from *new tank syndrome*. It happens when fish are exposed to toxic nitrogen compounds in their water: ammonia, nitrite, and nitrate (in decreasing order of toxicity). In an established aquarium, certain kinds of bacteria first convert ammonia to nitrite, a process called *nitrification*. Then, different bacteria convert nitrite into nitrate; not surprisingly, this process is called *denitrification*.

Ammonia in aquarium water exists in two forms, un-ionized free ammonia (NH_3) and ammonium ion (NH_4^+). Free ammonia is the much more toxic of the two. The total of the two together is often referred to as *total ammonia nitrogen* (TAN). The ratio of the two forms for any value of TAN depends both on the water's temperature and its pH.

For this activity you will use the tables and graph on the next page to answer some questions related to new tank syndrome.

How can I prevent ammonia toxicity in a new aquarium?

Materials
none

QUESTIONS

1. Ammonia is a water-soluble waste product produced when fish metabolize protein. Very little of this ammonia is excreted in either urine or feces. Suggest how fish primarily excrete ammonia.

2. Why is ammonia toxicity a problem in an aquarium but not in the wild?

3. Using the data in Tables 1 and 2, state the relationships between the fraction of TAN that is free ammonia versus temperature and pH.

4. According to the data, for any given amount of TAN, which tank conditions present the least risk of new tank syndrome? the most risk?

5. Nitrifying and denitrifying bacteria occur everywhere. Once a tank is set up, bacteria will naturally begin to grow in it. Graph A shows the amount of ammonia, nitrite, and nitrate in an aquarium over time. Explain what is happening in Graph A on the basis of bacterial growth.

TABLE 1
1.0 PPM AMMONIA TOXICITY CHART

pH	Temperature (°C/°F)					
	20/68	22/72	24/75	26/79	28/82	30/86
7.0	0.0039	0.0046	0.0052	0.0060	0.0069	0.0080
7.2	0.0062	0.0072	0.0083	0.0096	0.0110	0.0126
7.4	0.0098	0.0114	0.0131	0.0150	0.0173	0.0198
7.6	0.0155	0.0179	0.0206	0.0236	0.0270	0.0310
7.8	0.0244	0.0281	0.0322	0.0370	0.0423	0.0482
8.0	0.0381	0.0438	0.0502	0.0574	0.0654	0.0743

TABLE 2
2.0 PPM AMMONIA TOXICITY CHART

pH	Temperature (°C/°F)					
	20/68	22/72	24/75	26/79	28/82	30/86
7.0	0.0078	0.0092	0.0104	0.0120	0.0138	0.0160
7.2	0.0124	0.0144	0.0166	0.0192	0.0220	0.0126
7.4	0.0196	0.0228	0.0262	0.0300	0.0346	0.0396
7.6	0.0310	0.0358	0.0412	0.0472	0.0540	0.0620
7.8	0.0488	0.0562	0.0644	0.0740	0.0846	0.0964
8.0	0.0762	0.0876	0.1004	0.1148	0.1308	0.1486

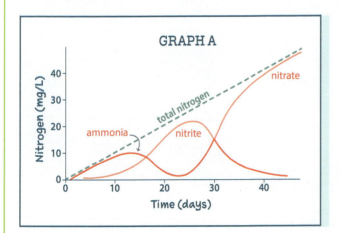

GRAPH A

6. On the basis of your answer to Question 5, suggest a strategy for introducing fish into a new aquarium setup.

7. Nitrate is an essential plant nutrient, but the plants in an aquarium are generally too few to have an appreciable effect on the amount of nitrate in the aquarium. Suggest a strategy for reducing the amount of nitrate in an established aquarium.

8. Fish from Lake Tanganyika in East Africa are popular aquarium residents. Do some internet research on Lake Tanganyika. Why must an aquarium keeper be especially mindful of new tank syndrome if keeping fish from Lake Tanganyika?

18.2 AMPHIBIANS

Classifying Amphibians

Unlike nearly all fish, many members of the class Amphibia, known as **amphibians**, live part of their lives on land and part in water. This fact is reflected in the class name, which comes from the Greek words meaning "both kinds of life." The most aquatic amphibians, such as the hellbender salamander, spend their entire lives in water. The most terrestrial amphibians, like the invasive cane toad, live most of their adult lives on land except during the mating season when they return to water to lay eggs. Because most amphibians—along with most reptiles, birds, and mammals—have four limbs, they are known collectively as **tetrapods**, a group that includes all vertebrates other than fish. Amphibians are divided into three orders, shown in the photo gallery on the next page.

?

Why do amphibians need to stay wet?

Questions

What is an amphibian?

How does an amphibian function and reproduce?

According to a biblical worldview, what should we do to conserve amphibians?

Terms

amphibian
tetrapod
caecilian
frog
salamander
cloaca
central nervous system
peripheral nervous system
estivation
hibernation
amplexus

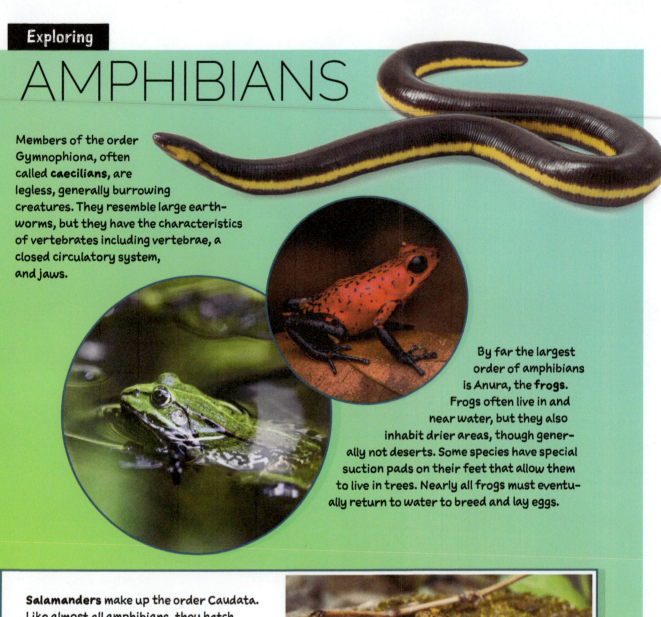

Exploring

AMPHIBIANS

Members of the order Gymnophiona, often called **caecilians**, are legless, generally burrowing creatures. They resemble large earthworms, but they have the characteristics of vertebrates including vertebrae, a closed circulatory system, and jaws.

By far the largest order of amphibians is Anura, the **frogs**. Frogs often live in and near water, but they also inhabit drier areas, though generally not deserts. Some species have special suction pads on their feet that allow them to live in trees. Nearly all frogs must eventually return to water to breed and lay eggs.

Salamanders make up the order Caudata. Like almost all amphibians, they hatch from eggs in water as larvae that typically breathe with gills. They go through metamorphosis as they transition to a terrestrial or semiaquatic life, but there are many variations of this general lifestyle.

Amphibian Structure

The most obvious feature of all amphibians is their thin, moist, highly permeable skin. In many amphibians the skin is thin enough to function in respiration. Most amphibians do possess lungs, but they are small, and much of their respiration occurs through the skin, especially when the animal is underwater. As blood flows through the vessels of the skin or lungs, carbon dioxide diffuses out and oxygen diffuses into the blood to be carried back to the animal's heart and then on to the rest of its body. Since moist skin is a much more effective respiratory organ than dry skin, amphibians secrete mucus when they are on land, giving their skin its moist look. Amphibians lack a diaphragm, so they use the lower part of their mouths to pump air into their lungs. We can easily observe this buccal pumping when looking at a living frog.

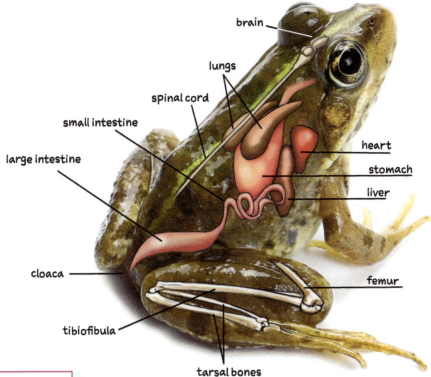

- brain
- lungs
- spinal cord
- small intestine
- large intestine
- heart
- stomach
- liver
- cloaca
- femur
- tibiofibula
- tarsal bones

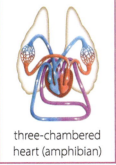

three-chambered heart (amphibian)

Amphibians have a three-chambered heart with two atria and only one ventricle. One atrium receives freshly oxygenated blood from the lungs and skin, and the other atrium receives deoxygenated blood from the veins. While the two atria pump both oxygenated and deoxygenated blood into the single ventricle, they contract at different times, so the two types of blood mix very little. The deoxygenated blood is pumped to the lungs and skin to be oxygenated, and the oxygenated blood is pumped to the body. As the blood circulates, it picks up nitrogenous waste from the body's cells and carries it to the kidneys to be removed and released as urine. Because water easily passes through amphibians' skin, their kidneys produce a very dilute urine when the animals are in the water. While amphibians are on land, they produce much less urine, allowing them to reduce water loss.

Amphibian larvae are generally herbivorous, but adults are carnivorous, generally eating anything that will fit into their mouths. After passing down the esophagus, the food is digested in the stomach before moving into the small intestine where nutrients are absorbed into the bloodstream. What cannot be digested passes through the large intestine and out through the **cloaca**. Both the digestive tract and the urogenital tract end in the cloaca, so solid wastes, urine, eggs, and sperm all pass out of the body through it.

As in all tetrapods, an amphibian's nervous system can be divided into two parts. The **central nervous system** is composed of the brain and spinal cord, enclosed and protected by the skull and vertebrae. From the central nervous system, all the nerves forming the **peripheral nervous system** run through the animal's body, allowing the brain to direct the muscles and sensory organs to send messages to its brain. If a frog by the lakeside hears a hiker walking by, the frog will jump into the water to hide—this simple action involves hundreds of nerve cells all working together with the frog's muscles and other organs.

Because amphibians are ectothermic, they are not able to adjust their body temperatures internally when the weather becomes too hot or too cold. Their moist skin can easily dry out, making them even more vulnerable to temperature extremes. As a result, many amphibians hide away and enter a state of torpor to avoid summer heat or winter cold. The animal's systems reduce activity to a minimum with very low metabolic and respiratory levels. Frogs that live in hot places often **estivate** (become dormant in summer) in the mud at the bottom of lakes. Others, such as the wood frog, live in regions where they must deal with winter cold. Wood frogs survive by **hibernating** (becoming dormant in winter) in the ground.

tadpoles

eggs

FROG LIFE CYCLE

skin grows over gills

hind legs appear

adult frog

front legs appear

Amphibian Reproduction

As the days in spring lengthen and temperatures increase, male frogs gather at creeks, ponds, and marshes and burst into their mating choruses. Each species has its own mating call, and females are attracted to the call of the males of their own species. When a pair mates, the male typically grasps the female from on top and behind, a position known as **amplexus**. Most anurans exhibit external fertilization, so this position allows the male to fertilize the eggs as the female lays them. Mating typically occurs in or near water where the eggs are laid; the water can range from a permanent lake to a temporary stream to a pocket of water held in the leaves of a plant. Salamanders, on the other hand, find each other by scent, and some perform courtship rituals before amplexus and mating. Unlike frogs, salamanders exhibit internal fertilization, with some species laying eggs and others bearing live young.

When the eggs hatch, the larvae often look very different from their parents, especially in the case of frogs. Amphibian larvae typically are aquatic and often have external gills. Frog larvae, or tadpoles, lack legs and swim with their large muscular tails. Salamander larvae do possess legs in addition to finned tails. The time from hatching to completing metamorphosis varies according to species. Spring peepers emerge as adult frogs only a few weeks after hatching, while American bullfrogs may take over a year to complete metamorphosis in the colder parts of their range.

The amount of parental care that amphibians provide for their young varies. Some amphibian species leave their eggs as soon as they are laid, some take pains to hide their eggs before leaving, and others actively guard their eggs or even their young. Some amphibians exhibit unusual or even bizarre forms of parental care. The male common midwife toad carries the eggs of his mate on his back until they are ready to hatch. Just before the larvae emerge, the male enters water and allows the tadpoles to swim away. The eggs of the common Surinam toad become embedded in the skin of the female's back during mating. The eggs hatch, and the tadpoles metamorphose into young frogs before they leave. The female horned marsupial frog takes this strategy one step further and houses the eggs and larvae in a pouch on her back. When the young frogs have completed metamorphosis, they emerge to begin life on their own.

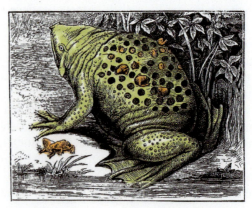

Surinam toad with emerging froglets

Amphibians in the Environment

Beginning in the 1980s, scientists began noticing that frogs were dying all over the world. For a while, these mysterious amphibian deaths baffled them. Then they discovered that the fungus *Batrachochytrium dendrobatidis* had begun to infect the frogs and kill them. This fungus, called the *chytrid fungus*, is now common in frog habitats throughout much of the world. Scientists estimate that as many as 30% of the amphibians in the world have been affected. But some frogs appear to be immune to the fungus because they carry the bacterium *Janthinobacterium lividum*, which produces an antifungal agent. Researchers are working on a way to introduce *J. lividum* to other frog populations to help them resist the fungus. Could fungus-resistant frogs be the key to curing their fellow amphibians?

Scientists continue to study the chytrid fungus, searching for possible ways to stop its destruction of amphibians. Since this research will likely cost a great deal of time and money, Christians must wrestle with questions of value. Is it worth spending large amounts of resources to save frogs? The answer to that question is not always simple. It is possible to be so concerned about the frogs that we hinder human well-being. Humans are more important than frogs, and we should always keep that priority in mind.

Taking care of the frogs can provide great value to humans and the frogs' habitats as well. Because their lifestyles often encompass both aquatic and terrestrial habitats, amphibians are an important part of both the food chain and their ecosystems. Tadpoles provide food for many fish, and adult amphibians consume large numbers of insects and other small invertebrates. Since many of their prey species carry diseases and are otherwise pests to humans, their consumption by amphibians benefits people. And because they are sensitive to changes in their environments, amphibians also serve as important indicator species. (See the case study on page 425.)

When God gave mankind dominion over the earth in Genesis 1:28, He gave him the responsibility to care for and rule over the animals of the world. Working to find a way to save amphibians from the deadly effects of the chytrid fungus fits within that responsibility as long as it does not threaten the welfare of humans.

18.2 SECTION REVIEW

1. Why are caecilians not a typical tetrapod?

2. In what amphibian order would the spring peeper, a common frog of eastern North America, be placed? Explain.

3. In most animals, a three-chambered heart that allows blood from the lungs to mix with blood from the body systems would be very inefficient. Why does it work well for amphibians?

4. What body system keeps an amphibian from becoming overhydrated in water and dehydrated on land? Explain.

5. Why would you not expect salamanders to make loud calls in the spring as frogs do?

6. Is a bullfrog more likely to lay its eggs in a permanent or a temporary body of water? Explain.

7. Should humans invest resources into trying to cure amphibian populations of the chytrid fungus? Explain.

18.3 REPTILES
Classifying Reptiles

Questions

What are the different orders of reptiles?

How does a reptile function and reproduce?

According to a biblical worldview, how should protecting endangered species be balanced with the well-being of humans?

Terms

amniote
turtle
scute
keratin
squamate
crocodilian
tuatara
septum
Jacobson's organ
pit viper
salt gland

Hot, dry desert sand. Wet, salty ocean sand. You're not going to find amphibians in either of these places, but you can find reptiles there. Amphibians' moist, permeable skin generally makes them unsuited for life in the desert or the ocean. But a reptile's scaly skin allows it to live in both environments. Whether turtles, lizards, snakes, or crocodilians, reptiles possess a wide variety of forms, colors, and lifestyles. But all reptiles have scaly skin and are ectothermic. Their young do not go through a metamorphic stage but instead look like smaller versions of their parents. The young are hatched from amniotic eggs. Animals that do so are **amniotes**, and this includes birds and mammals in addition to reptiles.

Are all reptiles poisonous?

the AMNIOTIC EGG

The embryo of an amniote is surrounded by a membrane, the *amnion*. This membrane holds the fluid within the *amniotic sac*, which protects and cushions an embryo.

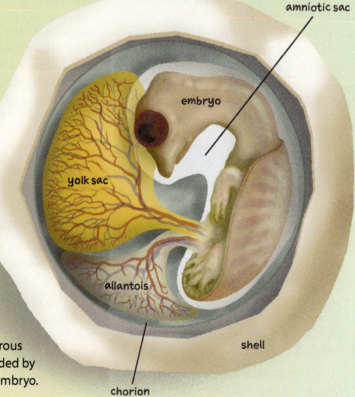

All *amniotic eggs* also contain a *yolk sac* that holds the yolk that nourishes the embryo. As the yolk is used, it is absorbed into the embryo's body.

amniotic sac

embryo

yolk sac

allantois

chorion

shell

Metabolic wastes are stored in the *allantois*, and the embryo and other structures are surrounded by the *chorion*. In oviparous species the entire egg is surrounded by a shell that further protects the embryo.

REPTILES

Living reptiles are divided into four orders: turtles, lizards and snakes, crocodilians, and tuatara.

TURTLES

Turtles, in order Testudines, are protected by a hard shell formed by bony plates fused to their ribs and vertebrae along their backs and to their ribs underneath. On top of the plates, the shell adds an additional layer of protection by being covered by large scale-like **scutes** made of **keratin**, a protein found in the external structures of amniotes. Horns, nails, hair, scales, and claws are all made of keratin. Turtles range in size from the three-inch speckled cape tortoise of South Africa to the eight-foot leatherback sea turtle.

scutes

LIZARDS & SNAKES

Most living reptiles are in the order Squamata, typically divided into the lizards in suborder Sauria and the snakes in suborder Serpentes. **Squamates** are distinguished by their loosely attached jaws. In snakes this attachment is so loose that they can swallow prey much larger than their own heads. Snakes can be very long, with the reticulated python surpassing twenty feet in length. People often fear snakes because some are infamously venomous, but 90% of species are nonvenomous, and most are perfectly harmless.

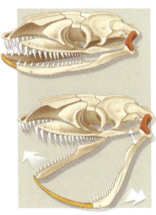

Unlike snakes, most lizards have legs, and all have ear openings and movable eyelids. Although their jaws are movable, they are less loosely attached than the jaws of snakes. Lizards such as the Komodo dragon can reach ten feet in length, but most are much smaller. Among lizards, the beaded lizard and the Gila monster are known to be venomous, along with the Komodo dragon.

REPTILES *(continued)*

CROCODILIANS

The unrivaled giants of the reptile world in terms of weight and ferocity are the **crocodilians** in the order Crocodylia. The largest of these massive creatures, the saltwater crocodile of Australia and Southeast Asia, can reach twenty feet long and weigh over a ton. Crocodiles in the family Crocodylidae are generally distinguished from alligators of the family Alligatoridae by their narrower bodies and snouts, by their more aggressive dispositions, and by the lower tooth on each side of their mouths that fits into a groove on the outside of their upper jaws.

In contrast to the crocodile, an alligator's lower teeth fit inside its mouth.

The lesser known but very large gharials of the family Gavialidae have the narrowest snouts of all crocodilians and primarily eat fish. The most distinguishing feature of the gharials is the bulbous growth on the end of the mature male's snout called a *ghara*. Its exact function is unknown.

TUATARA

The two living species in Rhynchocephalia are **tuatara** (s. tuatara). They inhabit islands north and south of North Island, New Zealand.

Tuatara often share burrows with seabirds such as the fairy prion. While early observers thought of the tuatara as protectors of the nesting birds, their relationship seems to be more complex than that, ranging from commensalism to predation.

scutes

scales

Reptile Structure

Like all animals, the reptile's organ systems work together to maintain the animal's homeostasis. Since reptiles often live in arid or aquatic environments, they need a way to keep water from diffusing through their skin in either direction. Reptiles are covered with an armor of scales or scutes. While these structures look similar and are both made of keratin, scales are produced by the epidermis while scutes are produced by the dermis. Squamates and tuatara have scales while crocodilians have scutes. Turtles typically have scutes on their shells and scales on the rest of their bodies. While the difference may seem small, it determines how the reptile replaces its skin. Animals with scales grow a new epidermis under the old and then shed the old one, either gradually in pieces, as turtles do, or all at once, as in the case of snakes. Animals with scutes simply add layers of keratin to the inner layer of their scutes as the outer layers are worn away. So you will never find a shed turtle shell or alligator skin.

Like amphibians, most reptiles have a three-chambered heart, but the ventricle has a wall called a **septum** that partially divides it. Although the deoxygenated blood from the body and the oxygenated blood from the lungs could potentially mix in the ventricle, the blood from the two atria generally remains separate due to the flow patterns within the chamber. Crocodilians are unusual among reptiles in having completely divided ventricles, making them the only ectothermic animals with four-chambered hearts.

As the blood flows through the lungs, carbon dioxide from the body's cells diffuses from the blood into the lungs, and oxygen diffuses from the lungs into the blood to be carried to the body's cells.

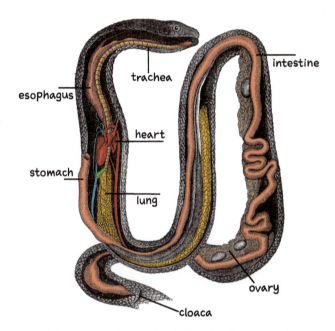

esophagus
trachea
intestine
heart
stomach
lung
ovary
cloaca

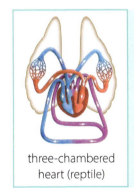

three-chambered heart (reptile)

Unlike mammals, reptiles lack a diaphragm and must use other means to expand their lungs. While amphibians pump air into their lungs, reptiles contract their rib muscles to allow their lungs to expand. When they increase the volume of their lungs, air pressure in the lungs decreases, drawing in air. The lungs are the only major respiratory organ in most reptiles.

Reptiles generally have more complex brains than amphibians and fish. Most reptiles see quite well and are believed to be able to perceive color better than many mammals. However, their sense of hearing is poor, though some have an ability for vocal communication. Some reptiles, such as snakes and tuatara, lack external ears entirely and so can hear sounds only within a very limited range of frequencies.

heat-sensing pit

Pit vipers use their pits to detect the heat of their endothermic prey.

Jacobson's organ

Squamates are known for flicking their tongues. They sample the air, bringing scent particles back into their mouths to be processed by the **Jacobson's organ**, a sensory organ in the roof of the mouth. This organ, found in many animals, more than makes up for squamates' rather poor sense of smell. Additionally, certain venomous snakes, such as the timber rattlesnake, are called **pit vipers** because of the heat-sensing pit located on each side of their heads between the eye and nostril. While turtles and crocodilians lack these types of specialized organs, they do have a relatively good sense of smell.

Pit vipers, like all snakes, are carnivorous, as most reptiles tend to be. Exceptions do exist among turtles and lizards. Not all reptiles possess teeth, and the teeth of those that have them are less suited for chewing than for grabbing and holding prey or, in the case of venomous snakes, for injecting venom into their prey. As a result, much of a reptile's food is swallowed whole or in large chunks. Salivary glands aid in swallowing, and digestion occurs primarily in the stomach. From there, the reptile's food moves to its small intestine, where nutrients are absorbed. Wastes pass through the large intestine and into the cloaca to be expelled.

The reptilian excretory system varies according to the habitat of the species. Aquatic species generally do not need to worry about conserving water, so they typically release diluted ammonia. However, the bodies of terrestrial species must conserve water, and even a small amount of ammonia is highly toxic. So terrestrial reptiles typically excrete nitrogenous wastes as uric acid, which is almost insoluble in water. This allows reptiles to remove the nitrogenous wastes in a harmless, nearly solid form with very little water loss. Marine species such as the green sea turtle and the marine iguana of the Galápagos Islands must deal with the buildup of salt in their systems. Reptile kidneys do not remove enough salt from their blood to prevent the salt concentration from reaching toxic levels; a **salt gland** near each eye removes excess salt and excretes it as a concentrated salt solution.

A marine iguana's salt glands allow it to eat salty marine algae by filtering salt from its blood and releasing it through the nostrils.

THOSE TERRIBLE LIZARDS

Move over, *Tyrannosaurus rex*! Meet the titanosaur *Dreadnoughtus*. This enormous plant eater probably lived up to its name and feared very little, but would it have feared humans? We often imagine dinosaurs as extremely ferocious, especially the carnivorous ones, but today even extremely large carnivorous mammals such as the grizzly bear will tend to avoid a confrontation with humans if given the choice. This fits with God's promise to Noah that animals would fear His image-bearers (Gen. 9:2), so it seems logical that dinosaurs would share this innate fear.

The term *dinosaur* (literally, "monstrous lizard") was coined by English paleontologist and biologist Sir Richard Owen in 1842, about the time of the first fossil finds of large reptiles. Although the term is normally used to describe the extremely large, presumably extinct reptiles, most species were about the size of horses, and some were as small as chickens. Technically, the term applies only to the land-dwelling reptiles that walked with their legs beneath them.

The biblical account of Creation says that God made the land animals on the sixth day, the same day that He made Adam and Eve. Therefore, dinosaurs and men must have lived together. Evolutionists believe that dinosaurs died out sixty-five million years ago and over fifty million years before humans and chimpanzees diverged. For an evolutionist, humans and dinosaurs never lived together.

As a result of these different models, creationists and evolutionists tend to interpret data differently. However, even holding the correct model does not guarantee that a scientist will always correctly *interpret* the evidence. Many creationists long believed that fossilized tracks near Texas's Paluxy River showed dinosaur and human tracks together. Evolutionists rejected this interpretation because it did not fit their model. As creationists have continued to study the prints, most have come to agree that these fossils show only dinosaur tracks. When the truth of the Bible is not in question, creationists should always be willing to humbly reexamine their interpretation of scientific evidence.

eggs

adult

hatchling

SEA TURTLE LIFE CYCLE

juvenile

Reptile Reproduction

Reptiles have a variety of reproductive strategies. All turtles, crocodilians, tuatara, and some lizards and snakes lay eggs, but many lizards and snakes bear live young. All reptiles exhibit internal fertilization since the oviparous species have eggs surrounded by a calcium-based shell that makes external fertilization impossible. With the exception of turtle eggs, the amount of calcium in reptile eggs is generally lower than in bird eggs, so most reptile eggs have a leathery feel.

The populations of a few lizards and at least one species of snake, the brahminy blindsnake, are entirely female, and as a result they reproduce asexually through parthenogenesis. Increasingly, captive females in other reptile species, including Komodo dragons, have laid eggs that hatch without any genetic input from a male. Some scientists think that this is a normal method of reproduction when the females cannot find a suitable mate. Others believe that this is merely a rare instance of incomplete meiosis that mimics parthenogenesis.

Reptiles generally bestow little parental care on their young. The female will often lay her eggs in a hole, cover them, and then leave. The primary exception to this rule is the crocodilians. Female crocodilians guard their nests from predators, and when the young hatch, the female breaks open the nest to allow them to escape. Females of some species care for their young for a year or more.

In crocodilians, tuatara, many turtles, and some lizards, the temperature of the nest determines the sex of the young. In some species, lower temperatures result in a large number of females, while higher temperatures produce primarily males. In other species, the opposite is true. In still other species, the effect of temperature on sex is even more complex with high and low temperatures resulting in females and in-between temperatures producing males. The temperature of the eggs affects the production of hormones that determine the sex of the hatchling. However, the sex of young snakes along with some young turtles and lizards is determined by sex chromosomes.

Crocodilians exhibit a level of maternal care unusual among reptiles. This mother is carrying her baby in her mouth, not eating it!

Reptiles in the Environment

Some reptiles are at the top of the food chain. Crocodilians and the largest snakes, such as the green anaconda, have few predators, such as large cats, as adults. However, even these reptiles are vulnerable before they mature. Young reptiles and adults of smaller species often fall prey to predators, including other reptiles. Birds and mammals also take a heavy toll. Tuatara on isolated small islands have historically been safe from predators. On islands where rats have been introduced and become established, the tuatara have declined and even died out entirely due to rats preying on tuatara eggs. On several of these islands, conservationists have exterminated the rats and reintroduced the tuatara, which seem to do very well in the absence of the invasive rodents.

It remains to be seen whether efforts to save the Komodo dragon will be similarly successful. Despite their protected status, the dragons still face a variety of threats. Logging, habitat fragmentation, the exotic pet trade, and the expansion of human settlements are persistent problems. Now scientists are examining the possible effects of climate change on dragon survival as well. Reduced populations also become susceptible to the effects of inbreeding and genetic drift. Any proposed solutions for these complex problems must take the needs of people into account. It is people, not giant lizards, that bear the image of God. At the same time, conserving an endangered species is a worthy goal and one that fits within the scope of the Creation Mandate.

18.3 SECTION REVIEW

1. Identify the three membranes in an amniotic egg marked in the diagram.

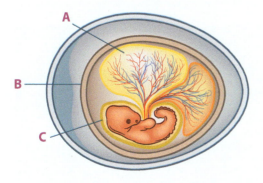

2. Which structure of egg-laying amniotes makes external fertilization impossible? Explain.

3. Which reptilian order contains species without legs?

4. Create a T-Chart showing which of the orders of reptiles have scales and which have scutes.

5. Members of which reptilian order have armored scutes and regularly prey on large mammals such as deer?

6. What circulatory structure is more complete in crocodilians than in other reptiles? Explain.

7. If crocodilians could survive near the Arctic Circle, what aspect of sex determination would prevent a population from being established?

8. Is the life of one person more important than the survival of an entire animal species? Explain.

18.1 CHORDATE INTRODUCTION AND FISH

- Chordates are animals that have four characteristics at some stage of their lives: a dorsal tubular nerve chord, a notochord, pharyngeal pouches, and a post-anal tail. Most chordates are vertebrates, that is, animals with backbones.

- Ectothermic vertebrates regulate their body temperature through their behavior rather than by internal means. Fish, amphibians, and reptiles are ectothermic vertebrates.

- Fish are classified into three classes that include cartilaginous fishes, bony fishes, and jawless fishes.

- Fish use fins for movement, and most have scales for protection. A row of sensory pits along the body—the lateral line—is an important part of a fish's nervous system.

- Fish have gills to extract oxygen from water and a closed circulatory system featuring a two-chambered heart.

- Many bony fish can maintain neutral buoyancy using a swim bladder filled with gas. The oily liver of a cartilaginous fish performs the same function.

- Many fish reproduce by spawning, using external fertilization, but other reproductive strategies are also used, including bearing live young and even parthenogenesis.

Terms
pharyngeal pouch • nerve cord • notochord • post-anal tail • cartilaginous fish • bony fish • jawless fish • scale • operculum • swim bladder • spawning

18.2 AMPHIBIANS

- Amphibians—caecilians, frogs, and salamanders—spend at least part of their lives in water. Most amphibians are tetrapods.

- An amphibian's thin, moist skin aids in respiration. Larval amphibians also respire through gills; adult amphibians have lungs, which they fill through buccal pumping.

- An amphibian's circulatory system includes a three-chambered heart.

- An amphibian's urinary, digestive, and reproductive tracts pass products through a common opening—the cloaca.

- Frogs utilize external fertilization, while salamanders exhibit internal fertilization. Eggs are typically laid in water. Some salamanders give birth to live young.

- Because of their sensitivity to environmental toxins, amphibians are often important indicator species.

Terms
amphibian • tetrapod • caecilian • frog • salamander • cloaca • central nervous system • peripheral nervous system • estivation • hibernation • amplexus

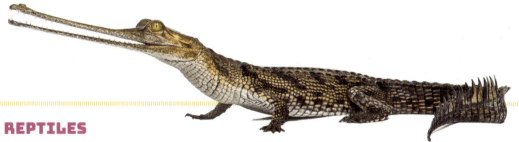

18.3 REPTILES

- Reptiles, as well as birds and mammals, are amniotes, that is, animals whose embryos are surrounded by a membrane—the amnion—and nourished by yolk.

- Reptiles include turtles, squamates (lizards and snakes), crocodilians, and tuatara.

- A reptile's skin is covered in either scutes or scales; both are made of keratin but produced by different layers of a reptile's skin.

- A reptile's three-chambered heart includes a ventricle that is partially divided by a septum, which, together with the blood flow pattern within the heart, helps prevent mixing of oxygenated and deoxygenated blood. Crocodilians have four-chambered hearts.

- Squamates sample scent particles with their tongues and process them with their Jacobson's organs. Pit vipers can detect heat with pits on each side of their heads.

- Most reptiles are carnivorous, though some lizards are herbivores.

- All reptiles exhibit internal fertilization, and most lay eggs, though some squamates bear live young.

Terms
amniote • turtle • scute • keratin • squamate • crocodilian • tuatara • septum • Jacobson's organ • pit viper • salt gland

Chapter Review Questions

RECALLING FACTS

1. What four characteristic physical features distinguish chordates from other animals?

2. Which class of fish is not really a vertebrate? Explain.

3. What is the function of the lateral line?

4. Which order of amphibians lacks the basic characteristic of a tetrapod? Explain.

5. How do amphibians survive periods of intense heat or cold?

6. How does an amphibian's diet change during metamorphosis?

7. What three structures function as respiratory organs during some time in a typical amphibian's life? Explain.

8. Which groups of vertebrates are not amniotes?

9. List three membranes found within an amniotic egg and describe the function of each.

10. Name three differences between a legless lizard and a snake.

11. What is the primary difference between reptilian scales and scutes?

12. What two sense organs does a timber rattlesnake have that a crocodile does not have?

13. What is unusual about the parental care that crocodilians exhibit compared with other reptiles?

UNDERSTANDING CONCEPTS

14. Tunicates, also known as sea squirts, are soft-bodied marine animals with no backbone or spinal cord, yet they are nevertheless classified as chordates. What can you deduce from this classification?

15. Identify each fish as cartilaginous, bony, or jawless.

    a. stickleback   **b.** blacktip reef shark   **c.** brook lamprey

16. Explain the difference between the gill structures of bony fishes and cartilaginous fishes.

17. Is the blood that leaves a fish's heart oxygenated or deoxygenated? Explain.

18. Compare cartilaginous fishes with bony fishes using a T-Chart. Include the jaws, gill structure, skeletons, fins, and organs used for buoyancy in your comparison.

19. What type of reproduction and fertilization would a fish that produces a mermaid's purse (shown on the right) exhibit?

20. How might shark fins be collected for shark fin soup in a responsible manner?

21. If a deaf female frog and a deaf female salamander were both searching for a mate, would either be able to find one? Explain.

22. How might bacteria be the answer to the amphibian die-off caused by the chytrid fungus?

23. Place each animal in the correct reptilian order.

    a. Nile crocodile   **b.** arboreal ratsnake   **c.** box turtle

24. Create a hierarchy chart showing the levels of classification of the ectothermic vertebrates. Include the following categories: amphibians, bony fishes, caecilians, cartilaginous fishes, Chordata, crocodilians, fish, frogs, jawless fishes, reptiles, salamanders, squamates, tuatara, turtles, and vertebrates.

25. How is an amphibian's method of filling its lungs different from the way that a reptile does it?

26. How might an unusually warm summer affect the sex of the snakes that hatch during that time? Explain.

27. When is a large reptile most likely to fall prey to a predator?

CRITICAL THINKING

Use the case study on the right to answer Questions 28–30.

28. Butterflyfish like the blue-cheeked butterflyfish feed on coral polyps. Why would this make them a good indicator species?

29. Would a species that lives in a wide variety of habitats make a good indicator species? Explain.

30. What are some potential problems that could arise when choosing an indicator species?

31. Some evolutionists see the progression from the two-chambered hearts of fish to the three-chambered hearts of amphibians and the three- and four-chambered hearts of reptiles as an example of evolutionary development. Refute this idea by considering how the different heart designs function with the animals' respiratory systems.

32. A terrestrial ectothermic animal often basks in the sun to raise its body temperature. At what time of a summer day in the desert is a lizard likely to bask?

33. Many male vertebrates fight over females. Were venomous snakes such as the king cobra to bite each other during such combats, they could easily kill each other. Instead, they engage in wrestling. The snake that pins its opponent's head to the ground, usually the larger one, wins. How does this behavior help ensure the survival of the species?

34. Write a short paragraph that explains the privileges and responsibilities of human beings in relation to animals.

case study

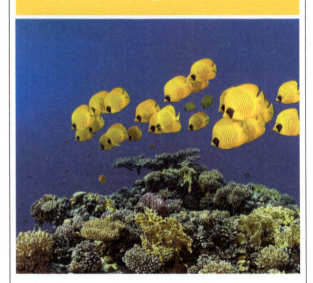

INDICATOR SPECIES

When scientists survey the health of an ecosystem, they can't sample all the organisms in the ecosystem since one community can contain thousands of individual species. Instead, scientists employ *indicator species*. These organisms are chosen because their abundance is tied to the health of the whole ecosystem, so they can often give early warning if the ecosystem is not doing well. In addition, indicator species can also help ecologists locate the boundary between one ecosystem and another.

19 ENDOTHERMIC VERTEBRATES

The Misfit

Beneath the dry grasslands of East Africa, naked mole-rats live out their existence without ever seeing the light of the sun. Despite its hideous appearance, scientists love studying this rodent. It's the only known ectothermic mammal, and it forms social groups similar to colonies of termites, each one complete with a queen. Scientists found that the naked mole-rat has the ability to survive on little oxygen, fight the effects of aging, and ward off cancer. Scientists scrambled to sequence its genome, hoping to unlock the secrets of one of the healthiest animals on Earth. Researching the naked mole-rat could lead to medical breakthroughs, but the research is laced with evolutionary thinking. Does that matter? How should someone with a biblical worldview respond to this research?

19.1 BIRDS

Endothermic Vertebrates

Most known birds and mammals (other than the naked mole-rat) are endothermic. Instead of relying on behaviors such as basking to maintain their body temperatures as ectotherms do, endotherms use heat energy produced during cellular respiration to keep their bodies at the right temperature. They often have complex internal regulation systems that keep them from becoming too hot or too cold. Because they rely on metabolic energy, endothermic vertebrates often need to eat much more than ectothermic vertebrates. Their bodies also contain layers of fat or special structures that insulate them from heat loss. Birds have feathers, while most mammals grow hair.

?

Why do birds migrate?

Classifying Birds

The **feathers** of birds are their most distinguishing feature. No other living animal has feathers. Without them, birds would not be able to fly. Although they have only two legs, they have four limbs including their wings, so they are considered tetrapods. Unlike fish, amphibians, and reptiles, which can each be easily divided into three or four large groups, birds are classified into more than thirty different orders. However, it is possible to loosely group them according to habitat, behavior, and appearance.

Questions

What is a bird?

How does a bird function and reproduce?

How are birds able to fly?

How are birds designed to live in different environments?

How do birds interact with other birds and the environment?

Terms

feather
keel
air sac
syrinx
crop
gizzard
courtship
albumen
incubation
migration

Endothermic Vertebrates   **427**

MASTERS of FLIGHT

Seabirds

The European herring gull (left) and similar birds spend much of their lives by the ocean. Most of them nest on land and fly over the beach and water looking for food, but a few, such as the shy albatross shown on the right, spend much of their lives either flying over the ocean or riding its waves, coming to land primarily to breed and nest.

Wading Birds

The great egret (left) and many other birds in this group are equipped with long legs for wading through the water. Many of them have sharp bills for spearing fish, though some, like the roseate spoonbill (above), have sieve-like bills for straining out the crustaceans and small vertebrates they eat.

Shorebirds

Birds like this spotted sandpiper spend most of their lives near water but rarely swim. They search for food along the shore in the company of other shorebirds, such as the sanderling. As many birders can attest, distinguishing the various species is a major challenge.

Raptors

Raptors like the red-tailed hawk (shown above) are carnivorous birds equipped with hooked beaks and strong grasping talons. Raptors are some of the largest birds and can carry half their own body weight while flying. Although they fly primarily by gliding on rising currents of warm air, they often take their prey in a high-speed dive. A peregrine falcon (above right) can dive at speeds in excess of 300 kph.

Waterfowl

The hooded merganser (above), like all ducks, is designed for water. Its webbed feet are much more suited for swimming than for walking, and its oily feathers and low-density bones make it naturally buoyant. However, even the largest waterfowl, such as the trumpeter swan, are also very capable flyers.

Game Birds

Game birds, such as ruffed grouse (right), quails, and doves, are a rather loose group of birds, commonly hunted for sport and food. Some species, including the ring-necked pheasant (above), have been introduced throughout the world outside their native Asia.

MASTERS of FLIGHT (continued)

Hummingbirds

Hummingbirds, such as the rufous hummingbird (left), are famous for their small size, scrappy personalities, and buzzing, hovering flight. They also have an extremely high metabolism that requires them to feed almost constantly on flower nectar. The bee hummingbird of Cuba (above right) measures less than 6 cm from beak to tail, making it the smallest bird known. But don't let its size fool you. Even though this bird is the size of an insect, it can live up to seven years!

Flightless Birds

Although all birds have feathers and wings, some, like the emu (left), do not fly. In fact, many of these flightless birds lack a *keel*, a structure on the breastbone where flight muscles attach in flying birds. However, the chinstrap penguin (below), like all penguins, has a keel that allows it to use its wings for swimming.

Perching Birds

Perching birds, such as the indigo bunting (above), are some of the most beloved birds due to their songs, their sometimes very bright colors, and their willingness to live near human habitation. A perching bird's feet automatically lock while perching until the bird releases them, so an eastern bluebird, for example, can sleep on a branch without falling off.

Avian Structure

Flying is not easy. As many engineers and designers have found over the years, flight requires much more than just wings. A bird takes wing because all its systems work together to achieve flight. In the early sixteenth century Leonardo da Vinci studied the flight of birds and bats in order to design machines that he thought might enable humans to fly. Four hundred years later, Wilbur and Orville Wright studied the flight of birds and realized that several of the commonly held beliefs about flight at that time were wrong. As a result of this insight, the Wright brothers accomplished the first powered aircraft for human flight.

Flying requires structural support that is both lightweight and strong. A lack of either characteristic spells doom for flight, but God has given birds special skeletons that meet their needs. The most obvious difference between the avian skeleton and other vertebrate skeletons is the presence of a **keel**, a large, flat structure on a bird's breastbone that provides a place for wing muscles to attach. Many of a bird's bones, including most of its vertebrae, are also fused to give the skeleton needed rigidity. They are hollow, yet strong. In fact, the skeletons of many birds are so light that they weigh less than the bird's feathers.

In addition to a lightweight, rigid skeleton, flight requires lift and thrust to carry birds upward and forward, and feathered wings provide both these forces for a bird. As a bird flaps its wings, the wings propel it both up (lift) and forward (thrust). Additionally, feathers cushion and insulate birds and reduce air resistance as they fly through the air. The feathers that you can see are *contour feathers*, but birds also have an insulating layer of *down feathers* next to their skin. Generally, the only parts of a bird that do not have feathers are its lower legs and feet, which are covered in scales.

keel

◀ Internal support structures inside a bird's bones make its skeleton lightweight and strong.

The Feather

A feather grows from a follicle deep in the skin and is made of a central shaft ➊ and a flat *vane* ➋ consisting of hundreds of rows of barbs ➌.

Each barb carries multiple pairs of barbules ➍ that overlap and connect with barbules on other barbs by means of tiny hooks. These hold all the barbs together as part of the vane of the feather.

BIRD WINGS

A bird's wings are suited to its lifestyle. A woodpecker could not maneuver with the wings of an albatross, and a falcon could never catch prey with the wings of a grouse. Generally, wings fall into the three general categories shown: high-speed, soaring, and elliptical.

High-speed wings have a long, thin, tapered shape that generates little drag in the air. The Himalayan white-throated needletail shown on the left uses its high-speed wings to reach level-flight speeds of over 100 km/h.

Elliptical wings are short and wide and provide for quick takeoffs and landings, low-speed flight, and maximum maneuverability. The elliptical wings of the great spotted woodpecker permit it to change direction quickly in its forest habitat.

Soaring wings are designed to give birds the ability to stay aloft while minimizing energy expenditure. Some birds, such as the great frigatebird shown above, have soaring wings that are long and narrow. This wing shape relies on the prevailing winds found near the ocean to generate lift. Frigatebirds can soar for days without landing.

Other raptors have a broader wing shape that provides for a tremendous amount of lift even at low speeds. The gaps between the long contour feathers at the ends of the wings enable the black kite (below) to soar by riding rising columns of hot air, changing course with subtle movements of its wing tips.

But feathers, a lightweight, strong, and fairly rigid skeleton, and wings suited to flight are not all that a bird needs to keep it flying. Muscles need sufficient oxygen to keep from tiring; a bird would not be able to fly very long without a constant supply of oxygen. Birds have no diaphragm and their lungs don't expand very much, but they have the most efficient respiratory system known. In addition to their lungs, they have nine **air sacs** that fill up much of their body cavity and even fill the hollow spaces in some of their larger bones. As the air enters and exits through the trachea, it passes through the **syrinx**, which the bird uses to make the characteristic songs and calls that endear them to so many people.

A bird's air sacs also help it maintain its internal temperature. As an endothermic animal, a bird builds up heat from its metabolism, but it also needs a way to release heat when it becomes overheated. As air passes through the sacs, it picks up heat from the tissues and carries it out of the body. Some birds pant when they become overheated to increase this cooling effect.

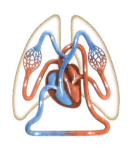

As the air passes through the bird's lungs, oxygen diffuses into the blood. The blood then travels to the four-chambered heart, entering the left atrium, then moving into the left ventricle, which pumps it to the rest of the body. When the blood returns to the heart, laden with carbon dioxide from the body's cells, it enters the right atrium and moves to the right ventricle, which pumps it to the lungs to release its load of carbon dioxide and pick up a new supply of oxygen. The avian heart is large for the size of the body, and its heartbeat is very fast. Some smaller birds have heart rates of 1000 beats per minute or faster during times of activity. The size and rate of a small bird's heart causes its blood pressure to be quite high.

As the blood circulates through the body, it also picks up nitrogenous wastes from the body's cells and carries them to the kidneys to be filtered out. Unlike many other tetrapods, birds do not have a bladder for storing urine. Instead, the kidneys release wastes as a white, sticky substance called *uric acid* directly into the cloaca. Marine birds have special salt glands in their heads that filter salt out of their blood and release it as a concentrated salt solution through the nostrils.

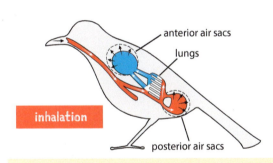

When a bird inhales, most of the air moves into the posterior air sacs and then is exhaled into the lungs. When the bird inhales again, the air moves out of the lungs and into the anterior air sacs, while air from the atmosphere rushes to refill the posterior air sacs.

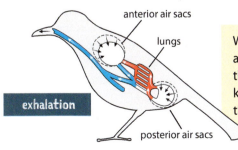

When the bird next exhales, oxygenated air from the posterior air sacs rushes into the lungs again and deoxygenated air in the anterior air sacs is released from the body. This system keeps oxygenated air and deoxygenated air from mixing, so the bird has a steady supply of oxygenated air.

Endothermic Vertebrates **433**

All birds have a beak made of keratin that they use to capture food. Birds feed on a wide variety of organisms, from plants to animals sometimes larger than themselves, and a bird's beak is shaped in a way that is suited to its food. While birds may use their feet or beaks to tear apart larger prey, most food is swallowed whole since birds have no teeth. After the food passes through the mouth and esophagus, it enters the **crop**, a sac-like enlargement of the esophagus that stores the food until it can be digested. It then moves to the two-part stomach. The first part, the *proventriculus*, produces digestive juices, and the second part, the muscular and thick-walled **gizzard**, grinds the food. Some birds swallow small stones and grit to help grind the food and mix it with gastric juices. The food then passes into the intestine where nutrients are absorbed into the bloodstream, while the digestive wastes are released into the cloaca with the uric acid from the kidneys. Because weight is an important factor in flying, most birds empty their cloaca just before takeoff and at other frequent intervals.

Flying takes a lot of energy and being endothermic requires a high metabolic rate. As a result, birds must eat a lot, even though the expression "eat like a bird" is used to describe someone with a poor appetite. Some smaller birds such as hummingbirds can eat their own weight in food each day.

In addition to the energy and oxygen that their muscles need to fly, birds require excellent navigation skills to dodge obstacles, avoid predators, and land safely. This kind of maneuverability is most needed by birds that live in the woods and must fly around tree trunks and through branches.

proventriculus

gizzard

crop

case study

CALIFORNIA CONDOR

Some of the least appreciated birds are the vultures. Most people find them mildly repulsive because of their predominately black feathers, bare heads, and habit of eating carrion. But by consuming carcasses, vultures help prevent disease and speed up the decomposition process.

In the past, not everyone understood the important ecological role of the vulture. For example, farmers seeing vultures eating a dead farm animal would mistakenly assume that the vultures had killed it. In the United States one common species, the black vulture, does occasionally kill calves and lambs, and since juveniles of several other vulture species look similar to the black vulture, all vultures were considered guilty. As a result, many vultures were killed, and one, the California condor, the largest bird in North America, was reduced to only twenty-two individuals

in the wild. As people came to understand the important role of vultures, the indiscriminate killing subsided and vulture populations have rebounded. The California condor was the subject of an intensive captive breeding program, and after being reintroduced into the wild, the number of condors increased. Today, scientists continue to study ways to help the condor population grow.

God has given birds the tools they need to traverse the air safely. Birds have very large eyes, and their sight exceeds even that of humans. Their color vision tends to be very good as well, and many can even see ultraviolet light, which is invisible to most animals. Their hearing also tends to be quite good, but many birds have poor or even no sense of smell or taste. But this deficiency allows the great horned owl to prey on skunks since the owl's poor sense of smell renders it immune to the skunk's noxious spray.

Avian Reproduction

During the spring birds sing in the air, in the trees, and on the ground looking for a mate. Birdsong is an example of a **courtship** behavior—any behavior aimed at attracting a mate. A male's courtship typically involves visual or vocal displays, or a combination of both, to attract a female. Since the male must attract his mate and defend the pair's territory from intruders and predators, he is often brightly colored. The female's typically duller colors camouflage her as she incubates her eggs. Courtships can be very elaborate affairs involving steep dives, like the sky dance of the golden eagle, or structure building, as in the case of the greater bowerbird. Shortly following courtship, the male and female birds mate. In most species a pair remains together for less than a year, pairing up with new mates the following year. However, a few birds, like the Canada goose, remain with their mates for life.

Although most vertebrate females have two sets of ovaries and oviducts, female birds develop only one set. After internal fertilization, the ovum and sperm unite in the *oviduct*. The fertilized egg, consisting of a tiny embryo and a large amount of yolk, passes down the female's reproductive tract, which adds a protective layer of **albumen** (the white of the egg) and shell. The egg then passes out of the body through the cloaca.

Most birds build nests to serve as nurseries for their developing young, each species building its own type. Eggs will hatch only if they are kept warm—the process of **incubation**—and most birds incubate their eggs by sitting on them. The females of many birds incubate the eggs while the male's activity ranges from leaving completely, to feeding the female while she sits on the nest, to actively assisting her in incubating the eggs and young. Mute swans (below) typically mate for life, and the males help care for their offspring, assisting with both the incubation process and raising the chicks once the eggs have hatched.

The male northern cardinal has brilliant red feathers, while his mate is primarily brown. These birds exhibit sexual dimorphism—a difference in appearance between males and females of the same species.

Birds in the Environment

As the second largest group of vertebrates, birds live in nearly every biome of the world, and many species are quite the globe trotters. A habitat that provides perfect conditions for rearing young in the summer may be very inhospitable in the winter, so many birds deal with this problem by spending the summer nearer the poles and moving to warmer climates during the winter. Some **migrations** are relatively short, like that of the American white pelican, which flies from the northern United States and southern Canada to Florida and Mexico. Others have far more dramatic migratory routes. The arctic tern breeds in the Arctic and winters in the Antarctic, a round trip of nearly 40,000 km.

Migration also makes conserving some species difficult. While nations like the United States have attempted to protect avian habitats, conservation laws in countries where birds travel to in the winter are often either poorly enforced or entirely absent. As a result, many organizations are seriously concerned about declining bird populations. However, the case of the eastern bluebird (right), which made a strong comeback in the late twentieth century after many people began setting up bluebird houses, suggests that even small efforts on the part of concerned individuals may help our feathered friends.

With all their variety, birds can occupy a number of niches in ecosystems, including some very unusual ones. Atlantic puffins spend most of their time during the winter riding the waves at sea since their wings are more suited to swimming than flying. When they do fly to land to breed, they nest in burrows dug in cliffs overlooking the sea (see page 53). Breeding and nesting in caves or crevices, the edible-nest swiftlet builds a nest almost entirely from its own saliva. These nests are harvested for use in bird nest soup, a Chinese delicacy.

19.1 SECTION REVIEW

1. Why can birds such as the dark-eyed junco live in snowy areas through the winter without hibernating like amphibians and reptiles?

2. You find a bird with naturally oily feathers and webbed feet. What type of bird is this likely to be?

3. Which bone structure found in most birds would an ostrich lack? Explain.

4. Arrange the following bird structures in order of largest to smallest: barb, barbule, feather, and hook.

5. What would happen if bird had a less efficient respiratory system?

6. In birds, what three body systems empty into the cloaca?

7. What evidence suggests that birds were designed for flight rather than having evolved the ability over time?

8. How do many species of birds survive the cold winters of their breeding grounds?

Use the case study on page 434 to answer Questions 9–11.

9. What role do vultures play in an ecosystem?

10. What led to the decrease of California condors?

11. How do Christians show good stewardship by helping the condor population increase?

MAMMALS

Classifying Mammals

Just as birds are distinguished by their feathers, animals in class Mammalia are set apart by their hair. Most mammals are covered by two layers of fur that protect and insulate them, and many of them possess specialized hairs that serve a variety of purposes. Some mammals, primarily aquatic ones, have only a few bristles. Notice some of the main groups of mammals.

? Why are blue whales classified as mammals and not fish?

Questions

How are mammals classified?

How does a mammal function and reproduce?

How do mammals interact with their environment?

How should we respond to useful research pursued by evolutionary scientists?

Terms
uterus
placenta
ungulate
rumen
prehensile tail
blubber
urine
diaphragm
larynx
mammary gland
gestation

Exploring

MAMMALS

Monotremes are quite different from the other groups of mammals. They are the only mammals that lay eggs. The platypus and four species of echidnas (spiny anteaters) are the only members of this group. Like other nonmammalian animals, monotremes have a cloaca, which provides a single opening for their digestive, excretory, and reproductive systems.

MONOTREMES

Kangaroos, koalas, and opossums are all marsupials. They are known for a unique pouch that functions to nourish and protect their developing young. After conception, a marsupial embryo begins to develop inside the mother in an organ called the **uterus**. Shortly after birth, the tiny offspring, essentially still an embryo, crawls to its mother's pouch and attaches to one of her nipples. In the largest marsupial, the red kangaroo, a young animal will continue to develop in its mother's pouch for as long as seven months.

MARSUPIALS

MAMMALS
(CONTINUED)

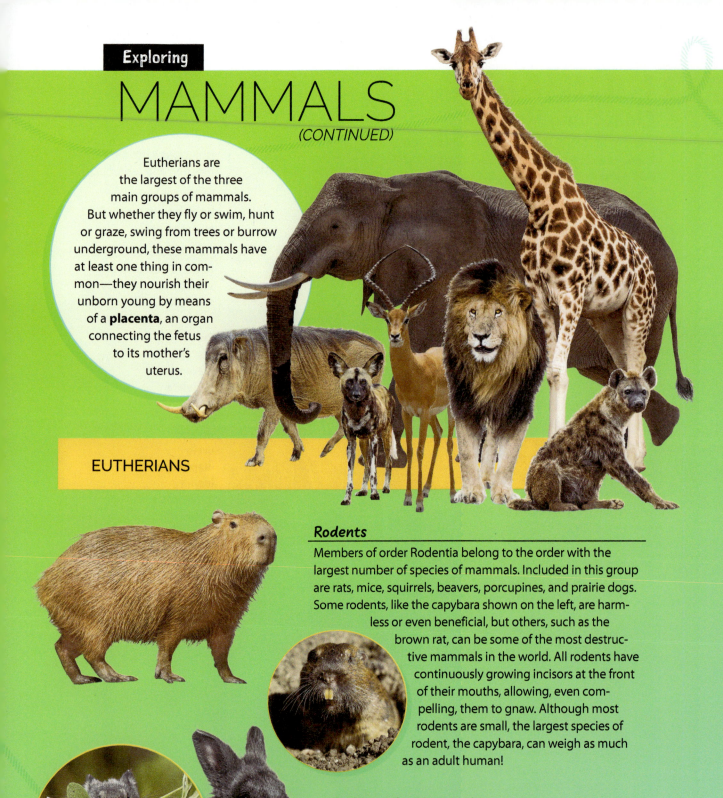

Eutherians are the largest of the three main groups of mammals. But whether they fly or swim, hunt or graze, swing from trees or burrow underground, these mammals have at least one thing in common—they nourish their unborn young by means of a **placenta**, an organ connecting the fetus to its mother's uterus.

EUTHERIANS

Rodents

Members of order Rodentia belong to the order with the largest number of species of mammals. Included in this group are rats, mice, squirrels, beavers, porcupines, and prairie dogs. Some rodents, like the capybara shown on the left, are harmless or even beneficial, but others, such as the brown rat, can be some of the most destructive mammals in the world. All rodents have continuously growing incisors at the front of their mouths, allowing, even compelling, them to gnaw. Although most rodents are small, the largest species of rodent, the capybara, can weigh as much as an adult human!

Rodent-Like Mammals

Belonging to the order Lagomorpha, rodent-like mammals are similar to rodents, but they have an additional pair of incisors located in their upper jaw. Species in this order are famous for their ability to reproduce quickly, leading to the expression "multiply like rabbits." Although rabbits and the bulkier, larger-eared hares are often difficult to tell apart, the third group of lagomorphs, the pikas (far left), appear distinctive because of their smaller, rounded ears.

Carnivores

Belonging to the order Carnivora, meat eaters like this lion are equipped with large canine teeth for tearing meat. While bears, dogs, and large cats often come to mind, carnivores include many diverse animals, including seals, weasels, and raccoons.

Hoofed Mammals

Ungulates are hooved mammals. *Even-toed ungulates* belong to the order Artiodactyla. They have two or four functional hoofed toes. This large order includes pigs, hippopotamuses, camels, giraffes, cattle, and deer. Some of these animals have a multiple-chambered stomach that allows them to digest cellulose in the plants they eat. Bacteria in the **rumen**, the first chamber of the stomach, break down the cellulose. The semi-digested food or cud is then brought back up to be chewed again.

Odd-toed ungulates belong to the order Perissodactyla. They have only one or three toes. A much smaller order than the even-toed ungulates, odd-toed ungulates include the tapirs, the rhinos, and the equids—horses, donkeys, and zebras.

MAMMALS
(CONTINUED)

Insect-Eating Mammals

Mammals in the order Insectivora are equipped with pointed snouts that allow them to search the ground for insects and burrow tunnels. Members of this order include elephant shrews (left), moles, and hedgehogs. Many species live most of their lives underground and have a good sense of smell but tiny eyes. Most species spend most of their time hunting insects and other small arthropods, which make up the bulk of their diet.

Trunked Mammals

The African bush elephant belongs to the order Proboscidea and is the largest land mammal in the world. The different species of elephants are distinguished from other mammals by a flexible and highly useful trunk, which is essentially an elongation of the nose and upper lip. Many also have two large tusks, which sadly may invite death at the hands of poachers involved in the ivory trade.

Primates

The western lowland gorilla and the other great apes make up the order Primates. This order also includes smaller primates such as monkeys, tarsiers, and lemurs (right). While some can walk erect on the ground, they prefer to walk on all fours and generally spend most of their time in trees. Some monkeys even have **prehensile tails** that they use almost like a fifth limb.

Strange-Jointed Mammals

Mammals in the order Xenarthra, meaning "strange joints," are classified according to the unique joint structure in their backbone. Nine-banded armadillos, anteaters (above), and sloths (right) are members of this group. They used to be known as toothless mammals, but armadillos and sloths have teeth, so they are now grouped together on the basis of their unusual joints.

Flying Mammals

Bats, the only true flying mammals, are members of the order Chiroptera. The members of this order are not the monsters that many people think they are. Many bats use echolocation to navigate and find insects in the dark. A single bat can eat several hundred thousand insects in the course of a year. Other bats, like the gray-headed flying fox (left), feed on fruit, while others may eat fish or frogs. And, yes, there are three species of vampire bats that do feed on blood, but they rarely bite humans.

Aquatic Mammals

Mammals that live in the water belong to the orders Sirenia or Cetacea. Although the clever and popular bottlenose dolphin can reach 4 m in length, it is small for a cetacean. The blue whale, the largest animal known to have ever lived, can reach 32 m in length and tip the scales at over 180,000 kg. Although cetaceans cannot live out of the water, they are mammals and must come to the surface to breathe. Several cetacean species are very social and have a relatively complex communication system in addition to echolocation that helps them navigate through the water.

The West Indian manatee (right) is a member of the order Sirenia along with three other species of sea cows. They are tropical or subtropical herbivores that live in the coastal waters of the ocean and sometimes move into estuaries and rivers. Although they are quite large, they have a reputation for being very gentle.

Mammalian Structure

Although mammals come in a great variety of shapes and sizes, they share many characteristics that enable them to maintain homeostasis in their various habitats. Like birds, mammals are endothermic, so most of them have two layers of hair for insulation, cutting down on the amount of heat that they lose to the environment. Hair grows from follicles deep in the skin, and while the root is living, the hair itself is made of dead cells filled with keratin. Many mammals, especially those that live in water or cold regions, have a layer of fat, called **blubber**, under the skin that provides an additional level of insulation. Mammals also need a way to avoid becoming overheated. Many animals pant, releasing heat through their mouths. Some mammals, like elephants and swift foxes, that live in hot places have large ears that release heat from the blood into the atmosphere. Predictably, arctic foxes of the far north have very small ears that work well to conserve heat.

Insulation is useful only if there is a source of heat, and in endothermic animals heat is produced by metabolism, so mammals must eat frequently. Mammals have various ways of finding food, but they all benefit by having a larger brain compared with their body weight than do other vertebrates. Aided by three bones that transmit sound in the middle ear, their hearing is very good (most other tetrapods have only one such bone). With the notable exception of primates, the mammalian sense of smell tends to be extremely keen, as demonstrated by the remarkable tracking ability of bloodhounds.

Mammalian eyesight tends to be weaker in comparison with birds' eyesight, especially in identifying color. While most mammals do have cells in the eye called *cones* that enable them to perceive color, they typically have relatively few of them. Furthermore, most mammals have only two types of cones instead of the three found in humans and the four found in birds, so they cannot distinguish between some colors. Many mammals, though, are able to see very well at night. As a rule, herbivores have eyes set in the sides of their heads, allowing them to see in almost all directions, a useful ability if you are potential dinner for large carnivores. Carnivores, on the other hand, tend to have their eyes in front of their heads, giving them the better depth perception needed for attacking prey. Compare eye placement in the two mammals shown here.

Whether a mammal is a carnivore chasing its prey or an herbivore traveling to a new grazing area, mammals must be able to move. Like most tetrapods, they generally have an endoskeleton complete with four limbs. However, the amount of variation is large. The forelimbs of bats form wings, and whales have flippers for forelimbs but no hind limbs.

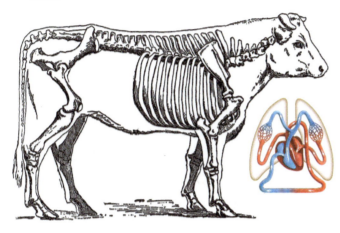

Once a mammal bites into its food, be it plant, invertebrate, or another vertebrate, the food enters the mouth and goes down the esophagus to the stomach. After being broken down by the stomach's acid and churning action (and by microorganisms and by being chewed multiple times in ruminants), the food moves into the small intestine where nutrients are absorbed. Unabsorbed nutrients then move into the pouchlike extension of the large intestine called the *cecum*, where salt and electrolyte absorption takes place. The cecum is especially important in nonruminant herbivores like horses, which don't have a rumen to help digest cellulose. Some mammals that feed primarily on insects lack a cecum altogether, and carnivores usually have very small ones. What is left of the mammal's food continues down the large intestine until it is expelled through the anus.

As the nutrients from the mammal's food are absorbed in the small intestine, they are carried along in its bloodstream to all the cells of the body. Along the way, the blood picks up nitrogenous wastes and carries them to the kidneys where they are removed and released as a diluted solution of urea called **urine**. Unlike other vertebrates, mammals other than monotremes have separate openings for their digestive and excretory systems.

As the blood flows through the body, it also picks up carbon dioxide produced by cellular respiration. Eventually, the blood reaches the right atrium of the four-chambered heart. From there, it moves into the right ventricle, which pumps it to the lungs. There, carbon dioxide diffuses into the lungs to be exhaled into the atmosphere, and oxygen diffuses into the blood and binds with a molecule called *hemoglobin* found in *red blood cells*. The blood then moves back to the heart and enters the left atrium. At this point, the blood empties into the left ventricle to be pumped to the rest of the body.

Mammals lack a bird's air sacs, but their lungs are far larger and more flexible. The **diaphragm**, a large muscle, separates the thoracic cavity holding the lungs and heart from the abdominal cavity holding the digestive organs. When the diaphragm contracts, the lungs expand, causing oxygen-rich air to rush into them. When the diaphragm relaxes, the lungs shrink, pushing carbon-dioxide-laden air back into the atmosphere. As the air exits the lungs, it flows through the **larynx** or voice box that is responsible for many of the sounds that mammals make.

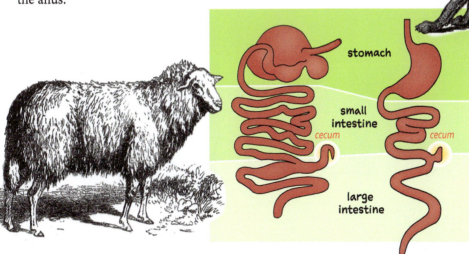

stomach

small intestine

cecum

cecum

large intestine

MINI LAB

Comparing Uric Acid and Urea

Birds have the ability to release nitrogenous waste by converting the ammonia produced from the waste to a less toxic substance known as *uric acid*. Mammals, on the other hand, release nitrogenous waste by converting ammonia to a different compound called *urea*. Both substances are less toxic than ammonia to the animals, but due to their difference in water solubility, one substance doesn't dissolve in water as easily and forms a solid, while the other easily dissolves to form a liquid. Let's investigate how this happens!

What's the difference between uric acid and urea?

Materials
urea, 4 g • uric acid, 4 g • distilled water, 60 mL • test tubes (2) • graduated cylinder • glass stirring rod • balance

PROCEDURE

A Place 30 mL of distilled water into two test tubes. Label the test tubes.

B Add 4 g of urea to one test tube and 4 g of uric acid to the other test tube.

C Stir each test tube for 5 min.

D Observe the test tubes. Record which substance was more easily dissolved in water.

CONCLUSION

1. What observations did you make when analyzing the two test tubes?

2. Explain what you think caused the differences that you observed in Question 1.

3. State what could be a possible benefit of animals producing uric acid to excrete waste.

GOING FURTHER

4. We've learned that mammals produce urea to rid the body of nitrogenous wastes that build up, but mammals also produce small amounts of uric acid. In certain mammals, as well as many reptiles, a condition called gout may occur if too much uric acid is in the blood. What do you think will happen if too much uric acid is in the body? Research and describe ways that gout can be prevented.

Mammalian Reproduction

Mammals have a set of structures called **mammary glands** that produce milk to nourish their young. In fact, the mammary glands are so unique among the animal kingdom that mammals derive their name from the Latin word *mamma*, meaning "breast."

As a rule, mammals are viviparous, although the five monotreme species are the exception to this and several other rules. In marsupials and eutherians the egg is fertilized by sperm inside the mother's body, and the zygote is nourished by yolk until it begins to attach to its mother's uterus.

Once the yolk sac is depleted, embryos become attached to the wall of the uterus by a placenta made partly of the mother's cells and partly of the embryo's cells. The placenta allows the mother to provide nourishment and oxygen to the embryo and remove wastes until the young mammal's systems are ready to function on their own. As a result, eutherians tend to have a long **gestation**, which is the period of time between conception and birth, producing young that are fairly well developed at birth.

On the other hand, marsupials have a very short gestation—they are born as soon as the yolk sac is depleted. Exceedingly tiny and poorly developed, the young marsupial makes the perilous journey from the uterus to its mother's pouch where it

continues development. In fact, red kangaroo joeys lack hind legs at birth. While marsupial gestation periods are much shorter than those of eutherians, the length of time female marsupials nurse their young is much longer.

Monotreme embryos remain in the uterus for an even shorter time than marsupial embryos. Often less than two weeks after they are conceived, the embryos are encased in a shell and laid as eggs. Mother echidnas guard their eggs in a pouch while platypuses lay their eggs in a burrow. While female monotremes produce milk for their young like all other mammals, the young lap the milk off their mother's fur since monotreme mammary glands lack nipples.

Mammals in the Environment

Mammals are extremely diverse and are found in almost every habitat, occupying many different niches from herbivore to top-level carnivore. They also serve many different purposes for humans within the niches they occupy. Horses, cattle, and camels are used as beasts of burden, dogs are used for hunting and companionship, and cats are kept as pets and to control rodents.

Many mammals are *generalists*; that is to say, they are able to survive in a wide range of habitats. These often outcompete specialists, especially in the aftermath of a change in the local area. Sometimes these differences can be seen in two very similar animals. For example, the Canada lynx (top, right) is rarely found in areas with a large population of humans, while the slightly smaller but very similar bobcat (bottom, right) thrives near humans. One factor that could give the bobcat the edge over the lynx is that bobcats are generalists that eat a wider variety of prey than do lynxes. Omnivores such as bears and foxes also thrive near human dwellings, as do coyotes, much to the chagrin of many sheep farmers.

Many large wild herbivores are generalists that are often hunted for food. With well-managed hunting seasons, these mammals, such as moose, can maintain large populations in the wild. Even the American bison is beginning to slowly recover from near extinction in the nineteenth century.

Large wild carnivores, on the other hand, are often scarce in developed countries. In the United States wolves were feared and nearly hunted to extinction because they were perceived as a threat to livestock and people, especially children. The reintroduction of wolves into Yellowstone National Park in the mid-1990s has resulted in wolves expanding their range into surrounding areas.

Though there have been no reports of wolves killing a human within the range of the introduced packs, human safety has been a concern. While this move remains controversial, the reintroduction of wolves has resulted in positive ecological changes in Yellowstone. Wolves have reduced the elk population, allowing tree species that the elk often eat to grow to full height. As a result of the increased tree growth, beavers have increased in numbers. The beaver dams have benefited numerous other species. The effect that one animal has on organisms throughout an ecosystem is called a *trophic cascade*.

Although many of the scientists involved with the Yellowstone project likely are evolutionists, they have had great success in managing God's world. Science is based on data, and data is available to all scientists. Whether or not a believer, a good scientist is able to collect data and use it for good purposes. A person's worldview still affects what data is chosen to examine and how information is interpreted, so it is important that we use discernment when we analyze data.

19.2 SECTION REVIEW

1. Which group of mammals are not viviparous? Explain.

2. What structure distinguishes marsupials from other mammals? Explain.

3. What three characteristics do all groups of mammals have?

4. What respiratory structure sets mammals apart from most other vertebrates?

5. What structure is important for digesting cellulose in nonruminant herbivores? Explain.

6. What structures do mammals use to nourish their young?

7. Arrange the following groups of mammals in order of the length of gestation from shortest to longest: eutherian, marsupial, monotreme.

8. Koalas feed almost exclusively on leaves from a few species of eucalyptus trees. Are koalas generalists or specialists?

19 CHAPTER REVIEW
Chapter Summary

19.1 BIRDS

- Birds are endothermic, have feathers, and are oviparous. Their feathers are their most distinctive feature.

- A bird's skeleton is light because of its hollow bones. Most birds are equipped with a keel, which provides a place for wing muscles to attach.

- To support its muscles' need for oxygen during flight, a bird has air sacs that supply it with a constant flow of oxygen.

- Birds have a crop to store food before digestion and a gizzard to grind the food and make it easier to digest.

- During avian courtship, males produce visual or vocal displays to attract females. After fertilization, an egg is produced, containing an embryo, yolk, albumen, and a protective shell.

- Many species of birds migrate to find hospitable conditions for living.

Terms
feather · keel · air sac · syrinx · crop · gizzard · courtship · albumen · incubation · migration

19.2 MAMMALS

- Mammals are generally endothermic and are set apart by the presence of hair.

- The three main groups of mammals are monotremes, marsupials, and eutherians. Monotremes are the only mammals that lay eggs. Marsupials are equipped with pouches, and eutherians have placentas that nourish and protect a fetus.

- Many mammals have a layer of fat that provides insulation, an efficient digestive system to break down food, and kidneys that remove nitrogenous wastes from the body.

- Mammals have mammary glands that produce milk for nourishing their young. They are generally viviparous, with the exception of monotremes.

- After fertilization, eutherian embryos attach to the mother's uterine wall by a placenta. Eutherians tend to have longer gestation periods. After a short gestation period, marsupials travel to their mother's pouch to grow and develop.

- Many mammals are generalists and can survive in a variety of different habitats.

Terms
uterus · placenta · ungulate · rumen · prehensile tail · blubber · urine · diaphragm · larynx · mammary gland · gestation

19 CHAPTER REVIEW
Chapter Review Questions

RECALLING FACTS

1. What special structures provide insulation in birds and mammals?

2. What purpose do layers of fat have in many endotherms?

3. How can perching birds sleep in trees?

4. What breastbone structure do some flightless birds lack?

5. Which of the two bones shown below belongs to a bird?

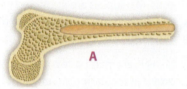 

A   B

6. Which painted bunting in the photo shown on the right is the male? How can you tell?

7. What two layers of a bird's egg protect the embryo?

8. In which direction is a bird in the Southern Hemisphere likely to fly as winter comes?

9. What negative consequences would result if vultures were to become extinct?

10. What structures distinguish ungulates?

11. Which two groups of mammals echolocate?

12. Ungulates that chew the cud have what stomach structure?

13. A mammal with a small cecum probably has what kind of diet?

14. Which of the following diagrams shows a mammalian heart?

A   B   C

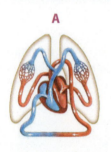 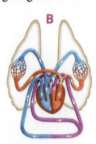 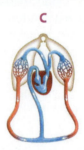

15. Why must marsupials nurse their young longer than eutherians?

16. How are mammals different from other vertebrates in the way that they nourish their young?

17. Why are animals that are generalists better able to survive than specialists when the local environment changes?

UNDERSTANDING CONCEPTS

18. Explain why endotherms need insulation.

19. Describe the uses of the three types of wings in birds.

20. Create a T-Chart classifying the following steps of a bird's respiration as either inhalation or exhalation: air enters the lungs, air enters the posterior air sacs, air leaves the body, air enters the anterior air sacs.

21. Contrast the ways that most birds and mammals expel digestive and nitrogenous wastes.

22. Some birds regurgitate food for their young. Where are they likely to store this food?

23. Describe the structural evidence that supports birds' design for flight.

24. Describe the process of courtship in some birds.

25. Explain why some birds migrate.

26. Explain how a bird's beak, wings, and feet relate to its environment.

27. Make a T-Chart comparing the reproduction differences between eutherians, monotremes, and marsupials.

CRITICAL THINKING

28. The shape of a bird's beak is suited to its lifestyle. Which of the birds shown on the right has a bill appropriate for probing in the mud for its food? Explain.

29. Why do you think that governments regulate hunting seasons for large herbivores?

30. The scientists who conduct research on naked mole-rats are often committed to evolutionary philosophy. Explain why the research on this creature could still be useful.

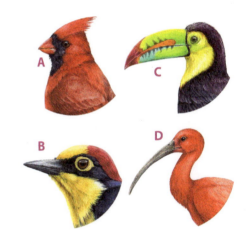

Use the graph on the right to answer Questions 31–36.

31. In what year between 1900 and 1930 was the hare population peak the greatest?

32. In what year in the nineteenth century did the hare population reach its lowest point?

33. In what year did the lynx population peak after the peak in the hare population in Question 31?

34. In what three 10-year periods listed on the graph did the lynx population reach its lowest points?

35. If the hare population peaks again in 2037, do you think that the lynx population is more likely to peak in 2035 or 2039? Explain.

36. On the basis of the pattern shown in the graph, if the hare population peaks in 2035, do you think that it will likely peak again around 2045 or 2050? Explain.

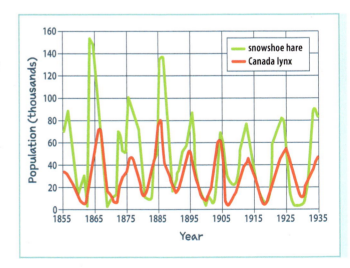

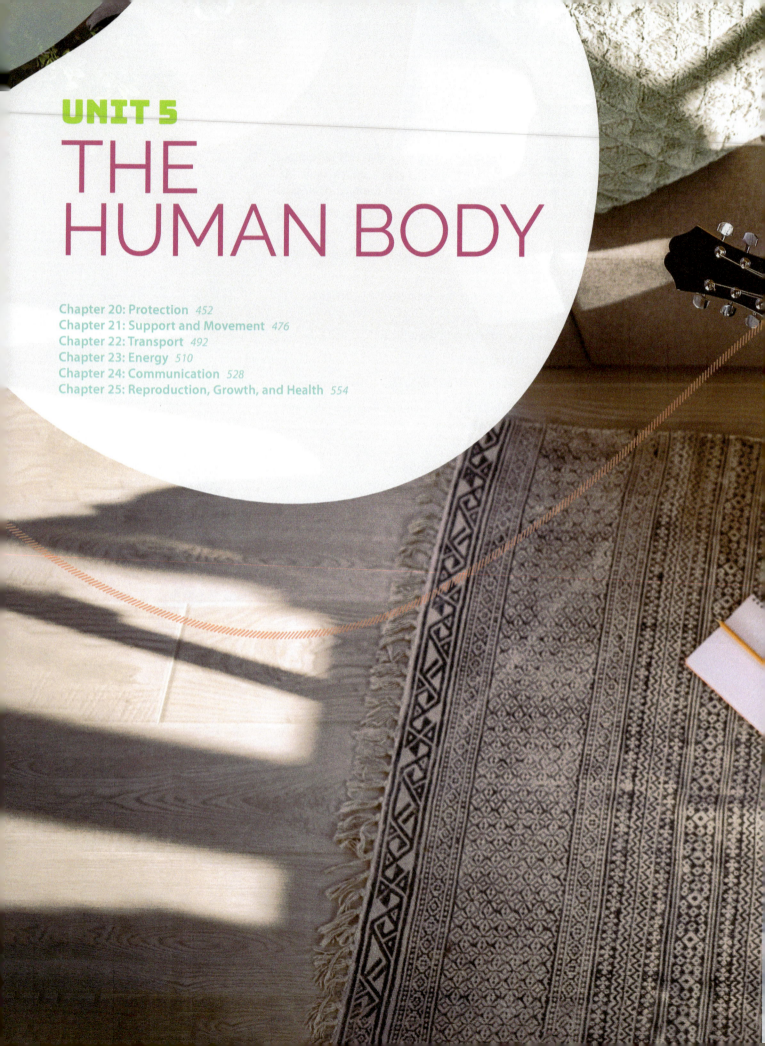

UNIT 5
THE HUMAN BODY

20 PROTECTION

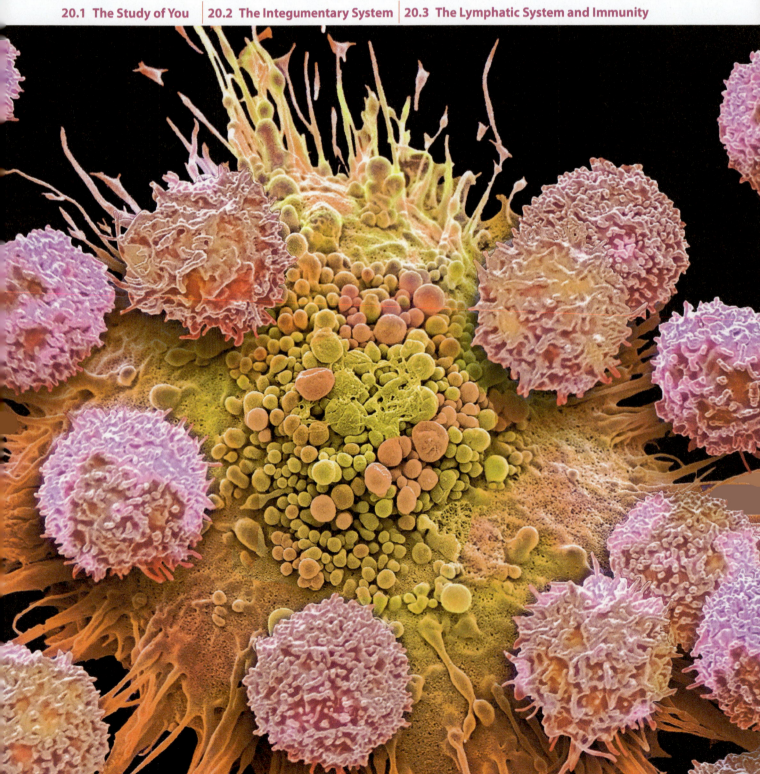

Looking for a Fight

Ah, the joys of spring—budding flowers, warm sun, and sneezing! Spring means pollen, and pollen means allergies. Even if you don't have them yourself, you probably know someone who does. In fact, allergies are the most common health problem among children in the United States, and they seem to be getting more common. About 20% of people in developed countries have allergies. And allergies can be far more serious than just having an occasional sniffle. Some people have severe reactions to foods or environmental toxins that could endanger their lives.

Some scientists suggest that allergies may happen for an odd reason—we are too clean. Lymphocytes (facing page) are white blood cells, your body's secret service agents that target foreign invaders called **antigens**. Because we live in sanitized environments, these scientists hypothesize that lymphocytes don't have much to resist, so they go out looking for a fight. This makes an immune system hypersensitive. Something small, like a pollen grain, can trigger an all-out war in the body. As more lymphocytes join in, allergy symptoms are triggered. What can we do about allergies?

20.1 THE STUDY OF YOU
The Essence of Humanity

We have been exploring a whole world full of life, but we've saved the best until last. There is one creature that has much in common with the animals yet is special among all the living things that God made. In God's hierarchy of life, humans are higher in value and authority than plants, invertebrates, arthropods, and ectothermic and endothermic vertebrates, primarily because of one distinguishing feature: people are made in God's image. We are not just bags of chemicals; we are spiritual as well as physical and rational beings capable of and responsible for making ethical decisions. Though animals and plants are worth studying, **human anatomy**, the study of our physical makeup and structure, and **human physiology**, the study of human life processes, are just a little different from the other parts of biology.

We study the remarkable complexity and design in the human body because we have ethical decisions to make to help people solve health problems, such as the problem of allergies. As we study human anatomy and physiology, it is amazing to see just how well God has designed our bodies to protect us from antigens and to carry out life-sustaining processes. For example, doctors and researchers have learned how the human immune system responds to antigens, and they have developed a way to train the immune system to ignore certain stimuli to avoid allergic reactions. This process is called *immunotherapy*. It involves giving an allergy patient a series of injections with increasing amounts of a customized mixture of all the things he or she is allergic to. If the treatment is successful, the patient's allergic reactions to these stimuli lessen over time.

The work, time, and money it takes to relieve human health problems are worthwhile because people are worthwhile. Studying human anatomy and physiology is one of the most effective ways to obey God's commands to wisely use His Earth and care for His image-bearers. We are accountable to our Creator to understand and wisely use our bodies for His glory. As we work to do this, we marvel at the infinite creativity of our loving Designer.

Why should we study the human body?

Questions

What makes the study of humans different from the rest of biology?

What is the human body made of?

What does each body system do?

How should we view the study of human anatomy and physiology?

Terms

antigen
human anatomy
human physiology
nervous tissue
connective tissue
matrix
muscle tissue
epithelial tissue

Human Tissues and Organs

It is astounding to consider how much action is going on in your body as you read this paragraph. It all starts with cells. We learned in Chapter 5 that living things are made of cells. These cells make up tissues, groups of similar cells that serve a specific purpose. Our bodies are made of four types of tissues—nervous, connective, muscle, and epithelial. Tissue of one or more of these four kinds make up organs, and organs make up body systems. Smooth muscle tissue teams up with epithelial and connective tissue to make an organ such as a blood vessel. Blood vessels work together with the heart to make the circulatory system function. Let's look at the different types of tissues in the human body.

Exploring

HUMAN TISSUES

We have **nervous tissue** in our brain, spinal cord, and nerves. Without it we couldn't move, get information about our environment, or think and feel. Nervous tissue is made of cells that receive sensory information from other sources and transmit electrical signals back to those sources through a series of chemical reactions.

Connective tissue includes bone, blood, lymph, tendons, fat, and cartilage. There is more of this type of tissue in our bodies than any other type. Connective tissue supports, connects, and protects other structures in the body. Between the cells in connective tissue is a substance called a **matrix** that holds the tissue together. The matrixes in various connective tissues differ, depending on the functions of the tissues.

Muscle tissue is made of cells that can contract. Muscles tighten to move either the body or substances through the body. The human body has three kinds of muscle tissue—skeletal, smooth, and cardiac (heart).

STRUCTURAL HIERARCHY

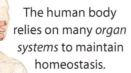

The human body relies on many *organ systems* to maintain homeostasis.

Matter is made of atoms.

Atoms combine to make up molecules.

Different *organs* work together to serve a specific purpose as an organ system in the human body.

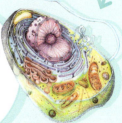

Molecules make up the cells' organelles, cytoplasm, DNA, and membranes.

Different kinds of *tissue* make up organs.

Similar *cells* make up a tissue.

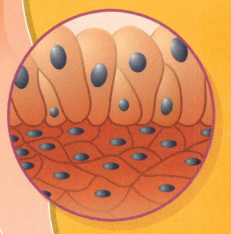

Epithelial tissue is made of layers of cells that cover or line surfaces, making up everything from the lining of our blood vessels to our skin. The cells in epithelial tissue are bound tightly together to protect surfaces.

Human Organ Systems

Organs and organ systems, like the blood vessels, heart, and circulatory system, mentioned in the previous subsection, work together to maintain homeostasis in an organism. We've studied how animals use organ systems to maintain homeostasis, such as the open circulatory system in arthropods and the closed circulatory system in earthworms. Let's consider the human organ systems that we'll study in this unit.

Exploring
HUMAN SYSTEMS

Integumentary System
This body system is made of skin, protecting the body from antigens.

Skeletal System
This body system is made of a variety of bones that support and protect the body and allow it to move.

Muscular System
Muscles make up this body system, powering the motion of the bones and joints.

Digestive System
The stomach and intestines in this system help the body digest food, absorb nutrients, and eliminate wastes.

Lymphatic System

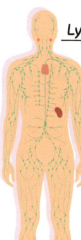

Lymph nodes in this system protect the body from disease-causing antigens that make it past the skin. The lymphatic system is an important part of the body's *immune system*—the organs and tissues from several different systems that defend the body from antigens.

Urinary System

This body system, like the digestive system, eliminates wastes and uses the kidneys and bladder to maintain water balance in the body.

Endocrine System

Body functions are controlled and regulated by hormone-secreting glands in this system, like the thyroid.

Circulatory System

This body system involves the heart, which pumps blood through blood vessels to deliver nutrients throughout the body.

Respiratory System

Lungs and all the tubes that bring in outside air allow the body to breathe, exchanging gases between the blood and air.

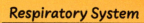

Nervous System

The brain, spinal cord, and a network of nerves coordinate and control movement and process sensory information from the body's environment.

Reproductive System

Males and females have the appropriate sex organs for reproduction, or procreation.

A Biblical View of the Body

Your body is important to you, and it is important to God too. Evolutionists believe that the human body is a product of completely random, natural processes, but believers of God's Word know that God created our bodies. We can't just use our bodies any way we want. Because God created the human body, it belongs to Him (1 Cor. 6:18–20). We are accountable to God to wisely use and care for the bodies that He has given us. We also need to help other people do that.

Though our bodies are created by God, they are also fallen. We bear God's image, but that image is twisted by sin. One of the consequences of the Fall is the shame that comes with nakedness (Gen. 3:7–10). But doctors and textbooks inspect the uncovered body to develop our understanding of human anatomy and physiology. While realizing that there is a proper time and place to study the human body, don't allow it to dull your sense of modesty.

You might also find disturbing the sight of human blood and tissue. Though we need to be able to recognize parts of the human body so that we can understand their shape and function, we should also retain a sensitivity to violence to the human body. Viewing our bodies clinically helps us deal with this aspect of human anatomy and physiology.

Scientists sometimes misapply their knowledge of human anatomy and physiology. As scientists and doctors work to understand the human body, they learn things that they can use to advise people in how to get relief for diseases and problems. Those decisions are deeply affected by one's worldview.

You are probably aware of people who have made unethical decisions about their bodies —abortion, euthanasia, and assisted suicide—sometimes even following the advice of doctors.

It is possible to develop an understanding of anatomy and physiology in the context of a biblical worldview so that we can use this science in a way that glorifies God. You'll be amazed when you learn how complex and sophisticated God has made your body.

Though there may be things that you don't like about your body, remember that God formed you in your mother's womb the way you are (Ps. 139). You are special to Him.

20.1 SECTION REVIEW

1. Discuss the similarities between the human body and the bodies of animals.

2. How is mankind different from other organisms?

3. What makes one type of tissue different from another type?

4. Which one type of tissue is found most often in each of the following systems?

   a. nervous system

   b. skeletal system

   c. muscular system

   d. integumentary system

5. Give an example of a body system and two organs in that system.

6. List each of the eleven systems in the human body and summarize the purpose of each with one word.

7. Compare the purposes of the integumentary and lymphatic systems.

8. How can knowledgeable doctors sometimes give patients bad advice, as when they, for example, advise abortion?

9. From a biblical worldview, why is immunotherapy a good tool of anatomy and physiology?

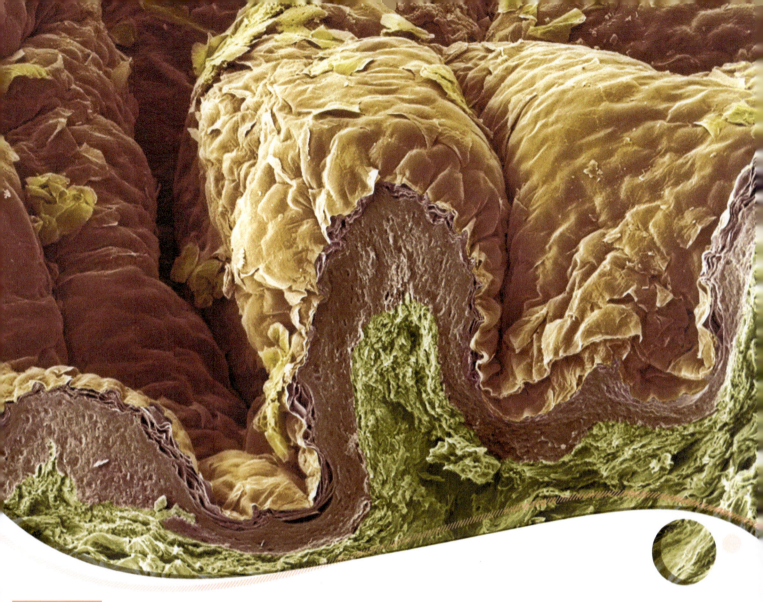

20.2 THE INTEGUMENTARY SYSTEM

The Structure of the Integumentary System

Let's study the human body, starting with the outside. In fact, the largest organ in your body is right there—skin! Your skin, hair, and nails cover an area of almost 2 m². Skin has varying thicknesses over your body, averaging 2.5 mm, and plays a vital part in maintaining your body's homeostasis.

Skin consists of two layers. The top layer of skin is the **epidermis**, made of epithelial tissue. The bottom layer of skin is the **dermis**, made of connective, muscular, epithelial, and nervous tissues, so it is the more complex layer. Under the dermis is the **subcutaneous layer**, a layer of cells made of connective tissues. Although the subcutaneous layer isn't a layer of skin, it is part of the integumentary system.

How does my skin help keep me alive?

Questions

What is skin made of?

What does your skin do for you?

How does your skin work with the rest of your body?

Terms
epidermis
dermis
subcutaneous layer
hair follicle
melanin
sweat gland
sebaceous gland

THE INTEGUMENTARY SYSTEM

One of our five senses is touch, and the skin has a lot to do with that! Receptor cells made of nervous tissue collect texture, pressure, and temperature information from the environment to send to the rest of the body.

epidermis

The epidermis, the top layer of skin, is completely replaced about every twenty-five days. Old cells build up keratin and die, to be sloughed off and replaced by new cells generated beneath the epidermis. When we get a cut, cells along the cut can regenerate to heal the wound in the same kind of process.

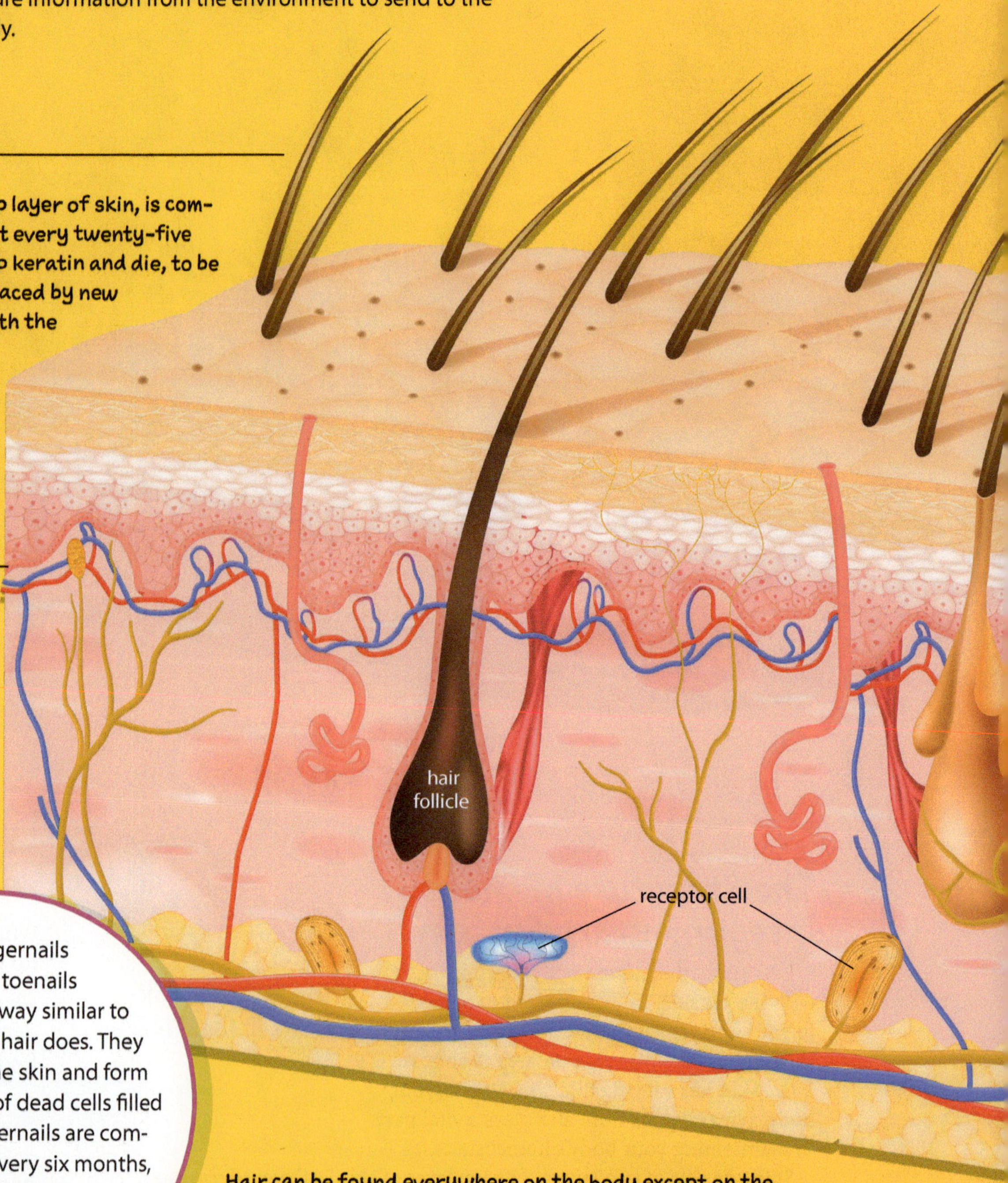

hair follicle

receptor cell

Fingernails and toenails grow in a way similar to the way that hair does. They develop from the skin and form a layer composed of dead cells filled with keratin. Fingernails are completely replaced every six months, while toenails grow more slowly, taking a year to replace.

Hair can be found everywhere on the body except on the palms and the soles of the feet. It grows from a root called the **hair follicle** buried deeply in the dermis. Hair follicles with elliptical openings produce curly hair, while those with round ones produce straight hair. The visible part of hair is called the *shaft*. It is made of a collection of dead, protein-filled cells. That's why it doesn't hurt to get a haircut!

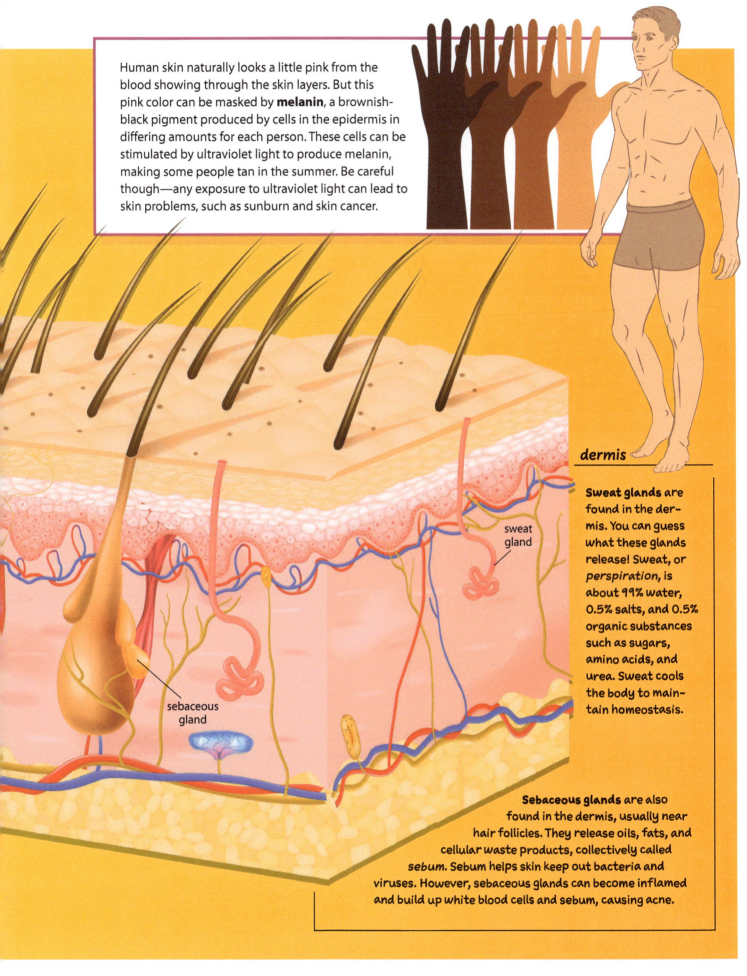

Human skin naturally looks a little pink from the blood showing through the skin layers. But this pink color can be masked by **melanin**, a brownish-black pigment produced by cells in the epidermis in differing amounts for each person. These cells can be stimulated by ultraviolet light to produce melanin, making some people tan in the summer. Be careful though—any exposure to ultraviolet light can lead to skin problems, such as sunburn and skin cancer.

dermis

sweat gland

sebaceous gland

Sweat glands are found in the dermis. You can guess what these glands release! Sweat, or *perspiration*, is about 99% water, 0.5% salts, and 0.5% organic substances such as sugars, amino acids, and urea. Sweat cools the body to maintain homeostasis.

Sebaceous glands are also found in the dermis, usually near hair follicles. They release oils, fats, and cellular waste products, collectively called *sebum*. Sebum helps skin keep out bacteria and viruses. However, sebaceous glands can become inflamed and build up white blood cells and sebum, causing acne.

MINI LAB

Skin Tone

We learned in this section that the pigment melanin is responsible for the color of human skin. But how can a single pigment account for the great variety of skin tones seen in people around the world?

?

How can a single pigment cause two people to have different skin tones?

Materials
pencil • plain paper

PROCEDURE

Create a sketch of the two people shown on the right using only a pencil. Your sketch should be made up of dots only—not lines! Have fun!

1. How did you recreate the different tones in the image?

2. How does this activity mimic what is happening with melanin in human skin?

3. Relate this activity to the biblical view of different "races."

The Function of The Integumentary System

One of the main functions of the integumentary system is to protect the human body from antigens, including bacteria, viruses, and foreign objects. These all can wreak havoc on the processes necessary to sustain life in the human body, causing illness and sometimes even death. The fact that people who lose large portions of skin have difficulty surviving testifies to the protective power of skin.

Another important function of skin is to keep the body's temperature under control. Why is this important? When we get too hot or too cold, our bodies aren't able to carry on the processes that they need to survive. Keeping the body at just the right temperature is an essential part of homeostasis. Skin helps keep us warm by providing insulation in the fats found in the subcutaneous layer. It keeps us cool as sweat glands release perspiration, which cools us off as it evaporates. On a hot day and with strenuous activity, a person can lose almost two gallons of water!

The same process that cools us down also helps with another part of homeostasis—ridding the body of harmful waste products. Both sebaceous and sweat glands help with this function of the integumentary system.

The neural receptor cells in the dermis also provide valuable feedback from the environment. We might not think that feeling pain is a good thing, but pain is the body's way of warning us that there is a problem somewhere in the body. Some people are born with a mutation in a gene that causes them to not feel pain. This is extremely dangerous, and many people with this disorder develop injuries without even knowing it. Touch receptors help us avoid parts of our environment that can harm the body. Even considering the discomfort of pain, we can see how God's design is for our good.

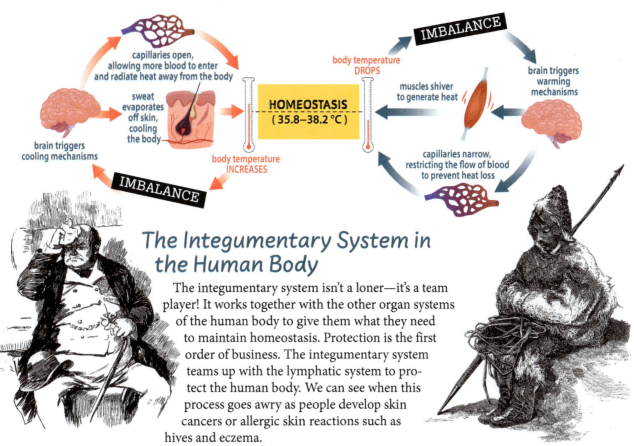

capillaries open, allowing more blood to enter and radiate heat away from the body

sweat evaporates off skin, cooling the body

brain triggers cooling mechanisms

HOMEOSTASIS (35.8–38.2 °C)

body temperature INCREASES

IMBALANCE

body temperature DROPS

muscles shiver to generate heat

IMBALANCE

brain triggers warming mechanisms

capillaries narrow, restricting the flow of blood to prevent heat loss

The Integumentary System in the Human Body

The integumentary system isn't a loner—it's a team player! It works together with the other organ systems of the human body to give them what they need to maintain homeostasis. Protection is the first order of business. The integumentary system teams up with the lymphatic system to protect the human body. We can see when this process goes awry as people develop skin cancers or allergic skin reactions such as hives and eczema.

Temperature regulation is the second main way that the integumentary system works with other systems in the body. It feeds neural information from the nerve receptors in the dermis to the nervous system. This can include information about temperature, causing the body to sweat or stimulating the muscular system to shiver to warm up the body. Fats in the subcutaneous layer provide insulation.

The integumentary system also works with several other systems. Sweat glands and sebaceous glands team up with the urinary system to help release waste products from the body, but skin can't manage this on its own. The kidneys (part of the urinary system) eliminate most waste products. Skin also uses sunlight to produce vitamin D, which provides nutrients for the skeletal system to make bones and for the digestive system to absorb calcium from food.

20.2 SECTION REVIEW

1. List the three layers of the integumentary system in order, starting with the outer layer.

2. How is the structure of the dermis different from the other layers of the integumentary system?

3. Review the types of tissue from the Human Tissues and Organs subsection (pp. 454–57). Suggest the purpose of each layer of the integumentary system on the basis of the tissues(s) that compose it.

4. How does the integumentary system help the human body maintain homeostasis?

5. Draw a concept map that shows the interactions of the integumentary system with other systems.

6. Moles, clumps of melanin-producing cells, sometimes become cancerous. On the basis of what you learned about cancer on pages 184–85, what do you think causes skin cancers such as melanoma?

20.3 THE LYMPHATIC SYSTEM AND IMMUNITY

The Structure of the Lymphatic System

Questions

What is the lymphatic system?

How does the lymphatic system keep you healthy?

What are some of the good and bad reactions of the immune system?

Terms

inflammatory response
antibody
lymph
lymph node
macrophage
lymphocyte
B cell
T cell
passive immunity
active immunity
cell-mediated immunity
humoral immunity
autoimmune disease
vaccination

Why is there a new flu shot available every year?

You're on the soccer team, and you have been playing for a few weeks when one day after the trees have started to bloom your eyes become red, itchy, and teary. Your body is undergoing a type of **inflammatory response**—your body's reaction to an allergen, toxin, bacteria, or other threat. The inflammation that you are experiencing is a signal to the immune system for an attack. In this case your body is having an allergic response, probably to the pollen from the blooming trees. But an inflammatory response can also be triggered by an infection. That's a good thing—this reaction rallies your immune system to eliminate an antigen.

The lymphatic system is an important part of the immune system, the parts of the body that protect it from antigens. The structures of the human immune system are difficult to spot—many are tiny and are parts of other systems that carry out one step of the defense process. Since the battle against pathogens happens wherever they invade the body, much of the immune system involves chemicals and blood cells that can easily be transported throughout the body. The immune system produces proteins called **antibodies** to fight antigens and keep them from disturbing the body's processes of life.

The lymphatic system uses a clear fluid called **lymph** to transport antigen-fighting cells and proteins through a network of vessels similar to blood vessels. Lymph comes from blood and is identical to plasma. It contains about 90% water along with other dissolved chemicals that supply nourishment to body tissues. Let's explore how the structure of the lymphatic system defends your health.

THE LYMPHATIC SYSTEM

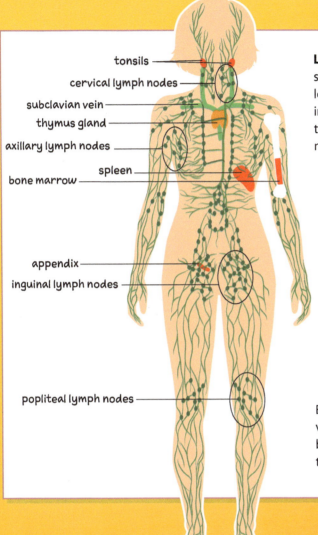

tonsils

cervical lymph nodes

subclavian vein

thymus gland

axillary lymph nodes

spleen

bone marrow

appendix

inguinal lymph nodes

popliteal lymph nodes

Lymph nodes are made of groups of cells surrounded by connective tissue. Lymph follows pathways through the body, starting out in tiny lymphatic capillaries all over the body, then going to lymphatic vessels, then to lymph nodes located in the body's trunk.

Two organs associated specifically with the lymphatic system are the thymus and the spleen. The *thymus* is a gland located near the heart where *T cells* mature and learn what to fight and what to leave alone. We are born with a fully developed thymus, but by our early teens it is gradually replaced by fats. The *spleen*, the largest organ in this system, stores *B cells* that make antibodies to eliminate bacteria and viruses.

Eventually, all lymph travels to one of the large veins near the collar bone on both sides of the body, which are called the *subclavian veins*. At this point, lymph rejoins the blood stream.

There are two types of warriors in the lymphatic system.

Macrophages fight antigens in the lymph nodes by grabbing and engulfing antigens.

Lymphocytes are white blood cells that are produced inside bone marrow.

There are two main types of lymphocytes: **B cells** and **T cells**. Lymphocytes stimulate the production of antibodies, proteins that latch onto and attack antigens.

The Function of The Lymphatic System

The main function of the lymphatic system is to protect the body from antigens. This is why it is such an important part of the immune system. There are two ways that our bodies can get protection. When someone is given antibodies from another organism, the person experiences **passive immunity**. This kind of immunity doesn't last long since the body hasn't produced these antibodies. But when the body makes its own antibodies to protect itself, it acquires **active immunity**.

To study two kinds of active immunity, let's put a different spin on it. Imagine that you are an evil flu virus that has entered a human body through inhalation. The whole scenario you trigger is just one way that the human body uses the lymphatic system to protect itself from antigens. This process, which uses mostly T cells to fight antigens that infect cells, is called **cell-mediated immunity**. Macrophages *generally* identify antigens, while T cells and B cells *specifically* identify them. Another way that our bodies can fight antigens is through **humoral immunity**. Instead of activating T cells, this process uses mostly B cells to fight antigens; it can occur when helper T cells capture antigens as prisoners and signal B cells to start making antibodies.

CELL-MEDIATED
IMMUNITY

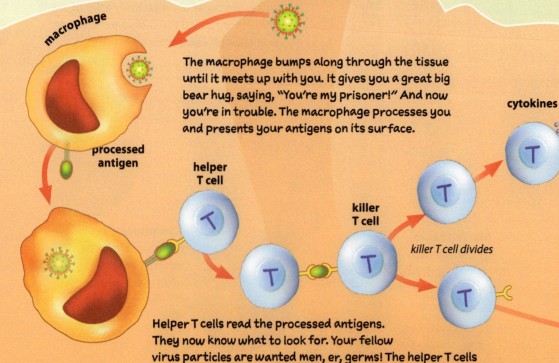

Here you are, the evil flu virus. You go up into the nose. At this point, you find a healthy cell to infect and make copies of yourself. You are free to wreak havoc in the body—until a macrophage finds you!

virus (antigen)

macrophage

The macrophage bumps along through the tissue until it meets up with you. It gives you a great big bear hug, saying, "You're my prisoner!" And now you're in trouble. The macrophage processes you and presents your antigens on its surface.

cytokines

processed antigen

helper T cell

killer T cell

killer T cell divides

Helper T cells read the processed antigens. They now know what to look for. Your fellow virus particles are wanted men, er, germs! The helper T cells use the processed antigens to activate killer T cells, which then replicate themselves. Some killer T cells produce chemical messengers called *cytokines*. Different cytokines can help T cells mature or direct them to the site of infection.

HUMORAL IMMUNITY

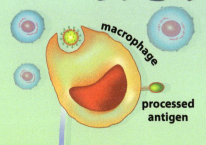

macrophage

processed antigen

Now you're the evil flu virus again. This time, before you've had the chance to infect a healthy cell, the immune system has detected you freely circulating between cells. Macrophages put you in a headlock, process you, and activate helper T cells.

helper T cell

The helper T cells activate B cells, which begin replicating. Some of these B cells begin producing antibodies. These antibodies bind to antigens on your fellow viruses. Your buddies have been tagged and marked for disposal by white blood cells called *phagocytes*.

B cell

antibodies

memory B cell

Other activated B cells remain in your body as memory B cells. These watchdogs remain watchful in case you turn up again. If you do, these B cells will be ready for you!

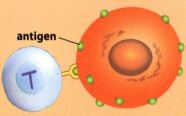

infected cell

antigen

Helper T cells then activate *suppressor T cells*, which calm down T cells after the antigen is taken care of. This is important because there are many diseases, called **autoimmune diseases**, in which a person's immune system can't distinguish the body's cells from foreign invaders. Autoimmune diseases include lupus, arthritis, diabetes, and allergies.

Other killer T cells hunt down your antigens on infected cells and induce cell death, or lysis. Cytokines call in phagocytes to clean up the mess.

cell death

Cell-mediated immunity takes place when an antigen is spotted on the surface of a cell by the immune system. Humoral immunity takes place in between cells and relies on fluids, or *humors*, to circulate antibodies to different parts of the body. Both of these kinds of active immunity are called a body's *primary response* to an antigen just like you, the evil flu virus.

So what happens when this person is exposed to the flu again? If you as the evil virus haven't experienced any mutations, the person's memory T cells and memory B cells are prepared to fight the virus quickly. This is called the *secondary response* since the body has seen you and your copies before, and it knows just what to do.

Sometimes doctors penetrate the defensive layer of skin by giving a **vaccination**, which exposes a person to an antigen in the form of a pathogen, in a safe way, to train the body's immune system to respond properly to it. The pathogen could be dead, or it could be live but weakened in some way. A vaccine can be administered as a shot or orally. Recall the polio vaccine from Chapter 12, which can be administered in both ways.

The secondary response is the primary reason why people are vaccinated for serious diseases such as smallpox, measles, and whooping cough. Doctors use vaccines to stimulate the body to produce antibodies or to activate T cells to provide just as much protection as those formed from actually having the disease. For some diseases, people need booster shots to remind the body to produce memory cells and antibodies in order to maintain immunity.

Immunity in the Human Body

The lymphatic system is the interstate for your body's immune defense. It works with other organs in other systems to help protect you from disease.

The first line of defense is the skin, which keeps out most antigens and pathogens. There are other barriers in the body besides skin that section it off and contain any antigens that penetrate the skin's defenses. One example of a barrier is the blood-brain barrier, which keeps blood and nervous tissue separate.

When an antigen gets past the skin, the body keys up the immune response, perhaps triggering inflammation near an antigen. When you get a fever, your brain is firing up the immune response to an antigen.

Organs in other systems support the work of the lymphatic system. Muscles push lymph through lymphatic vessels; there is no pump equivalent to the circulatory system's heart. Bone marrow in the skeletal system produces lymphocytes, both B cells and T cells. B cells mature in bone marrow. The lymph system also works closely with the circulatory system to get lymphocytes where they need to go to fight antigens. There are even lymphatic tissues in your tonsils and appendix, all part of your digestive system. The digestive system also relies on the lymphatic system to help it remove fats.

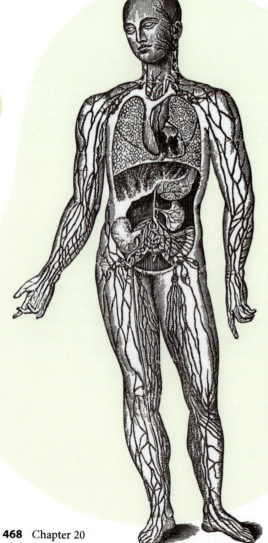

God has fearfully and wonderfully made our bodies (Ps. 139:14), providing for our needs even in a fallen world where there are many diseases to plague us. Though God has given us bodies having amazing abilities to protect us, He sometimes in His wisdom allows us to experience sickness, disease, and suffering. We can trust the Creator of life on Earth—including human life—to do what is best for us. Through our knowledge of human anatomy and physiology we can not only help people who are sick but also worship a Creator who put us together and knew every cell of our bodies even before we were born.

20.3 SECTION REVIEW

1. You get a fire ant bite that turns red, swollen, and itchy. What is your immune system doing?

2. Outline the path that lymph travels, starting at the extremities of the body and moving toward the center.

3. Some breast-fed infants have immunity to chicken pox that bottle-fed infants do not. What type of immunity is this and how is it acquired?

4. Why would the breast-fed infants in Question 3 still need a chickenpox vaccination later in life?

5. How does the lymphatic system help to maintain homeostasis in the body?

6. Classify cell-mediated immunity and humoral immunity as either active or passive immunity. Explain your answer.

7. Create a T-Chart comparing T cells and B cells. Include their places of origin and maturation, type of immunity they contribute to, and the different cells they activate.

8. Why is it important that lymphocytes not stay in the lymphatic vessels all the time?

9. Compare the cells involved in cell-mediated immunity and humoral immunity.

10. Compare the locations of the antigen in cell-mediated immunity and humoral immunity.

11. How do the integumentary and lymphatic systems work together to protect the human body from antigens?

12. Why is the presence of muscular tissue necessary in lymphatic vessels?

13. On the basis of what you have read in this chapter, suggest a reason why there is currently no vaccine for HIV, the virus that causes AIDS.

20.1 THE STUDY OF YOU

- Human anatomy is the study of the structure of the human body, while human physiology studies the body's processes.

- The human body is made of nervous, connective, muscle, and epithelial tissues organized into organs and organ systems.

- The body's organ systems work together to maintain homeostasis.

- Understanding how the body works helps believers glorify God and serve others.

Terms
antigen • human anatomy • human physiology • nervous tissue • connective tissue • matrix • muscle tissue • epithelial tissue

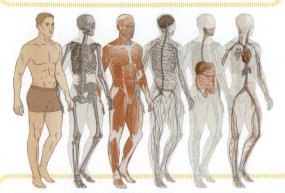

20.2 THE INTEGUMENTARY SYSTEM

- The main organ of the integumentary system is the skin, composed of an outer layer, the epidermis, and an inner layer, the dermis. Beneath these is a subcutaneous layer of connective tissue that is part of the integumentary system but not of the skin itself.

- Cells in the epidermis contain a pigment, melanin, that determines skin color. The epidermis is constantly being replaced as older cells on its surface are sloughed off.

- The dermis is composed of all four kinds of tissues along with sweat glands, sebaceous glands, and hair follicles.

- Important functions of the integumentary system include protecting the body from antigens, regulating body temperature, providing sensory information about the environment, excreting some waste products, and producing vitamin D.

Terms
epidermis • dermis • subcutaneous layer • hair follicle • melanin • sweat gland • sebaceous gland

20.3 THE LYMPHATIC SYSTEM AND IMMUNITY

- The lymphatic system consists of a clear fluid—lymph—that flows through lymphatic capillaries to lymph nodes. The thymus and spleen are also part of this system.

- The main function of the lymphatic system is to protect the body from antigens as part of the body's immune system.

- Passive immunity is of short duration and is acquired by receiving antibodies from another organism. Longer-lasting active immunity is produced by the body in response to an infection.

- The two types of active immunity are cell-mediated, involving mainly T cells, and humoral, using mostly B cells.

- Immunity can be acquired by exposing a person to an antigen in the form of a vaccine.

- The lymphatic system relies on the muscular system to transport lymph and on the skeletal system to produce lymphocytes.

Terms
inflammatory response • antibody • lymph • lymph node • macrophage • lymphocyte • B cell • T cell • passive immunity • active immunity • cell-mediated immunity • humoral immunity • autoimmune disease • vaccination

Chapter Review Questions

RECALLING FACTS

1. List the four kinds of tissues found in the human body.

2. What is the single purpose of almost every organ system in the human body?

3. Which two body systems are mostly responsible for movement?

4. Which two body systems transport needed nutrients around the body?

5. Which two body systems are most closely related to immunity?

6. Which two body systems are most closely related to water intake, eating, and nutrition?

7. Which two body systems help the body control itself in its environment?

8. List which one of the ten other body systems the integumentary system most closely works with for each of the following functions.

   a. sense of touch

   b. shivering

   c. immunity

   d. sweating

9. What is the purpose of the lymphatic system in the body?

10. Name two organs of the lymphatic system.

11. List which one of the ten other body systems the lymphatic system most closely works with for each of the following functions.

    a. keeping antigens out of the body

    b. the production of lymphocytes

    c. the onset of fever

UNDERSTANDING CONCEPTS

12. Review the discussion of animal body plans on page 351. What kind of symmetry, if any, does the human body display?

13. Which kind of tissue lines the small intestine?

14. Is "skin" just another name for the integumentary system? Explain.

15. What part of the skin does sunburn most often affect? Why is this?

16. Draw and label the three layers of the integumentary system.

17. Trace the path of lymph through the body from the extremities to the bloodstream by putting the following steps in order.

1 Lymph between cells in the finger migrates to a nearby lymphatic capillary.

2 Lymph in lymph nodes travels through lymphatic vessels to the subclavian vein.

3 Lymph in the subclavian vein enters the blood stream.

4 Lymphatic capillaries merge in lymphatic vessels.

5 Lymphatic vessels bring lymph to collection stations in lymph nodes.

18. Create a T-Chart comparing cell-mediated and humoral immunities. Include the types of lymphocytes involved, the location of antigens, the use of macrophages, the cells produced, and the occurrence of lysis in your comparison.

19. A person develops an allergic response when an antigen from the environment called an allergen is identified by the immune system. Helper T cells initiate the inflammatory response. What kind of immunity do allergies involve?

20. What is the main difference between a primary and secondary response to an antigen?

CRITICAL THINKING

21. How should a Christian view the study of human anatomy and physiology, and why is such knowledge important?

22. There are three severities of burns: first, second, and third degree. Suggest how these would relate to the layers of the integumentary system.

23. In hatcheries, many fish are confined in a contained area. Fish have immune systems that are slow to respond to pathogens. Hatchery managers use broad spectrum antibiotics to head off an epidemic rather than take the time to figure out exactly what is causing the problem. How would this compare with human immune systems and handling epidemics?

24. The tonsils and the appendix are part of the lymphatic system, both producing antibodies. However, they were once deemed vestigial organs, evolutionary relics with no specific purpose. In light of this example of premature conclusion, how do you think one's worldview affects the progress of science?

25. Blood vessels are organs made mainly of epithelial, muscle, and connective tissues, with very little nervous tissue. Why don't blood vessels need much nervous tissue?

case study

SMALLPOX

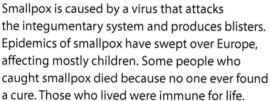

Smallpox has killed people for centuries. Ramses in Egypt, Mozart in Austria, George Washington in the United States, and Joseph Stalin in the Soviet Union all got smallpox in different places and at different times in history.

Smallpox is caused by a virus that attacks the integumentary system and produces blisters. Epidemics of smallpox have swept over Europe, affecting mostly children. Some people who caught smallpox died because no one ever found a cure. Those who lived were immune for life.

But smallpox doesn't kill people anymore. In 1958 the World Health Organization organized a campaign to eliminate smallpox. A key part of this was vaccinations, which have been administered for smallpox for centuries, though safely only for decades. Smallpox is one of the few human diseases to be eliminated worldwide.

Use the case study above and the graph on the right to answer Questions 26–30.

26. When was the highest worldwide incidence of smallpox?

27. Relate the answer that you gave in Question 26 to the World Health Organization's campaign to exterminate smallpox.

28. Suggest a reason why there were no reported cases of smallpox after 1980.

29. Why don't we vaccinate for smallpox today?

30. How is the development of a safe smallpox vaccine a good application of science?

Use the information in the ethics box on the next two pages to answer Questions 31–32.

31. Compare the responses from each worldview.

32. Are there any conclusions in the bioethics strategy that seem to conflict with biblical teaching? If so, what are they?

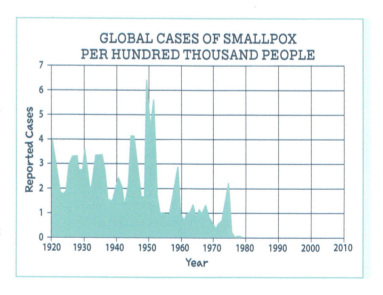

ETHICS

PUBERTY BLOCKERS

If you pay attention to social media or the news, you are sure to be aware of the issue of gender dysphoria. This is a condition in which a person feels anxiety and discomfort with his or her gender. One option for treatment is a puberty blocker, which is a class of drugs that do exactly what its name implies—it delays the onset of puberty. Puberty blockers were originally intended to treat a condition called *precocious puberty*, the onset of puberty at an unusually early age. Precocious puberty is often completely normal in all respects except for age, but sometimes it is caused by underlying medical conditions. Some of these conditions have harmful long-term effects. Using puberty blockers to treat such cases is consistent with exercising godly compassion for others. But what about gender dysphoria? Should puberty blockers be used to help those who believe they are the wrong gender?

Below is a typical response to this question using the principles of bioethics.

1. What information can I get about this issue?

Puberty blockers delay the onset of puberty by preventing the development of secondary sexual characteristics such as menstruation in girls, facial hair in boys, and the development of both male and female sexual organs. Many transgender-identifying children who receive puberty blockers eventually go on to hormone therapy and sometimes even sex reassignment surgery. There is debate about the long-term effects of puberty blockers even if a child does not take other measures to complete a transition between sexes. Puberty blockers stunt a child's growth and also significantly lessen bone density as a child approaches adulthood.

2. How are the people involved in this issue respected and given the freedom to choose?

Informed consent must be obtained, but because puberty blockers by definition are administered to children, this process is complicated. Depending on the laws in a given location, consent can involve parents or guardians of the child. According to the principles of bioethics, a child (and guardian) must be informed of the function, benefits, and complications of taking puberty blockers. Bioethicists would say that ultimately the needs and desires of the patient must be considered regardless of who is responsible for making the decision.

3. How are individuals protected from harm or injury?

As with almost any medical treatment, there are possible side effects when administering puberty blockers. Some of the potential side effects include diminished bone density and sterility. Additionally, numerous people come to regret their decision to transition sexes; some even have surgery to return to their birth sex. Secular bioethicists would say that alleviating gender dysphoria and allowing the patient to develop psychologically and emotionally outweighs any potential side effects.

4. How are people helped by the action taken?

Those experiencing gender dysphoria often struggle emotionally and psychologically and believe that puberty blockers will help them. Studies range in their conclusions about how helpful puberty blockers have proved to be to youth experiencing gender dysphoria. Other people feel that using the blockers helps by reducing the stress and anxiety in those with gender dysphoria and is therefore beneficial.

5. How is the action taken just and fair?

Puberty blockers are available in most places in the United States. There may be limits on the availability of puberty blockers to underage children without parental consent in some states, especially considering the uncertain causes of gender dysphoria, the experimental nature of some of these drugs, and their possible long-term effects. Some advocates argue that these treatments should be available without parental consent in all states to be fair and equitable.

6. What action do the principles of bioethics recommend?

According to the principles of bioethics, puberty blockers should be offered to youth after careful assessment of an individual's situation (e.g., diagnosis of gender dysphoria). The decision-makers (patient, parent, guardians) must be informed of the benefits as well as possible complications before they give their consent.

Below is a response to this question from a biblical perspective.

1. What information can I get about this issue?
See the facing page.

2. What does the Bible say about this issue?

For much of human history people have recognized what the Bible clearly teaches, that there are only two genders: male and female. This truth is taught in the very first chapter of the Bible in Genesis 1:27, and it is repeated elsewhere (Gen. 5:2, Matt. 19:4). Moreover, the Bible teaches that each person is a creation of the sovereign God who alone decides whether one is to be male or female (Is. 29:16). Our proper response to God is gratitude for how He created us. One common element associated with gender dysphoria, same-sex attraction, is repeatedly forbidden in Scripture (Lev. 18:22; 20:13; Rom. 1:26–27; 1 Tim. 1:10). The possibility of any gender(s) other than male or female is not even broached in Scripture.

3. What are the acceptable and unacceptable options?

The reality is that some people do feel a disconnection with their gender, and coping with gender dysphoria may very well be a trial for a believer. We can accept that we are fearfully and wonderfully made and accept how God created us. We can rely on the promises in the Bible that God will be faithful to help those who struggle with temptation (1 Cor. 10:13; James 4:7). Or we can wrongly rely on feelings rather than truth. Those who struggle with gender dysphoria should seek medical help for any underlying condition that may be impacting their struggles in this area. They should also seek biblical counsel on how to cope with their feelings.

4. What are the motivations of the acceptable options?

Believers can be motivated to exercise faith in God when faced with gender dysphoria. Every Christian has different struggles, so someone struggling with gender dysphoria can learn to hope in God's promises of help and deliverance. A believer could be motived to love God and others by seeking to live out his God-given identity. Other believers can support those that struggle with gender dysphoria by supporting their need for possible medical treatment and recommending biblical counsel.

5. What action should I take?

Recognizing the authority of God's Word and the one who inspired it, the believer should seek to honor God by being thankful for his gender and living it out according to the instructions found in Scripture (e.g., Eph. 5:21–31).

Scripture teaches that God made each of us in His image. In His infinite knowledge and wisdom, He created us as we are. We are to live a God-honoring life as we were created. While acknowledging that some struggle with feelings of gender dysphoria, the use of puberty blockers is not a biblical response.

21 SUPPORT & MOVEMENT

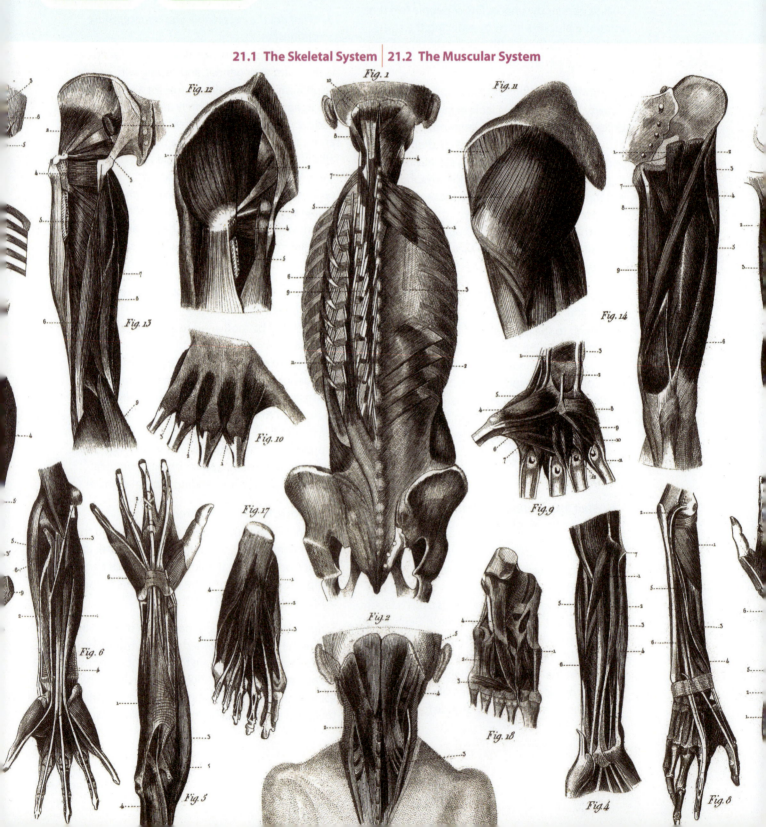

Fig. 12
Fig. 1
Fig. 11
Fig. 13
Fig. 14
Fig. 10
Fig. 9
Fig. 17
Fig. 2
Fig. 6
Fig. 18
Fig. 5
Fig. 4
Fig. 8

21.1 THE SKELETAL SYSTEM
The Structure of the Skeletal System

Claire Lomas, the paraplegic woman who walked the 26.2 mile London Marathon course in seventeen days, used her ReWalk exoskeleton to enable her to stand and walk. The exoskeleton also replaced the function of her leg muscles in helping her walk. While both bones and muscles have multiple functions, their most obvious functions—support and movement—are intimately connected. Knowing how bones and muscles function enables us to help those among God's image-bearers whose bodies do not function properly as a result of a birth defect or injury.

An infant's skeleton can contain up to 300 bones made of a firm and flexible connective tissue called **cartilage**. But as a child grows, bones undergo **ossification**, a process where new bone material is being laid down, resulting in bones that are harder and more rigid. As the bones ossify, some of them begin to fuse, so an adult has only around 206 bones, though the exact number may vary slightly from one person to another.

While most of the cartilage in a child's bones ossifies, some cartilage remains at the very ends of the bones to cushion them where they meet at a **joint**. Strong bands of connective tissue called **ligaments** attach to the **periosteum**, the bone's outer covering, to hold the bones together. A connective tissue sheath called the *joint capsule* covers the ends of the bones forming the joint. The *synovial membrane* lines the inner surface of the joint cavity and produces *synovial fluid*, which lubricates the joint and absorbs shocks. Synovial membrane tissue is also found in sac-like structures called *bursas*, which are located between moving parts of joints, such as ligaments and bones, and reduce friction between the structures as they move.

?

Are bones alive?

Questions

What bones make up the axial and appendicular skeletons?

What are the parts of a bone?

Where are the locations of different joints?

How is the skeletal system connected to other parts of the body?

Terms

cartilage
ossification
joint
ligament
periosteum
axial skeleton
appendicular skeleton
osteoblast
spongy bone
compact bone
osteon
osteocyte
Haversian canal
growth plate
osteoclast

ligament

articular cartilage

synovial fluid

periosteum

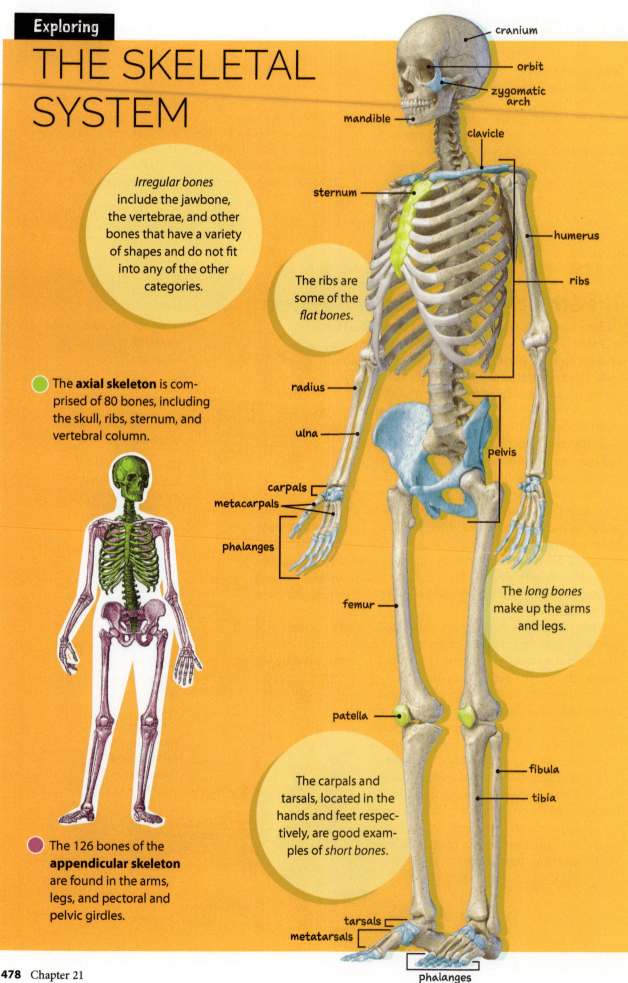

THE SKELETAL SYSTEM

Irregular bones include the jawbone, the vertebrae, and other bones that have a variety of shapes and do not fit into any of the other categories.

The ribs are some of the *flat bones*.

• The **axial skeleton** is comprised of 80 bones, including the skull, ribs, sternum, and vertebral column.

• The 126 bones of the **appendicular skeleton** are found in the arms, legs, and pectoral and pelvic girdles.

The *long bones* make up the arms and legs.

The carpals and tarsals, located in the hands and feet respectively, are good examples of *short bones*.

cranium
orbit
zygomatic arch
mandible
clavicle
sternum
humerus
ribs
radius
ulna
pelvis
carpals
metacarpals
phalanges
femur
patella
fibula
tibia
tarsals
metatarsals
phalanges

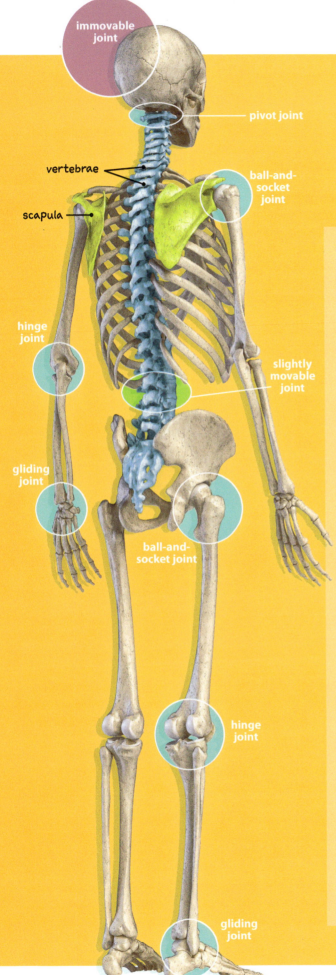

immovable joint

pivot joint

vertebrae

ball-and-socket joint

scapula

hinge joint

slightly movable joint

gliding joint

ball-and-socket joint

hinge joint

gliding joint

JOINTS

Joints are often classified according to how the two bones move in relation to one another. Let's take a look at the different types.

Immovable Joints

As a person grows, some adjoining bones fuse. An infant's skull bones are joined by soft tissue that eventually hardens into bone, uniting the skull. These immovable joints are called *sutures*.

Slightly Movable Joints

The vertebrae have pads of cartilage between them that allow only limited movement.

Freely Movable Joints

Pivot Joints. The top two vertebrae fit together so that the lower one fits through the upper one to form a pivot for the skull. This arrangement allows the head to rotate and swivel.

Gliding Joints. The wrist and ankle bones have slightly convex and concave surfaces that fit together to allow limited lateral and vertical movement.

Ball-and-Socket Joints. A ball-shaped head moves within a hollow socket. This design creates rotating movement and allows free movement in all directions. The hip and shoulder joints are ball-and-socket joints.

Hinge Joints. The bones at these joints fit together so that they bend in only one direction. The knee and elbow are hinge joints.

The Function of the Skeletal System

The sutures in the skull form as a person's skull bones grow together. While we often think of bones as dead, the bones of a living person are living and changing. As we learned in Chapter 20, the cells of connective tissues such as bone are embedded in a matrix that differs from one type of connective tissue to another. In bones, the collagen matrix contains a large amount of calcium salts that make bones hard and strong.

During ossification, cells called **osteoblasts** begin replacing the disintegrating cartilage with spongy bone. Despite its name, **spongy bone** should not be thought of as soft. It is actually quite hard and strong. Its name relates to its irregular, pitted appearance.

As some osteoblasts are creating spongy bone to replace cartilage, other osteoblasts located just under the periosteum are laying down **compact bone**. Unlike the irregularly shaped spongy bone, compact bone is organized into circular units called **osteons**. An osteon consists of circular layers of hardened matrix called *lamellae* (s. lamellum). Bone-forming cells called **osteocytes** live between the lamellae. Blood carrying nourishment to the osteocytes flows through vessels in a tiny passageway called the **Haversian canal** at the very center of the osteon.

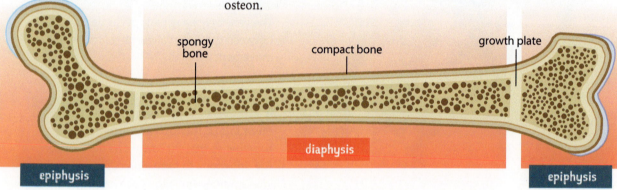

A long bone is divided into the shaft or *diaphysis* and the two ends or *epiphyses* (s. epiphysis). While the diaphyses contain much of the compact bone in our bodies, an epiphysis remains spongy bone. Bone itself doesn't really grow; instead, a small layer of cartilage called the **growth plate** remains between the diaphysis and each of the two epiphyses. This layer of cartilage continually divides, and the older layers are ossified to increase the length of the bone. When a person has reached full height, the growth plate becomes bone entirely, halting growth.

While the growth plate is increasing the length of the bone, the bone must also increase in diameter. A long, thin bone would be in constant danger of breaking. Cells called **osteoclasts** destroy the inner spongy bone of the diaphysis. Osteoblasts and osteocytes form new compact bone using the material that the osteoclasts release. In fact, bones continue to be in a constant state of remodeling even after a person has finished growing. Bones may develop stress points that would weaken the entire bone, but osteoclasts remove the bone material to allow osteoblasts and osteocytes to build new, stronger bone. Bones that constantly undergo stress from exercise become denser, so a runner's femur bones tend to be stronger than those of the average person.

The Skeletal System in the Human Body

Protection is one of the main functions of the skeletal system. Skull bones fuse into one structure to form a single plate to protect the brain. Since the central nervous system controls most of the other organ systems, it is essential that it be protected. The spinal cord runs the length of the torso and is protected by the surrounding vertebrae. The ribs form a protective cage around the heart and lungs, and the pelvic girdle provides some protection to the intestines and reproductive organs.

The skeleton is also a major storage site for certain minerals that our bodies need. As the osteoclasts break down bone material for the osteoblasts and osteocytes to use, some of that material can be removed from the bone altogether if the body needs it. Calcium is especially important to the body since the muscular, nervous, and reproductive systems all use it. The bones store calcium in the form of calcium phosphate, so when calcium is released, phosphate is simultaneously released. Phosphate forms part of ATP and the nucleic acids, connecting the skeletal system to both cellular respiration and information storage. Moving calcium and phosphate to or from the bones is controlled by hormones from the endocrine system. If a person's diet does not include enough calcium, the body may be forced to remove calcium from the skeletal system, leading to brittle bones that fracture easily.

On the other hand, if the calcium in the diet is greater than the body needs, the excess can be stored in the bones until it is needed.

The skeletal system is also connected to the circulatory system. At the center of bones lies a soft tissue called *bone marrow*. At birth, all the bone marrow is *red* bone marrow, which makes red blood cells, platelets, and several types of white blood cells. Eventually, most of the bone marrow in the diaphysis becomes *yellow* bone marrow, which can produce cartilage and bone. The yellow bone marrow also functions as a storage site for fat. The red bone marrow remains in the spongy bone of the epiphyses of long bones and in some of the flat bones. If the body's supply of blood becomes severely depleted, some of the yellow bone marrow can transform back into red bone marrow to increase blood production.

Skeletal systems perform many roles, but the most obvious function of the skeletal system is support. The skeleton allows a person to stand upright and hold his or her shape. It also provides a rigid framework for the muscles to attach to, and the bones and muscles together allow the whole body to move.

21.1 SECTION REVIEW

1. What two tissues form the outside covering of bones?

2. Which type of joint are the knuckles? Explain.

3. Are long bones found more in the axial skeleton or in the appendicular skeleton? Explain.

4. What bone connects the scapula to the sternum?

5. What types of cells are found in compact bone?

6. What type of bone is found in the epiphyses?

7. What essential element is stored in the bones?

8. What type of bone marrow is most common in a sixty-year-old man? Explain.

9. How does the skeletal system affect the circulatory system?

Questions

What are the different kinds of muscles?

How do muscles move?

How does the muscular system interact with other systems in the body?

Terms

skeletal muscle
tendon
fascicle
myofibril
myosin
actin
muscle fiber
smooth muscle
cardiac muscle
sarcomere
antagonistic pair

The muscular system is intimately connected with the skeletal system. In fact, without the muscles, the bones wouldn't even hold their positions, and we'd all be stuck lying on the ground unable to move. Many muscles connect two or more bones, one of which moves while the others remain essentially stationary. The place on the stationary bone where the muscle attaches is called its *origin*. The muscle's attachment location on the movable bone is called its *insertion*. Practically your entire body is covered with muscles, but you are probably not aware of many of them until you strain one.

How do the different parts of the muscular system keep the body moving?

Exploring

THE MUSCULAR SYSTEM

The name of an individual muscle may reflect its size, shape, function, location, or number of origins.

Size

major: large muscle

minor: small muscle

maximus: largest muscle in a region

Shape

deltoid: triangular-shaped muscle

trapezius: trapezoidal-shaped muscle

oblique: muscle with fibers arranged obliquely (in a slanting direction)

Location

sternocleidomastoid: muscle with origin at the sternum and clavicle and insertion at the mastoid process

Number of Origins

biceps brachii: muscle with two origins (*bi-*, meaning "two"; *-ceps*, "head")

Function

adductor group: muscles that adduct (draw in) the thigh

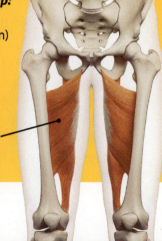

adductor group

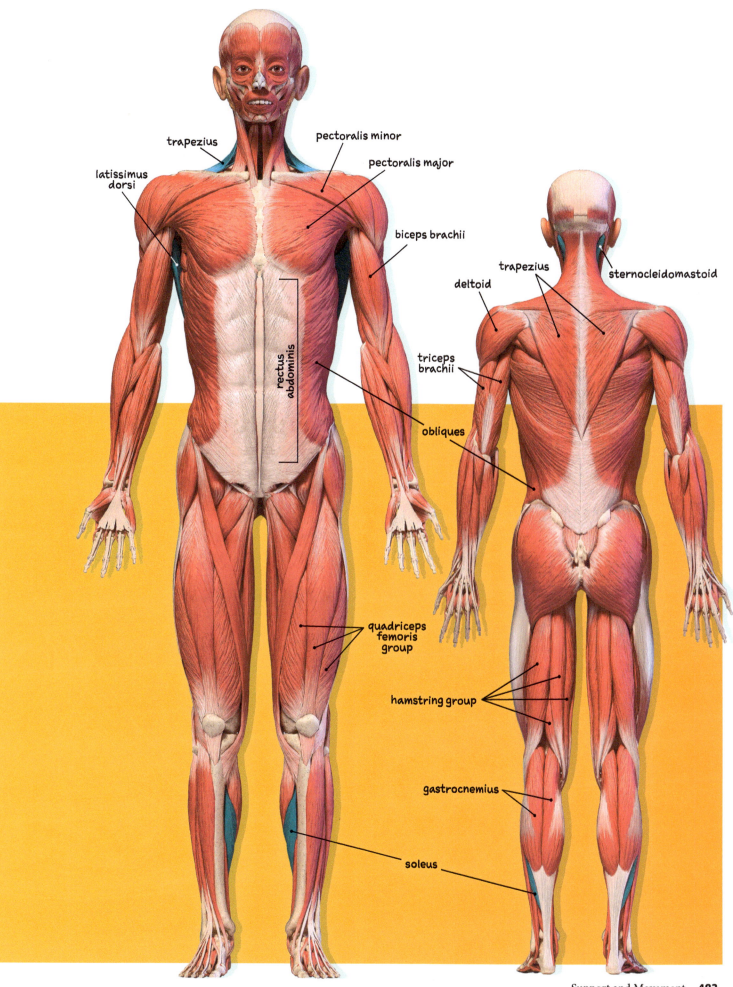

trapezius

pectoralis minor

pectoralis major

latissimus dorsi

biceps brachii

rectus abdominis

quadriceps femoris group

soleus

deltoid

trapezius

sternocleidomastoid

triceps brachii

obliques

hamstring group

gastrocnemius

MUSCLE TYPES

Skeletal muscle is so named because it is usually attached to bones. It is also described as *striated* because of the dark and light stripes in its cells. These striations are filaments of protein in the muscle cells. Skeletal muscle is also known as voluntary muscle tissue because it is controlled primarily by conscious thought. We move our arms by just thinking about it. Our brain sends signals to the muscles that control the arm, such as the trapezius and the pectoralis major.

Exploring

THE SKELETAL MUSCLE TISSUE

Tendons are an extension of the connective tissue that surrounds the entire muscle. They connect the muscle to the bone by intertwining with the periosteum.

Multiple **fascicles** together form a muscle. Each is surrounded by a layer of connective tissue and contains anywhere from 10 to 100 muscle cells.

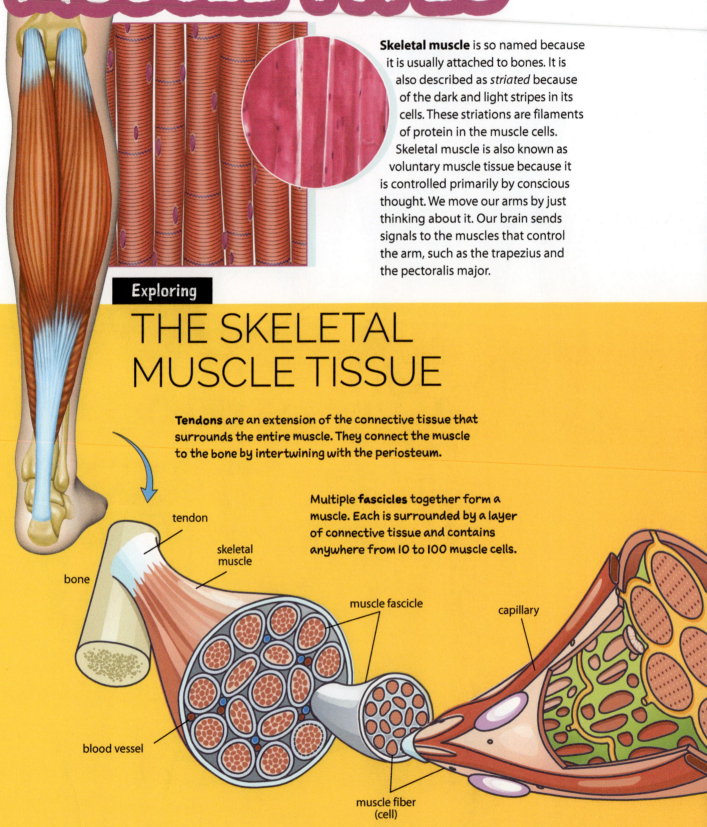

tendon

skeletal muscle

bone

muscle fascicle

capillary

blood vessel

muscle fiber (cell)

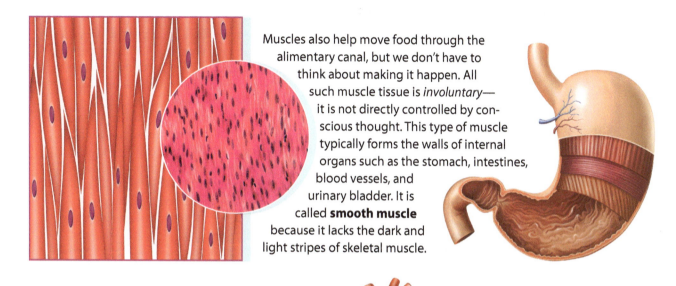

Muscles also help move food through the alimentary canal, but we don't have to think about making it happen. All such muscle tissue is *involuntary*—it is not directly controlled by conscious thought. This type of muscle typically forms the walls of internal organs such as the stomach, intestines, blood vessels, and urinary bladder. It is called **smooth muscle** because it lacks the dark and light stripes of skeletal muscle.

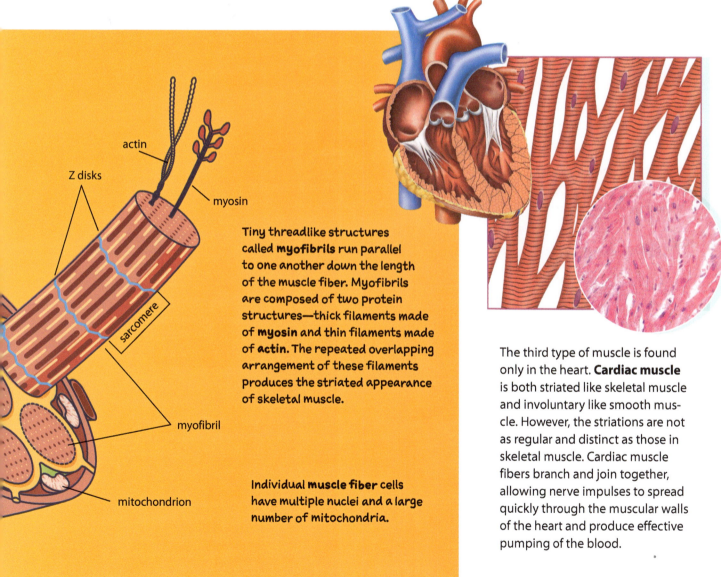

actin

Z disks

myosin

sarcomere

myofibril

mitochondrion

Tiny threadlike structures called **myofibrils** run parallel to one another down the length of the muscle fiber. Myofibrils are composed of two protein structures—thick filaments made of **myosin** and thin filaments made of **actin**. The repeated overlapping arrangement of these filaments produces the striated appearance of skeletal muscle.

Individual **muscle fiber** cells have multiple nuclei and a large number of mitochondria.

The third type of muscle is found only in the heart. **Cardiac muscle** is both striated like skeletal muscle and involuntary like smooth muscle. However, the striations are not as regular and distinct as those in skeletal muscle. Cardiac muscle fibers branch and join together, allowing nerve impulses to spread quickly through the muscular walls of the heart and produce effective pumping of the blood.

actin

myosin

The Function of the Muscular System

It is the myosin and actin filaments that produce muscle contractions. The individual actin filaments, which look like a strand of twisted beads, are connected at one end by a protein structure called the *Z line*. Z lines form the edges of a **sarcomere**, the functional unit of muscle contraction. Actin filaments alternate with myosin filaments so that they partially overlap. The myosin filaments have tiny structures called *heads* that resemble the head of a golf club.

At the beginning of the *cross-bridge cycle*, a nerve impulse reaches the muscle and calcium ions prepare the actin to bind to the myosin. ATP binds to the myosin heads and allows them to bind to the actin filament and bend toward the center of the sarcomere, pulling the actin filaments. The myosin head releases the actin filament and then, with the addition of another ATP molecule, repeats the process. As the myosin filaments pull on the actin filaments, the area where actin and myosin overlap increases, while the sarcomere unit shortens. The myosin filaments in a sarcomere are connected to each other at their midpoints by a protein structure called the *M line*. M lines allow the myosin filaments in a sarcomere to act together as a unit.

cross-bridge cycle

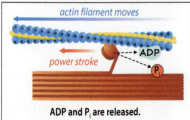

ADP and P$_i$ are released.

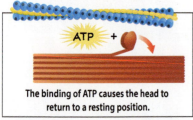

The binding of ATP causes the head to return to a resting position.

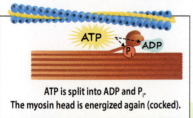

ATP is split into ADP and P$_i$. The myosin head is energized again (cocked).

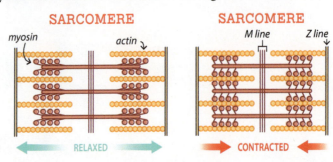

A contraction is caused by myosin filaments pulling actin filaments inside the sarcomeres, which are found in the myofibrils of the muscle fibers. A strong muscle contraction occurs when many of the fibers in a muscle contract, while a more gentle contraction occurs when only a few of the fibers contract. When the nerve impulse ends, the myosin filaments cease pulling on the actin filaments, and the sarcomere returns to its original size.

A myosin filament uses a lot of ATP as it pulls on its neighboring actin filaments. Muscle fibers contain a high number of mitochondria to supply the myosin filaments with the ATP they need. As we learned in Chapter 7, normal cellular respiration is an aerobic process—it takes oxygen. When we exercise vigorously, our muscles begin to use more ATP than the mitochondria can supply with the available oxygen. As the available oxygen runs low, our muscle cells begin to perform lactic acid fermentation. The reduction in oxygen in the muscle cells causes us to breathe heavily, creating a feeling often described as being out of breath.

After a person finishes a strenuous activity, the mitochondria in muscle cells begin to regenerate ATP using the oxygen that is becoming available. The lactic acid that was produced during lactic acid fermentation is carried by the blood to the liver where

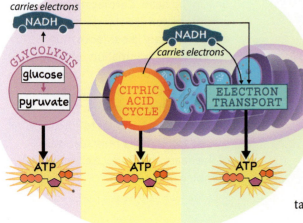

it is converted into glucose. However, this process takes more ATP, requiring even more oxygen. The additional need for oxygen resulting from strenuous activity is called *oxygen debt*. While exercising does not increase the number of muscle fibers in a person's muscles, it does increase the number of filaments. In other words, a strong person doesn't have more muscle cells—just larger ones. Exercising also increases the number of mitochondria in a person's cells so that the muscle fibers can sustain strenuous activity for a longer period of time before they switch to fermentation.

Notice that myosin filaments can only *pull* on the actin filaments; they cannot *push*. Since this is true of myosin filaments, it is also true of myofibrils, muscle fibers, and whole muscles. In order to have movement in two directions at a hinge joint, muscles are arranged in **antagonistic pairs**. One muscle or group of muscles pulls in one direction, while another muscle or group of muscles pulls in the other direction. For example, when the biceps brachii muscle contracts, the triceps brachii muscle relaxes, allowing the arm to bend at the elbow. When the triceps contracts, the biceps relaxes, and the arm straightens. Additionally, muscles often contract together. The main muscle is called the *prime mover*, and other muscles that assist the prime mover by contracting with it are called *synergists*.

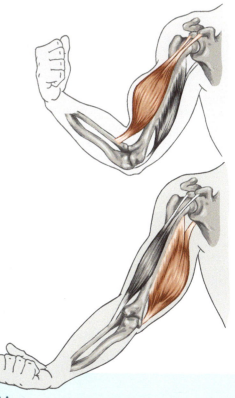

The Muscular System in the Human Body

Like any system of the body, the muscular system interacts with the other systems in order to function properly. Without the respiratory system, muscles would not have the supply of oxygen they need. Without the digestive system, they would not have a supply of glucose to make ATP. Likewise, the respiratory system would not be nearly as effective without the diaphragm. This muscle pushes down on the organs in the abdominal cavity and allows the lungs to expand, drawing in air. Thus, a muscle aids the respiratory system in obtaining oxygen for the muscles and all the other cells of the body.

The hardest-working muscle in the body is the heart. During an average person's lifetime the heart may beat several billion times, pumping blood through all the blood vessels of the body and carrying nutrients and oxygen to all the cells of the body while removing their metabolic wastes.

Muscles are also involved with the integumentary system, primarily in reacting to cold. Muscles in the skin cause shivering, which generates heat.

Other tiny muscles called *arrector pili* muscles are attached to hair follicles and contract, causing goosebumps to form on the skin. When this occurs, the hair from the follicle stands erect, holding a layer of air next to the body, creating an additional degree of insulation.

Muscles are also dependent on the nervous system to function properly. Claire Lomas, who completed the London Marathon in the ReWalk exoskeleton, was crippled by a spinal cord injury in a riding accident. Although her bones and muscles were not directly affected, the injury to her central nervous system limited her skeletal and muscular systems. All the systems of the body work together, so if one is severely damaged, the rest are affected. By understanding how the skeletal and muscular systems work, people in multiple areas of science and medicine are able to make devices such as the ReWalk exoskeleton possible. As a result, people who have been crippled for years may be able to walk and have their quality of life improved.

MINI LAB

Muscle Trick

Take a moment and command your arm to straighten. Did it do what you wanted it to do? This voluntary movement was performed because your brain created a command, which was delivered to the muscles and caused your arm to move. But do your arms involuntarily move without being prompted? In most cases they don't, but there is one popular trick that can demonstrate our arms' involuntary capabilities.

Can my arms rise on their own?

Materials
doorway • timer

PROCEDURE

A Stand in the doorway with the backs of your hands pressing against each side of the doorframe for forty seconds.

B After forty seconds, step out from the doorway and relax your arms. Observe and record what occurs.

QUESTIONS

1. What happened after you stopped pushing against the doorframe and relaxed your arms?

2. Explain how your arms felt after stepping out of the doorway.

3. Calcium is a very important mineral that the body needs to make muscles contract. After contracting the arms for forty seconds, calcium still remains in the muscles. How do you think this remaining calcium causes your arms to lift?

GOING FURTHER

4. Try the activity again, only this time, after pressing on the doorframe, step out and concentrate on keeping your arms close to your sides. Did this suppress the involuntary motion of your arms moving up? Explain.

5. Conduct more research by trying these procedures with different muscles in the body and different contraction times. Record your observations.

21.2 SECTION REVIEW

1. Judging from their names, which do you think is larger, the teres major or the teres minor?

2. Identify each muscle as skeletal, smooth, or cardiac muscle.

   a. wall of the right atrium

   b. soleus

   c. pyloric sphincter between the stomach and small intestine

3. Arrange the following structures from largest to smallest: fascicle, muscle, muscle fiber, myofibril, and myosin filament.

4. One end of which type of filament is attached to the structure at the edge of the sarcomere? Explain.

5. Which myofibril structure contains heads?

6. If the amount of oxygen available to muscle cells runs low, what substance is produced as an indirect result of ATP production?

7. Which systems of the body does the muscular system rely on for oxygen and glucose to make ATP?

8. How can a Christian use knowledge of the skeletal and muscular systems to help others?

21 CHAPTER REVIEW
Chapter Summary

21.1 THE SKELETAL SYSTEM

- There are 80 bones in the axial skeleton and 126 bones in the appendicular skeleton.

- Ligaments attach to the periosteum of bones and hold bones together.

- Joints are classified as immovable, slightly movable, and freely movable. The freely movable joints include ball-and-socket joints, gliding joints, hinge joints, and pivot joints.

- Bones elongate when the growth plates between the diaphysis and two epiphyses continually divide and ossify.

- Bones widen when osteoblasts and osteocytes form new compact bone.

- The skeletal system provides support for the body, stores minerals, and contains bone marrow.

Terms
cartilage · ossification · joint · ligament · periosteum · axial skeleton · appendicular skeleton · osteoblast · spongy bone · compact bone · osteon · osteocyte · Haversian canal · growth plate · osteoclast

21.2 THE MUSCULAR SYSTEM

- The name of a muscle can reveal its size, shape, function, location, or number of origins.

- The three main muscle types are skeletal muscle (striated), smooth muscle (involuntary), and cardiac muscle (heart).

- Skeletal muscle tissue is made up of tendons, fascicles, muscle fiber cells, and myofibrils.

- During contraction, myosin filaments inside the sarcomeres pull actin filaments.

- The muscular system interacts with the other systems of the body to help carry out everyday functions.

Terms
skeletal muscle · tendon · fascicle · myofibril · myosin · actin · muscle fiber · smooth muscle · cardiac muscle · sarcomere · antagonistic pair

Chapter Review Questions

RECALLING FACTS

1. Explain the role of cartilage in a joint.

2. Explain the role of ligaments in a joint.

3. Which of the two main divisions of the human skeleton protects many of the vital organs? Explain.

Use the image at right to answer Questions 4–6.

4. Which letter identifies the ulna?

5. What bones are labeled *D*?

6. What is the name of the only labeled bone that is part of the axial skeleton?

7. What type of joint connects the femur to the pelvis?

8. What type of cells dissolve bone as part of the remodeling process?

9. What two types of bones typically contain some red bone marrow in an adult?

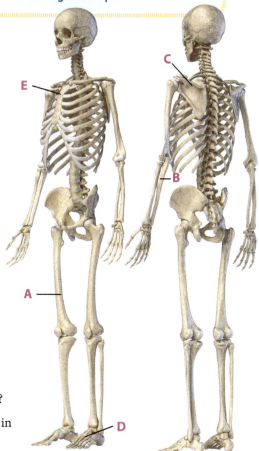

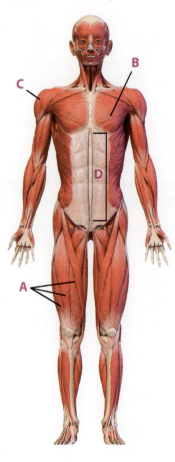

Use the diagram at left to answer Questions 10–13.

10. Which letter identifies the pectoralis major muscle?

11. Which muscle is labeled *A*?

12. Which muscle is labeled *C*?

13. Which muscle is labeled *D*?

14. How many origins does the quadriceps have?

15. What structure connects muscle to bone?

16. Which structure of the myofibril binds to ATP?

17. What determines the strength of a muscle contraction?

UNDERSTANDING CONCEPTS

18. Explain how an infant's bone structure can begin with 300 bones but decrease to 206 by adulthood.

19. Classify each of the the following movable joint descriptions as a pivot, ball-and-socket, hinge, or gliding joint.

   a. neck

   b. ankle bones

   c. hip

   d. elbow

20. In which section of a bone would you expect to find osteons? Explain.

21. Explain the role of the skeletal system in the human body.

22. The biceps brachii muscle connects the scapula and the radius. Which of these two bones is the insertion of the biceps? Explain.

23. Under what conditions do muscle cells produce lactic acid?

24. Describe two ways that the muscular system interacts with other parts of the human body.

CRITICAL THINKING

25. Following an injury that occurred during a game of basketball, thirteen-year-old Matt was diagnosed with a fracture on the growth plate in his ankle. What could be a possible long-term effect of this injury?

26. Muscle disorders can be caused by mutations within the sarcomeres. Explain how this could affect the functioning of muscles.

27. Claire Lomas can't walk because of a spinal cord injury. Injuries to the spinal cord are difficult to treat because nerve cells do not regrow as bones and muscles do. A lot of research is being done to find ways to trick the spinal cord into healing itself. While such long-term solutions are being researched, shorter-term solutions such as the ReWalk system are being developed. Both take a lot of research money to develop. What is the balance between developing both short- and long-term solutions?

Use the case study below to answer Questions 28–31.

28. The most common medication used to treat osteoporosis is bisphosphonate. Astronauts aboard the ISS have also benefited from this medicine. It works by stopping osteoclasts from dissolving bone. How does this slow bone loss?

29. A rare side effect of bisphosphonates is a break in the diaphysis of a femur. Given what you know about the role of osteoclasts, why do you think this might occur?

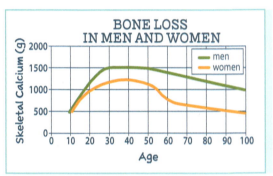

30. Osteoporosis-related fractures occur in 50% of women over the age of fifty but in only 20% of men of the same age. This disparity is the result of a drop in hormone levels that occurs in women around age fifty. Using the graph at right, determine the other reason that osteoporosis occurs more in women than in men.

31. Would female astronauts, most of whom are under age fifty, likely experience a higher rate of bone loss in space than male astronauts? Explain.

case study

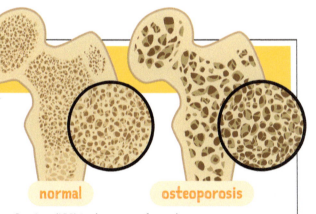

BONE DENSITY

Our bones are permanently in remodeling mode. Osteoblasts and osteocytes continuously build bone while osteoclasts continuously dissolve it. This remodeling is necessary for healthy bones, and as long as the two processes balance each other, all is well. However, many older people, especially women, experience *osteoporosis*, the result of osteoclasts dissolving bone faster than it can be built. Many people do not even know they have it until a bone breaks as a result of a minor injury. In many cases the femur breaks, usually in the epiphysis that forms a joint with the pelvis.

Astronauts experience similar loss of bone density in space because the microgravity environment doesn't put enough stress on their bones to encourage osteoblasts and osteocytes to continue building bone. The International Space Station (ISS) is thus a perfect place to test potential treatments for osteoporosis.

Exercise is a major guard against loss of bone density, but it has its limits. For those on the ground, it is most effective when a person has exercised throughout life. In space, exercise can only slow bone loss; it cannot stop it. In addition to regular exercise some osteoporosis sufferers need medication and a balanced diet high in calcium and vitamin D, a vitamin that helps the body absorb calcium.

22 TRANSPORT

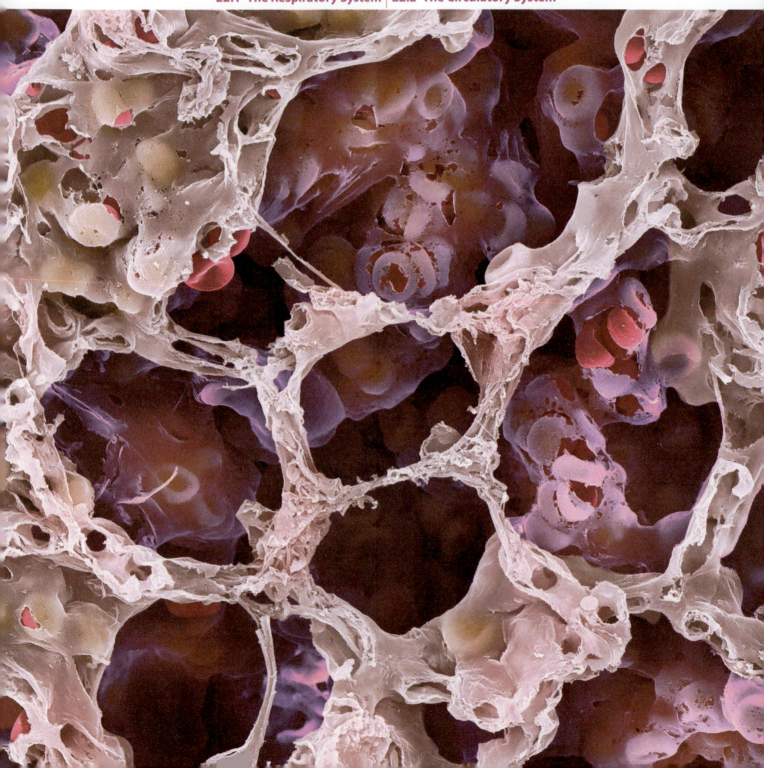

A group of tourists gathers in the crowded airport of Kathmandu, Nepal. They are from all over the world, but they have one mission: to climb Mount Everest. Just getting to one of its two base camps is a challenge—they will hike for one to two weeks just to reach that goal—let alone the summit. Why so much time, when they flew to Nepal in a day or less? It's because of Everest's elevation.

One of Mount Everest's base camps is at an elevation of 17,600 feet, higher than all but five of the mountains in North America. The human body has to gradually adjust, or *acclimatize*, to air that is thinner. It is 20% oxygen, just like the air at sea level—but there is less of it up there. Nobody can live permanently above 20,000 feet, and most people need supplemental oxygen to climb Mount Everest. If hikers ascend to base camp too quickly, they can develop possibly fatal altitude sickness that can cause the brain and lungs to swell. In fact, a person instantaneously brought from sea level to the summit of Everest could lose consciousness and die within minutes.

22.1 THE RESPIRATORY SYSTEM

The Structure of the Respiratory System

Everyone knows that people need to breathe, but do you know why? As we learned about cell processes in Chapter 7, cells need to get oxygen and the body needs to eliminate carbon dioxide in order to get energy from food through cellular respiration. Just as the well-being of people who climb Mount Everest is affected by their ability to obtain sufficient oxygen, the health of individual cells in the body is likewise affected. By understanding the human body, we are better able to explore and wisely use the world that God has placed us in.

Oxygen moves from the air into the lungs through the respiratory system. But to make it available to cells, the circulatory system transports the oxygen from the lungs via the blood to the entire body. The body eliminates carbon dioxide by the reverse process, traveling from the body to the blood, and then to the lungs so that it can be exhaled. Let's explore how the respiratory system gets oxygen from the air into the lungs.

?

Why is it so hard to breathe at high altitudes?

Questions

What are the parts of the respiratory system?

How does oxygen get from the air to the blood?

How does breathing work?

How does the respiratory system work with the rest of the body?

Terms

nasal cavity
pharynx
epiglottis
trachea
bronchus
lung
bronchiole
alveolus
capillary
lung capacity
hemoglobin

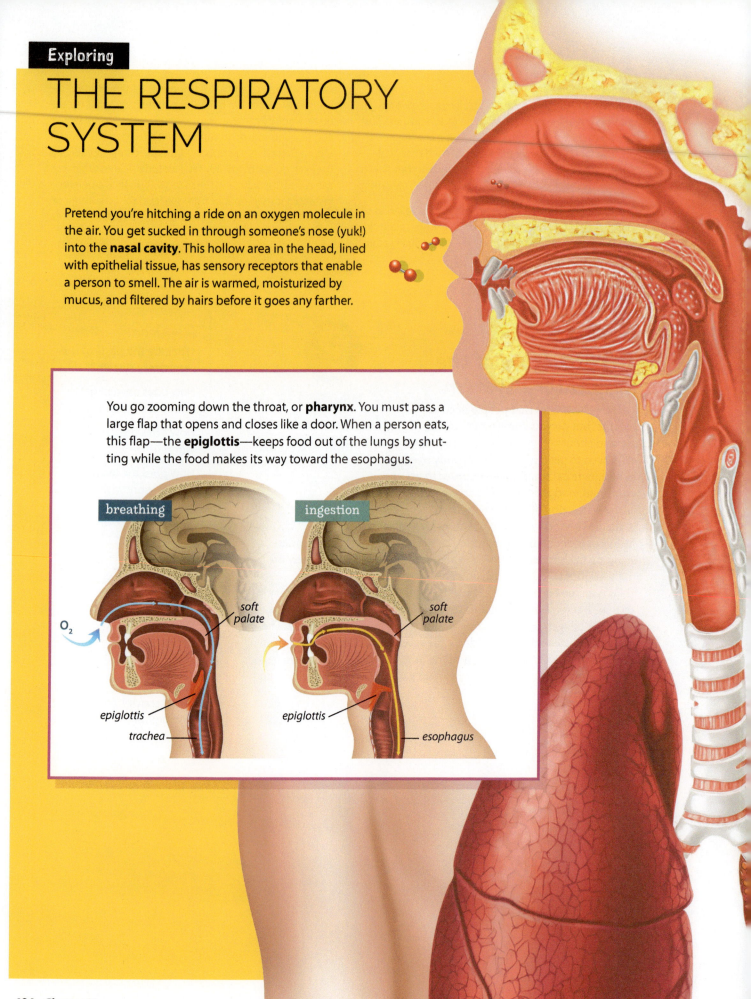

THE RESPIRATORY SYSTEM

Pretend you're hitching a ride on an oxygen molecule in the air. You get sucked in through someone's nose (yuk!) into the **nasal cavity**. This hollow area in the head, lined with epithelial tissue, has sensory receptors that enable a person to smell. The air is warmed, moisturized by mucus, and filtered by hairs before it goes any farther.

You go zooming down the throat, or **pharynx**. You must pass a large flap that opens and closes like a door. When a person eats, this flap—the **epiglottis**—keeps food out of the lungs by shutting while the food makes its way toward the esophagus.

breathing

ingestion

O₂

soft palate

soft palate

epiglottis

epiglottis

trachea

esophagus

After passing the epiglottis, you go through a pair of flaps into the *larynx*, or voice box. (When a person speaks or sings, air travels over these flaps (called *vocal folds*) in the opposite direction, causing them to vibrate and produce sound, like a bow on the strings of a violin.) Then you move into the **trachea**, or windpipe, which filters and moisturizes the air one last time with cilia and mucus before the air enters the lungs. Cartilage gives flexible support to the trachea and epiglottis in the respiratory system.

vocal folds open

vocal folds closed

larynx

trachea

With a big whoosh, like going downhill on a rollercoaster, you pass down the trachea through one of two branching tubes, called **bronchi** (s. bronchus) **1** , into the **lungs 2** . Then the bronchi branch into smaller tubes called **bronchioles 3** . You come to a dead end in a bubble-like sac called an **alveolus** (pl. alveoli) **4** . Very small blood vessels called **capillaries 5** wind around the alveoli, and you pass through the alveoli into the capillaries. Now you're surfing the bloodstream!

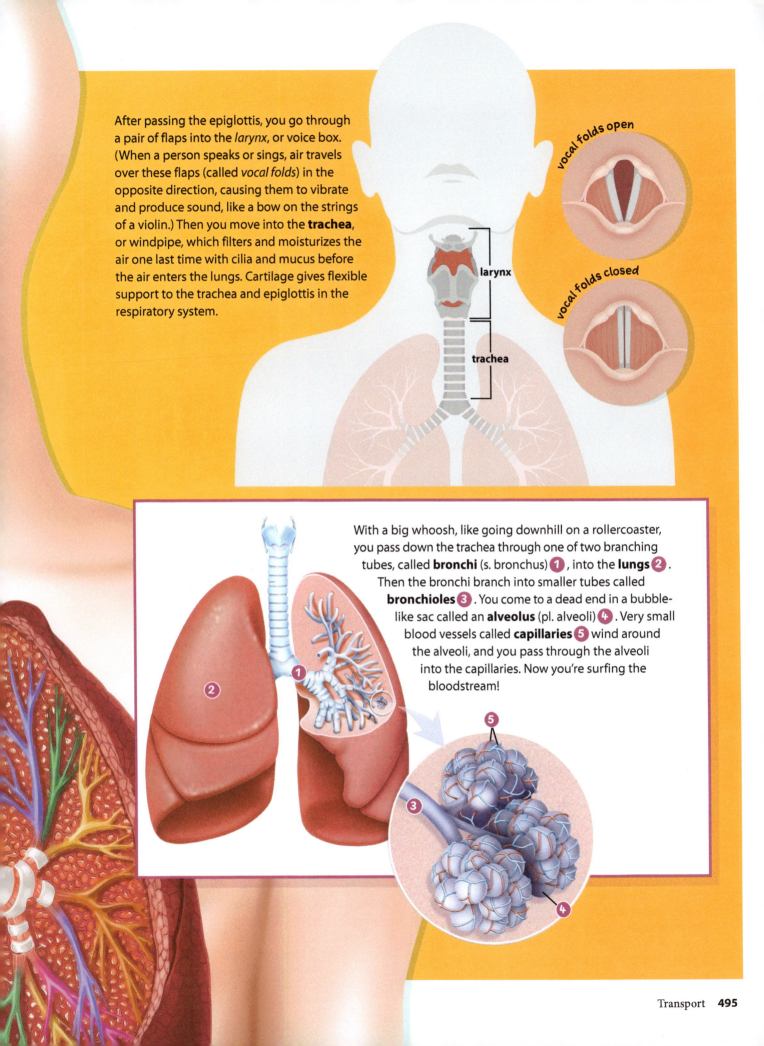

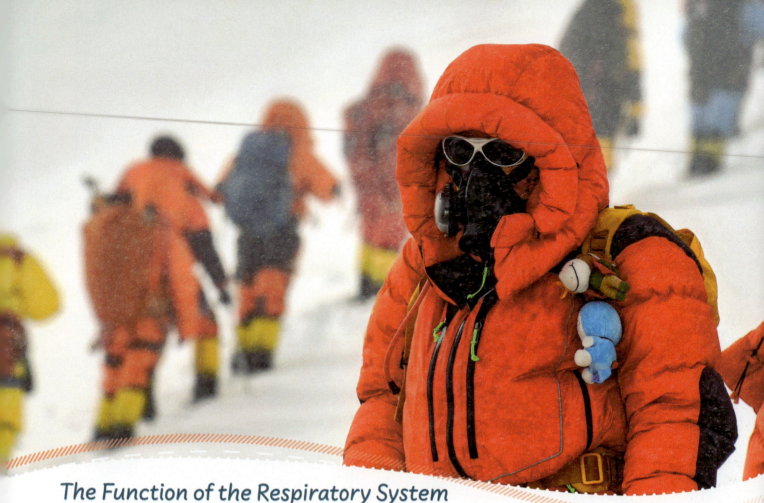

The Function of the Respiratory System

Now how do all those organs of the respiratory system work together to help us breathe? There are two phases of breathing: inhaling, or *inspiration*, and exhaling, or *expiration*.

When we breathe in (inspiration), a muscle under the lungs—the *diaphragm*—contracts, moves down, and becomes flatter. At the same time, muscles between the ribs contract to increase the volume in the chest cavity. The air pressure in the lungs becomes less than the pressure outside the body, and air rushes through the nasal cavity, pharynx, and trachea down into the lungs. It's similar to what happens when a syringe is used to siphon a liquid.

The opposite process happens when we breathe out (expiration). The diaphragm and the muscles between the ribs relax, reducing the volume in the chest cavity. This causes the air pressure in the lungs to become greater than the pressure outside the body, forcing air out of the lungs.

When breathing in and out normally, the body takes in and gives out about 0.5 L of air. This quantity is known as the *tidal volume* of the lungs. But when a healthy adult male takes in a really deep breath, like when he's getting ready to dive into the deep section of a pool, he can breathe in about 3.3 L more. This is his *inspiratory reserve volume*. If he deliberately forces that deep breath

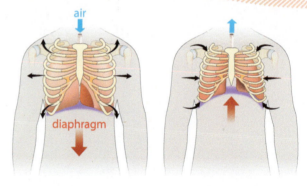

out, he can release the 3.8 L he breathed in plus an *expiratory reserve volume* of about 1.1 L of air. Of course, he can't force all the air out of his lungs. The fraction he can't force out—his *residual volume*—is approximately 1.2 L. Adding up all these numbers gives about 6.1 L, the measure of an average man's total **lung capacity**. An average adult woman's total lung capacity is about 4.2 L.

People with lung disease or lung problems like asthma, which constricts the bronchioles, have smaller lung capacities. On the other hand, people who are healthy and exercise regularly often have a greater lung capacity. A few climbers on Mount Everest with unusually large lung capacities have been able to summit Everest without supplemental oxygen.

The Respiratory System in the Human Body

OXYGEN EXCHANGE

Once air gets into the lungs, why does oxygen transfer from the alveoli into the capillaries that wrap around them? This happens because there is a higher concentration of oxygen in the air than in us! It's a matter of diffusion from a region of high concentration to a region of low concentration. To make this diffusion possible, the walls of both the alveoli and the capillaries that surround them are incredibly thin—less than 1 μm. These thin structures allow for efficient gas exchange but also make lung tissue susceptible to damage by airborne particles such as dust or soot.

But how are oxygen and carbon dioxide transported in the body? Do they travel as tiny bubbles alongside blood cells? Actually, gases in the blood are either dissolved in or chemically combined with blood substances. Almost all the oxygen in the blood (98%) is combined with **hemoglobin**, the oxygen-carrying protein found in red blood cells. Hemoglobin is what makes blood bright red—this protein has iron in it, which binds with the oxygen to form *oxyhemoglobin*. Blood that carries oxygen is called *oxygenated blood*.

To move oxygen to the rest of the body, the respiratory system relies on the circulatory system. Oxygen (O_2) diffuses through the walls of the alveoli. Red blood cells carry oxyhemoglobin through the capillaries, which provide tissues with blood. In these capillaries oxygen leaves the oxyhemoglobin to go to the body cells because there is a greater concentration of oxygen in the blood cells than in the cells of the body. The blood absorbs the carbon dioxide (CO_2) produced by cellular respiration for the same reason. Hemoglobin can carry carbon dioxide as well, even carrying it at the same time as oxygen since these two molecules have different bonding sites on hemoglobin. However, hemoglobin holds carbon dioxide with a stronger bond than it holds oxygen.

So oxygen goes through the respiratory system, to the bloodstream, and then to the heart, which pumps the blood through vessels to cells in tissues. The body uses the oxygen in tissues to release energy from glucose. This shows the three stages of respiration—*external respiration* (inhaling air), *internal respiration* (supplying cells with oxygen), and *cellular respiration* (using oxygen to produce energy from glucose). Many body systems team up with the respiratory system to make this happen, especially the muscular and circulatory systems.

oxygen transport cycle

Oxygen from alveoli binds to hemoglobin.

Oxygen is released to tissue cells.

ARTERY

Red blood cells carry oxygen from lungs to cells.

CONTROL OF BREATHING

But what controls this whole process? You think about breathing when you speak, sing, or swim. But what about when you're sleeping? The nervous system takes over in response to different chemicals produced by the body and environmental stimuli to control breathing. The nerve cells in the part of the brain that controls breathing make up what is called the *respiratory center*. It monitors the level of carbon dioxide in the blood by keeping an eye on the blood's pH, which is affected by both dissolved oxygen and dissolved carbon dioxide. When too much carbon dioxide builds up in the blood, the muscles involved in breathing are stimulated to increase the respiration rate to raise oxygen and lower carbon dioxide in the blood. This can happen when you exercise or when you climb to Mount Everest Base Camp! It's another example of a feedback mechanism for maintaining homeostasis in the body. The lungs can respond to these signals by expanding alveoli to increase oxygen in the blood and by involving more alveoli to exchange gases with the blood.

When climbers traveling to Mount Everest Base Camp acclimatize, their bodies get used to a different atmospheric pressure. This involves helping their brains to adjust how deeply they breathe and to determine how many alveoli are involved in gas exchange. These changes affect how much oxyhemoglobin is in the blood and how quickly blood circulates. More red blood cells carry oxygen and pass more rapidly through the blood vessels. Almost makes you want to climb a mountain!

air

diaphragm

inhalation

exhalation

22.1 SECTION REVIEW

1. Which kind of tissue—nervous, connective, epithelial, or muscular—lines the nasal cavity, pharynx, larynx, and bronchioles? Suggest the primary purpose of this tissue.

2. Give five examples of organs in the respiratory system and indicate the function of each.

3. What happens when you breathe in and out?

4. List two factors that can affect a person's lung capacity.

5. Through what process does oxygen leave the lungs and enter the bloodstream?

6. Suggest why the quality of air that a person breathes, which is eroded by pollution and smoking, affects his or her overall health.

7. The level of which dissolved gas in the blood triggers a new breath cycle?

8. Study the diagram on page 494. Suggest why people cough and sneeze and describe how this works.

9. British mountaineer George Mallory, who died during a 1924 attempt to climb Mount Everest, was once asked why he wanted to climb Everest. Mallory responded, "Because it's there." Is exploration like this justified from a biblical worldview? Consider both the risks and benefits.

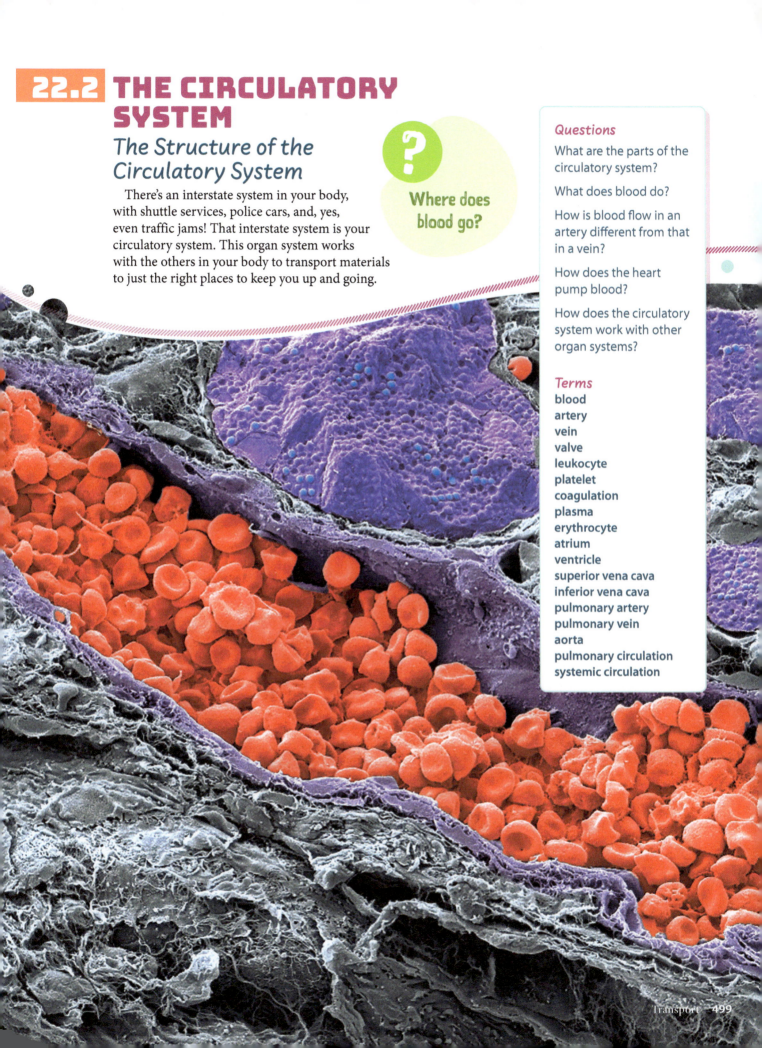

22.2 THE CIRCULATORY SYSTEM

The Structure of the Circulatory System

There's an interstate system in your body, with shuttle services, police cars, and, yes, even traffic jams! That interstate system is your circulatory system. This organ system works with the others in your body to transport materials to just the right places to keep you up and going.

?

Where does blood go?

Questions

What are the parts of the circulatory system?

What does blood do?

How is blood flow in an artery different from that in a vein?

How does the heart pump blood?

How does the circulatory system work with other organ systems?

Terms
blood
artery
vein
valve
leukocyte
platelet
coagulation
plasma
erythrocyte
atrium
ventricle
superior vena cava
inferior vena cava
pulmonary artery
pulmonary vein
aorta
pulmonary circulation
systemic circulation

THE CIRCULATORY SYSTEM

When you think of the circulatory system, you probably think of blood, the current of travelers on the body's internal interstate system. **Blood** is the marvelous red river that transports oxygen, nutrients, and hormones to all body cells and helps in removing waste products from the body. A healthy adult has almost 5 L of blood. You can lose about a quarter of your blood without any serious problems.

The heart is the muscle that pumps blood through the circulatory system. All blood traffic goes through the heart. It is a hollow organ about the size of your fist and has four chambers made of muscular tissue. And what a muscle! The average person's heart contracts 70–80 times per minute, 100,000 times per day, 37 million times per year. Every heartbeat pushes about 80 mL of blood from the heart, adding up to about 8000 L per day.

The "roads" in your circulatory system are blood vessels, and there are three kinds—arteries, veins, and capillaries. **Arteries** carry blood away from the heart, **veins** carry blood toward the heart, and capillaries connect the two. Veins have flaps called **valves** that act like double doors that open only one way to regulate flow to the heart. Valves in the veins open when muscles around them contract. Nearly every cell in the body is located next to a capillary so that cells can exchange nutrients, oxygen, and waste products with blood.

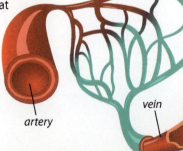

capillaries

artery

vein

The Function of the Circulatory System

Now that you know the basic organs of the circulatory system, let's look at each of these parts in depth and see what they do to move substances around in your body. Let's start by taking a close look at blood.

BLOOD

Leukocytes, or white blood cells, are the police officers of the internal interstate system. Lymphocytes like T cells and B cells are types of leukocytes, and they are about twice as big as erythrocytes. They migrate from the lymphatic system to rid the body of antigens through phagocytosis. In a healthy person there is about one leukocyte for every six hundred erythrocytes in blood. When a person has an infection, a blood sample will show a greater ratio of leukocytes to erythrocytes.

LEUKOCYTES AND THROMBOCYTES
1%

Platelets, or *thrombocytes*, are the smallest components of blood. Platelets are cell fragments that lack a nucleus and are less than half the size of an erythrocyte. When you get a cut or bruise, broken blood vessels must be sealed to prevent blood loss. Platelets clump together in a complex process with several feedback mechanisms to form a *clot* (magnified at right) with a microscopic mesh that entangles red blood cells. This is called **coagulation** and may take just minutes! If the injury is minor, platelets release a chemical that makes the smooth-muscle walls of the blood vessels clamp down to stop blood loss.

Plasma is a yellow fluid and is the main constituent of blood. It is 90% water and is almost identical to lymph. The remaining 10% is made of proteins, dissolved gases, minerals, vitamins, nutrients, hormones, and waste substances. The components of plasma can change after you've finished a big meal or drunk a lot of water.

PLASMA
55%

Though plasma is yellow, blood, of course, is red! The interaction between the iron in hemoglobin and oxygen that binds to it makes blood red. Red blood cells, or **erythrocytes**, aren't really cells at all; they are disk-shaped structures that hold hemoglobin. This shape gives them a greater surface area to do their job—carrying oxygen. Erythrocytes are the most common component in blood besides plasma. If the circulatory system is like an internal interstate system, erythrocytes are the driver service that delivers hemoglobin and oxygen.

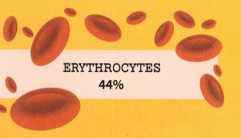

ERYTHROCYTES
44%

The Heart

Now let's consider the muscle that pumps this blood. The *heart* has several parts that work together, using two different loops, to keep blood moving to both the lungs and the rest of the body.

The heart has four chambers, or rooms. The top two small chambers are called *atria* (s. **atrium**). The bottom two larger chambers are called **ventricles**.

The *septum* is the muscular wall that separates the heart into a right and left side, so designated according to the perspective of a person. The left side of the heart is on a person's left side, that is, the right side to an observer.

The heart is held within a strong, double-walled sac called the *pericardium* (not shown). The heart itself is made of muscular tissue. The thickest part of the heart wall is called the *myocardium*. It contracts to pump blood.

right side *left side*

right atrium

valve

left atrium

septum

left ventricle

myocardium

right ventricle

Valves regulate blood flow through the heart. There are valves between the atria and ventricles and at the major arteries that connect to the heart. The "lub-dub" sounds of the heart are caused by closing valves. Heart valves don't move by themselves—they move in response to the flow of blood.

Blood enters the heart in several places.

Deoxygenated (oxygen-depleted) blood from the upper body above the diaphragm enters the right atrium of the heart through the **superior vena cava** ❶.

Deoxygenated blood from the lower body enters the right atrium of the heart through the **inferior vena cava** ❷.

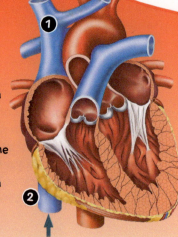

IN A HEARTBEAT

What happens during a heartbeat? There are two parts to a heartbeat, or a *cardiac cycle*. The heart muscle goes through one complete contraction and one complete relaxation cycle. Both sides of the heart are working at the same time to relax and contract—and the heart does this about seventy to eighty times per minute!

1 A nerve pulse at the right atrium activates the heart into contracting. This nerve pulse, a natural pacemaker, sets the pace for how fast the heart beats.

2 The left and right atria contract first, sending oxygenated blood into the left ventricle and deoxygenated blood into the right ventricle, respectively. About 0.1 s later, the impulse reaches the ventricles after they have filled with blood.

3 Oxygenated blood is squeezed out of the left ventricle into the aorta to supply the body with oxygenated blood. Deoxygenated blood is squeezed out of the right ventricle into the pulmonary arteries to take on oxygen from the lungs.

4 The heart relaxes, and deoxygenated blood flows from the body through the vena cava into the right atrium. Oxygenated blood flows from the pulmonary veins into the left atrium.

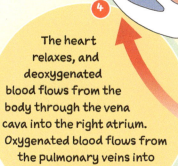

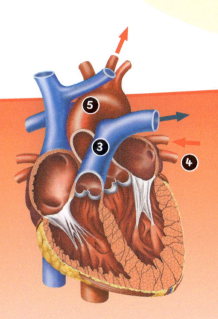

Blood leaves the heart through the **pulmonary arteries** ❸. These arteries deliver deoxygenated blood to the lungs to exchange carbon dioxide for oxygen.

Pulmonary veins ❹ deliver oxygenated blood to the heart, entering at the left atrium.

Oxygenated blood leaves the left ventricle through the **aorta** ❺, one of the strongest arteries in the body. It arches around and forks to provide the upper and lower parts of the body with blood.

MINI LAB

Heart Rate

We saw earlier that the human heart rate is about 70–80 beats per minute (bpm), but what do you suppose the *average* rate is? Is it 75 bpm? Let's investigate!

How fast does the average human heart beat?

Materials
small piece of clay • toothpick • stopwatch

PROCEDURE

A Form the piece of clay into a flat disk about 1 cm in diameter and insert the toothpick into its center.

B Lay one arm flat on your desk with your wrist facing up. Find your pulse, then set the clay atop that spot on your wrist. You should be able to see the toothpick twitch slightly with each heartbeat.

C Have your partner use the stopwatch to measure off one minute as you count the number of beats. Record your result on a piece of paper, then repeat the procedure to measure your partner's heart rate.

D Record the results from the rest of your classmates to form a larger pool of data.

ANALYSIS

E Use your class dataset to create a bar graph (histogram) for heart rate in beats per minute. Group the rates into 5–10 bpm increments.

QUESTIONS

1. According to your graph, what is the average heart rate for your class?

2. How does your answer in Question 1 compare to the value given on page 500 (70–80 bpm)?

3. Would it be proper to conclude that the average heart rate for your class is the average for all people? Explain.

The Circulatory System in the Human Body

We've seen what happens when the heart beats. But where does the blood go? There are a whole lot of organs in the body that need nutrients and oxygen from the blood!

The heart is the hub of four loops. A separate loop goes to each of the lungs and back. The flow of blood between the lungs and heart through these two loops is called **pulmonary circulation.** Deoxygenated blood is carried to the lungs for gas exchange and then returned. This involves about 0.5 L of blood at any given time and requires only a few seconds to cover the loop. The flow of oxygenated blood from the left ventricle to the extremities through the remaining two loops is called **systemic circulation**. One loop delivers blood to the head and upper extremities, the other to the abdomen and lower extremities. The systemic system involves much more blood and takes longer since it must cover more distance.

All the organs in your body need oxygenated blood, even your heart and lungs! Though oxygen is the most crucial substance that your body needs, it must also obtain nutrients and eliminate waste products. Carbon dioxide, a waste product of cellular respiration, is transported in the blood to the lungs so that it can be exhaled. Capillaries near the intestines pick up nutrients absorbed from food to distribute to the body via the bloodstream. The kidneys and liver filter the blood to remove waste products, toxins, and old red blood cells to be excreted in urine. Of course, the heart also relies on the brain and nervous system to control and monitor the cardiac cycle. And the plasma in blood regularly interacts with the lymphatic system.

It's easy to see why it's important to take good care of the bodies that God has given us, eating well and exercising to maintain a healthy heart and lungs. Our bodies don't belong to us; they belong to God. As we develop an understanding of the complexities of the human body, we can better help people to maintain good health so that they can have the strength to explore and wisely use God's world.

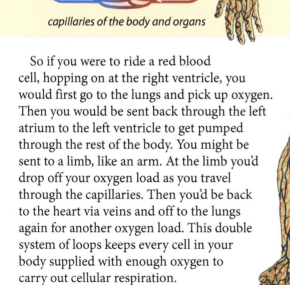

lung capillaries

pulmonary arteries

pulmonary veins

pulmonary

aorta

vena cava

systemic

capillaries of the body and organs

So if you were to ride a red blood cell, hopping on at the right ventricle, you would first go to the lungs and pick up oxygen. Then you would be sent back through the left atrium to the left ventricle to get pumped through the rest of the body. You might be sent to a limb, like an arm. At the limb you'd drop off your oxygen load as you travel through the capillaries. Then you'd be back to the heart via veins and off to the lungs again for another oxygen load. This double system of loops keeps every cell in your body supplied with enough oxygen to carry out cellular respiration.

VAPING

Your family's car rolls to a stop at a red light. The window of the car next to you is rolled down, and out of it issues a large cloud of what looks like smoke, but it's not. The car's driver is using an *electronic cigarette*, also called an *e-cigarette*. Instead of burning tobacco to release the addictive drug nicotine, e-cigarettes use electricity to heat a liquid mixture. The liquid consists of *carriers*, usually a mix of propylene glycol and glycerin, containing nicotine and flavorings. The heated liquid produces a vapor that the user inhales. The practice is widely referred to as *vaping*.

The first non-tobacco cigarette was patented in 1965, but the modern e-cigarette wasn't introduced until 2003. Since then, the popularity of vaping has steadily increased. Ironically, vaping has often been used to help wean people *away* from smoking, but some research suggests that vaping may also serve as a gateway drug that leads to smoking tobacco. Because vaping has been around for only two decades, its long-term health effects are not known, but the practice isn't without risks. Hard questions are being asked about the variety of liquids used as carriers and the presence of toxins in the vapor, including heavy metals. And, of course, the dangerous effects on health stemming from nicotine addiction are well-known. Not all vaping liquids contain nicotine, but a recent CDC study found that 99% of them do.

22.2 SECTION REVIEW

1. What are the main parts of the circulatory system, and what is the function of each?

2. Describe the basic structure of the heart in your own words.

3. How are veins different from arteries?

4. How do both blood vessels and the heart use muscular tissue?

5. Identify the three particles found in blood plasma and give the main purpose of each.

6. List the blood vessels that enter the heart.

7. List the blood vessels that leave the heart.

8. How does the circulatory system work with the respiratory system?

9. Order the following events in the circulation of carbon dioxide and oxygen in the circulatory system by letter, beginning with the letter *a* statement.

   a. Oxygenated blood leaves the left ventricle.

   b. Deoxygenated blood leaves the right ventricle.

   c. Deoxygenated blood drops off carbon dioxide and picks up oxygen in the capillaries near the alveoli.

   d. Oxygenated blood travels through the aorta.

   e. Deoxygenated blood travels through the inferior vena cava.

   f. Oxygenated blood travels to the intestines and kidneys to provide them with oxygen, pick up nutrients, and filter out waste products.

10. Which of the steps in Question 9 are involved in pulmonary circulation?

11. Which of the steps in Question 9 are involved in systemic circulation?

Use the case study above to answer Questions 12–13.

12. What is the link between vaping and the circulatory system?

13. How should a Christian view vaping?

22 CHAPTER REVIEW
Chapter Summary

22.1 THE RESPIRATORY SYSTEM

- The respiratory system helps maintain homeostasis in the body by extracting from air the oxygen needed for cellular respiration and ridding the body of carbon dioxide.

- In the lungs, oxygen and carbon dioxide are exchanged with the atmosphere through diffusion.

- Breathing is controlled by the brain's respiratory centers in response to the concentration of dissolved carbon dioxide in the blood.

- A breath cycle begins with inspiration as the diaphragm contracts, expanding the lungs and drawing air into the lungs. In the lungs oxygen diffuses from the air into the blood, and carbon dioxide diffuses from the blood into the air.

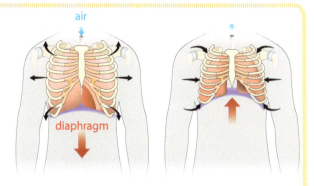

- Expiration occurs as the diaphragm relaxes, reducing the volume of the lungs and forcing air out.

Terms
nasal cavity • pharynx • epiglottis • trachea • bronchus • lung • bronchiole • alveolus • capillary • lung capacity • hemoglobin

22.2 THE CIRCULATORY SYSTEM

- The circulatory system transports dissolved gases, nutrients, wastes, and other substances throughout the body.

- The heart pumps blood throughout the body through a network of blood vessels. The four chambers of the heart are arranged so that oxygenated and deoxygenated blood do not mix.

- A heartbeat consists of two phases, a relaxation and a contraction. Blood enters the atria during the relaxation phase. During a contraction, blood first moves from the atria to the ventricles and is then forcefully expelled from the ventricles to either the lungs or extremities.

- Blood is carried away from the heart through arteries, and blood is returned to the heart through veins. Capillaries connect arteries and veins.

- Blood is a mixture of erythrocytes, leukocytes, platelets, and plasma.

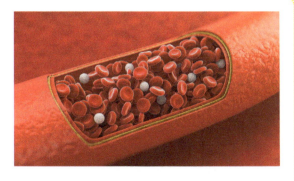

- During pulmonary circulation, blood travels between the heart and lungs to drop off carbon dioxide and pick up oxygen. Systemic circulation moves blood to and from the body's extremities.

Terms
blood • artery • vein • valve • leukocyte • platelet • coagulation • plasma • erythrocyte • atrium • ventricle • superior vena cava • inferior vena cava • pulmonary artery • pulmonary vein • aorta • pulmonary circulation • systemic circulation

RECALLING FACTS

1. What are the three major organs of the respiratory system located in the head and throat? Describe the function of each.

2. What causes air to move into your lungs?

3. How does lung disease affect lung capacity?

4. Where do the gases move between the respiratory and circulatory systems?

5. How does the body sense when it is time to breathe?

6. What structure in the heart prevents the mixing of oxygenated and deoxygenated blood?

7. Identify the chambers of the heart. Indicate which pump oxygenated blood and which pump deoxygenated blood.

8. What are the only veins that carry oxygenated blood? Explain.

9. What are the two major divisions of the circulatory system?

UNDERSTANDING CONCEPTS

10. Order the following structures according to the sequence in which air passes through each: alveoli, bronchi, bronchioles, larynx, nasal cavity, pharynx, trachea.

11. Locate the following structures by associating them with the letters on the bronchial tree: alveolus, bronchiole, bronchus, larynx, trachea.

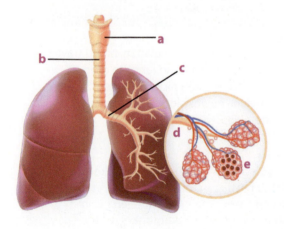

12. Is it correct to say that gas exchange happens only in the lungs? Explain.

13. A friend says that oxygen and carbon dioxide travel in the blood as tiny bubbles. Is he correct? Explain.

14. How does oxygen move from the air in the lungs into the bloodstream?

15. Is breathing voluntary or involuntary? Explain.

16. Compare arteries, veins, and capillaries.

17. How are platelets different from blood cells?

18. What is the purpose of plasma in the blood?

19. While chopping vegetables, you slice your finger! Describe the component(s) in blood that help(s) you recover.

20. Locate the following structures by associating them with the letters on the diagram below: aorta, left atrium, right atrium, heart muscle, pulmonary artery, pulmonary vein, septum, inferior vena cava, superior vena cava, left ventricle, right ventricle.

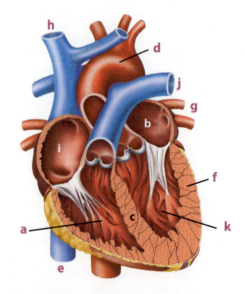

21. How does the body eliminate the carbon dioxide that is produced as a byproduct of cellular respiration?

22. How do pulmonary and systemic circulations differ?

23. Why is it important to know about and take care of our bodies?

CRITICAL THINKING

24. Use a table to compare inspiration and expiration. Include in your comparison the movement of air, the relative size of the chest cavity, the pressure in the lungs, and the state of the diaphragm.

25. What are some of the advantages that a multi-loop circulatory system has over a single loop?

26. Why doesn't drinking carbonated beverages elevate the levels of carbon dioxide in your blood?

27. Most people have three lobes in their right lung and only two in their left. Hypothesize as to why God may have designed our lungs in this manner.

28. Do some research to determine what makes breathing difficult for people with asthma.

29. Blood clots are helpful to prevent blood loss during an injury. Do some research to determine a few major problems that can develop when blood clots develop in the wrong places.

case study

THE EKG

Doctors investigate heart health with a special test called an *EKG*, or *electrocardiogram*. An electrocardiograph records the electrical activity of the heart, measuring its electrical pulses and muscle contractions. Each cycle on an EKG reading represents one complete cardiac cycle. A normal heartbeat (top EKG) includes three waves. The P wave represents the atria contracting, the QRS complex shows the ventricles contracting, and the T wave shows how the heart electrically recalibrates to prepare for the next contraction. This repeating cycle is called the *normal sinus rhythm*. Now you can be a student cardiologist!

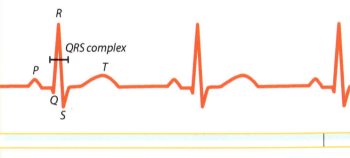

Use the case study above to answer Questions 30–34.

30. If the vertical axis represents the size of the heart muscle contraction, what does the horizontal axis represent?

31. What do the flatter parts of an EKG represent?

32. How can we determine the heart rate from an EKG?

33. Compare the second EKG with the normal EKG. What do you notice?

34. Do some research on the abnormal pattern shown in the second EKG and suggest what a cardiologist might do for a patient with this EKG reading.

23 ENERGY

Energy Crisis

In the United States there's an energy crisis. It's not a lack of fossil fuels or a nuclear meltdown—we actually have too much stored energy! Americans are eating too much of the wrong things, and it's killing them. Many Americans are overweight and as a result are developing diseases at a rate that has steadily risen over the past several decades. Most people gain weight because they eat more calories than they burn.

When many try to lose weight, they find it physically and emotionally difficult, and they often quit in discouragement. Yet caring for our bodies by learning to manage our caloric energy is an important part of glorifying God in all we do. Our bodies are the first area over which we have stewardship. How can we motivate and show love to people struggling to balance their food intake and activity level?

23.1 THE DIGESTIVE SYSTEM

The Structure of the Digestive System

Eating is both something we enjoy and something we need. Food tastes good! And our bodies are very efficient at extracting nutrients from food as it moves along the **alimentary canal**, which is sometimes called the *gastrointestinal (GI) tract*. Many **accessory organs** also contribute to the digestive process by releasing digestive chemicals into the alimentary canal. Let's take a look at the digestive system, including the alimentary canal and accessory organs.

How do I get nutrition from my food?

Questions

How does the body use food and water?

What parts of the body participate in digestion?

How does digestion produce energy?

Terms

alimentary canal
accessory organ
oral cavity
esophagus
bolus
peristalsis
stomach
chyme
small intestine
large intestine
rectum
salivary gland
appendix
liver
bile
pancreas
gallbladder
nutrient
amylase
pepsin
lipase

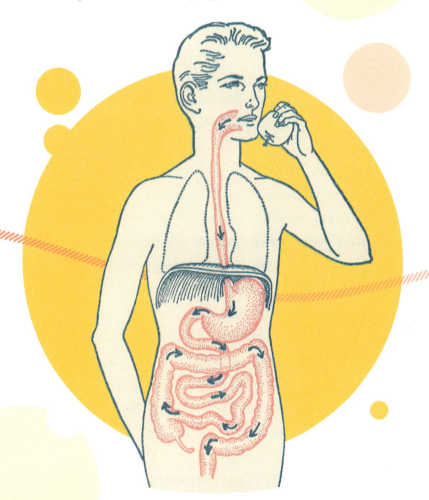

THE DIGESTIVE SYSTEM

Alimentary Canal

Food enters the alimentary canal through the mouth, or **oral cavity**. God has given us large *incisors* for cutting food and pointed *cuspids* for tearing it before the *premolar* and *molar* teeth grind it. This variety of teeth allows us to eat both meat and vegetables.

The tongue pushes food through the pharynx into the **esophagus**, a hollow and muscular tube, as a ball-like mass called a **bolus**. As food travels down the esophagus, it is propelled by **peristalsis**, muscular contractions that continue throughout the alimentary canal.

When a bolus of food reaches the end of the esophagus, it passes through the lower esophageal sphincter into the **stomach**. There the food is digested with a combination of mechanical grinding and chemical action of enzymes and hydrochloric acid. The digested food leaves the stomach through the *pyloric sphincter* as a thick, semiliquid paste called **chyme**.

Most of the digestive enzymes in the body are found in the **small intestine**. Tiny, finger-like *villi* (s. villus) lining the inside of the small intestine use diffusion and active transport to absorb most of the nutrients that the body needs.

By the time the chyme exits the small intestine, the body has already extracted most of the nutrients from it. The **large intestine** absorbs water and some remaining nutrients, including important vitamins. The undigested material passes through the **rectum**, the last section of the large intestine, and is expelled through the *anus* as feces.

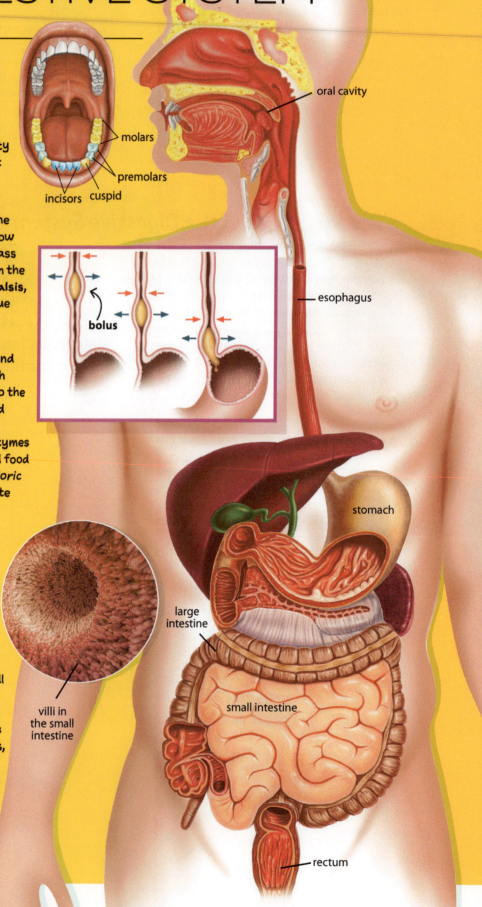

molars

premolars

incisors   cuspid

bolus

oral cavity

esophagus

stomach

large intestine

small intestine

villi in the small intestine

rectum

Accessory Organs

Saliva from the **salivary glands** mixes with food during chewing. It lubricates the food for swallowing and also contains some digestive enzymes.

The **appendix** attaches to the large intestine just below the place where it joins the small intestine. Its exact purpose is still a matter of debate. It is composed of lymphatic tissue, so it may be involved in helping lymphocytes recognize pathogens. Or the appendix could also be a reservoir for beneficial bacteria. It's also possible that the appendix serves a different, yet unknown function.

The **liver** is the largest internal organ in the body and has a multitude of functions. As we learned in Chapter 22, the liver filters toxins and wastes from the blood. It also produces **bile**, a greenish fluid that helps digest fats in the small intestine.

After the liver makes bile, the **gallbladder** stores it until it is needed in the small intestine. Bile contains a large amount of water, some of which the gallbladder removes, concentrating the bile. However, sometimes the solutes in bile precipitate, forming gallstones, which occasionally become so problematic that the gallbladder must be removed. A person without a gallbladder may find it more difficult to digest fat and may have to make dietary adjustments.

The **pancreas** is the primary producer of digestive enzymes. These different enzymes are carried along by *pancreatic juice* into the small intestine where they break down carbohydrates, fats, proteins, and nucleic acids. The pancreas also helps neutralize the pH of the extremely acidic chyme entering the small intestine from the stomach.

gallstones in a gallbladder

The Function of the Digestive System

As the food we eat passes through the GI tract, the digestive system extracts **nutrients**, the substances that our bodies need to stay healthy. We can divide these into six categories.

NUTRIENTS in FOOD

WATER

Water is essential to most of the chemical reactions in the body because they occur in an aqueous environment. Water also dissolves some foods for easier digestion, and it balances fluids, blood pH, and body temperature.

MINERALS

We learned in Chapter 21 that bones store calcium for added strength and for other uses in the body. Calcium is just one of the minerals that our bodies need. These inorganic substances are found naturally in soil and are absorbed by plants. Minerals are important for nerve and muscle function, blood coagulation, blood pH maintenance, enzyme function, and many other uses.

FATS

Fats are used to form cell membranes, protective membranes around nerve cells, and certain hormones. They also serve to cushion delicate structures and provide energy. The body stores excess energy as fats in *adipose tissue*, so a low-fat diet with too many calories can still cause a person to build up adipose tissue.

skin

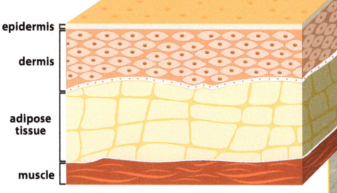

epidermis

dermis

adipose tissue

muscle

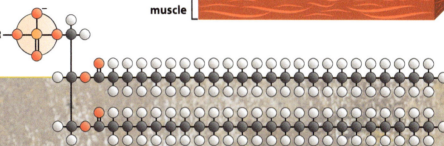

PROTEINS

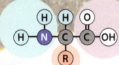

Proteins make up about 50% of the body's dry weight; they are found in all living cells and in nonliving structures such as hair, nails, and the matrix of connective tissues. Most enzymes and some hormones are also proteins, and while proteins are most important for growth and repair, they may also be used for energy production. Proteins are formed from chains of complex molecular structures called *amino acids*. Our bodies can make some of these amino acids, but others must come from our diet.

VITAMINS

Our bodies can't make most of these organic molecules, which they need to aid metabolism, so we must get them from our diet. Some vitamins serve as coenzymes that enable enzymes to perform their roles in the body while others regulate metabolic processes. Although they are necessary for health, our bodies need vitamins in much smaller amounts than most of the other nutrients.

CARBOHYDRATES

Carbohydrates include sugars, such as fructose and sucrose, and more complex molecules, such as starch and glycogen. Although the body can use several carbohydrates as energy sources, all are broken down to glucose before being used in cellular respiration.

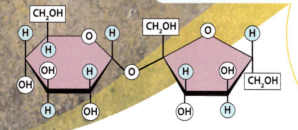

CHEMICAL DIGESTION

Mouth

The salivary glands release saliva.

Amylase (enzyme) breaks down maltose (disaccharide) and starch.

Stomach

The stomach lining releases gastric juices.

Pepsin (enzyme) breaks down protein fragments.

Small Intestine

The pancreas releases pancreatic juices.

Amylase (enzyme) breaks down maltose, sucrose, and lactose (disaccharides).

Lipase (enzyme) breaks down fatty acids and glycerol.

Trypsin and chymotrypsin (enzymes) break down proteins.

The intestinal lining releases intestinal juices.

Maltase, sucrase, and lactase (enzymes) break down monosaccharides.

Lipase (enzyme) breaks down fatty acids and glycerol.

Peptidase (enzyme) breaks down proteins.

The liver releases bile.

Bile breaks down emulsified fats.

Now that we've covered the digestive organs and the different classes of nutrients, let's discuss how the digestive organs function to extract nutrients. The digestive system uses a combination of mechanical and chemical digestion. *Mechanical digestion* consists of physically breaking down the food. *Chemical digestion* involves the breakdown of food by acids and enzymes.

Many organs use both mechanical and chemical digestion at the same time. The teeth grind food while **amylase** in saliva begins to break down starch, a large carbohydrate polymer. The stomach mashes and grinds the food while the enzyme **pepsin** begins to split proteins into smaller units of amino acids. The small and large intestines continue to mix chyme as it passes through, while bile begins to separate molecules of fat to make it easier to digest. Various enzymes all work to break food molecules into usable nutrient molecules.

Most digestive enzymes come from the pancreas and small intestine. The pancreas releases additional amylase to further digest carbohydrates and **lipase** to digest fats. The pancreas and small intestine release several enzymes to finish the digestion of proteins begun in the stomach by pepsin. In Chapter 2 we learned that enzymes have active sites that their substrates fit into. If the pH of the environment is not correct, the shape of the enzyme changes, so the substrate will no longer fit in the active site. Pepsin can function only in the extremely acidic environment of the stomach, while lipase cannot work there at all. This is one of the reasons why the pH of the chyme must be changed in the small intestine. Most of the digestive enzymes would not work on the chyme that comes from the stomach until the acid has been neutralized.

Once the small intestine has absorbed these nutrients, they can be used by the body. Carbohydrates serve many different purposes, but their primary purpose is to provide the energy necessary to make ATP during cellular respiration. All carbohydrates can enter glycolysis, though some enter at different points in the process. If the body runs short on glucose, the cells can use fats and amino acids in cellular respiration. Fatty acids, glucose, and amino acids are converted to *acetyl CoA*, which then enters the citric acid cycle. Other amino acids are fed into the citric acid cycle later in the process.

Under normal circumstances, fats and amino acids are used throughout the body. The liver processes fats and creates the various molecules needed by the body. Amino acids serve many different purposes, but they carry a danger with them—nitrogen. When amino acids are used for energy or converted to fats for storage, they release ammonia (NH_3). We will learn how this issue is remedied in the section on the urinary system.

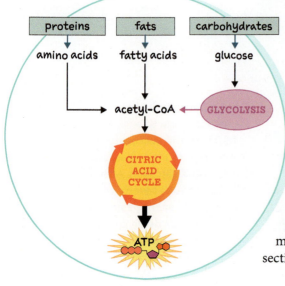

Modeling Digestion

If you walk down a medicine aisle in a drug store, you'll notice that there are many types of pills in many different shapes and sizes. Some of those pills appear to have special coatings, kind of like some chocolates have a thin candy shell. But what does that coating do? Does it affect digestion?

Will the coating on pills impact digestion?

Materials
uncoated aspirin (1 pill) • enteric-coated aspirin (1 pill) • clear plastic cups (2) • vinegar, 8 oz • water, 4 oz • baking soda, 6 g • pinch of salt

A Fill one clear plastic cup with vinegar and the other cup with baking soda, a pinch of salt, and a half cup of water.

B Drop one uncoated aspirin and one enteric-coated aspirin into the cup with vinegar. Record your observations.

C After three minutes have passed, transfer any remaining aspirin from the vinegar to the cup with baking soda. Record your observations.

1. Since the enzymes and hydrochloric acid in the stomach are responsible for the further break-down of food, which of the two cups, vinegar or baking soda with salt and water, would resemble the digestion that occurs in the stomach? in the small intestine?

2. Which pill dissolved the quickest in the cup filled with vinegar? Explain why you think this happened.

3. Describe what you observed after you placed the enteric-coated aspirin in the baking soda, salt, and water solution. Explain why you think this happened.

CONCLUSION

4. How does this activity demonstrate digestion in the human body?

GOING FURTHER

5. Research the differences between uncoated and enteric-coated aspirin. State any benefits you find.

The Digestive System in the Human Body

Like the other systems of the body, the digestive system works in conjunction with other systems. The digestive system wouldn't function at all without the muscles that form sphincters and line the alimentary canal and provide peristalsis. The digestive system is comprised of smooth muscle, which we learned in Chapter 21 is controlled involuntarily.

As part of the body's ability to maintain homeostasis, the small intestine is equipped with bacteria that help digest and absorb food and protect the organ against harmful bacteria. Once the small intestine absorbs the nutrients from our food, distribution of those nutrients relies on other systems. Carbohydrates, amino acids, minerals, vitamins, and water are transported by cells in the villi of the small intestine to the bloodstream. The circulatory system then carries these nutrients throughout the body. The small intestine treats fats very differently than it does other nutrients. Fat molecules are absorbed by the villi just like other nutrients, but they are transported to lymphatic vessels called *lacteals*. From there, the fat molecules are carried through the lymphatic system until it empties into the subclavian vein. The circulatory system then distributes the fat molecules to the rest of the body.

The digestive system also interacts with the integumentary system. The intestines cannot absorb calcium without sufficient vitamin D. While some vitamin D must come from food, some of it also comes from the skin. Ultraviolet light causes the epidermis to form vitamin D. The newly formed vitamin is transported by the circulatory system, and some reaches the intestine where it increases the absorption of calcium. But before you rush out to catch some rays and maximize your calcium absorption, remember that the same ultraviolet rays also cause skin cancer. Moderation is best when it comes to being in the sun.

Finally, the digestive system takes in nutrients to build our cells and tissues and to provide the energy we need, but when we take in more energy than our bodies can use, the excess is stored as fat. Many people in the western world, especially in the United States, struggle with consuming more energy than they use. This problem is due not only to the typical American diet but also to insufficient exercise. Exercise increases the rate of cellular respiration and improves respiratory and circulatory efficiency.

Believers have a responsibility to glorify God with their bodies (1 Cor. 6:20). This includes eating a healthy diet and exercising so that their bodies will be useful for the Lord's service.

case study

EXERCISE

Exercising is one way to increase the number of calories that you burn. Different activities burn calories at different rates. Running for half an hour takes more energy than walking for the same amount of time. Various websites and apps are available to help you calculate the approximate number of calories that you will burn while doing an activity for a specified length of time.

It might seem like common sense to pick the exercise form that burns the most calories, but it's not that simple. One of the main problems many of us have with exercising is consistency. It's hard to keep up an exercise routine, so it's better to choose an exercise that we enjoy. We are more likely to continue a routine that we like than one we dislike. Sometimes injury can also lead a person to find a less stressful exercise, even though it may burn fewer calories per hour than some others.

So what's the point of finding out how many calories you burn during an activity? It gives you an idea of how much time you should spend doing that activity in order to burn the targeted number of calories. If your favorite activity doesn't burn a lot of calories, you will need to spend a longer time each week doing that activity, or you could find a second, more energy-consuming activity.

CALORIE EXPENDITURE

ACTIVITY	CAL/LB/H BURNED	IF YOU WEIGH 130 LB	IF YOU WEIGH 180 LB
Baseball	2.27	295	409
Basketball	3.63	472	654
Cycling (moderate)	3.63	472	654
Flag Football	3.63	472	654
Hiking	2.72	354	490
Jogging	3.63	472	654
Martial Arts	4.54	590	817
Rollerblading	5.45	708	981
Running (7.5 mph)	6.13	797	1103
Soccer	4.54	590	817
Swimming	2.72	354	490
Ultimate Frisbee	3.63	472	654
Walking	1.50	195	270

23.1 SECTION REVIEW

1. Describe peristalsis.
2. Make a T-Chart that divides nutrients into organic and inorganic substances.
3. Mechanical digestion takes place primarily in which two digestive organs?
4. What is the role of bile in the small intestine?
5. Which two accessory organs produce amylase?
6. A bacterium that normally lives in the mouth moves to the stomach. Is it likely to cause an infection? Explain.
7. Which type of nutrient is *not* transported *directly* by the small intestine to the circulatory system? Explain.

Use the case study and table above to answer Questions 8–10.

8. Which activity burns the most calories per hour?
9. Jordan, a 130-pound teenager, wants to increase his calorie expenditure, but he doesn't enjoy running. However, he does like cycling. How many hours a week would he have to ride his bike to use the same number of calories that he would use running three hours a week?
10. Rachel weighs 130 lb and wants to burn an extra 1500 calories a week. She exercises only about three hours a week, and her schedule doesn't allow her to play organized sports. She used to run, but she developed plantar fasciitis, a medical condition resulting in pain in her heel. What new exercise would you suggest for her?

23.2 THE URINARY SYSTEM
The Structure of the Urinary System

Questions

What are the parts of the urinary system?

What do the kidneys do?

What other organs excrete waste?

How much water do you need to keep healthy?

What are the views on artificial nutrition and hydration?

Terms

kidney
ureter
urinary bladder
urethra
metabolic waste
nephron
glomerulus
Bowman's capsule
loop of Henle

In the last section we learned that the body produces ammonia when it catabolizes amino acids. Ammonia is highly toxic, but the liver changes it to urea, a much less harmful chemical. But even urea must be removed from the blood or it will eventually poison the body. This important task is performed by the urinary system. While all the organs below are integral to the urinary system, the kidneys are the primary and most structurally complex of them.

Why is it so important for me to drink water?

Exploring

THE URINARY SYSTEM

The two **kidneys**, which look like large kidney beans, filter urea from the blood to be excreted from the body as urine.

Each kidney connects to a tube called a **ureter**, which uses peristalsis to conduct the urine away from the kidney.

Urine exits the ureters into the **urinary bladder**, which collects the urine until it can be excreted.

Urine exits the urinary bladder through the **urethra**, which conducts the urine outside the body. The urethra has two sphincters. One opens involuntarily, and the other is voluntary. (The anus has a similar design.)

The Kidneys

1 Cone-shaped masses of tissue called *renal pyramids* make up the *renal medulla*.

2 The *renal cortex* forms the outer layer of the kidney. Extensions of the renal cortex called *renal columns* lie between the renal pyramids.

3 The *renal artery* carries **metabolic wastes** from the liver into the kidneys.

4 The *renal vein* conducts the filtered blood out of the kidney and back into the rest of the circulatory system.

5 The *renal pelvis* collects the urine from the renal pyramids and releases it into the ureter.

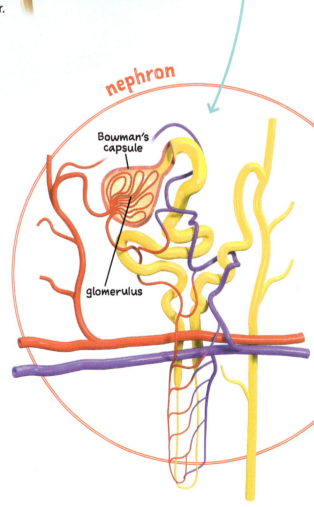

nephron

Bowman's capsule

glomerulus

The Function of the Urinary System

To rid the body of wastes, the kidney performs three major functions: *filtration*, *reabsorption*, and *secretion*. A renal artery carries waste-laden blood into a kidney and then divides into several branches that further divide multiple times to carry blood to the nephrons. The **nephron** is the functional unit of the kidney, and a single kidney contains on average a million of them. Much of the nephron lies in the renal cortex, and the remainder forms part of the renal pyramids. When the blood reaches a nephron, it passes into a collection of capillaries called the **glomerulus**, which is surrounded by a cup-shaped structure called the **Bowman's capsule**. Because blood flowing into the glomerulus is under higher pressure than the fluid in the Bowman's capsule, plasma moves from the glomerulus into the Bowman's capsule. This plasma carries with it smaller molecules such as urea, vitamins, salt, and glucose, while blood cells and larger molecules like whole proteins remain in the blood vessel. This process is called filtration. The fluid and substances that enter the Bowman's capsule are called the *filtrate*.

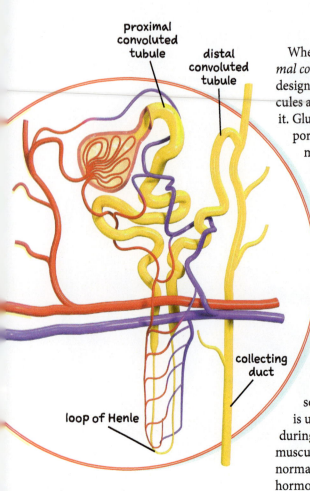

proximal convoluted tubule

distal convoluted tubule

collecting duct

loop of Henle

When the filtrate flows from the Bowman's capsule into the *proximal convoluted tubule*, it enters the reabsorption phase. The tubule is designed to contain the wastes and ensure that the beneficial molecules and minerals are reabsorbed into the capillaries that surround it. Glucose, sodium, potassium, and calcium ions are actively transported back into the capillaries. As the concentrations of these molecules increase in the blood vessels, most of the water in the proximal convoluted tubule and the subsequent **loop of Henle** diffuses back into the capillaries by osmosis. Some reabsorption continues to occur in the *distal convoluted tubule*.

In the distal convoluted tubule, additional substances move from the blood to the tubule in the secretion phase. Substances such as urea, various ions, drugs, and other waste products are secreted. Hydrogen ions may also be secreted into the tubule to adjust the pH of the blood. As the fluid flows from the distal convoluted tubule into the collecting duct, water may be reabsorbed into the bloodstream, especially if the body is becoming dehydrated. The remaining fluid that flows from the *collecting ducts* of the renal pyramids into the renal pelvis is urine.

Normal urine is about 95% water, with the rest being solid solutes. Urea, formed in the liver from protein catabolism, is usually the most common organic solute. Uric acid, formed during nucleic acid breakdown, and creatinine, produced during muscular contraction, are also common substances in urine. A normal urine sample also contains very small quantities of pigments, hormones, enzymes, vitamins, and various inorganic substances.

The Urinary System in the Human Body

The urinary system is part of a larger body system—the *excretory system*—that removes wastes. In addition to the organs of the urinary system, the excretory system includes the liver, which creates urea from ammonia produced by the breakdown of amino acids. The skin also serves as a minor excretory organ since it releases a limited amount of urea in perspiration. Finally, the lungs release carbon dioxide as part of respiration. Since CO_2 is produced as a part of cellular respiration, this makes the lungs part of the excretory system as well.

As we learned in the previous section, water is an essential nutrient. In fact, water makes up over half the body weight of most adults. Water forms much of the cytoplasm in the cells, blood plasma, and lymph. Since we are constantly losing water through urination, sweat, and exhalation, we need to replace it to avoid dehydration. Three-fifths of the water that our bodies need comes from the fluids we drink; most of the rest comes from the food we eat. While our exact fluid needs depend on many different factors, including height, weight, and gender, we need to drink lots of fluids. If a person consistently passes clear urine, then he is sufficiently hydrated. However, since what is "clear" is somewhat subjective, many sources suggest drinking between 64 and 100 ounces of fluids a day. While this may seem like a lot, drinking just an 8-ounce glass of water every two hours between 8:00 a.m. and 10:00 p.m. would add up to 56 ounces, very close to the recommended amount. Any type of fluid helps meet this requirement; water just has the added benefit that it has no calories, unlike many other beverages.

Our bodies must have a way of regulating water levels. Too much water in the blood plasma causes water to enter our cells by osmosis, resulting in swollen cells. If this were to continue, the cells might even start bursting. Too little water in the blood plasma causes water to leave cells via osmosis, causing them to shrivel. The urinary system regulates the balance between too much and too little. If blood vessels have too little volume because of a lack of water, the urinary system releases a hormone called *antidiuretic hormone* (ADH). ADH causes the nephrons to reabsorb more water during the reabsorption phase. If the body has too much water, then the body does not release ADH, reducing the amount of water that the nephrons reabsorb and allowing more water to be released in urine.

Drinking plenty of water is also another way to maintain good health. Since our bodies are roughly 60% water and our blood is nearly 90% water, hydration is extremely important. Drinking enough water daily is essential for everyday bodily functions. Being properly hydrated gives our bodies energy, helps us to think more clearly, regulates body temperature and blood pressure, helps with weight loss, and prevents kidney damage. Water is needed in order to keep us healthy and maintain homeostasis so that we can serve the God who made us for His glory.

23.2 SECTION REVIEW

1. Which two organs of the urinary system are essentially tubes?

2. A person is having trouble with urea not being removed from the blood. Which organ(s) of the urinary system is (are) likely to be failing? Explain.

3. If a blood clot is preventing blood from entering a kidney but is not affecting blood flow in other vessels, which vessel most likely contains the clot?

4. Place the following organs and tissues in the order urea passes through them on its way to being expelled from the body: collecting duct, loop of Henle, renal artery, renal pelvis, ureter, urethra, urinary bladder.

5. Which structure of a nephron is located between the proximal convoluted tubule and the distal convoluted tubule?

6. Which one of the three major functions of the kidney is responsible for ensuring that the body does not lose beneficial molecules, such as glucose and sodium?

7. Name one of the three excretory system organs that are not part of the urinary system, along with the organ system that it is part of.

8. How does water affect homeostasis in the body?

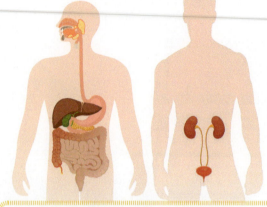

23.1 THE DIGESTIVE SYSTEM

- Organs along the alimentary canal extract nutrients from food. These organs include the mouth, esophagus, stomach, small intestine, and large intestine.

- Accessory organs of the digestive system release chemicals into the alimentary canal. These organs include the salivary glands, liver, gallbladder, pancreas, and appendix.

- Our bodies need six nutrients: vitamins, minerals, water, fats, carbohydrates, and proteins.

- The digestive system uses a combination of mechanical and chemical digestion to breakdown food.

- Amylase, pepsin, lipase, and peptidase are enzymes that break down foods chemically.

- The digestive system works with other systems of the body including the muscular, circulatory, lymphatic, and integumentary systems.

- Bacteria and other systems of the body work together with the digestive system to maintain homeostasis.

Terms
alimentary canal · accessory organ · oral cavity · esophagus · bolus · peristalsis · stomach · chyme · small intestine · large intestine · rectum · salivary gland · appendix · liver · bile · pancreas · gallbladder · nutrient · amylase · pepsin · lipase

23.2 THE URINARY SYSTEM

- The urinary system is part of the excretory system and rids the body of wastes by filtration, reabsorption, and excretion.

- The kidneys filter waste from the renal artery to the nephron where it is then filtered and sent to the Bowman's capsule.

- Reabsorption occurs when the filtrate is reabsorbed and water diffuses into the capillaries.

- The urinary system works with other systems of the body, including the digestive, respiratory, and integumentary systems.

- Water is an important nutrient for life and helps to maintain homeostasis by regulating many bodily functions.

Terms
kidney · ureter · urinary bladder · urethra · metabolic waste · nephron · glomerulus · Bowman's capsule · loop of Henle

Chapter Review Questions

RECALLING FACTS

1. Which organs does chyme pass through?

2. Classify each substance into one of the six classes of nutrients.

   a. inorganic phosphate (PO_4^{3-})

   b. lactose (milk sugar)

   c. lean meat

3. What nutrient is a human's dry weight mostly composed of?

4. Which part of the alimentary canal has the lowest pH?

5. Indicate the nutrient that each enzyme digests.

   a. amylase

   b. lipase

   c. pepsin

6. What organ system are the lacteals part of?

7. Why must the urinary system convert ammonia to urea before removing it from the body?

8. Match each function listed below with one of the four main organs in the urinary system.

    a. connecting tube

    b. filter

    c. exit

    d. holding tank

9. If a single nephron ceased functioning, would it make much of a difference in the kidney's function? Explain.

10. What structure serves as the functional unit of a kidney?

11. Match each part of a nephron listed below with one of the following processes: filtration, reabsorption, secretion.

    a. collecting duct

    b. glomerulus

    c. loop of Henle

UNDERSTANDING CONCEPTS

12. What sections of the alimentary canal would carry secretions from only one set of accessory organs? Explain.

13. What is the difference between mechanical and chemical digestion?

14. Rory lives in Whitehorse, Yukon Territory, Canada, and his friend Eduardo lives in Mexico City. During winter, which vitamin would Rory need to be more concerned about getting from his diet than Eduardo? Explain.

15. Which of the following would never be found in the urine of a healthy person: urea, water, or red blood cells? Explain.

16. Describe what occurs in a nephron.

17. Other than the kidneys, identify the organs that are involved in the functioning of the excretory system. Describe how they participate in excretion.

18. Kidney stones are mineral deposits that form in the kidneys. While many different factors can lead to kidney stones, one of the more common causes is overly concentrated urine. What could a person do to ensure that his urine does not become too concentrated?

CRITICAL THINKING

19. You are a nutritionist, and one of your clients wants to lose weight by reducing his daily calorie intake from 2300 to 2000 calories a day. He suggests that he should go on a completely fat-free diet so that he can reduce the amount of fat in his body. Why won't this work?

20. Why is urinalysis a common way of testing a person for drug use?

21. You are a doctor working in an emergency room. A patient is brought in with a case of alcohol overdose. Since you know that alcohol blocks ADH, what major problem should you check your patient for?

22. If ADH were to be continuously released for a long period of time, what might result?

Use the case study below to answer Questions 23–31.

23. How many tortillas are in one unopened package of this product?

24. John ate three tortillas for lunch. How many grams of protein did he eat?

case study

NUTRITION FACTS LABELS

The Food and Drug Administration requires all food companies to place a label on their food packages listing the amount of fat, carbohydrates, and proteins in a recommended serving. Additionally, the label also shows what percentage of the recommended daily allowance of fats, carbohydrates, cholesterol, and sodium the product contains. A second part of the Nutrition Facts label contains information on what percentage of the daily recommended dose of specific vitamins and minerals the food product contains. The very bottom part of the label includes recommended daily allowances of fat, cholesterol, sodium, and carbohydrates.

Nutrition Facts

Serving Per Container 2
Serving Size 2 Tortillas (51g)

Amount Per Serving

Calories 110   Calories from fat 10

	% Daily Value*
Total Fat 1g	2%
Saturated Fat 0g	0%
Trans Fat 0g	
Cholesterol 0mg	0%
Sodium 30mg	1%
Total Carbohydrate 30g	7%
Dietary Fiber 2g	9%
Sugars 0g	
Protein 3g	

Vitamin A 0%	•	Vitamin C 0%
Calcium 2%	•	Iron 4%

*Percent Daily Values (PDV) are based on a 2,000 calorie diet. Your daily value may be higher or lower depending on your calorie needs.

	Calories	2,000	2,500
Total Fat	Less than	65g	80g
Sat Fat	Less than	20g	25g
Cholesterol	Less than	300mg	300mg
Sodium	Less than	2,400mg	2,400mg
Total Carbohydrate		300g	375g
Dietary Fiber		25g	30g

25. The 30 mg of sodium is said to be 1% of the daily recommended allowance. According to this, what is the recommended daily allowance of sodium?

26. According to the bottom panel, what is the daily recommended allowance of sodium?

27. Is the label misrepresenting the amount of sodium in the product?

28. Given that cholesterol is produced by animal cells, is this product more likely made with vegetable oil or lard (pig fat)?

29. The recommended daily allowance for vitamin A is 900 mg for men over age fourteen and 700 mg for women of the same age. How many milligrams of vitamin A would a thirty-three-year-old woman need to consume to achieve her daily recommended allowance after eating two tortillas? Explain.

30. How could a person seeking to glorify God with his or her eating habits use the nutrition facts label?

31. A friend cites Psalm 90:10 and tells you, "The Bible says that we are going to live only about seventy years anyway. We don't have to measure every little thing we eat and spend our life exercising." How would you respond?

Use the ethics box on this and the next page to answer Questions 32–33.

32. Use the process modeled in Chapter 20 to deconstruct the response from the principles of bioethics and formulate a biblical position on artificial nutrition and hydration.

33. Are the principles of bioethics enough to use when analyzing ethical issues?

ARTIFICIAL NUTRITION & HYDRATION

ETHICS

ISSUE

Providing health care often involves many decisions. At times, these decisions can be difficult. One such difficult situation families may encounter is whether to give a loved one artificial nutrition and hydration (ANH) as supportive measures in prolonging life.

After being diagnosed with a terminal illness, Irene's condition gradually worsened. Eventually, she entered into a nonresponsive state and could no longer advocate for herself. Irene tasked her son Collin with making medical decisions for her. As her illness progressed, Irene began to have difficulty taking in food and water. Doctors proposed using ANH in an effort to treat Irene's lack of nutrition, but Collin was unsure whether such medical intervention would be wise in her situation. Imagine you are thinking Collin's situation through with him.

Below is a typical decision-making process according to the principles of bioethics.

1. What information can I get about this issue?

Artificial nutrition and hydration is a form of medical care in which patients are given nutrition and fluid intravenously or through a feeding tube. This is usually given if the patient is unable to take food and water by mouth. In an effort to sustain life, caregivers may choose to use this method of care if the patient has indicated this desire and doctors agree that it would be to his or her benefit. Personal decisions regarding ANH may be made on the basis of religious or moral beliefs. Decisions made by

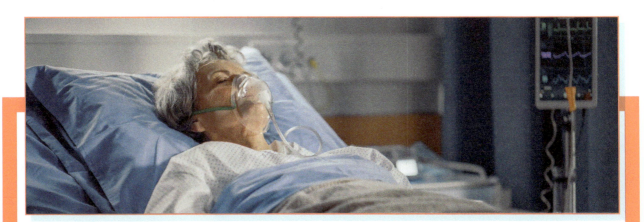

caregivers may be made on the basis of believing that withholding nutrition and hydration could increase a patient's pain. Although it is known that people will die without food and water, it is debated whether ANH is truly beneficial to a patient who is likely going to die from a terminal illness.

2. How are the people involved in this issue respected and given the freedom to choose?

Collin, as Irene's caregiver, has a responsibility to do what is in her best interest. Doctors suggest ANH to strengthen his mother with sufficient nutrition and hydration. But Collin is hesitant to go against the natural process of dying. He knows that Irene is suffering and believes that it may do more harm than good to provide ANH. Since Irene is unable to make a decision about her care, the principle of autonomy would indicate that Collin, as Irene's authorized caretaker, has priority in the decision-making process over doctors' suggestions. Regardless of his choice, doctors must respect the autonomy of the patient that has been transferred to Collin.

3. How are individuals protected from harm or injury?

Collin is struggling because he knows that his mother is starving without ANH and wishes to prevent that. However, he also believes that providing nutrition might prolong her life and lead to worse suffering from other symptoms of her illness. Collin must determine which decision he thinks is most likely to lead to the least amount of harm.

4. How are people helped by the action taken?

If the decision is made to give Irene ANH, her life may be slightly prolonged with the additional nutrition and hydration. It could also make her death less painful because she will at least be fed and hydrated, but there is conflicting evidence on this topic. Alternatively, Collin might think that allowing Irene to progress through the natural process of death would be most beneficial to her in her current condition.

5. How is the action taken just and fair?

According to the principle of justice, patients should be given equal opportunities in the administration of different forms of treatment. ANH is neither an expensive nor exclusive treatment and therefore justice and fairness do not have an impact on whether or not to administer ANH.

6. What action do the principles of bioethics recommend?

Considering only the principles of bioethics, Collin could decide either way. Justice is not a factor in this decision. Help or harm could be viewed either way. It ultimately boils down to a question about autonomy. Irene would have the right to make the decision for herself if she were aware and able to, but since she cannot, Collin is free to make the decision that he thinks is in Irene's best interest.

24 COMMUNICATION

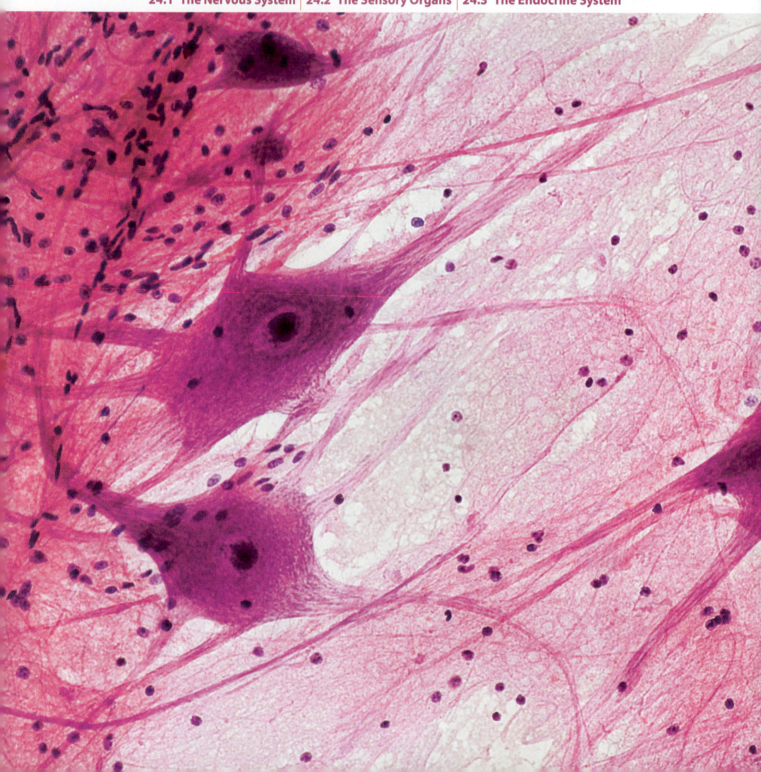

A World Without Pain

Ever burn your hand on a hot stove or cut it on a knife? For some people in the world, touching something hot or cutting their fingers wouldn't register at all—they wouldn't feel a thing! These people aren't considered blessed to be free of the pain of cuts or burns; in fact, this lack of pain can be life-threatening. Worldwide, one of the most common causes of this insensitivity is damage to the nervous system—*neuropathy*—resulting from leprosy. The leprosy bacteria destroy nerve cells, depriving the sufferers of their sense of touch. As a result, they may injure themselves without realizing it, leading to major health issues or even death. Neuropathy may also occur in developed countries due to diseases such as diabetes, though in these cases it generally results in pain, tingling, and numbness. How can science be used to improve the quality of life for God's image-bearers who suffer from neuropathy?

24.1 THE NERVOUS SYSTEM
The Structure of the Nervous System

The nervous system forms a massive communication network that runs throughout the body. It receives information from the sensory organs in the nose, tongue, ears, eyes, and skin, and it processes that information at speeds that would make a supercomputer designer green with envy.

How does my body respond in an emergency?

Questions

What are the parts of the nervous system?

How does the nervous system work?

What does your brain do?

How does information get from one neuron to another?

How does information get from your body to your brain and back again?

Terms

meninges
cerebrospinal fluid
neuron
cell body
dendrite
axon
cerebrum
thalamus
hypothalamus
cerebellum
brain stem
resting potential
action potential
synapse
neurotransmitter
threshold
sensory neuron
interneuron
motor neuron
parasympathetic nervous
 system
sympathetic nervous
 system

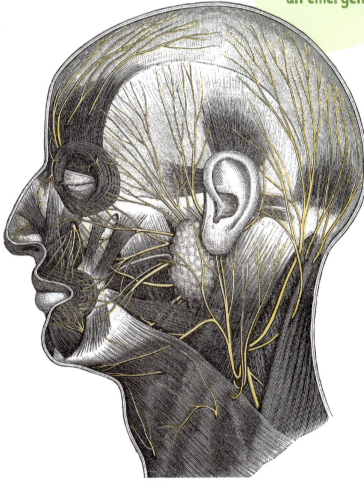

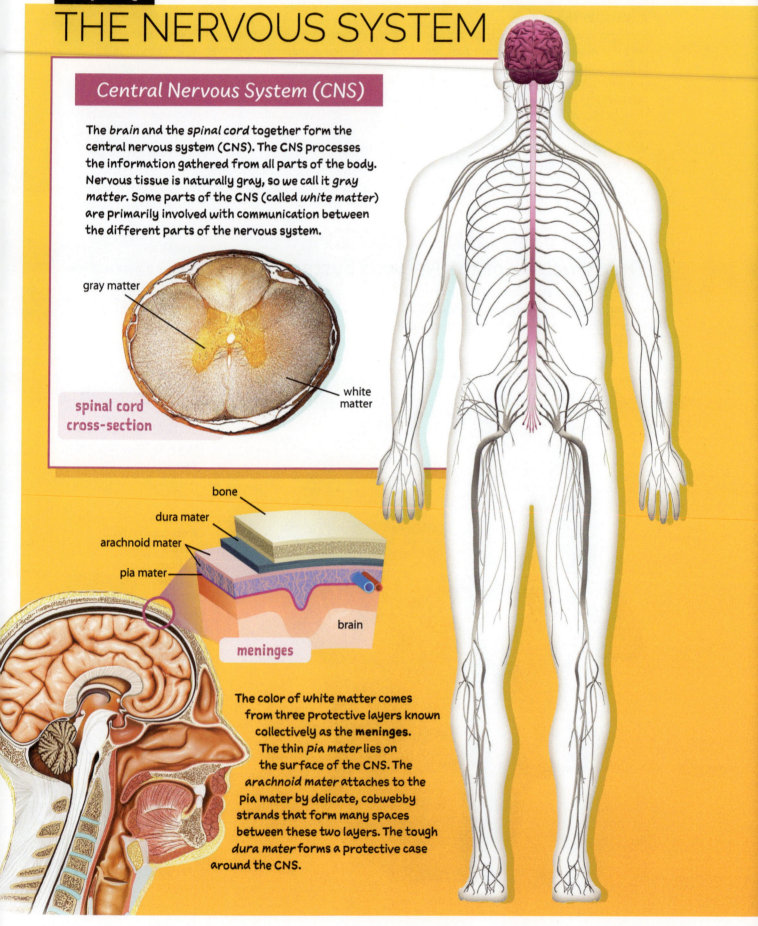

THE NERVOUS SYSTEM

Exploring

Central Nervous System (CNS)

The *brain* and the *spinal cord* together form the central nervous system (CNS). The CNS processes the information gathered from all parts of the body. Nervous tissue is naturally gray, so we call it *gray matter*. Some parts of the CNS (called *white matter*) are primarily involved with communication between the different parts of the nervous system.

gray matter

spinal cord cross-section

white matter

bone

dura mater

arachnoid mater

pia mater

brain

meninges

The color of white matter comes from three protective layers known collectively as the **meninges**. The thin *pia mater* lies on the surface of the CNS. The *arachnoid mater* attaches to the pia mater by delicate, cobwebby strands that form many spaces between these two layers. The tough *dura mater* forms a protective case around the CNS.

Special capillaries in the pia mater form **cerebrospinal fluid** by allowing only specific substances to move from the blood into the spaces between the arachnoid mater and the pia mater. The fluid bathes the CNS, providing nutrition to the cells, cushioning the structures, and removing wastes. Another set of structures in the arachnoid mater above the brain allows excess cerebrospinal fluid to reenter the bloodstream.

cerebrospinal fluid

Peripheral Nervous System (PNS)

The peripheral nervous system (PNS) forms the other half of the nervous system. Its sensory division gathers information from the sensory organs and carries it to the CNS to be processed. Its motor division then relays information from the CNS to *effectors*—usually muscles, glands, and organs. The part of the motor system that activates skeletal muscles and can be controlled voluntarily is called the *somatic nervous system.* The *autonomic nervous system* controls smooth and cardiac muscle and functions involuntarily.

neuron

dendrite

cell body

nucleus

myelin sheath

axon

The **neuron** is the functional unit of the nervous system. Neurons have several different forms, but they all have three main parts: a single **cell body**, usually multiple **dendrites**, and a single **axon**. Axons with their protective *myelin sheaths* form the white matter of nervous tissue. The gray matter of nervous tissue is usually formed by cell bodies and dendrites.

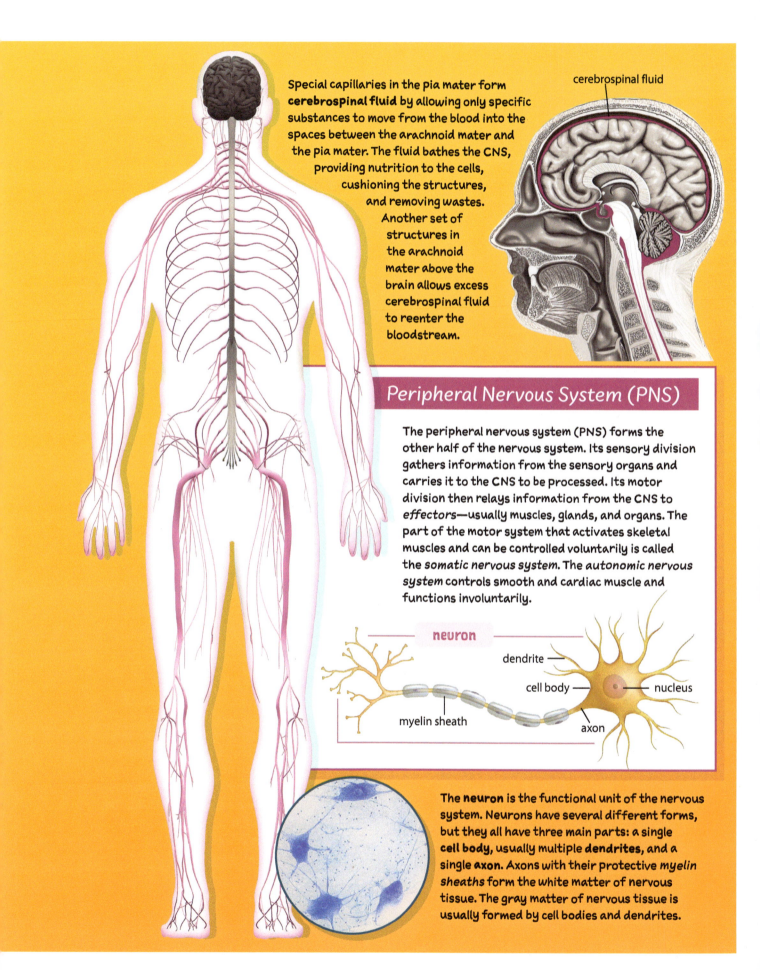

The central and peripheral nervous systems are essential to each other. We will study the PNS in more detail in the next section. For now, let's focus on the CNS.

THE BRAIN

The **cerebrum** ❶ is by far the largest region of the brain and is responsible for conscious activities. The depressed areas, called *fissures*, and ridges, called *gyri* (s. gyrus), increase the surface area of the cerebrum.

The **thalamus** ❷ receives general sensations and relays the important ones to the cerebrum. The thalamus also helps keep us awake and alert.

The **hypothalamus** ❸ is a tiny part of the brain, weighing only 4 g, but it regulates many involuntary activities, including body temperature, blood volume, and fluid balance. It also has some control over appetite and emotional expression. Finally, and perhaps most importantly, it is the link between the nervous system and the endocrine system.

The **brain stem** ❺ connects the rest of the brain to the spinal cord and controls involuntary respiration, blood vessel diameter (blood pressure), and heart rate. It also controls responses to sight and sound as well as eyeball movement. The brain stem controls body movements and posture; it controls but does not initiate sneezing, coughing, swallowing, and vomiting. Because many basic homeostatic processes are controlled by the brain stem, an injury to this part of the brain is often fatal.

Located behind and beneath the cerebrum, the **cerebellum** ❹ controls body coordination. Although it initiates no voluntary movements on its own, it monitors and adjusts activities and movements that are stimulated by other brain regions. When a movement occurs, the cerebellum regulates its quickness and force, ensuring that it performs the desired action.

THE CEREBRUM

The cerebrum is divided longitudinally into two hemispheres. Because the nerves coming from the cerebrum cross to the opposite side of the body, the right hemisphere controls the left side of the body and the left hemisphere the right side of the body. Each hemisphere is further divided into lobes, each controlling different functions.

The *frontal lobe* performs mental functions such as reasoning, planning, and memorizing. It also controls verbal communication and starts the commands for voluntary body movements.

The *parietal lobe* receives sensations such as pain, pressure, touch, and temperature. It then directs this information to the frontal lobe, which determines what to do about it. The parietal lobe also uses muscle tension to maintain a sense of the position of the body.

The *occipital lobe* is involved primarily in vision and memory of objects and symbols. A severe blow to the head may cause the sensation of "seeing stars" because it stimulates the neurons leading from the eyes to the occipital lobe. Nerve impulses from these fibers are interpreted as visual impulses by the occipital lobe.

The *temporal lobe* perceives the sensations of hearing and smell. It also provides the ability to remember the pronunciation of words and the melody of songs and stores memories of both sight and sound.

The spinal cord continues from the brain stem down most of the spinal column. It relays messages between the brain and the PNS, which relays messages to and from the rest of the body. The spinal cord is a cylindrical mass of nervous tissue composed of thirty-one segments with one pair of spinal nerves originating from each segment. The spinal cord is made of white matter in the outer regions and gray matter in the center. The neuron cell bodies that make up much of the central gray matter are connected primarily with the PNS, and the axons that form the outer white matter carry signals to and from the brain. Some axons carry messages to the brain; others carry messages only away from the brain. For example, pain signals pass along only certain axons, while signals to muscles pass along other axons.

The Function of the Nervous System

Neurons carry messages using electrochemical signals that travel down the length of the neuron. Neural cells have proteins in their membranes called *sodium-potassium pumps* that actively transport sodium cations out of the cell and potassium cations into the cytoplasm. The concentrations of several types of anions also build up inside the cell so that the membrane has a net positive charge on the outside and a net negative charge on the inside. This difference in charge across the membrane forms the neuron's **resting potential**.

When a neuron receives a signal, protein channels in the membrane open, causing an influx of sodium ions into the cell, an event called an **action potential**. The action potential causes other protein channels to open, passing the action potential down the length of the neuron. When the sodium ions rush in, other protein channels open, allowing potassium cations to rush out, reestablishing the resting potential. The sodium-potassium pumps then begin transporting sodium cations back out of the cell and potassium ions back into the cell.

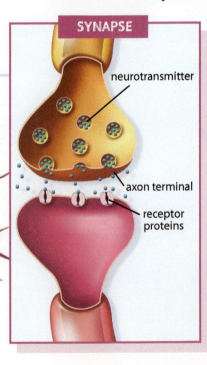

SYNAPSE

neurotransmitter

axon terminal

receptor proteins

A neuron typically receives a signal at one of its dendrites, and the action potential moves down the axon until it reaches the *axon terminal*. The axon terminal is part of the junction between two neurons known as a **synapse**. The *synaptic cleft* is a gap between the axon terminal of one neuron and the dendrite of the next neuron. The action potential cannot cross the synaptic cleft; instead, it causes the axon terminal to release molecules called **neurotransmitters**, which stimulate receptor proteins in the other neuron. Neurotransmitters can also carry an action potential between a neuron and a muscle or gland. Enzymes in the synaptic cleft quickly inactivate the neurotransmitters so that receptor proteins are ready for the next action potential.

So how do neurons avoid setting off action potentials unnecessarily? For a stimulus to cause an action potential in a neuron, it must be significantly strong enough. This level is called a **threshold**. A stimulus weaker than the threshold will not cause an action potential.

Notice that an action potential starts at the dendrites and ends at the axon terminal. It never goes the other way. Therefore, we need nerves that travel in each direction to take messages from sensory organs to the brain and back to an effector. Neurons that carry information about our surroundings toward the brain are called **sensory neurons**. Their dendrites may detect sensory input such as heat or cold, or they may be connected to a special sensory organ such as the eye. Sensory stimuli cause an action potential to begin in the dendrites and proceed up the axon until it reaches the axon terminal. The axon terminal then hands the signal to an **interneuron** in the CNS. Interneurons direct the impulse to the brain where the impulse is interpreted.

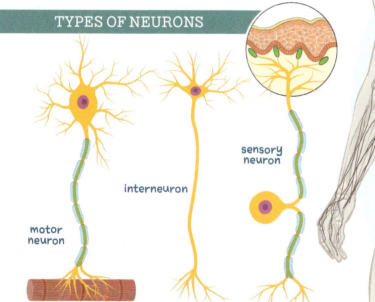

TYPES OF NEURONS

motor neuron

interneuron

sensory neuron

Let's say that sensory nerves in your skin that detect cold perceive that the temperature in the house is too cold. When your brain receives the message, another signal is sent down the spinal cord to multiple interneurons in the spinal cord that then pass the impulse to **motor neurons**. Motor neurons have cell bodies in the spinal cord and axons that conduct the action potential to effectors all over the body. Depending on what you decide to do about the cold, you may start shivering, move to a warmer location, put on more clothing, turn up the thermostat, or perform all these actions.

The speed of an action potential is incredibly fast. But a few stimuli require action before a signal has time to reach the brain and return. When you touch a hot object, you need to move your hand away before thermal energy does too much damage to your tissues. The pain receptors send an action potential up a sensory nerve. When the impulse reaches an interneuron in the spinal cord, it takes a shortcut called a *reflex arc*. The interneuron sends the impulse to one or more motor neurons. So by the time you are even aware that you are touching a hot surface, your hand has already jerked back. This reflex is entirely involuntary. You don't even have to think about it!

The Nervous System in the Human Body

The nervous system intricately controls the body, but even more amazingly, the nervous system performs tasks that we cannot explain. How does a cell or group of cells in the brain remember a beautiful mountain landscape seen two years ago? Or, for that matter, how does it store the concept and word *mountain*? How does a person think? What is a dream? What is an emotion? The brain controls all these experiences thousands of times each day. Scientists continue to explore these questions, and we may never know for sure how the brain does all that it does. But what we continue to learn fits much better with the worldview that says that our brain was designed by God than with one that says that it is a product of time and chance.

So far we have discussed the nervous system under normal circumstances. During such times the **parasympathetic nervous system**, part of the autonomic nervous system, is operating. The nerve impulses travel through neurons located in the brain stem and at the very bottom of the spinal cord. They stimulate proper digestion, absorption of food, and elimination of wastes. They also maintain normal heart and breathing rates as well as lower blood pressure. These functions are performed best when a person relaxes in a pleasant, peaceful environment. The parasympathetic system controls all body functions while we sleep.

During times of danger the other part of the autonomic nervous system—the **sympathetic nervous system**—takes over. Neural impulses pass through neurons located in the middle portions of the spinal cord. They help the body gear up in order to survive and be successful in stressful situations, such as a natural disaster. The impulses direct the adrenal glands, part of the endocrine system, to release a rapid surge of the hormone *epinephrine*, which stimulates a quick increase in heart and breathing rates and causes a rise in the blood sugar needed for muscle action. The pupils of the eyes dilate quickly so that the person can see where he or she is running. The digestive system is shut off, and the entire nervous system is put on emergency standby. This is often called the *fight-or-flight response*.

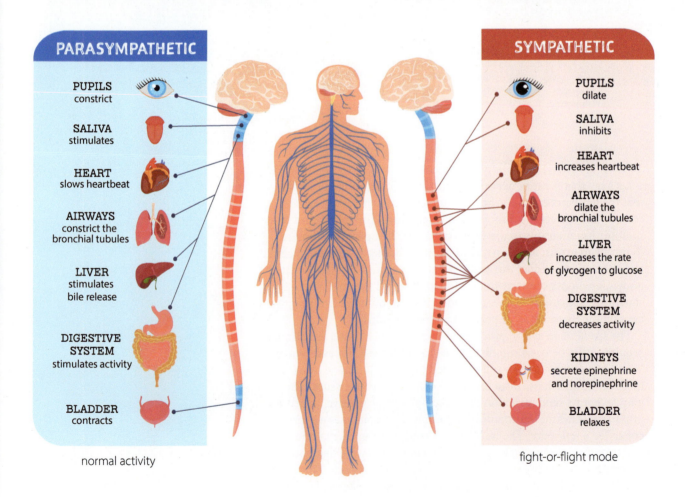

PARASYMPATHETIC

PUPILS
constrict

SALIVA
stimulates

HEART
slows heartbeat

AIRWAYS
constrict the
bronchial tubules

LIVER
stimulates
bile release

DIGESTIVE
SYSTEM
stimulates activity

BLADDER
contracts

normal activity

SYMPATHETIC

PUPILS
dilate

SALIVA
inhibits

HEART
increases heartbeat

AIRWAYS
dilate the
bronchial tubules

LIVER
increases the rate
of glycogen to glucose

DIGESTIVE
SYSTEM
decreases activity

KIDNEYS
secrete epinephrine
and norepinephrine

BLADDER
relaxes

fight-or-flight mode

MINI LAB

Reaction Time

Many activities require a person to react quickly to a stimulus. Some of those activities are simply for fun, like hitting a fastball during a baseball game. Others may be matters of life or death, like stepping quickly on a brake pedal when a person moves unexpectedly into the path of a car. How fast can you react? Let's find out!

How fast can I react to a stimulus?

Materials
computer with internet access

PROCEDURE

A Go to the website provided by your teacher. Follow the instructions for taking the reaction time test. Record your reaction time for each trial.

B Determine your average reaction time.

QUESTIONS

1. What was your average reaction time?

2. How does your average reaction time compare with that of others in your class?

3. A few hundred milliseconds may not be a lot of time, but it's still time. Do you think it would be possible to achieve a reaction time close to 0 ms? Explain.

GOING FURTHER

4. Professional athletes rarely play beyond the age of forty. Several factors contribute to this, such as the aging human body's slower healing in response to injury. Is reaction time also a factor? Do an internet search using the keywords "reaction time by age," then record your findings.

24.1 SECTION REVIEW

1. Identify each of the following structures as part of the CNS or the PNS.

   a. the cerebellum

   b. an axon connected to a pain receptor in the skin

   c. an interneuron

2. Which part of the brain is the link between the nervous system and the endocrine system?

3. Which cation is concentrated outside a neuron when it's at its resting potential?

4. Explain how an action potential crosses a synaptic cleft.

5. List the three types of neurons involved in a reflex arc in the order in which an action potential travels during a reflex.

6. Which glands release epinephrine as part of the response when the sympathetic nervous system takes over?

7. Why might someone who is constantly overly excited have problems with high blood pressure and poor digestion?

8. Create a concept map using the following terms: *autonomic nervous system*, *central nervous system*, *nervous system*, *parasympathetic nervous system*, *peripheral nervous system*, *somatic nervous system*, and *sympathetic nervous system*.

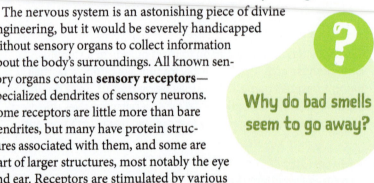

24.2 THE SENSORY ORGANS
The Structure of the Sensory Organs

The nervous system is an astonishing piece of divine engineering, but it would be severely handicapped without sensory organs to collect information about the body's surroundings. All known sensory organs contain **sensory receptors**—specialized dendrites of sensory neurons. Some receptors are little more than bare dendrites, but many have protein structures associated with them, and some are part of larger structures, most notably the eye and ear. Receptors are stimulated by various external and internal conditions and inform the body of changes. A single receptor is usually sensitive to only one type of stimulus. The sensory organs also contain structures and tissues that support and assist the receptors.

We perceive our world primarily aurally and visually, with sight being the predominant sense. Our sense of hearing allows others in close proximity to communicate verbally with us, and our eyes detect a host of color, distance, and texture variations in our surroundings. Let's take a closer look at these structures.

?

Why do bad smells seem to go away?

THE EAR

① Outer Ear

The visible part of the ear is the *auricle*, a flap of cartilage designed to collect sound waves. The shape of the auricle directs the sound vibrations into the **external auditory canal** toward the **tympanic membrane**, or eardrum. The skin lining the auditory canal contains *ceruminous (wax) glands*. The skin continuously sheds dead surface cells, which combine with earwax and gradually move out of the ear, carrying away any foreign matter.

② Middle Ear

Three tiny bones, the **malleus**, or hammer, **incus**, or anvil, and **stapes**, or stirrup, pick up vibrations from the tympanic membrane and carry them to the inner ear. The *eustachian tube* allows air to enter from the pharynx to equalize pressure between the middle ear and atmosphere. If the pressure in the middle ear is unable to equalize with the atmospheric pressure, the difference causes the tympanic membrane to bulge outward or inward. This decreases the effectiveness of the eardrum and could even burst it if the pressure difference were to become too great.

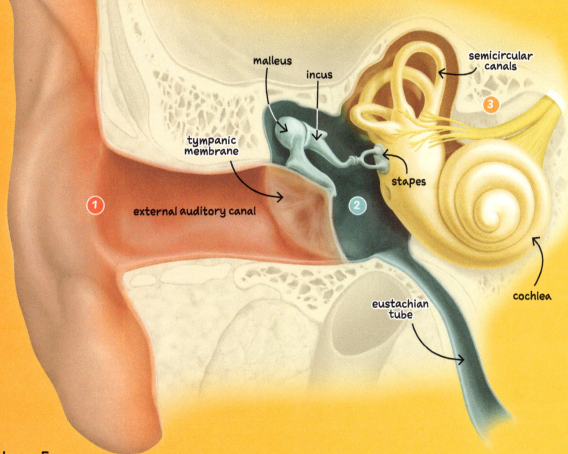

malleus

incus

semicircular canals

③

tympanic membrane

stapes

① external auditory canal

②

cochlea

eustachian tube

③ Inner Ear

A series of channels and cavities within the skull form the *bony labyrinth* of the inner ear. The *membranous labyrinth* closely duplicates the shape of the bony labyrinth in a tube-within-a-tube arrangement. The **cochlea** receives vibrations from the stapes and transforms them into nerve impulses to be sent to the brain. The **semicircular canals** along with the *utricle* and *saccule* (both located between the semicircular canals and the cochlea) provide a sense of balance.

THE EYE

The outer layer of the eye, known as the **sclera**, is often called the *white of the eye*. This white, fibrous tissue maintains the shape of the eyeball. The **cornea** is a transparent portion of the sclera at the front of the eye that allows light to enter the eyeball.

The middle layer, the **choroid**, is fragile and thin and contains many blood vessels for nourishing the innermost part of the eye. The pigmented **iris**, lying behind the cornea, is a muscular portion of the choroid that surrounds the **pupil**, which allows light into the eyeball. The muscles in the iris change the diameter of the pupil, regulating the amount of light entering the eye.

The third and innermost layer of the eye, the **retina**, contains thousands of specialized neurons and their fibers. These neurons are **photoreceptors** that can be stimulated by light. The impulses from the photoreceptors are transmitted to the **optic nerve**.

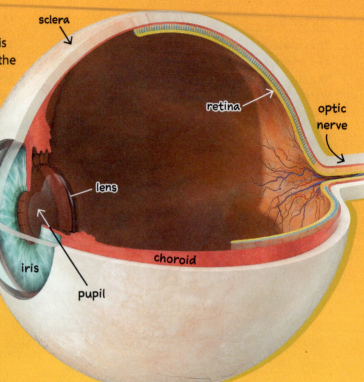

The **lens** is a double convex, semisolid structure supported by the *ciliary muscles* and *suspensory ligaments*, which can change the lens's shape.

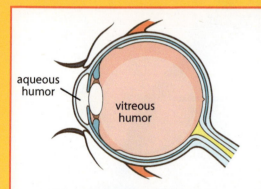

The eye contains two cavities. The *aqueous humor*, a transparent, watery fluid, diffuses from blood vessels located near the ciliary muscles and fills the chamber between the lens and cornea. The larger chamber of the eyeball lies behind the lens and contains the *vitreous humor*, a clear, jellylike substance that provides support for the eye.

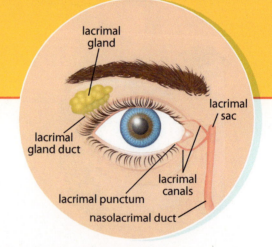

The *lacrimal (tear) gland* lies in the upper lateral region of each eyelid. Every day it secretes about 1 mL of fluid, which is spread evenly over the surface of the eyeball each time you blink. The fluid lubricates the eyelid and moistens and cleans the cornea. It also contains the enzyme *lysozyme*, which kills bacteria.

Because humans rely on their hearing and sight for most of their information, the other senses are often called *minor senses*. This is not to suggest that they are unimportant. Our chances of survival would be greatly reduced without our minor senses, and certainly life would be a lot less pleasant.

THE MINOR SENSES

Cutaneous receptors sense touch, cold, heat, pain, and pressure. Each sensation has different receptors, and certain receptors are more numerous in certain areas. Touch receptors lie near the skin's surface and are especially numerous in the fingertips and palms of your hands and the soles of your feet. Heat and cold receptors are both buried deep in the dermis. Pressure receptors lie below the skin, and pain receptors can be found in nearly every tissue of the body. All the receptors create a nerve impulse that other nerves carry to the brain, which interprets the impulses and responds to them.

free nerve ending (pain)

Merkel ending (pressure)

root hair plexus (hair deformation)

Ruffini ending (heat)

Meissner corpuscle (touch)

Pacinian corpuscle (vibration)

The sense of smell, or *olfactory sense*, is one of the least understood senses. Humans apparently can distinguish several thousand different odors. The molecules in the air stimulate *olfactory receptors*, which consist of numerous cells in the mucous membranes that line the upper region of the nasal cavities. After a short period of stimulation, the olfactory receptors become insensitive to that specific stimulus (smell). So you may notice a strong smell when you first walk into a room, but you will probably not notice it after several minutes. This is called *accommodation*, and while it occurs with most nerve sensations other than pain, it usually does not occur with these as quickly as with smell.

olfactory bulb

olfactory nerve fiber

We have about 10,000 taste buds, primarily on the papillae of the tongue, but some are found on the inner surface of the cheek, roof of the mouth, and throat structures. Each taste bud holds numerous taste cells. These taste cells have chemoreceptors called *taste hairs* that stimulate a nerve impulse when they bind to one of five particular molecules or ions. As a result, the tongue perceives five basic taste sensations—sweet, sour, bitter, salty, and umami (savory, meaty taste).

taste bud

taste cell

The Function of the Sensory Organs

Cutaneous, olfactory, and taste receptors all create nerve impulses that are carried along axons to the CNS. The photoreceptors in the eyes and the auditory receptors in the ears send impulses in the same manner. Since the structures of the eyes and ears are better understood and are especially important to our existence, most of our discussion will concern sight and hearing.

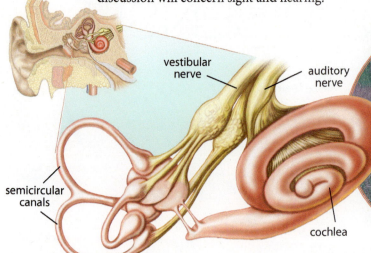

HEARING

Sound waves travel through the air and enter the external auditory canal. When sound waves encounter the tympanic membrane, they cause it to start vibrating. The malleus, which is attached to the inside of the eardrum, carries those vibrations to the incus, which carries them to the stapes. The joints between these bones are flexible, and they amplify the strength of sound vibrations about ten times as they carry them from the eardrum to the inner ear.

The stapes is attached to the inner ear through a membrane-covered hole in the bony labyrinth called the *oval window*. When the stapes vibrates the oval window, the fluid inside the cochlea vibrates along with it. Running the full length of the cochlea's inner surface is the *organ of Corti*, containing thousands of receptor cells called **hair cells**. The vibrations in the cochlear fluid cause the cilia of the hair cells to bend as the fluid moves around them. Different hair-cell cilia will bend, depending on the frequency of the vibrations, allowing the cochlea to distinguish between different frequencies. The bending of the hair cells produces a nerve impulse that is transferred to nerve cells. These impulses are carried by the auditory nerve to the temporal lobe of the cerebrum, where they are perceived as different sounds.

BALANCE

While the cochlea is processing vibrations, the rest of the bony and membranous labyrinths are helping us keep our balance. The utricle and saccule control our sense of *static equilibrium*, our sense of body position when we're not moving. The utricle and saccule are lined with sensory hair cells embedded in a jellylike substance that contains crystals of calcium carbonate. When the head moves, the calcium carbonate crystals slide, pulling the jelly. This movement bends the hair cells, stimulating nerve impulses that are sent to the brain via the *vestibular nerve*.

The body's ability to respond automatically to positional changes when we are moving is called *dynamic equilibrium*. The three semicircular canals control this. Each semicircular canal collects data for a different plane of three-dimensional space. Like the utricle and saccule, the semicircular canals are lined with hair cells embedded in a gelatinous layer, but they also contain fluid. During walking, running, and other dynamic activities, the fluid flows over the gelatinous material, bending the hair cells. The movement of the hairs stimulates nerve impulses that travel along a branch of the auditory nerve to the temporal lobe and cerebellum. Impulses from the cerebellum adjust muscle actions, producing coordinated movements for each position change.

RETINA

cone cell rod cell

SIGHT

Some objects, such as the sun and light bulbs, emit light. Most objects, including the moon, houses, and our bodies, reflect light. When light strikes an object, the object's atoms and molecules absorb certain wavelengths of light and reflect others. Our brain interprets different wavelengths as colors.

When light strikes the eye, it passes through the transparent cornea and aqueous humor and enters the pupil. It must then pass through the lens. By nature, the lens is nearly spherical, but the suspensory ligaments attached at the upper and lower edge pull it nearly flat. The ciliary muscles contract to reduce the tension of the suspensory ligaments, allowing the lens to become more convex. A more convex lens focuses light from nearby objects on the retina, allowing us to clearly see the objects. When focusing on distant objects, the ciliary muscles relax, allowing the suspensory ligaments to pull the lens into a flatter shape. The ability of our eyes to focus on objects at different distances is called *visual accommodation*.

Once the now-focused light has passed through the lens, it traverses the vitreous humor and strikes the retina at the back of the eye. The retina of each eye contains about 130 million photoreceptors. Most of these photoreceptors, called **rods**, are scattered over the retina. They are sensitive to low-intensity light but cannot distinguish color. As a result, color is difficult to see in low-light situations.

The other type of photoreceptors, the **cones**, detect color. We have three types of cones. Each one detects a range of wavelengths of light, which our brain interprets as different colors. Each type of cone is most stimulated by the wavelengths that we perceive as one of the primary colors. So one type of cone is most stimulated by the wavelengths

that we interpret as red, one is most stimulated by the wavelengths that we interpret as green, and the third is most stimulated by the wavelengths that we interpret as blue. Of course, we can see many more colors than just those three. When light waves such as those that we see as yellow strike our retinas, they stimulate both red and green cones. The brain receives the signals from both of the stimulated types of cones and interprets the object as yellow. The exact shade of color depends on which type of cone is most stimulated by that particular wavelength. The cones are especially concentrated in the *fovea*, a small depression in the central region of the retina. Therefore, we see the sharpest color image of an object when we are looking directly at it in a well-lit environment. In dim light an image is less clear because there are no rods at the fovea.

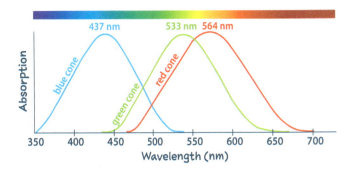

Both rods and cones contain light-sensitive pigments, and when the photoreceptors are stimulated by light, the pigments decompose. These substances initiate a complex biochemical pathway that changes light energy into a nerve impulse, which is carried to the cerebrum's occipital lobe by the optic nerve. The brain then interprets the sensory input as images and colors.

The Sensory Organs in the Human Body

Our sensory organs work with our nervous system to keep us aware of our surroundings. We can see potential danger, hear the voice of a family member, or feel whether an object is hot or cold without actually touching it. Most of a baby's sensory organs are fully functional at birth, although eyesight continues to develop over the first five months of life. But babies lack a framework to process data; this comes from experience. A growing child learns to incorporate new information with previously learned information.

A newborn can feel pain, and while we often think of pain as a bad thing, it serves a very useful purpose. It alerts us when something is wrong. For this reason, pain receptors are located in nearly every tissue of the body, unlike other sensory organs, which are usually concentrated in a single organ. However, some tissues have more pain receptors than others.

People who suffer neuropathy from a disease such as leprosy often suffer injuries or infection without realizing it. Even though antibiotics can cure leprosy, they cannot restore the function of the pain receptors destroyed by the leprosy bacteria. If any pain receptors survive the leprosy attack, they are usually located deep in the tissue. As a result, the person may not feel any pain from an infection until it has reached these deep tissues. This kind of infection is much harder to treat than one that is recognized quickly. And though an injury may not cause pain, it still causes tissue damage, which can lead to infection, possibly requiring amputation of a limb.

Those living with neuropathy caused by conditions such as diabetes often suffer continual pain, tingling, and numbness. In addition to making life difficult, neuropathy can create the same problems as it does for leprosy sufferers. Injuries may go unnoticed, and infection may become very bad before the victim seeks treatment.

The main difficulty in treating neuropathy lies in the fact that nerve tissue, unlike other tissue, typically does not regenerate. Although learning to trick neurons into regenerating would be a great long-term goal, several strategies can now be used to limit the damage done by neuropathy.

Currently, the best way to prevent neuropathy is to address its root cause. For neuropathy caused by leprosy, this means diagnosing and treating the disease quickly before it has time to destroy the patient's neurons. But this is often difficult since most cases of leprosy occur in places with little access to health care. The need for medical workers is often great. Several Christian organizations have established medical clinics that work with a local church or missionary, seeing to both the physical and spiritual needs of people.

This same *control-the-underlying-cause* approach also works best for treating neuropathy caused by type 2 diabetes or other non-contagious, chronic conditions.

Type 2 diabetes can be managed by changing one's diet and exercise habits. Pain relievers can also relieve some of the discomfort caused by neuropathy. But treating the underlying cause works only when we know what it is. Some neuropathy sufferers are diagnosed with *idiopathic neuropathy* because physicians cannot find a cause. In these cases, the treatments currently available help the individual manage the symptoms without curing the disease.

1. Indicate whether each of the following structures is part of the outer ear, middle ear, or inner ear.

   a. incus

   b. auricle

   c. saccule

2. The cornea is a transparent section of what layer of the eye?

3. Order the names of the following structures from largest to smallest: taste bud, taste cell, taste hair.

4. List the following structures in the order that a vibration would pass through them: auditory nerve, cochlea, eardrum, external auditory canal, hair cells, stapes.

5. Which structures of the inner ear control our sense of dynamic equilibrium?

6. Visual accommodation describes the function of what structure of the eye?

7. You are walking after dark, and you see the silhouette of an object up ahead. Which type of photoreceptors are your eyes using? Explain.

8. Some people with leprosy-induced neuropathy can pick up a hot dish without feeling the heat. Why might this be dangerous to them?

24.3 THE ENDOCRINE SYSTEM

The Structure of the Endocrine System

Where do hormones come from?

Remember the adrenal glands that are responsible for the fight-or-flight response? They are part of the endocrine system. Glands such as the salivary glands secrete their products through a duct, a tube that connects the gland to the organ where its product is needed. However, endocrine glands have no ducts. Instead, they release their products, chemical messengers called *hormones*, directly into the bloodstream. As we learned in Chapter 15, a hormone can affect cells far from the hormone's source.

The nervous system can cause a change in the body very quickly, but its effect is usually short-lived. To even flex your arm requires many nerve impulses, one right after another. The endocrine system usually works more slowly, but its effects are longer lasting. Various endocrine glands are located throughout the body.

Questions

What's the difference between the nervous system and the endocrine system?

How do hormones tell the body what to do?

How does a gland know when to produce a hormone?

How do hormones help a child grow into an adult?

Terms

pituitary gland
growth hormone
pineal gland
gonad
androgen
estrogen
thyroid
parathyroid gland
adrenal gland
insulin
glucagon
target cell
steroid hormone
nonsteroid hormone

THE ENDOCRINE GLANDS

As we learned in Section 24.1, the hypothalamus is the connection between the nervous system and the endocrine system. Its neurons produce hormones that are released at the ends of the axons in much the same way that neurotransmitters are released by most other neurons. Many of these hormones control the pituitary gland by function-ing as *releasing hormones*, like a chemical green light, and *release-inhibiting hormones*, like a chemical red light.

pituitary gland

pineal gland

When the hypothalamus's hormones reach the marble-sized **pituitary gland**, they affect the front half of the gland, the *anterior lobe*. The anterior lobe secretes several different groups of hormones that control other endocrine glands. It also releases **growth hormone**, which largely determines an individual's height and size. The gland's back half, the *posterior lobe*, provides a place for the hypothalamus's neurons to release two other important hormones. One of these, antidiuretic hormone, we learned about in Chapter 23 in connection with the urinary system. The other, *oxytocin*, stimulates smooth muscles to contract.

A small oval structure, the **pineal gland**, lies deep in the brain between the cerebral hemi-spheres near the thalamus. The pineal gland secretes the hormone *melatonin*, which is believed to affect a person's circadian rhythms. The pineal gland releases melatonin in response to darkness and stops releasing it in response to light.

From birth, the **gonads**, or reproductive organs, produce small amounts of sex hormones. In males the testes produce hormones called **androgens**, the most familiar of which is *testosterone*. In females the ovaries produce hormones called **estrogens** and another hor-mone called *progesterone*. During the early teen years the anterior lobe of the pituitary gland begins to produce *gonadotropins* that cause the gonads to begin producing sex hor-mones in far greater amounts.

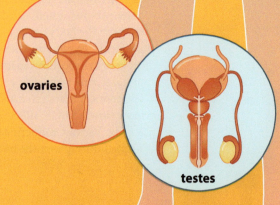

ovaries

testes

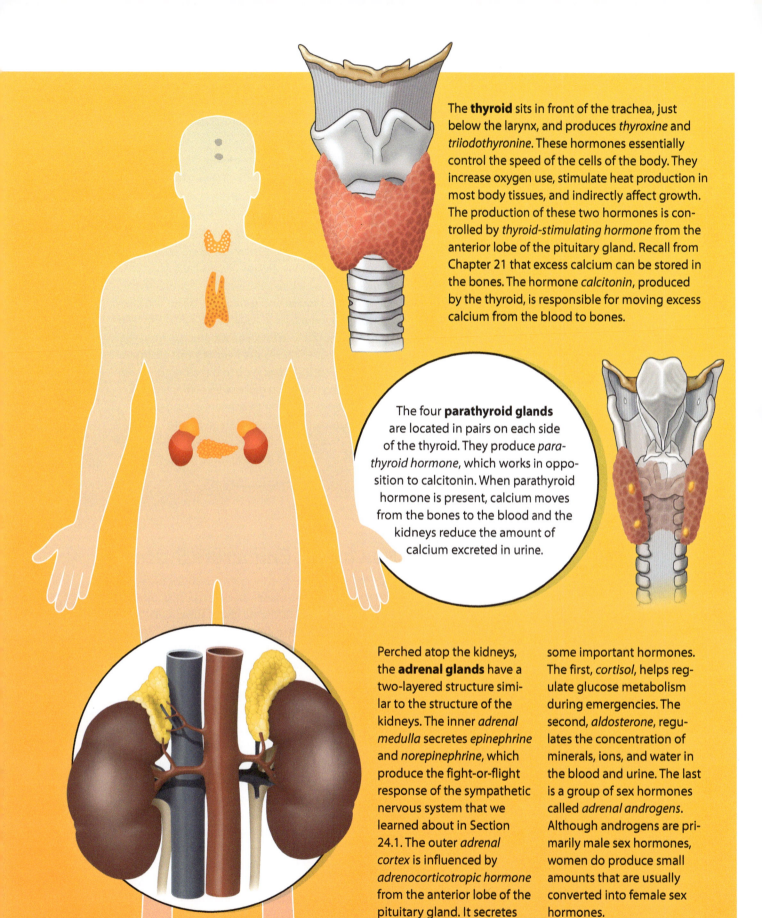

The **thyroid** sits in front of the trachea, just below the larynx, and produces *thyroxine* and *triiodothyronine*. These hormones essentially control the speed of the cells of the body. They increase oxygen use, stimulate heat production in most body tissues, and indirectly affect growth. The production of these two hormones is controlled by *thyroid-stimulating hormone* from the anterior lobe of the pituitary gland. Recall from Chapter 21 that excess calcium can be stored in the bones. The hormone *calcitonin*, produced by the thyroid, is responsible for moving excess calcium from the blood to bones.

The four **parathyroid glands** are located in pairs on each side of the thyroid. They produce *parathyroid hormone*, which works in opposition to calcitonin. When parathyroid hormone is present, calcium moves from the bones to the blood and the kidneys reduce the amount of calcium excreted in urine.

Perched atop the kidneys, the **adrenal glands** have a two-layered structure similar to the structure of the kidneys. The inner *adrenal medulla* secretes *epinephrine* and *norepinephrine*, which produce the fight-or-flight response of the sympathetic nervous system that we learned about in Section 24.1. The outer *adrenal cortex* is influenced by *adrenocorticotropic hormone* from the anterior lobe of the pituitary gland. It secretes some important hormones. The first, *cortisol*, helps regulate glucose metabolism during emergencies. The second, *aldosterone*, regulates the concentration of minerals, ions, and water in the blood and urine. The last is a group of sex hormones called *adrenal androgens*. Although androgens are primarily male sex hormones, women do produce small amounts that are usually converted into female sex hormones.

THE ENDOCRINE GLANDS
(continued)

We learned in Chapter 23 that the pancreas releases several different digestive enzymes in pancreatic juice. The pancreas also contains groups of cells called *islets of Langerhans*, which release two hormones into the bloodstream: **insulin** and **glucagon**. When the blood glucose level rises, for instance after a meal, insulin causes the body's cells to absorb glucose from the bloodstream and causes the liver to store glucose as glycogen. Both actions reduce the blood glucose level. When the blood glucose level drops, glucagon causes the liver to convert glycogen to glucose and causes fats to be converted to glucose, raising the blood glucose level.

islets of Langerhans

steroid hormone

receptor protein

hormone-receptor complex

protein synthesis

DNA

TARGET CELL

receptor protein

finished product

second messengers

activated enzyme

nonsteroid hormone

starting substance

TARGET CELL

The Function of the Endocrine System

Unlike the nervous system, which directly connects to the organs and tissues that it controls, endocrine glands can control structures far away from their location. Once a hormone enters the circulatory system, it can travel to its **target cells** in any part of the body. Target cells have specific receptors that recognize and bind specific hormones. As a result, hormones pass by nontarget cells without affecting them.

Hormones come in two basic forms, and each type binds to the receptors in its target cell differently. **Steroid hormones** have a high lipid content and can easily diffuse through the target cell's cell membrane and enter the cytoplasm. Once inside the cell, the steroid hormone binds with one or more receptors to form a *hormone-receptor complex* that alters the cellular processes of the target cell. Often the hormone-receptor complex binds to the DNA and activates specific genes. The gene is then transcribed into mRNA, which is then translated into proteins that cause the desired effect in the target cell.

Nonsteroid hormones lack a high lipid content and cannot diffuse through the cell membrane of their target cells. Instead, they bind to receptors embedded in the cell membrane of their target cells, causing the release of other chemicals, called *second messengers*, into the cytoplasm. These second messengers begin a chain of cellular processes that produce the desired changes in the target cell.

The Endocrine System in the Human Body

Hormones influence or control many metabolic processes throughout the body, including the water balance in blood, reproduction, growth, and blood pressure. They also prepare the body for important life changes. The sex hormones are steroid hormones that, during *puberty* (Ch. 25), activate genes that have been inactive since birth. As these genes produce proteins, major changes begin to happen. The reproductive organs begin to grow and the body begins to change.

As we can see, hormones are powerful. So how does the body control them? Hormone production is often controlled by negative feedback. For instance, the blood glucose level is part of the negative feedback loop for both insulin and glucagon. A low blood glucose level signals the pancreas to stop producing insulin, and a high blood glucose level halts glucagon production. This type of control allows the different hormones to maintain homeostasis, instead of disrupting it as they would do if they were allowed to operate unchecked. Additionally, many endocrine glands control or at least influence the production of others. The pituitary gland affects many of the other glands, and the adrenal glands initiate the production of sex hormones from the reproductive organs, to name just two examples.

If hormones remained in the bloodstream, they would build up until they caused all the different body processes to run out of control. Homeostasis would be destroyed, and we would all die very quickly. However, our loving Lord has designed our bodies to prevent this from occurring. The liver breaks down hormones, and many are removed by the kidneys to be excreted in urine.

The endocrine system seems to be interconnected with our thoughts and emotions. The exact relationship between these aspects of our minds is not known, but it is important to remember that humans are spiritual as well as physical creatures and that though the endocrine system is powerful, it is not all-powerful.

In spite of what is sometimes thought, Christians who are walking with God can feel depressed. Many people struggle with mood swings or just "feeling down," at least partially as a result of changes in their hormone levels. But while these types of mood changes are normal, we can easily fall into the trap of mistaking these feelings for reality. Even when believers are feeling down, God's grace is sufficient to enable them to do what they should do. In addition to seeking appropriate medical care, it's important to remember during dark times that Christ has redeemed all those who trust Him. He has promised to always be with His people, even when they don't *feel* that it's true. He has not called us to always feel happy, but He has called us to trust and obey Him.

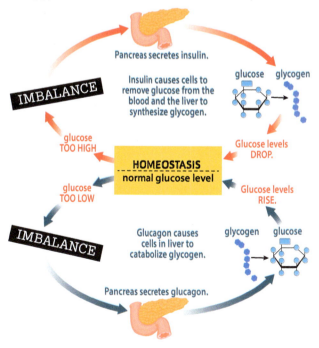

TYPE 2 DIABETES

Diabetes is a disease in which the blood glucose level is consistently too high, and yet the body's cells do not receive enough glucose. There are several causes, but one of the common forms, type 2, is caused by the cells becoming insensitive to insulin, a condition called *insulin resistance*. Although sufficient insulin is present, the target cells do not respond as they should.

Because the body's cells do not absorb glucose, they cannot perform cellular respiration, and the blood glucose level continues to rise. Over time, the high concentration of glucose can damage many different organs, including the kidneys, heart, eyes, and nerves. Because the cells are starving, the body begins to break down fats for energy, but this doesn't really help either. Instead, the breakdown of fats lowers the blood pH, a potentially fatal condition known as *acidosis*.

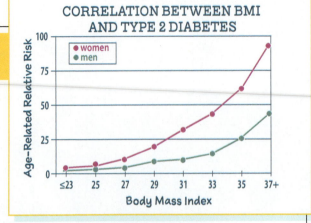

CORRELATION BETWEEN BMI AND TYPE 2 DIABETES

People who have diabetes must manage their condition. Since being overweight makes diabetes worse, being careful to maintain a healthy body weight is important. Exercise also improves blood glucose level, so a daily exercise routine will often help a diabetic person manage the condition. Diet is probably the most important aspect of a diabetes management policy. A diet low in sugars and high in protein helps keep blood glucose from spiking after each meal. Finally, some diabetics need to check their blood glucose level each day and give themselves injections of insulin.

24.3 SECTION REVIEW

1. Which structure is part of both the nervous system and the endocrine system?

2. Which endocrine gland releases hormones that control several other endocrine glands?

3. Which structures in a male begin to increase hormone production in response to gonadotropins?

4. Which part of the adrenal glands releases hormones in response to signals from the sympathetic nervous system?

5. Why is an endocrine system response to a stimulus slower than that of a nervous system response?

6. Explain why a steroid hormone does not need a membrane-bound receptor to influence its target cell.

7. A target cell is acted on by a hormone, but the mRNA levels in the cell remain the same. What type of hormone acted on this cell? Explain.

8. If a person's pancreas releases insulin, causing glucose to be removed from the bloodstream and converted to glycogen, why doesn't all the blood glucose get converted to glycogen?

9. The sex hormones are steroid hormones. How do these hormones cause the physical changes of puberty?

10. What two organs are especially important in removing hormones from the bloodstream?

11. Human beings are both physical and spiritual beings. What does that tell us about our emotions and our responses?

Use the case study above to answer Questions 12–13.

12. According to the data in the study's chart, what is something that teens can do to reduce their risk of developing type 2 diabetes?

13. While type 2 diabetes often strikes in adulthood, type 1 typically shows up early, often during childhood. In type 1 diabetes the pancreas reduces or even ceases producing insulin, often because the body's immune system mistakes the islets of Langerhans as foreign cells. Type 2 diabetes patients can sometimes manage their diabetes without insulin. Could type 1 diabetes patients do the same?

24 CHAPTER REVIEW
Chapter Summary

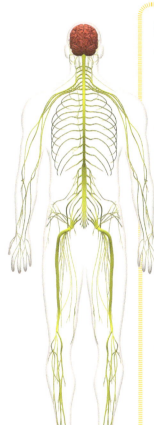

24.1 THE NERVOUS SYSTEM

- The nervous system consists of the central nervous system, composed of the brain and spinal cord, and the peripheral nervous system.

- Neurons are the functional units of the nervous system. Each neuron consists of a cell body, axon, and dendrites. Neurons send and receive electrical signals in response to stimuli.

- The brain controls most of the body's activities, both conscious (e.g., mental functions) and unconscious (e.g., blood pressure and heart rate). The exceptions to this rule are reflex arcs, which do not involve a nerve signal being processed by the brain.

- A nerve impulse is generated by the difference in electric charge between a nerve cell's resting and action potentials. A nerve impulse is transmitted from one neuron to the next by chemical neurotransmitters.

- Sensory neurons transmit sensory information to the brain, while motor neurons act on signals sent by the brain. Interneurons transmit signals between neurons.

- Under normal conditions, the body's functions are controlled by the parasympathetic nervous system. In times of stress, additional responses are generated by the sympathetic nervous system.

Terms
meninges • cerebrospinal fluid • neuron • cell body • dendrite • axon • cerebrum • thalamus • hypothalamus • cerebellum • brain stem • resting potential • action potential • synapse • neurotransmitter • threshold • sensory neuron • interneuron • motor neuron • parasympathetic nervous system • sympathetic nervous system

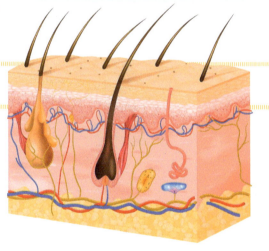

24.2 THE SENSORY ORGANS

- Sensory organs contain sensory receptors, providing the body with information about its environment. The two most important human senses are hearing and vision.

- The human ear consists of three divisions. Sound waves are collected by the outer ear, transferred to the bones of the middle ear by the tympanic membrane, and then converted into nerve impulses by the cochlea in the inner ear. The inner ear's semicircular canals also provide the body's sense of balance.

- The human eye passes light through its cornea, then focuses images onto the sensory cells of the retina by means of a lens. Rods detect low-intensity light, while three types of cones distinguish colors.

- Other senses include touch, smell, and taste. Each of these depends on specialized receptors located in the skin, nasal cavities, and mouth.

Terms
sensory receptor • external auditory canal • tympanic membrane • malleus • incus • stapes • cochlea • semicircular canal • sclera • cornea • choroid • iris • pupil • retina • photoreceptor • optic nerve • lens • hair cell • rod • cone

24 CHAPTER REVIEW
Chapter Summary

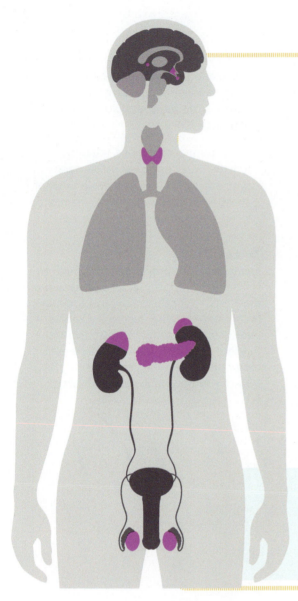

24.3 THE ENDOCRINE SYSTEM

- Unlike the nervous system, the endocrine system regulates body functions by means of chemical messengers called hormones.

- The hormones produced by various endocrine system glands regulate growth and development, govern daily cycles of activity, control the amounts of chemical substances in the body, and play important roles in behavior and emotions.

- The hormones produced by a gland are carried to target cells by the circulatory system. Steroid hormones can pass easily through cell membranes, while nonsteroid hormones bind with receptors in a cell membrane, triggering the release of secondary messengers.

- Some hormones work in pairs within a negative feedback loop to regulate the amounts of substances within the body. An example of this is insulin and glucagon produced by the pancreas to regulate the body's blood glucose levels.

Terms
pituitary gland · growth hormone · pineal gland · gonad · androgen · estrogen · thyroid · parathyroid gland · adrenal gland · insulin · glucagon · target cell · steroid hormone · nonsteroid hormone

Chapter Review Questions

RECALLING FACTS

1. What are the names of the three protective layers that surround the CNS? What are these collectively referred to as?

2. Describe the functions of cerebrospinal fluid.

3. Which part of your brain handles activities such as deciding to go for a walk?

4. What part of the brain includes control of the endocrine system among its functions?

5. Make a simple side-view sketch of the brain and label the frontal lobe (F), parietal lobe (P), occipital lobe (O), and temporal lobe (T).

6. Having decided to go for a walk, you step outside and realize how hot it is. Which lobe of your cerebrum receives the nerve impulses from your temperature receptors?

7. Indicate whether each of the following ions is inside or outside a neuron's cell membrane during a resting potential.

   a. potassium cations

   b. anions

   c. sodium cations

8. Which part of a neuron typically receives an incoming nerve impulse?

9. Which structure of the cochlea contains the hair cells?

10. Which structures of the membranous labyrinth are not involved in hearing?

11. In what order does light entering the eye pass through the eye's two humors?

12. Which endocrine gland releases a hormone when the blood sugar is too high?

UNDERSTANDING CONCEPTS

13. People can often survive an injury to the cerebellum, but one part of the brain is especially susceptible to fatal injury. Explain.

14. A speck of dust strikes the skin of a person with normal sensory perception, but the person does not feel it. What is the most likely explanation?

15. Jeff left school at 3:30, went to the gym and played basketball for a half-hour, showered, and walked home, arriving just before supper at 5:00. Which division of his autonomic nervous system was active during this time?

16. Rhonda is brought to an eye doctor. Her pupil is not changing size in response to changes in light intensity. It is determined that the issue is in a smooth muscle. What structure is likely at fault?

17. Describe olfactory accommodation.

18. Briefly describe how a sound wave is converted to a nerve impulse.

19. Which kind of receptors would the human eye need more of in order to see better at night? Explain.

20. Curtis can see some colors but not all of them. The most likely cause is a mutation in one of the genes that code for which eye structure?

21. What is the main obstacle to treating neuropathy?

22. What would happen if the body accumulated too much parathyroid hormone?

23. Hypoglycemia occurs when the blood glucose level becomes too low. Generally, it is caused when too little of a hormone is produced. Given what you have learned about hormones, which hormone does the body of a hypoglycemic person lack?

24. Place the following terms in order of the sequence that steroid hormones effect change in a cell: cell membrane, DNA, hormone-receptor complex, protein synthesis.

25. A newly discovered hormone is isolated and discovered to be made almost entirely of amino acids. What type of hormone is it? Explain.

CRITICAL THINKING

26. A friend says that an involuntary reflex is controlled by the PNS, not the CNS. Is your friend correct? Explain.

27. People sometimes blame poor behavior on overactive hormones. How should a Christian respond to such an assertion?

28. Many of the neurons in the CNS have myelin sheaths protecting the axons. In multiple sclerosis (MS) the immune system attacks the myelin sheaths and eventually destroys the nerves. What symptoms is MS likely to cause?

29. Although there is no cure for MS, sometimes there are ways to slow its progress. One of the most common classes of MS medicine is beta interferons, which are thought to inhibit some aspects of the immune system. How might this help?

25 REPRODUCTION, GROWTH &HEALTH

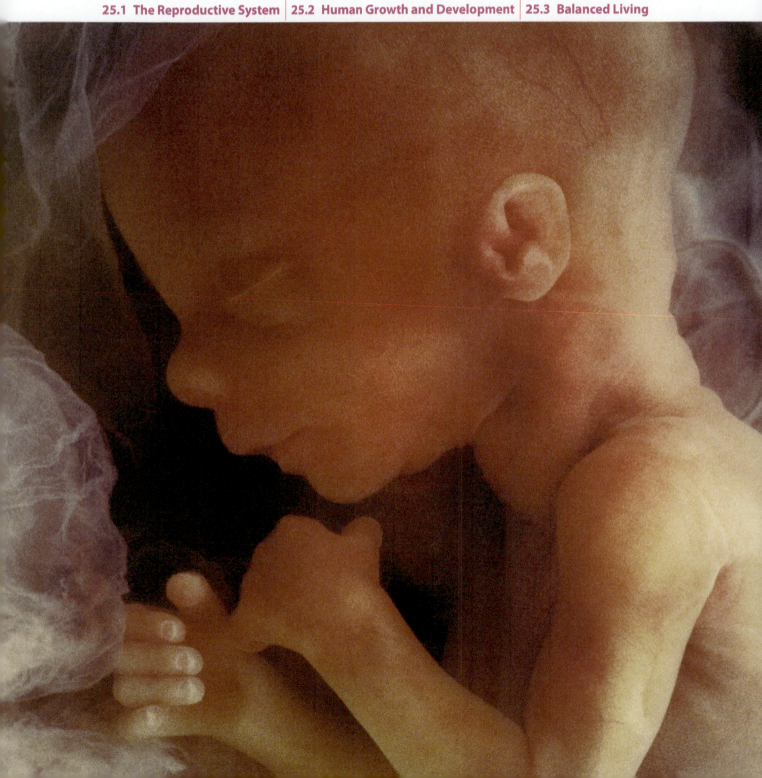

Missing Babies

For many people the birth of a child is a celebration of life. But for others a child may be considered unfashionable, burdensome, or perhaps even illegal. Countries in Europe and Asia have seen their populations remain static or even shrink as people stop having children. In places like China the government was so concerned about overpopulation and poverty that for many years it set limits to the number of children married couples could have in order to protect the country's resources. Abortions of baby girls have also been an issue in some countries. Because of this, there is a substantial gender gap in countries such as China and India that have a history of favoring male babies. This has left many men unable to find wives their own age or in their own country. This issue also affects the care and provision for the elderly, which are highly dependent on a growing, balanced population. When families stop having children or abort the ones they do conceive, society suffers.

But why is having children a good thing? Modern culture is questioning the very basics of family, gender, and sexuality. Who decides what is good for society?

25.1 THE REPRODUCTIVE SYSTEM

The Structure of the Reproductive System

?

How does the body produce new life?

The reproductive system is different from every other system that we've studied in the human body: *you can live without it*. In fact, the reproductive system works *against* all the other systems. Reproduction actually upsets rather than supports homeostasis. Just ask any pregnant mom—reproduction really does make life processes harder. An expectant mom has to breathe, eat, excrete, and drink for both herself and her growing baby, putting increased demands on her respiratory, digestive, excretory, and circulatory systems. However, though an *individual* can live without the reproductive system, the *human race* can't continue to exist without it.

Reproduction involves one aspect of the Creation Mandate that we haven't actually mentioned: the command to have children (Gen. 1:28). Reproduction is part of God's good and wise command for all of creation, and it extends to His highest creation—mankind. God made sex something that people can enjoy within God's design for marriage, but He also made it something productive for society. He is the only one who can determine what is ultimately good for our families, culture, society, and mankind.

Because of the Fall, reproduction is harder, just as the work of exercising good and wise dominion is. But that doesn't mean that we shouldn't reproduce. In fact, one part of having good and wise dominion means having children. Society and the world will bear the consequences if we do otherwise. On the other hand, we can enjoy the blessings of sex and children within the context of the biblical family. Let's take a look at the male and female organs that God has given us for our blessing and benefit. Follow the numbers in the diagrams that follow to see where sperm in a male and eggs in a female come from and how they unite to produce a baby.

Questions

What are the structures of the reproductive system?

How do an egg and sperm unite to become a new person?

What does the Bible say about sex?

Terms

testis
seminiferous tubule
epididymis
vas deferens
scrotum
seminal vesicle
semen
penis
prostate gland
ovary
ovum
fallopian tube
vagina
cervix
menstrual cycle
ovulation
corpus luteum
menstruation
cleavage
blastocyst
implantation
gastrulation

THE Male Reproductive SYSTEM

The **testes** (s. testis) are the most significant organs of the male reproductive system. Males have two testes, which produce tadpole-like sperm and the hormone testosterone. Sperm are the male reproductive cells (gametes) that fertilize the female's eggs.

A male's urinary and reproductive systems use many of the same vessels.

The urethra conducts both urine from the urinary system and sperm-containing semen from the prostate gland

Cells lining the **seminiferous tubules**, tiny tubes in the testes, undergo meiosis to produce millions of haploid sperm.

Sperm migrate to the tightly coiled **epididymis** to mature.

Sperm travel to the prostate gland through the **vas deferens**.

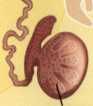

seminiferous tubules

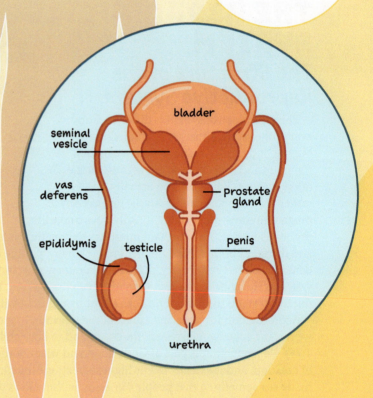

bladder

seminal vesicle

vas deferens

prostate gland

epididymis

testicle

penis

urethra

The **scrotum** is the sac of skin that contains the testes. It keeps the sperm at just the right temperature, about 2 °C lower than body temperature. The testes in a baby boy gradually descend from the pelvis into the scrotum during development before birth.

A pair of glands called **seminal vesicles** helps to produce **semen**, the fluid that surrounds, protects, and nourishes sperm migrating from the testes.

The walnut-sized **prostate gland** works with the seminal vesicles to produce the necessary fluid in semen. Sperm in semen migrate to the urethra.

The **penis** is the male appendage that deposits sperm. The penis also contains the urethra, as well as muscles, ligaments, and spongy material that control the angle of the penis. The muscles relax and blood flows into the spongy material when the male is sexually aroused, causing an *erection*. Sperm in semen are deposited in a process known as *ejaculation*, which takes place at the peak of a man's sexual experience, called an *orgasm*. If ejaculation happens during *sexual intercourse*, it may result in a fertilized egg.

THE Female Reproductive SYSTEM

The **ovaries** are the most significant organs of the female reproductive system; they produce the female's haploid eggs, called **ova** (s. ovum). Females have two almond-sized ovaries with thousands of immature ova. However, a sexually mature woman's ovaries usually release only one ovum each month. A developing ovum is nourished by surrounding cells called *follicles*.

Unlike in a male, the female's urinary and reproductive systems do not use the same vessels and are kept separate.

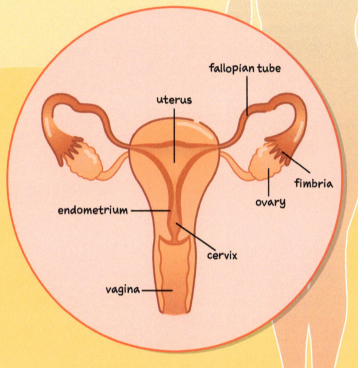

fallopian tube

uterus

fimbria

ovary

endometrium

cervix

vagina

Ducts known as the **fallopian tubes**, also called *oviducts*, carry the single, mature ovum away from the ovaries toward the uterus. This is where the male's sperm and the female's ovum unite to form a zygote, which develops into a human being.

Sperm from the male enter a female's body during sexual intercourse at the **vagina**, an elastic tube that leads from outside a woman's body to the cervix.

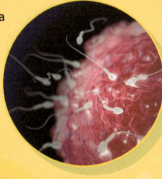

fertilization

Sperm migrate up the vagina through the **cervix**, a narrow opening into the uterus.

A pear-shaped organ—the *uterus*—is where a fertilized ovum ends its trip if the timing is right. Here a new life receives nourishment and develops. The uterus's epithelial lining, called the *endometrium*, thickens every month to prepare to nourish a fertilized ovum.

The Function of the Reproductive System

God's purpose for the function of the human reproductive system is to produce children within the context of a biblical family. But how does a woman's body prepare for and provide for a developing baby?

It begins with the **menstrual cycle**, the cycle of changes that happen to a woman's uterus to prepare to nurture a developing baby in the event that one of her ova is fertilized by a man's sperm.

The Menstrual CYCLE

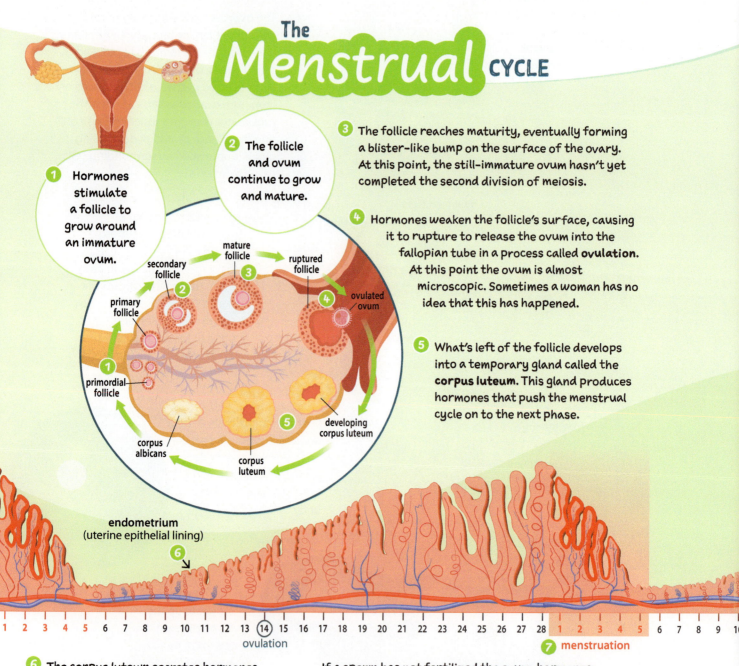

1 Hormones stimulate a follicle to grow around an immature ovum.

2 The follicle and ovum continue to grow and mature.

3 The follicle reaches maturity, eventually forming a blister-like bump on the surface of the ovary. At this point, the still-immature ovum hasn't yet completed the second division of meiosis.

4 Hormones weaken the follicle's surface, causing it to rupture to release the ovum into the fallopian tube in a process called **ovulation**. At this point the ovum is almost microscopic. Sometimes a woman has no idea that this has happened.

5 What's left of the follicle develops into a temporary gland called the **corpus luteum**. This gland produces hormones that push the menstrual cycle on to the next phase.

primordial follicle
primary follicle
secondary follicle
mature follicle
ruptured follicle
ovulated ovum
developing corpus luteum
corpus luteum
corpus albicans

endometrium
(uterine epithelial lining)
6

1 2 3 4 5 6 7 8 9 10 11 12 13 (14) 15 16 17 18 19 20 21 22 23 24 25 26 27 28 | 1 2 3 4 5 6 7 8 9 10
ovulation
7 menstruation

6 The corpus luteum secretes hormones that make the endometrium thicken in preparation for the attachment of an ovum that may have been fertilized in the fallopian tube after ovulation. A fertilized ovum migrates from the fallopian tube to the uterus.

If a sperm has not fertilized the ovum, hormones stimulate the uterus to shed the endometrium over the course of several days. This process is known as **menstruation** **7**, what many people refer to as a woman's period. Fluctuations in hormones during this time may affect a woman's mood, and muscle contractions in the uterine walls may cause cramps.

fusion of egg and sperm nuclei

zygote

2 Fertilization

A male's sperm must migrate up the female's reproductive system within a few days after ovulation for the ovum to be fertilized. At *fertilization* the haploid sperm and egg cells, which each contain twenty-three chromosomes, fuse together to form a diploid cell with forty-six chromosomes. The immature ovum nucleus divides, completing the second phase of meiosis to form a zygote. A new life has just begun.

3 Cleavage

The zygote undergoes mitosis to produce daughter cells in a phase of embryo development called **cleavage**.

2-celled zygote

4-celled zygote

MITOSIS

morula

8-celled zygote

1 Ovulation

During ovulation the mature ovum moves to the fallopian tube, surrounded by a few follicle cells for protection and nourishment.

Fertilization & Implantation

What happens if an ovum is fertilized during intercourse by one of millions of sperm? How would a woman's menstrual cycle be changed? A fertilized ovum sets in motion a whole series of events that ends with a baby being born after about nine months.

4 Blastocyst Formation

The rapidly dividing embryo takes three days to migrate to the uterus, and then floats free in the uterus for another three days. Cell division continues, producing a mass of cells in a fluid-filled sphere called a **blastocyst**.

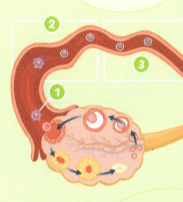

blastocyst

6 Gastrulation

During **gastrulation** the blastocyst forms the three germ layers that we learned about earlier—the ectoderm, mesoderm, and endoderm.

gastrula

5 Implantation

The blastocyst uses up its reserves of nutrients and attaches to the thickened endometrium in the uterus for nourishment— we call this **implantation**. From here, the baby's development accelerates.

The Reproductive System in the Human Body

The other body systems are deeply affected by the reproductive system, heightening the senses during sexual intercourse and involving the respiratory, circulatory, nervous, and integumentary systems. In addition, as mentioned previously, the reproductive system disrupts homeostasis.

A woman's organ systems maintain homeostasis by accommodating stresses on her body. The immune system plays a big part in this. If a woman becomes pregnant, her immune system is kept from attacking a developing embryo. There is a delicate balance between the body's ability to protect her life and the life of her baby from possible infections. Other body systems are involved too. The respiratory system carries oxygen and carbon dioxide for both mom and baby. The circulatory system transports nutrients to the baby and wastes from the baby through the placenta.

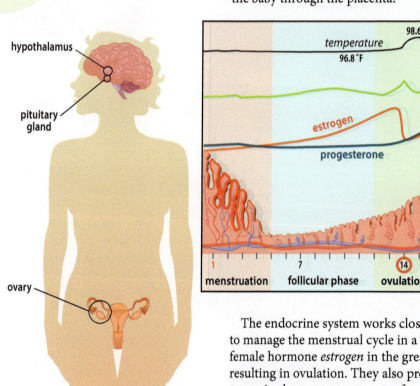

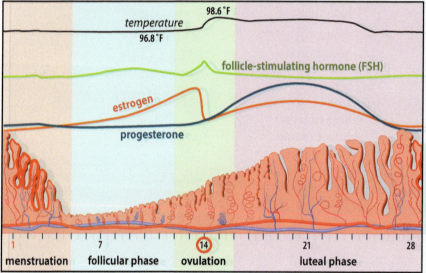

The endocrine system works closely with the reproductive system to manage the menstrual cycle in a woman. The ovaries produce the female hormone *estrogen* in the greatest amounts leading up to and resulting in ovulation. They also produce the female hormone *progesterone* in the greatest amounts after ovulation to thicken the uterus. Hormones offer a negative feedback mechanism similar to the way that the body monitors temperature. This negative feedback mechanism depends on whether an ovum is fertilized.

But the ovaries are not the only organs that produce hormones controlling the reproductive system. When a girl reaches sexual maturity, the hypothalamus signals the pituitary gland to begin secreting a gonadotropin called *follicle-stimulating hormone* (FSH). This hormone stimulates several follicles to develop each month to nourish a developing ovum in the ovaries. The pituitary gland is involved in producing the hormones that cause the corpus luteum to form in the ovaries.

When a boy reaches sexual maturity, the hypothalamus signals his pituitary gland to produce FSH to stimulate the development and nourishment of sperm in the seminiferous tubules of the testes.

Sexuality, Family, and a Biblical Worldview

God created sex, marriage, and the family. He created Adam and Eve as male and female and brought them together to experience union without shame (Gen. 2:24–25). The Bible uses the marriage relationship to illustrate Christ's love and care for the church (Eph. 5:23–27). Children are the natural and beautiful result of God's plan for marriage and the family.

The Fall has led to the perversion of God's good design, and modern society increasingly undermines the basics of sexuality. Not only do some husbands and wives forsake their God-ordained roles, but they break their marriage union through adultery and divorce. It is commonplace for unmarried couples to be immoral and to live together. Pornography is a thriving industry. Sexual abuse and rape threaten people's well-being. Homosexual marriage has been declared legitimate by the US Supreme Court. Our culture continues to wrestle with the basic question of gender identity. These things are not part of God's good and original plan for humanity, and they have the potential to unravel the fabric of society.

But God is working His grand plan of Redemption for the human race. He preserves society in spite of its fallen condition. When we live out God's good and perfect plan for the family, we can be salt and light in a dark world. At the end of all things, we will see God's perfect plan revealed in the complete redemption of His bride, the church.

You can make decisions *right now* that show that you believe that God's way is best. You can avoid behaviors like homosexuality and sexual immorality that fall outside God's design (Heb. 13:4). You can refuse to look at things that lead you to sin and that God calls wicked (Job 31:1). You can refuse to fantasize about sin (Matt. 5:28). Moral purity is not easy in our fallen world. But by God's grace it is achievable (2 Cor. 9:8). God's restoration of the human race needs to begin with the choices that we make every day. When we make choices that please and honor Him, our lives can be light in a dark world.

25.1 SECTION REVIEW

1. How does the function of the reproductive system fit with God's commands to mankind?

2. Order the following organs according to the formation and movement of sperm through the male reproductive system: epididymis, penis, prostate gland, seminal vesicle, seminiferous tubules, and vas deferens.

3. State the most significant organ of both the male and female reproductive systems and describe the purpose of each.

4. Arrange the following events in chronological order to describe how a woman's body prepares for and releases an ovum to be fertilized and nourished.

   a. An ovum is released to the fallopian tube during ovulation.

   b. A blastocyst attaches to the uterine wall during implantation.

   c. A fertilized ovum divides to form germ layers during gastrulation.

   d. A fertilized ovum migrates down the fallopian tube to the uterus.

   e. An ovum to be fertilized is nourished by developing follicles in the ovaries.

   f. An ovum unites with sperm in the fallopian tube to form a zygote.

5. What additional purpose does the female reproductive system serve that the male system does not?

6. Give an example not mentioned in Section 25.1 of how the Fall affected God's command to have children.

7. Give an example of how you can make decisions right now that show that you believe that God's plan for the family is best.

25.2 HUMAN GROWTH AND DEVELOPMENT

Questions

How does a baby develop and grow in the womb?

How does a child change with age?

How is a teen's body different from that of a child?

How does an adult change with age?

Terms
umbilical cord
fetus
amniotic fluid
puberty
basal metabolic rate

Prenatal Development and Birth

If you've ever seen a woman who is pregnant, then you've probably observed her eagerness to meet the baby forming inside her.

We've tracked how a woman's body produces an ovum that is fertilized in the fallopian tubes and implanted in the uterus where it is nourished. But what happens after that? It takes about forty weeks, or nine months, for a baby to mature from a fertilized egg into a squealing newborn. Doctors divide this time into three parts, called *trimesters*.

?

How have you changed as you have grown?

THE FIRST TRIMESTER

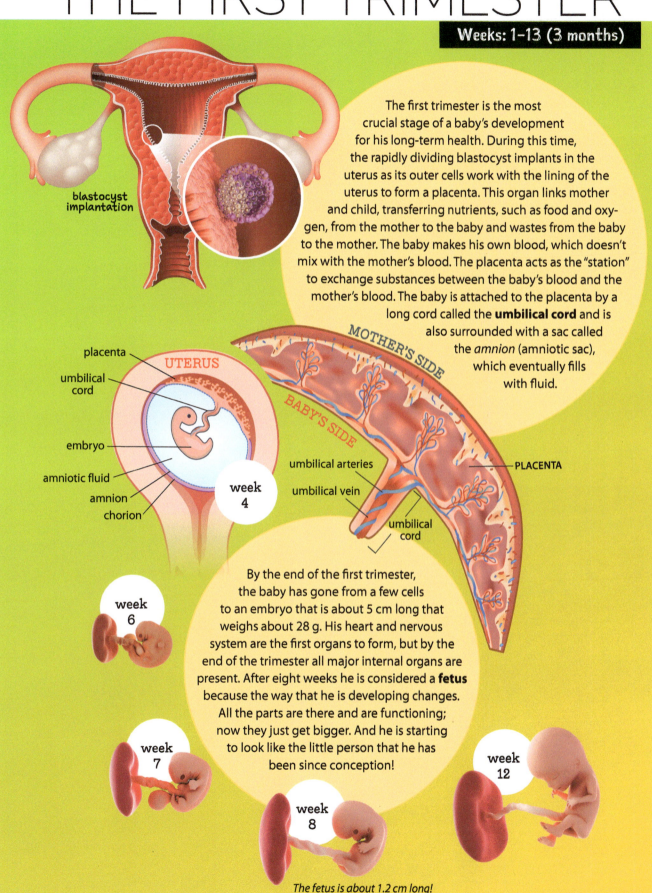

blastocyst
implantation

The first trimester is the most crucial stage of a baby's development for his long-term health. During this time, the rapidly dividing blastocyst implants in the uterus as its outer cells work with the lining of the uterus to form a placenta. This organ links mother and child, transferring nutrients, such as food and oxygen, from the mother to the baby and wastes from the baby to the mother. The baby makes his own blood, which doesn't mix with the mother's blood. The placenta acts as the "station" to exchange substances between the baby's blood and the mother's blood. The baby is attached to the placenta by a long cord called the **umbilical cord** and is also surrounded with a sac called the *amnion* (amniotic sac), which eventually fills with fluid.

placenta

umbilical cord

UTERUS

embryo

amniotic fluid

amnion

chorion

week 4

MOTHER'S SIDE

BABY'S SIDE

umbilical arteries

umbilical vein

umbilical cord

PLACENTA

week 6

By the end of the first trimester, the baby has gone from a few cells to an embryo that is about 5 cm long that weighs about 28 g. His heart and nervous system are the first organs to form, but by the end of the trimester all major internal organs are present. After eight weeks he is considered a **fetus** because the way that he is developing changes. All the parts are there and are functioning; now they just get bigger. And he is starting to look like the little person that he has been since conception!

week 7

week 8

week 12

The fetus is about 1.2 cm long!

THE SECOND TRIMESTER

Weeks: 14–26 (3 months)

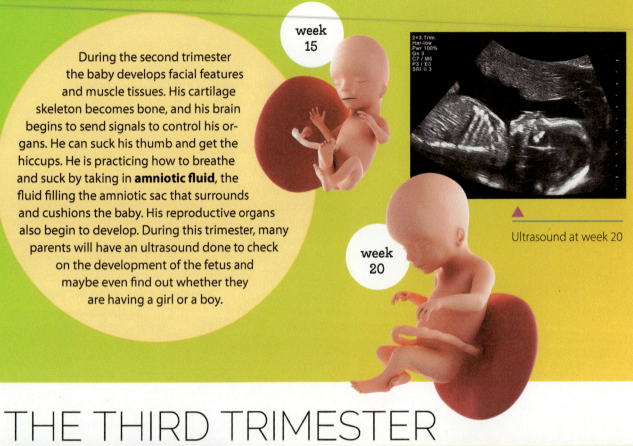

During the second trimester the baby develops facial features and muscle tissues. His cartilage skeleton becomes bone, and his brain begins to send signals to control his organs. He can suck his thumb and get the hiccups. He is practicing how to breathe and suck by taking in **amniotic fluid**, the fluid filling the amniotic sac that surrounds and cushions the baby. His reproductive organs also begin to develop. During this trimester, many parents will have an ultrasound done to check on the development of the fetus and maybe even find out whether they are having a girl or a boy.

week 15

week 20

2+3.Trim.
Har-low
Pwr 100%
Gn 3
C7 / M5
P3 / E3
SRI II 3

Ultrasound at week 20

THE THIRD TRIMESTER

Weeks: 27–40 (3 months)

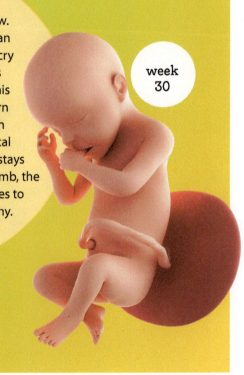

week 28

The baby is growing quickly now. He is moving so much that he can make mom uncomfortable! He can cry and open and close his eyelids—he's practicing for after he is born! Fuzz on his head is the beginning of hair. If he is born during the third trimester, he has a high possibility of survival with close medical supervision. However, the longer he stays inside his mother's womb, the better his chances to be born healthy.

week 30

A 3D ultrasound reveals many detailed characteristics of a growing baby.

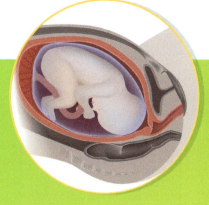

BIRTH

It's getting close to the time that baby first sees the world. The mother's uterus is stretched to its maximum size, and the fetus is too big for his cozy home. After a hormone, *oxytocin,* is released, the muscles in the uterus start to contract. Contractions become stronger and more frequent until they peak and occur at regular intervals. This is a sign that mom is in labor!

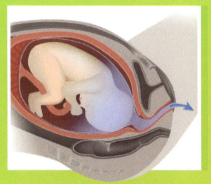

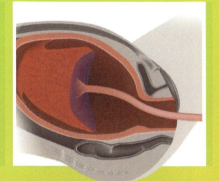

1 During labor, the mother's cervix becomes thin and opens (dilates). The amniotic sac breaks, releasing a gush of amniotic fluid. Eventually, these contractions push the baby's head into the vagina, or *birth canal.* Both the mother's pelvic area and the baby's head flex to allow it to push through the vagina.

2 Then the baby's head crowns (i.e., the top or "crown" becomes visible) and completely emerges, and his body follows. However, he's still attached to the placenta by the umbilical cord. This cord is pinched and is usually cut within one to three minutes after the baby is born. It's time for him to breathe on his own!

3 After a baby is born, the placenta and the rest of the umbilical cord, together called the *afterbirth,* are delivered. This normally happens within about twenty minutes after a baby is born.

It will take a mother about four to six weeks to recover from childbirth. In the meantime, the baby can get his nourishment from her in an entirely new way—by nursing from her breasts, which produce milk in a process called *lactation.* Hormones have already triggered her body to produce this supply.

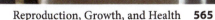

CHILDHOOD

If you've had a baby born in your family, you know that he doesn't stay little for long! During the first year, he changes dramatically, learning how to communicate, make decisions, and move about—perhaps even walk.

His development continues after his first year as he grows more teeth and learns how to regulate his body's functions. His body's proportions change as he matures, his head forming a smaller percentage of his body as he ages. A child has significant nutritional needs to fuel his growing body, such as calcium for his quickly growing bones and muscles.

ADOLESCENCE

As a child grows to become a teen, his body begins to dramatically change as the hypothalamus signals the pituitary gland in the endocrine system to produce a whole new batch of hormones. These hormones cause the child to begin to develop *secondary sex characteristics*. Boys begin to look trimmer as they lose the body fat of childhood and develop muscles. Vocal folds get larger, making both male and female voices deeper. External sex organs, or *genitals*, enlarge. Hair begins to grow in places other than on the head, such as under the arms, on the legs, and in the *pubic region*, the area of the body near the genitals. Teen girls experience similar changes and begin their menstrual cycle. This dramatic time of life in which a person physically transitions to adulthood is called **puberty**.

But the changes in a person's body during adolescence don't just involve his reproductive and endocrine systems. Huge swings in hormones also affect a person's nervous system, influencing his emotions and sense of balance as his brain tries to keep up with his quickly growing body. This can also affect a teen in social and spiritual ways.

Sometimes people struggle with their sexuality during adolescence. Some teens in this stage of life experiment with pornography, heterosexual immorality, and homosexuality. Some experiment with all of them, searching for fulfillment. Occasionally teens struggle with gender identity. In fact, some professionals make a distinction between *sex*, as determined by one's reproductive organs and secondary sex characteristics, and *gender*, one's masculinity or femininity as communicated in his or her social interactions.

God's Word tells all of us how to think about ourselves and our sexuality. The Bible prohibits all sexual activity outside of marriage (Exod. 20:14), and it limits marriage to one man with one woman (Gen. 2:18, 24). The Bible makes no distinction between a person's sex and gender (Matt. 19:4). While it is true that the Fall has affected every part of man, it does not follow that gender is undetermined and can be chosen by an individual. Men and women should accept their sex as God's will for them and should seek to act in keeping with that determination.

You are in this stage of life. Your body is approaching *physical* maturity, but you will spend the rest of your life acquiring wisdom and maturing *spiritually*. That process can occur only one way—by living in biblical wisdom and choosing God's way as the best way. A relationship with Jesus Christ is the only path to real fulfillment.

ADULTHOOD

So what stage comes after adolescence? When you reach your twenties, you will most likely experience your peak physical condition. It is at this stage of life that you have the potential to be strongest and most capable of procreating. Men begin to produce sperm at the onset of puberty. For a woman, ova have been present in her body since birth but now regularly mature to allow her to become pregnant. Women also experience an entirely different range of hormones during pregnancy, birth, and while nursing a baby.

After our twenties, however, both men and women begin to experience more changes. In our thirties and forties, we begin to lose muscle strength and ligament elasticity and hair eventually begins to turn gray as melanin production is diminished. As we age, all life processes begin to slow down and become less efficient. The amount of energy the body uses during resting, called the **basal metabolic rate**, changes as we grow older. This rate starts to decrease significantly in the thirties and forties, and weight gain may result. We also tend to get slightly shorter as we age. By the time a woman reaches her fifties, her hormone production has changed dramatically, and her menstrual cycle and ovum production stops during what is known as *menopause*. Men continue to produce sperm for the rest of their lives, though the number produced decreases.

The life expectancy for an American is now about eighty years. In our later years of life we can still have rich experiences, investing in our families and the people around us. We can still have a meaningful life, especially as we invest in younger people. Take a moment and think—what older people have invested themselves in you?

God knows, even if we don't, how many days He will give us in this life (Ps. 90:10). If He gives us seventy or even eighty years of life, we still will eventually die, and the wise thing to do is to make choices in light of that reality (Ps. 90:12). Death is not the end. The Bible teaches us that our bodies will be raised and that we will spend an eternity either in the presence of God or in the place of judgment. For Christians, the hope of resurrection is the hope of going to be with Jesus and being reunited with believing loved ones. Death is only the beginning of an eternity in heaven.

GENDER IDENTITY

It is likely that you have heard of issues regarding gender identity. Terms that tend to go along with gender identity include *nonbinary*, *transgender*, or *gender fluidity*. But what do these terms mean? Simply stated, these terms describe the gender identity that a person adopts. Trends in those claiming to be nonbinary, identifying as neither male nor female, are increasing. Why is this so?

Statistics have shown that *gender dysphoria*, the clinical term used to describe the desire to identify as another gender, can present itself in children as early as two to four years of age but more often manifests around the time of puberty. Studies have also suggested that there are a number of factors that influence the development of this condition. In some cases those who experience gender dysphoria have also experienced childhood trauma, though a causative link between the two has not been conclusively established. And sadly, people with gender dysphoria are also more likely to develop depression and anxiety, and even commit suicide.

In response to the desire to become a different gender, up to half of those with gender dysphoria decide to undergo hormone replacement therapy or gender reassignment surgery. But choosing such options is no guarantee of future happiness. Studies have suggested that some of those who have made gender modifications later regret their decision and decide to return to their original gender. The main reason people gave for this type of de-transitioning was simply that making the change didn't seem to help them feel any different. They came to realize that though they changed outwardly, their genetics remained the same. Currently, there seems to be a push to change the body on the basis of feelings of gender identity without taking into account that feelings are subject to change.

25.2 SECTION REVIEW

1. What is the difference between an embryo and a fetus?

2. What are the three stages of pregnancy? Describe the baby's development in each stage.

3. How is a child's body different from that of a baby?

4. List three of the secondary sex characteristics that develop with puberty.

5. What organ and organ system are responsible for the changes that happen to a person during puberty?

6. Describe three ways that a person's body changes in the fourth decade of life.

7. How should a believer view death from a biblical worldview?

Use the case study above to answer Questions 8–11.

8. According to the information given, what connections can you make regarding the presence of gender dysphoria and trauma?

9. What does the Bible say about gender identity?

10. According to what the Bible says about gender, what conclusions can we draw?

11. On the basis of these conclusions, how should a Christian respond to someone dealing with gender dysphoria?

25.3 BALANCED LIVING
What You Take In

Up to this point in life you have had many of your decisions made for you. But as you get older, you will become more independent. You'll need to make life decisions that affect your body. We need to consider what we take into and do with our bodies, recognizing that a person is more than just a body.

NUTRITION

We have already considered the kinds of nutrients that our bodies need to be healthy. There is a lot of disagreement over what the healthiest lifestyle is, and scientific research is beginning to indicate that what works well for one person may not work well for another. However, just about everyone agrees that we all need to drink a sufficient quantity of water, limit the amount of added sugars and processed foods in our diets, and increase the amount of time we spend exercising. This doesn't mean that a person can never have dessert again—just not at every meal!

To have a healthy body, we need certain amounts of nutrients from the food we eat, including water, carbohydrates, fats, proteins, minerals, and vitamins. But many Americans are overeating. Additionally, many Americans are not nearly active enough, so they are consuming far more calories than they need. People often find that it's easy to overload on calories by replacing water with sugary drinks such as sodas.

Although people have been dieting for years, many still struggle with being overweight. Those who lose weight while dieting often gain it back again soon after going off the diet. Dieting is often an attempt to quickly fix the underlying problems of an unhealthy lifestyle.

While it's a good idea to keep our weight in check to avoid the problems that come with being overweight, there's an even better reason. God lives in the bodies of believers, and it is our responsibility to maintain these bodies as the Holy Spirit's temple (1 Cor. 6:19–20). It's easy to think about this when considering drinking alcohol, taking drugs, or smoking cigarettes, but it applies to eating too. Being healthier can make a believer more useful for God's glory.

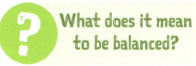

What does it mean to be balanced?

Questions

What outside factors affect the body's homeostasis?

How do exercise, sleep, and hygiene help maintain homeostasis?

How do your mind and spirit affect your body?

Is assisted suicide ethical?

Terms
drug
withdrawal
dopamine
REM sleep

Healthy eating is both a matter of what we eat and how much we eat.

▼

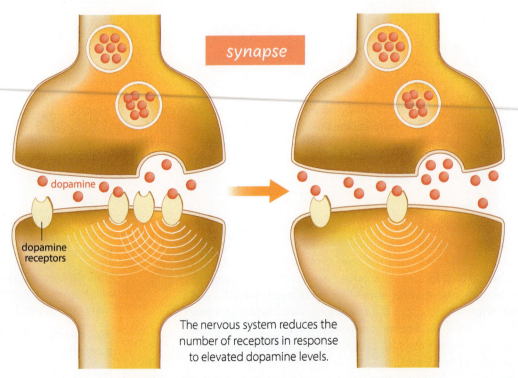

synapse

dopamine

dopamine receptors

The nervous system reduces the number of receptors in response to elevated dopamine levels.

DRUGS

A **drug** is any chemical that changes the function or structure of a living tissue. The same drug can be good or bad, depending on how much we take. People use drugs as medicines to prevent, treat, or cure a disease. Certain drugs, such as sleeping pills, affect body function. Aspirin and ibuprofen act as inhibitors to enzymes that cause us to feel pain. Vaccines give us immunity and antibiotics fight infections. Some substances, like antidepressants, can even affect our emotional state.

With some drugs, a person gradually becomes less responsive to the same amount of a drug, an effect called *drug tolerance*. An increased dosage is required to achieve the desired physical response. For some drugs like alcohol, caffeine, and morphine, a person can form a physical or emotional addiction. A person can experience **withdrawal**— a series of physical consequences that make it very difficult to be without the drug.

Why does this happen? Drugs that cause addictions affect neural synapses by producing a neurotransmitter called **dopamine**. Dopamine is one of several chemicals that help your brain feel happy. For example, when you exercise, your body naturally releases dopamine to produce a

"high." Substances as innocent as caffeine or as sinister as heroin work by causing the release of dopamine.

When drug tolerance occurs, an addict begins to need more of the drug to produce the same effect. This happens because the brain naturally reduces the number of dopamine receptors in the nervous system when there is an abundance of the substance. The body also produces much less dopamine. A person experiencing withdrawal feels depressed and can undergo significant symptoms, such as chills, vomiting, insomnia, and pain.

Medicines are good and useful for bringing people healing, but because we are sinners, we can use good things in evil ways. Many of the people who develop addictions to drugs are desperately seeking some kind of happiness, some kind of satisfaction that will last. Jesus Christ is the only one who can fill our greatest need and bring healing that lasts for eternity. For centuries people with addiction problems have found lasting freedom, joy, and satisfaction in life through Christ. That opportunity still exists for people today.

What You Do

EXERCISE

Though you're probably not too worried about your health right now, the decisions you make today can set you up for a healthy lifestyle. You can actually raise your basal metabolic rate and reduce your risk of cancer, heart disease, diabetes, arthritis, and depression in one fell swoop. And, by the way, you could also have a trimmer body. Sound like a miracle pill? You just need to get moving and keep going! Exercising is a healthy way to naturally increase your dopamine, and your body needs it to stay balanced.

Almost any movement will do—stretching, walking, gardening, bicycling, even doing chores around the house (won't your mom like that!). You don't have to hit the gym to get and keep moving. Consider taking up a sport, whether an individual or team sport. Whatever you do, make it fun so that you will be more likely to keep doing it. Start slow and enjoy the outdoors while you're at it. You can focus on *aerobic exercises*, which increase your rate of breathing, or focus on getting stronger with strength, weight, or resistance training. You can work on flexibility. Or you could do all of these. There's almost no bad way to exercise if you start slow and work up gradually.

SLEEP

You probably like to stay up late and enjoy texting or chatting on social media; just make sure that you don't have to get up at 6:00 the next morning! Your body needs sleep to refresh all your organ systems, especially your brain. During a research study done in 1963, sixteen-year-old Randy Gardner went without sleep for a record-breaking eleven days. He experienced hallucinations, short-term memory loss, and mood swings. Even though he completely recovered without any problems, it's clear that people need sleep!

There are two types of sleep—**REM** (rapid eye movement) **sleep** and *NREM (non-REM) sleep*. REM sleep is the sleep of dreams during which the brain's activity is the highest. NREM sleep is a deep, dreamless sleep during which the body physically rests. You could think of REM as active sleep for the brain, while NREM is quiet sleep for the brain. During the night, your body naturally goes through cycles of these two types of sleep. If your sleep gets disturbed, your whole body suffers!

Though different people at different ages have different sleep needs, right now you need about eight and a half to ten hours of sleep each day. Newborns can sleep up to eighteen hours a day.

The brain waves that a person presents during the different stages of sleep are displayed. The tighter the waves, the more active the sleeper is. Compare REM brain waves with the ones that a person has when awake.

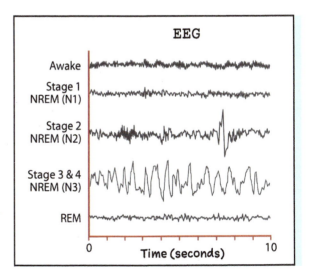

EEG

Awake

Stage 1
NREM (N1)

Stage 2
NREM (N2)

Stage 3 & 4
NREM (N3)

REM

0   Time (seconds)   10

HYGIENE

You're probably used to wearing clean clothes, taking a shower every day, brushing your teeth, and using deodorant so that you don't repulse everyone in your school! But there are other reasons for good hygiene beyond being socially acceptable. Good hygiene can help keep you and the people around you healthy by maintaining your body's homeostasis.

Brushing your teeth twice a day and flossing daily prevents tooth decay. You might not realize it, but cavities are contagious! There are even some scientific studies that link gum disease to heart problems and strokes.

Keeping your skin clean can also help prevent rashes and acne. Wear sunglasses on sunny days and sunscreen every day to protect your eyes and skin from the sun. Exposure to the sun over the course of your life can lead to skin cancers and speed up aging in your skin.

Washing your hands, especially before meals, protects your body from antigens that could cause infections and disease. Keeping your fingernails trimmed helps the skin in this high-traffic area of your body do its job to keep out invaders.

Showering every day and washing your hair regularly helps keep harmful bacteria at bay.

Who You Are

What makes a living, healthy person? It isn't just breathing and eating, though these things are necessary for life. You are not just a body—you are a rational, emotional, and social being made in God's image. Part of being alive and healthy is having healthy thoughts, emotions, and relationships.

Many people struggle with their mental health. Worldwide, roughly 264 million people suffer from depression. Mental problems incapacitate millions of people each year, causing them to flounder in their work and relationships. When people are stressed and depressed, it affects their bodies. They don't do what they should with nutrition, exercise, hygiene, and sleep. Some even take their own lives, despite the fact that help is available. We live in a world where some people have no hope.

Many people struggle with their relationships. Marriages end in divorce every day, and almost 300,000 Americans experience some kind of sexual abuse every year. These strains on relationships can sometimes affect an individual's physical and emotional health. The abortion of nearly 900,000 babies every year in America places mothers under an incredible burden of guilt. Divorce, abuse, and guilt bring stress and plunge people into depression that affects their bodies. We live in a world where people are hurting and feel insecure.

Science can't give us hope, and it can't give us security. God wants us to have abundant life, and that comes through one Man—Jesus (John 10:10). So we end where we began: the one who created the world and watched it fall became a man who brings us true, eternal life. Only Jesus restores the hope and security that was present in Eden and robbed from mankind at the Fall. When believers live in obedience to God's commands, they live out the work of redemption that Christ has begun in them. And He will bring it to completion one day, rising as the Sun of righteousness with healing in His wings for those who fear Him (Mal. 4:2).

MINI LAB

Researching the Impact of Our Thoughts

Philippians 4:8 is a popular verse of Scripture that Christians often quote when confronted with issues dealing with their thoughts. What some may not realize is that a vast amount of research has been done on this topic. And just like many other references in the Bible, much of this research supports God's original plan for our thought lives. God knew the significance of our thoughts well before it was researched. But how do our thoughts truly impact our daily lives?

Do our thoughts really impact our health?

Materials
computer with internet access

PROCEDURE

A Perform an internet search for peer-reviewed journals with keywords such as "how a person's thoughts affect his health." Record your findings.

B Research strategies that may help support and encourage the right forms of thinking. Record your findings.

C Create a plan for the implementing one of the thought strategies.

D Test your plan by putting it into practice. Over the next few days note any positive or negative effects that you have experienced during your testing.

QUESTIONS

1. Explain the research you found regarding the effects of a person's thoughts on his health.

2. What researched strategy did you find that may produce the greatest positive effect on health?

3. What thought strategy did you choose to implement? Explain your reason for choosing this strategy.

4. What other practices did you come across in your research that may lead to better health outcomes?

GOING FURTHER

5. Other than Philippians 4:8, list two other Bible verses that speak about a person's thoughts. Explain how the research supports the verses that you selected.

25.3 SECTION REVIEW

1. When considering calories, what must a person do to lose weight?

2. What substances, when ingested, affect the body's homeostasis?

3. Give an example of a drug and explain why your example is considered a drug.

4. When does a drug become harmful?

5. Give an example of an aerobic exercise.

6. During which type of sleep does the brain use the most oxygen and glucose from the body (see graph on page 571)? Explain your answer.

7. How does good hygiene help your body maintain homeostasis?

8. How do mental health and relationships affect the body's homeostasis?

9. How can a Christian enjoy a healthy mind and healthy relationships?

25.1 THE REPRODUCTIVE SYSTEM

- The male reproductive organs include the testes, seminiferous tubules, scrotum, epididymis, vas deferens, prostate gland, penis, and urethra.

- The female reproductive organs include the ovaries, fallopian tubes, uterus, cervix, and vagina.

- As part of a women's menstrual cycle, an ovum is produced after hormones stimulate a follicle to grow. At that time the endometrial lining in the uterus thickens in preparation for a fertilized egg.

- After intercourse, if an egg is fertilized, it rapidly divides as it travels through a fallopian tube to the uterus, where it will implant and continue developing.

- The reproductive system is controlled by the hormones released by the endocrine system.

- Man's sinful nature has altered God's original plan of marriage and family.

Terms
testis · seminiferous tubule · epididymis · vas deferens · scrotum · seminal vesicle · semen · penis · prostate gland · ovary · ovum · fallopian tube · vagina · cervix · menstrual cycle · ovulation · corpus luteum · menstruation · cleavage · blastocyst · implantation · gastrulation

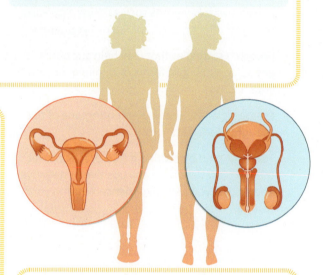

25.2 HUMAN GROWTH AND DEVELOPMENT

- After implantation, an embryo grows and is nourished by the mother through the placenta.

- Throughout the duration of three trimesters, or nine months, a fetus develops and grows until it is ready to be born.

- Development continues throughout the stages of childhood, adolescence, and adulthood.

- Puberty is a transition to adulthood in which hormones activate changes in a child's body. These changes affect an adolescent's reproductive, endocrine, and nervous systems.

- In the beginning of adulthood, men and women are at their peak physical condition. As they age, life processes begin to slow down, and the body becomes less efficient at using energy.

Terms
umbilical cord · fetus · amniotic fluid · puberty · basal metabolic rate

25.3 BALANCED LIVING

- What we take in, what we do, and who we are all affect the body's homeostasis.

- Drugs have the ability to prevent, treat, or cure diseases, but they can also become addictive and harmful.

- Exercise, sleep, and good hygiene are important ways to keep our bodies healthy.

- Having healthy thoughts, emotions, and relationships are also an important aspect of good health.

Terms
drug · withdrawal · dopamine · REM sleep

Chapter Review Questions

RECALLING FACTS

1. Where are the seminiferous tubules located in the male reproductive system?

2. Which organ of a woman's reproductive system nourishes a developing embryo?

3. How many ova do a woman's two ovaries produce each month?

4. List the following events in chronological order: blastocyst formation, cleavage, fertilization, gastrulation, implantation, ovulation.

5. Chronologically order the following events in the life of a baby.

    a. The baby is considered a fetus, with all organs present.

    b. The baby's cells begin to divide, and a blastocyst implants in the uterus.

    c. The baby can open and close his eyes and is growing hair on his head.

    d. The baby forms a placenta and an umbilical cord.

    e. The baby's reproductive organs move into place.

    f. The baby's heart and nervous system form.

6. Given that our life and health are ultimately in God's hands and not ours, what is a biblical reason for wanting to live a healthy life?

7. What happens when a person stops using a drug that he or she has become addicted to? Why does this happen?

UNDERSTANDING CONCEPTS

8. Explain how a man's body keeps sperm alive as long as possible.

9. Explain how a woman's reproductive system works with her endocrine system.

10. In one or two sentences, describe a biblical view of sex.

11. Describe how an embryo develops during the different stages of pregnancy.

12. Explain how puberty starts.

13. Describe the physical changes that occur during puberty.

14. Explain how adolescence is different from adulthood.

15. Describe what substances affect the body's homeostasis.

16. Explain how exercise affects the body's homeostasis.

17. Why do we say that a person is more than a body?

18. Other than what we put into our bodies, explain what contributes to a person's health.

CRITICAL THINKING

19. In a tubal pregnancy an embryo fails to migrate from the fallopian tube to the uterus and implants in the fallopian tube instead. This is an example of an *ectopic pregnancy*. Why do you think this is problematic?

20. Two months ago, Jared began taking a medicine to manage his chronic back pain. But more recently he has been unable to get his pain levels down. After a quick appointment with his doctor, Jared was prescribed a higher dosage of the same medicine. Explain what occurred when Jared began experiencing pain. What do you believe may happen with a higher dosage of the medicine?

21. Calculate your basal metabolic rate using the equations below.

    Women: BMR = 655 + 4.35 × (weight in lb) + 4.7 × (height in in.) – 4.7 × (age in years)

    Men: BMR = 66 + 6.23 × (weight in lb) + 12.7 × (height in in.) – 6.8 × (age in years)

22. Now multiply your BMR by the appropriate ratio below on the basis of your level of activity to calculate how many calories you need each day to provide the energy that your body uses.

    I almost never exercise: BMR × 1.2

    I exercise 1–3 days per week: BMR × 1.375

    I exercise 3–5 days per week: BMR × 1.55

    I exercise 5–7 days per week: BMR × 1.725

    I am undergoing intense physical training: BMR × 1.9

Use the graph at right to answer Questions 23–26.

23. Compare the various cycles of sleep each night.

24. After what type of sleep do people usually wake?

25. Which cycle of sleep is usually the longest?

26. What do you notice about Stage 4 NREM sleep?

Use the case study below to answer Questions 27–30.

27. Should Katja feel that the rape was her fault? Should she remain silent?

28. Is rape the only way that a person can experience sexual abuse? Explain.

29. What can Katja do now to make decisions that honor God?

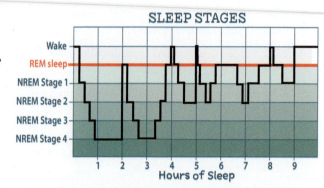

SLEEP STAGES

30. Sexual abuse doesn't affect just girls. A recent study on sexual abuse showed that 14% of women and 4% of men reported having been abused as children. Suggest why it might be difficult for men to deal with sexual abuse and get help.

case study

SEXUAL ABUSE

Katja had just graduated from high school. She was a Christian student attending a summer orientation for the state college that she was planning to attend. She enjoyed touring the campus, and when her roommates invited her to a party, she decided to go. However, when she got there, alcohol and drugs were everywhere. Katja knew better than to try any of it, but she also had no way of getting back to campus, which was miles away.

Then a guy came over to her and started talking. Surprisingly, Roger wasn't drunk or high. He said that he was a Christian too. When he offered to get her a soda, she thought nothing of it. But

soon, Katja became very drowsy. Before she knew what had happened, Roger had taken her to another room and raped her. He had used a date-rape drug to weaken her resistance. A few months later, Katja found that she was pregnant. What should she do?

Sadly, rape—when one person forces another to have sexual intercourse—is a common and devastating occurrence in our world. When someone forces or tricks someone to satisfy his sexual desires, it is a type of sexual abuse. About 20% of women are sexually abused at some point in their lives.

Use the ethics discussion on the next page to answer Questions 31–32.

31. Do some research on the topic of assisted suicide to answer the first question of both the bioethical decision-making process and the biblical triad: What information can I get about this issue?

32. The position according to the principles of bioethics is given on the next page. Use the process modeled in Chapter 20 to deconstruct and develop a biblical position on the issue of assisted suicide.

ASSISTED SUICIDE

ISSUE

Brittany Maynard was faced with a tough decision. When she was 29, a year after she was married, she found out that she had terminal brain cancer. She began undergoing treatments, but the cancer continued to grow. She began to form a plan for the time when her health would degenerate and death would come, but she decided that she wasn't interested in care that would simply make her comfortable while she was dying. She left California with her husband and moved to Oregon. Why? Oregon's Death with Dignity Act provides for assisted suicide, which allows doctors to prescribe a medication to a terminally ill patient that will end life. She could choose when she would die. There would be no surprises, no suffering, and a quiet death surrounded by her loved ones. She shared her story with the world to help get death-with-dignity legislation passed elsewhere. Her husband approved of her decision and has, since her death, been instrumental in getting such legislation passed in several states. But should people have the right to choose to die even when diagnosed with a terminal illness?

1. How are the people involved in this issue respected and given the freedom to choose?

Allowing people the ability to choose death when faced with a terminal illness gives them complete autonomy in the decision as to when they die. When a government agency restricts their access to life-ending prescriptions, patients' preferences in the timing of death are not considered.

2. How are individuals protected from harm or injury?

Giving people the choice of when they would like to die may protect them from the extended pain experienced with many terminal illnesses. Allowing a patient to choose to die could be viewed as giving them the option of avoiding harm. The patient is in the final stages of a terminal disease, so ultimately this is seen as an alternative and quicker way to avoid the painful process of death.

Brittany Maynard's family, including her mother shown here, is using Brittany's story to push for death-with-dignity legislation.

3. How are people helped by the action taken?

People are helped by being given a medication that allows them to peacefully die at a time of their choosing rather than experience the extended pain and discomfort of their terminal diagnosis. Families are allowed time to grieve with the patient and to avoid watching a loved one go through the sometimes-painful process of death. Assisted suicide also permits family members to visit just prior to death or to be with the loved one when they die.

4. How is the action taken just and fair?

According to the principle of justice, allowing every patient the ability to choose his or her time of death when diagnosed with a terminal illness is just and fair. Advocates would argue that individual states prohibiting assisted suicide are unjust and unfair to their citizens.

5. What action do the principles of bioethics recommend?

According to the principles of bioethics, patients who are given fewer than six months to live should be allowed to choose when they want to die. This would allow them maximum autonomy, they could avoid the pain and suffering of a prolonged death, and they could die in the presence of family and friends.

APPENDIX A

Apologetics and Evidence
Christian Apologetics

Differences in worldviews between people can devolve into heated arguments. But should Christians involve themselves in such disputes? The apostle Peter reminds us that we should always be ready to give a reason for our faith with respect and gentleness (1 Pet. 3:15). The formal word for the act of giving an account or explanation of the Christian faith is *apologetics* (from the Greek word for "speaking in defense").

Christians have often used apologetics

» to explain to others the reasons for their beliefs and practices, as the apostles did through their epistles and when standing before various authority figures;

» to try to persuade an unbeliever to become a Christian;

» to try to persuade another person who holds to a non-biblical position to change his or her mind; and

» to encourage believers that their beliefs are reasonable.

Like any kind of effective argumentation, apologetics relies on presenting credible evidence, building a logical presentation, and making a persuasive defense. Christian apologetics draws mainly from

» biblical evidence that comes from properly interpreted Scriptural passages;

» non-biblical evidence (e.g., historical or scientific);

» a well-constructed argument that follows accepted rules of logic and avoids logical fallacies; and

» a respectful and loving appeal to understand, to change one's belief, or to be strengthened in one's faith.

The Limits of Evidence as an Apologetic

A key part of convincing someone that your position is correct is presenting evidence to support your position. But this is where worldview comes into the picture. All evidence is filtered and interpreted through one's worldview. With the exception of evidence from the Bible, which Christians hold to be perfect and without error, no other kind of evidence is truly objective or certain, not even scientific evidence! But even biblical evidence, while perfect, is subject to interpretation by the reader.

A worldview guides one's interpretation of the significance, or even in some cases, the validity, of the evidence. And no one has absolutely logical or flawless reasoning. Our fallen minds can misinterpret, misunderstand, or draw illogical conclusions from evidence. Normally, if evidence doesn't fit into our view of things or makes us uncomfortable at some deep level in our minds, we either reject it or explain it away.

When using non-biblical evidence to present arguments for what we believe, we must be careful about its quality. Some non-biblical evidence that we might have used at one time is no longer considered valid. For example, many Bible believers, including Martin Luther, believed that the earth was the center of the universe. When Copernicus, Galileo, and Kepler built evidence for a heliocentric theory, many Christians rejected that idea as unbiblical. Some still do! But the Bible is written in a way that doesn't necessarily support a geocentric or a heliocentric theory. Today we can send probes out to the edge of the solar system to observe that our solar system is, in fact, best modeled by the heliocentric theory.

On the other hand, high-quality, non-biblical evidence can be effective in helping unbelievers question their presuppositions. For instance, geologists can describe a single stratum of sedimentary rock at least 1600 km across covering portions of the central United States. It is distinctive because it's mostly sandstone and shows evidence of deposition under rapidly flowing water (cross-bedding). Most geologists interpret this feature as sediments gradually deposited by a large system of meandering rivers during a vast period of time, but the evidence better fits deposition by a broad, fast-moving flood all at once in a short period of time.

Some disagreements resulting from differences between Christian and secular worldviews don't have clear non-biblical answers. The "starlight and time" problem in cosmology is an excellent example of this. If the universe is only about 7000 years old, then how can we see light from objects that are more than 13 billion light-years away if the speed of light is unchanging?

When trying to change the heart of an unbeliever through apologetics, the most concrete, impressive evidence will not work on its own (Luke 16:27–31). Only the Holy Spirit through the presentation of Scripture convinces and persuades (Rom. 10:8–17). Piling up non-biblical evidence will save no one, and the unbeliever can make his or her own pile of evidence to counter a believer's. The relative heights of the piles cannot determine truth!

Even a believer who bases some convictions on non-biblical evidence may be on shaky ground. A Christian's life ultimately must rest on faith, not sight (2 Cor. 5:7). Belief that is based on non-biblical evidence or improperly interpreted biblical evidence is unstable because the view of such evidence changes with time. For example, since the 1900s many Christians believed that the canopy theory was true. But recent atmospheric modeling and evaluation of the theory by qualified creationist atmospheric scientists have concluded that a vapor canopy could not possibly have been the source for the rains during the Flood and would have made life on Earth impossible before that event. A person who puts faith on evidences such as these could easily become disillusioned and perhaps bitter if the evidence is later shown to be invalid. Only sound biblical evidence obtained from proper hermeneutics (methods of interpreting the Bible) and accompanied by the ministry of the Holy Spirit will convince and persuade.

A Proper View of Evidence as an Apologetic

God is able to use all sorts of means to convince an unbeliever. And evidence and logical argument are two of those potential means, which Paul himself employed. He pointed to eyewitnesses (1 Cor. 15:5–6), and he used reason and persuasion (Acts 19:8–9).

But Paul knew, as we should, that no amount of evidence, on its own, is sufficient to convert someone. Only God's powerful Holy Spirit can do that because a person's ultimate problem is not in the brain or in the science textbook. Humanity's shared problem is that we are all born loving all sorts of things other than God.

Even when we do appeal to evidence, we shouldn't let our debate stay there. We should use evidence as a shovel to dig deeper, to question another person's presuppositions. The moral law is a good way to do that. Secular scientists have recently claimed that evolution can account for the rise of love, justice, and self-sacrifice as important emotions that help the survival of the species. But how do these work with the underlying evolutionary assumption that natural selection acts through untold millennia of survival of the fittest? This contradiction isn't easily explained, but the evolutionist will come up with an explanation because love, justice, and self-sacrifice all exist. They must be useful for survival. The only real way to compare worldviews is to ask, "Which one has the greater explaining power? Which worldview better accounts for morality and the existence of life?"

So we graciously push back: "Can you really live as if love and justice aren't right—just useful for preserving the species? How can a worldview that is based on natural selection and random chance account for our shared human conviction that some things are right and others wrong?"

A Christian worldview accounts for love and justice because the source of these values is an absolute, personal, triune God. Love exists because it is part of God's nature as a triune Being to love—the Father loved the Son and the Spirit, the Son loved the Father and the Spirit, and the Spirit loved the Son and the Father, from eternity past.

All of life is a journey of discovery. We're always learning and growing—or we should be! We may grow to find that some of our understanding of Scripture was wrong or perhaps incomplete. This is part of life in a fallen world. Even dedicated Christians are still fallible. They can't assume that they understand the Bible perfectly. But with the Spirit's help they can obey what they do understand and seek to refine that obedience and understanding over time.

When building up a believer's faith and tearing down strongholds of unbelief (2 Cor. 10:4–5), we begin with careful, consistent interpretations of Scripture. We then show how careful, consistent interpretations of scientific or cultural evidence can be understood in light of these biblical truths. But we remember that presuppositions play a role in everyone's evaluation of the evidence. We can't test the big bang or a 7000-year-old Earth with the scientific method. So biblical presuppositions must interpret all the evidence that we encounter. Science and human reasoning must never be our starting point; the fear of the Lord is the beginning of knowledge (Prov. 1:7).

APPENDIX B

Creating Graphic Organizers

There will be several times in this textbook when you will be asked to demonstrate that you understand a concept or a group of concepts by creating a graphic organizer. A *graphic organizer* is just a visual way to represent data. Use this appendix as a guide to help you create different kinds of graphic organizers to show how concepts in biology compare, how they are structured, or how they move through a process.

Table/T-Chart

A *table* is often used in science to organize data, but it can also be used to organize descriptions and other information. A table can be a helpful way to compare two or more concepts in biology. For example, the three classes of levers are compared in the table below. If you are comparing only two concepts, this kind of table using only two columns can be called a *T-Chart* because of its shape.

COMPARING CLASSES OF LEVERS

	FIRST-CLASS LEVER	SECOND-CLASS LEVER	THIRD-CLASS LEVER
Order of Parts	resistance, fulcrum, effort	fulcrum, resistance, effort	fulcrum, effort, resistance
Comparing Lengths of Effort Arm (EA) and Resistance Arm (RA)	varies	EA > RA	EA < RA
Mechanical Advantage	> 0	> 1	< 1

When you create a table, place the concepts that you are comparing in the top row. Put the characteristics under consideration in the leftmost column. Fill in the cells, considering how each characteristic or category differs for the concepts that you are comparing.

Hierarchy Chart

A *hierarchy chart* is a chart that shows the relationships between several concepts. For example, a company has an organizational chart that does the same thing. Hierarchy charts are especially useful in biology when studying classification.

When creating a hierarchy chart, think of the different categories that a larger category includes. The example below shows how all the objects in our solar system can be classified as the sun, planets, small solar system bodies, moons, or dwarf planets.

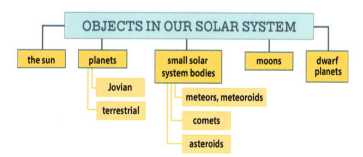

APPENDIX B

Concept Map

A *concept map* is a free-form graphic organizer that relates concepts, subconcepts, and their related characteristics. It's similar to an outline in visual form. Because concept maps are so flexible, they can be more difficult to create.

To make a concept map, start with the biggest concept and put it in a circle. Use lines or arrows to connect this main concept with smaller concepts. You can use linking words to help a viewer see the connection between two concepts and read the concept map like a sentence. The further you get from the main idea, the more specific the information should become.

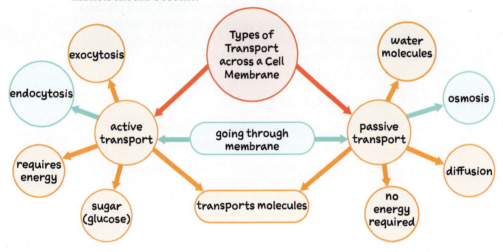

Process Map/Flow Chart

A *process map* illustrates a specific chain of events from the first event in a biological process. Because of its purpose, it almost always includes arrows to indicate the direction of the process.

When you create a process map, think of the process that you are illustrating and the different steps that are part of this process. Make sure that you put them in chronological order! You can show a process in a line of events such as a timeline, or if the process repeats itself, you can show it in a circle as a cyclical process that doesn't have a beginning or an end.

Concept Definition Map

The last type of graphic organizer is a *concept definition map*. This graphic organizer takes a concept and considers four questions: What is it? What is it not? What are its properties? What are some examples?

Think through these four questions for a concept when you create a concept definition map. Then put the concept in the middle of the map and organize these four questions in boxes labeled with these questions around the concept.

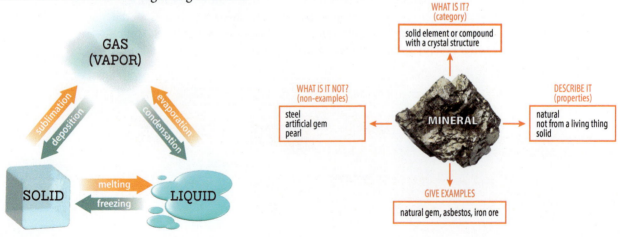

Combining Forms

You may find science a little intimidating because of all the long, unfamiliar words that scientists use. But you can often unravel the meaning of these words by breaking them down into simple parts that do have meaning to you. When you see a difficult scientific word, look at the entries in this appendix to help you understand a scientific term. These word parts may come at the beginning, at the end, or in the middle of the term, depending on the meaning.

For example, if you ran into the word *aerophyte*, you could separate it into two parts—*aero-* and *-phyte. Aero-* means "air," and *-phyte* means "plant." So an aerophyte is an air plant, that is, a plant that is able to pull water out of the air and doesn't have roots embedded in the soil.

Some roots can be used only as prefixes, others only as suffixes. Still others can be used as either. For example, *cardiologist* is the name of a heart doctor, while *tachycardia* refers to a rapid heart rate.

a, **an** (Gk.)—not, without

ab (L.)—away from

ac, **ad**, **ag** (L.)—to, toward

acanth, **acantha**, **acantho** (Gk.)—spine, thorn

acou (Gk.)—hear

aer, **aero** (Gk.)—air

alb, **albi**, **albid** (L.)—white

allelo (Gk.)—one another, pair

alter (L.)—change

amal (Gk.)—soft

amph, **amphi**, **ampho** (Gk.)—on both sides

ana (Gk.)—up, throughout, again

ant, **anti** (Gk.)—opposite, against

ante (L.)—before

ap, **apic** (L.)—tip, extremity

aqua, **aquari** (L.)—water

arthr, **arthro** (Gk.)—joint

ase—used for forming enzyme names

aster, **astero**, **astr** (Gk.)—star

audio (L.)—hear

aur, **aure** (L.)—gold, golden

aut, **auto** (Gk.)—self

bact, **bacter**, **bactr** (Gk.)—a rod

bacteri, **bacterio** (Gk.)—bacteria

bar, **baro** (Gk.)—weight, pressure

bi (L.)—two, twice, double

bio, **bios**, **biot** (Gk.)—life

bola, **bole**, **bolo**, **bolus** (Gk.)—a throw; clod, lump

calc, **calci** (L.)—calcium

calor (L.)—heat

card, **cardi**, **cardia**, **cardio** (Gk.)—heart

carn, **carneo**, **carni** (L.)—flesh

cat, **cata**, **cato** (Gk.)—down, downward

caud, **cauda** (L.)—tail

cell, **cella**, **celli** (L.)—chamber, cell

cent, **centi** (L.)—a hundred

centr, **centri**, **centro** (Gk.)—center

cephal, **cephala**, **cephalo** (Gk.)—head, brain

cereb, **cerebell**, **cerebr**, **cerebro** (L.)—brain

chem, **chemi**, **chemo** (Gk.)—juice; pour; chemistry

chlor, **chloro** (Gk.)—green

chondr, **chondro** (Gk.)—grain, corn; cartilage

chord, **chorda** (L.)—string, cord

chrom, **chroma**, **chromato**, **chromo** (Gk.)—color

chron, **chrono** (Gk.)—time

chrys, **chryso** (Gk.)—gold

cili, **cilia**, **cilio**, **cilium** (L.)—eyelash, eyelid, small hair

cis, **cide**, **cision** (L.)—to kill or cut

cline (Gk.)—sloping

co, **com**, **con** (L.)—with, together

coagul (L.)—drive together, curdle

crani, **crania**, **cranio**, **cranium** (Gk.)—skull

cret, **creta** (L.)—to separate

crist, **crista** (L.)—crest, tuft

cupr (Gk.)—copper

cut, **cutane**, **cuti**, **cutic** (L.)—skin

cyan, **cyane**, **cyani**, **cyano** (Gk.)—dark blue

cycl, **cycle**, **cyclo** (Gk.)—circle, wheel

cyst, **cystis**, **cysto** (Gk.)—bladder, pouch

cyt, **cyte**, **cytic**, **cyto**, **cytus** (Gk.)—hollow place, cell

dactyl, **dactylo**, **dactylus** (Gk.)—finger, toe

de (L.)—loss, removal

deci (L.)—tenth

derm, **derma**, **dermi**, **dermis**, **dermo** (Gk.)—skin

detrit (L.)—to wear off

di, **dis** (Gk.)—separate, apart; two, double

dia (Gk.)—across, through

dipl, **diplo** (Gk.)—double, two

div (Gk.)—apart

dors, **dorsi**, **dorso**, **dorsum** (L.)—back

duce, **duct** (L.)—to lead

dyna, **dynam**, **dynamo**, **dynast** (Gk.)—be able; power, energy

e (L.)—out, without, from

echin, **echino**, **echinus** (Gk.)—spiny, sea urchin

eco (Gk.)—house

APPENDIX C

ecto (Gk.)—outside, outer, external

electro (L.)—electricity

ell, elle (L.)—small

em (L.)—in, into

emia (Gk.)—blood

en, end, endo (Gk.)—within, inner

epi (Gk.)—upon, over, beside

equ, equa, equi (L.)—equal

erythr, erythro (Gk.)—red

eu (Gk.)—good, true, well

ex, exo (Gk.)—out, outside, without

extra (L.)—outside, more, beyond, besides

fasci, fascia (L.)—a bundle

fer, fera (L.)—to bear. *See also* **phor**

fibr, fibril, fibrin, fibro (L.)—fiber

fissi (L.)—split, divide

flagell, flagellum (L.)—whip

flam (L.)—fire

foli, folia, folium (L.)—leaf

fund (L.)—basis

fusi (L.)—to join

gamet, gamete, gameto (Gk.)—marriage, spouse

gast, gaster, gastero, gastr, gastri, gastro (Gk.)—stomach, belly

gen, genus (L.)—race, sort, kind

gene, genea, geneo, genesis, geno (Gk.)—birth, origin, family, beginning

genic, genous (Gk.)—producing

geo (Gk.)—earth

germ (L.)—bud, sprout, seed

gest (L.)—to carry

gli, glia, glio (Gk.)—glue

glob, globo, globus (L.)—ball, globe

gluc, gluco (Gk.)—sweet

glutin (L.)—glue

glyc, glycer, glyco (Gk.)—sweet

gon, gone, gonia, gonidi, gonium, gono, gony (Gk.)—seed, generation, offspring

grad (L.)—step, walk, slope

gran, grani, grano, granum (L.)—grain

graph, grapho, graphy (Gk.)—to write

grav (L.)—heavy

gymn, gymno (Gk.)—naked, bare

gyro (Gk.)—spinning

halo (Gk.)—salt, sea

hapl, haplo (Gk.)—single

hem, hema, hemato, hemia, hemo (Gk.)—blood. *See also* **emia**

hemi (Gk.)—half

herb, herba, herbi, herbo (L.)—grass

hered (L.)—heir

heter, hetero (Gk.)—other, different

hist, histo (Gk.)—web, tissue

homeo, homio, homo (Gk.)—like, same, resembling

homo (L.)—man

hybrid (L.)—offspring of different parents

hydr, hydra, hydri, hydro (Gk.)—water, fluid

hyper (Gk.)—over, beyond

hypo (Gk.)—under, beneath

ic (Gk.)—of, relating to

ichthy, ichthyo, ichthys (Gk.)—fish

il, im, in, ir (L.)—into, in; not

inter (L.)—between

intr, intra (L.)—within

is, iso (Gk.)—equal

ism (Gk.)—belief, process of

jug, jugo (L.)—*See* **zygo**

kary, karyo (Gk.)—nut, nucleus

kine, kinema, kinemato, kines, kinesi, kinet, kineto (Gk.)—move, moving, movement

lampr, lampro (Gk.)—shining, clear

later, lateral, latero (L.)—side

leuco, leuko (Gk.)—clear, white

libra (L.)—a balance

liga, ligam, ligat (L.)—bound; a band

lip, lipo (Gk.)—fat

log, logo, logus, logy (Gk.)—word, study of

lun, luna (L.)—moon

lys, lysi, lysio, lysis, lytic (Gk.)—loosening, break apart

macr, macro (Gk.)—large

magneto (Gk.)—magnetic

matr, matri, matro (L.)—mother

medi, media, medio (L.)—middle

mei, meio (Gk.)—less

melan, melano (Gk.)—black

mens, mensa (L.)—table

mer, mere, meri, mero (Gk.)—a part

mes, meso (Gk.)—middle

met, meta (Gk.)—between, with, after, change

meter, metr, metra, metri, metro, metry (Gk.)—measure

micro (Gk.)—small

mill, mille, milli, millo (L.)—one thousand

mit (L.)—to send

mito (Gk.)—thread

mono (Gk.)—one

morph, morpha, morpho (Gk.)—form, shape

mult, multi (L.)—many

muta (L.)—change

my, myo, mys (Gk.)—muscle

myc, myce, mycet, myceto, myco (Gk.)—fungus

nan, nani, nano, nanus (Gk.)—dwarf, one billionth

nast, nasto (Gk.)—pressed closed, solid

nem, nema, nemato, nemo (Gk.)—thread

nephr, nephri, nephron, nephrus (Gk.)—kidney

neur, neura, neuro (Gk.)—nerve

nomy (Gk.)—the science of

nuc, nucle, nucleo (L.)—central part

ocul, oculi, oculo, oculus (Gk.)—eye

oid (Gk.)—like, form

omni (L.)—all

oo (Gk.); **ov, ovi, ovo, ovum** (L.)—egg

opt, opti, opto (Gk.)—eye, vision

ora (L.)—mouth

organ (Gk.)—living

orth, ortho (Gk.)—straight, correct; upright, perpendicular

os, oss, osse, ossi (L.); **ost, oste, osteo, osteum** (Gk.)—bone

ose—used for forming sugar names

osmo (Gk.)—pushing

ox, oxy (Gk.)—oxygen

par, para (Gk.)—beside

parous (L.)—to give birth

pause (Gk.)—to stop

ped, peda, pede, pedi, pedo (L.)—foot

pend (L.)—hanging

per (L.)—through, by means of

peri (Gk.)—around, near

phag, phage, phago, phagy (Gk.)—to eat

pharm, pharmac (Gk.)—drug, poison

pheno (Gk.)—show, seem, appear

phil, phila, phile, phili, philo (Gk.)—beloved, loving

phob, phobia, phobo (Gk.)—fear

phon, phono (Gk.)—sound

phor, phora, phore, phori, phoro (Gk.)—to bear

phos, phot, phota, photi, photo (Gk.)—light

phyll, phyllo, phylum (Gk.)—leaf

physis (Gk.)—growth

phyt, phyto, phytum (Gk.)—plant

plan, plani (L.); **plat, plate, plati, platy** (Gk.)—broad, flat

plasm, plasma, plasmato, plasmo (Gk.)—something molded

plast, plasto (Gk.)—formed, molded

plur, pluri (L.)—more, several

pneumo, pneumon, pneumona (Gk.)—lungs

pod, podo, pody (Gk.)—foot

poly (Gk.)—many

post (L.)—after

poten, potent (L.)—powerful

pre (L.), **pro** (Gk.)—before, in front of

prot, prote, proto (Gk.)—first, original

pter, ptero, pterum, ptery (Gk.)—wing, feather, fin

pulmo, pulmon, pulmono (L.)—lung

purpur, purpure (L.)—purple

pus (Gk.)—foot

pyro (Gk.)—fire

radi, radia, radiat, radio (L.)—spoke, ray

re (L.)—back, again

ren, rena, reni, reno (L.)—kidney

reticul (L.)—network

retro (L.)—backward

rhiz, rhiza, rhizo (Gk.)—root

sacchar, saccharo (Gk.)—sugar

sal (L.)—salt

sarc, sarci, sarco (Gk.)—flesh

scientia (L.)—knowledge

scop, scope, scopo, scopy (Gk.)—to see, watch

scri, scrib, script (L.)—to write

secret (L.)—set apart

sect (L.)—to cut

seism (Gk.)—earthquake

semen, semin (L.)—seed

semi (L.)—half

sexu (L.)—sex

sis (Gk.)—the act of

sol (L.)—sun

soma, somat, somato, some (Gk.)—a body

son (L.)—sound

spec (L.)—see, look at

sperm, sperma, spermato, spermi (Gk.)—seed

sphere (Gk.)—ball, globe

spir, spira, spiro (L.)—coil or twist

spire (L.)—to breathe

spor, spora, spore, spori, sporo (Gk.)—seed

stas, stasi, stasis (Gk.)—to stand still

stat, stati, stato (Gk.)—standing, placed

stoichi, stoicho (Gk.)—element

stom, stoma, stomato, stomo (Gk.)—mouth

stria, striat (L.)—furrow, streak

stroma, stromato (Gk.)—anything spread out, coverlet

sub (L.)—below, under

super (L.)—above, over

sy, syg, syl, sym, syn, sys (Gk.)—with, together

tel, teleo, telo (Gk.)—end, complete, far

tele (Gk.)—distant

terra (L.)—earth

tetr, tetra (Gk.)—four

therm, thermos (Gk.)—heat

thesis (Gk.)—an arranging

thigm, thigma, thigmato, thigmo (Gk.)—touch

thora, thoraco, thorax (Gk.)—breastplate

thylac, thylaco (Gk.)—sack, pouch

tom, tome, tomi, tomo, tomy (Gk.)—to cut

ton, tono (Gk.)—tone or tension

top, topo, topus, topy (Gk.)—place

tot, tota, toti (L.)—all

toxic, toxicum (L.)—poison

tran, trans (L.)—across, through

tri (L.)—three

trop, tropae, trope, tropo (Gk.)—turn, change

troph, trophi, tropho (Gk.)—food, nourishment

turg (L.)—swell

typ, type, typi, typo (Gk.)—blow or strike; impression

uni (L.)—single, one

ur, ura, uro (Gk.)—tail

vacu (L.)—empty

valen, valent (L.)—strength; be worth

vari, vario (L.)—difference

vas, vasa, vaso (L.)—vessel, duct

vect (L.)—to carry

vent, venter, ventr, ventro (L.)—belly, underside

vert, verta, verte (L.)—to turn

volu (L.)—bulk, amount

vor, vora, vore (L.)—to devour, eat

xanth, xantho (Gk.)—yellow

zo, zoa, zoi, zoo, zoon (Gk.)—animal

zyg, zygo, zygus (Gk.)—yoke

zym, zyma, zymo (Gk.)—yeast, leaven, ferment

Reading Tables, Graphs, and Scientific Diagrams

Scientists often like to compare two or more groups of observations to see whether they are related in some way. Doing so is a type of *analysis*—the process of looking for patterns and relationships within sets of data. These patterns and relationships can often be better understood by presenting them in a visual form, such as a table, graph, or diagram. Not surprisingly, then, science textbooks normally contain many tables, graphs, and diagrams too. Knowing how to read them will greatly enhance your learning experience.

Tables

One of the simplest means of organizing data is to use a *table*. A table arranges data into rows and columns. Each row or column corresponds to a defined set of data or variables. For example, Table 1 shows the results of a stream survey conducted by fisheries biologists. The table has been set up to show at a glance *what kind* of fish were observed during the survey (row headings), as well as *how many* of each sex were observed (column headings).

TABLE 1 STREAM SURVEY

SPECIES	MALE	FEMALE	UNKNOWN
Chinook	3	5	1
Coho	11	9	2
Steelhead	1	0	0

Data tables are usually read from left to right across the table but may also be designed to include useful information if read in another direction as well. Reading from left to right, we can see that the biologists observed three male Chinook salmon, five female Chinook salmon, and one Chinook salmon whose sex could not be determined. Reading from top to bottom reveals that the biologists observed a total of fifteen male fish of all species. If you wanted to know how many female coho salmon were observed, you would first find the row with the coho heading, then scan over to where that row intersects the column for female fish. Doing so shows that nine female coho salmon were observed.

Table 1 displayed the results of a *survey*, but tables are often used to record the results of *experiments* as well. A group of students used Table 2 (facing page) to record the results of an experiment to test the relationship between the time needed for a car to travel a certain distance on a slant track and the angle at which the track was set.

In Table 2, the two changing quantities, track angle and time, are called *variables*. Track angle, recorded in the first column, is the *independent variable* of the experiment since the angle of the track is set by the experimenters and does not depend on the values for time. The time to travel down the track, measured in the next three columns, is the *dependent variable* because its value changes depending on the angle of the track. Notice that in the Trial column headings, the units used to make each measurement are given. When the units are given in the heading, it means that all the measurements in that column are in those units; for example, every trial time is measured in seconds.

TABLE 2
CAR TRACK SLOPE AND TIME OF TRAVEL

TRACK ANGLE	TRIAL 1 (s)	TRIAL 2 (s)	TRIAL 3 (s)	AVERAGE (s)
10°	3.25	3.10	3.30	3.22
20°	2.75	2.60	2.55	2.63
30°	2.25	2.30	2.20	2.25

Table 2 displays the results of three trials done at each of three track settings for a total of nine trials. The students also calculated the average speed at each track setting and recorded the result in a fourth column. Can you find the result for the second trial at a 20° track setting?

Graphs

Another way to graphically display the relationship between two variables is to plot their values on a *graph*. The values of the independent and dependent variables are called the *coordinates* of the data. Each point plotted on a graph represents an *ordered pair* of coordinates, where the first number in the pair is the independent data coordinate and the second is the dependent data coordinate. You may have graphed ordered pairs in the form (x, y) in a math class.

Useful graphs include a title describing the graph's purpose and labels identifying the quantities and units used on each axis. Numbered scales and a grid are usually included to help the reader estimate values of the variables plotted on the graph.

Scientists usually plot the independent variable on the horizontal axis of a graph, also known as the x-axis, with increasing values to the right. The dependent variables are plotted on the y-axis, the vertical axis, with the values increasing upward. These are not hard and fast rules, and many graphs are arranged differently to improve the clarity of the plot. One of the first rules of reading graphs is to notice what the two variables are.

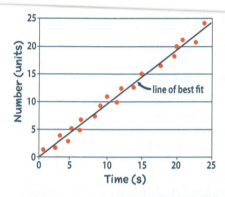

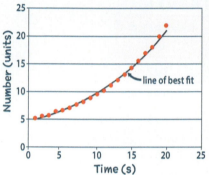

Graphs come in different forms. A simple plot of points on a graph is called a *scatterplot.* This is the starting point for many graphs. Scientists like to detect trends in the data and to create an equation that describes the trend. They draw a straight or curved line through the pattern of dots. The kind of line depends on how the pattern changes. We call the trend line a *line of best fit.* Graphs whose line of best fit is a straight line are said to show *linear relationships.* These are fairly rare in nature. Most trends in nature are slightly curved to really wavy! These graphs, logically enough, show *nonlinear relationships.*

One of the most important things to learn from a graph is the rate at which the dependent variable changes in comparison to the independent variable. This rate is shown by the *slope* of the graph. A rising, or *positive,* slope shows that things are changing in a connected way—each increase in the independent variable produces an increase in the dependent variable. A horizontal slope shows that there is no relationship between the variables—a change in one does not produce a change in the other. A dropping curve has a *negative slope,* meaning that each increase in the independent variable produces a decrease in the dependent variable.

Scatterplots and trend lines can be read to obtain information not directly measured in the data set. The trend line connecting two data points represents values that are estimated, not measured. Obtaining unmeasured data this way is called *interpolation.* In addition, scientists often try to predict the values of the dependent variable if the independent variable continues to change beyond the range of the measured data. This method of analysis is called *extrapolation.* Extrapolating data depends heavily on assuming that the trend will continue as it has in the actual measured data.

Another common kind of graph is a *bar graph.* A bar graph depicts the value of the dependent variable of a data set in the form of a bar

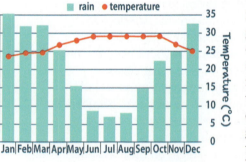

whose height (or length, depending on how the graph is set up) is determined by the variable's value. In the graph on the left, for example, the total rainfall for each month of the year in a rainforest can easily be seen. The shorter the bar, the less the amount of rainfall for a particular month. If you were asked to pick which months are the wettest in a rainforest, your eyes would quickly be drawn to the tallest bars on the graph. As this example shows, bar graphs are a good way to present data in which the independent variable has a nonnumerical value.

You will get to practice reading graphs in the Chapter Reviews of many chapters in this textbook. You will have a few opportunities to *make* graphs in exercises in this textbook and even more in the Lab Manual.

Scientific Diagrams

Although biologists do make measurements during their investigations and present their findings in graphs, much of the content of this textbook is *conceptual*, meaning that ideas, facts, principles, and models are presented without using a lot of mathematics. (Perhaps this is a relief to you!) Biological processes and models are frequently illustrated by *diagrams*. These are visual aids whose purpose is to help make a concept more clearly understood. Diagrams are usually intended to help clarify something presented in the text, so the text and diagram should be studied together in order to best understand the idea being presented. Diagrams are used in this textbook mainly to illustrate two things: structures and processes. Let's take a look at what to expect in each type of diagram.

STRUCTURAL DIAGRAMS

A structural diagram is intended to help you identify the parts of an organism. Keep in mind that the primary purpose of a structural diagram is to *inform* the reader, not to be an *exact representation* of the structure being discussed. Structural diagrams often make use of certain enhancements to make something clearer.

Structural diagrams can enhance color, zoom in on a structure, or cut away the exterior so that we can see inside. *Color coding* is often used to make the parts of a diagram more distinct, and in many cases makes use of colors not found in the original structure. *Exploded diagrams* separate the parts

of a structure so that each individual piece can be clearly defined. A *zoom-out* is a close-up view of one portion of a diagram, giving more focus and detail to that portion. A *cutaway* shows how a structure might look if we could see into its interior. Structure diagrams will use callouts or numbered bullets to label parts of the diagram. Each of these techniques is designed to make a diagram more information-rich. In the diagram below, notice the use of both the cutaway and zoom-out features to help you better understand the structure of a mushroom.

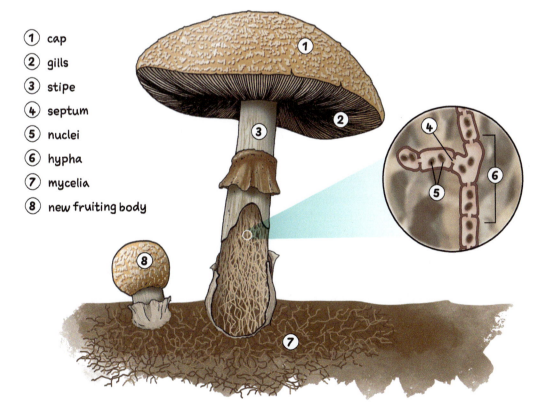

1. cap
2. gills
3. stipe
4. septum
5. nuclei
6. hypha
7. mycelia
8. new fruiting body

PROCESS DIAGRAMS

A process diagram is intended to show not how something *looks*, but how it *works*. Because some biological processes are very complex, diagrams of those processes can seem overwhelming at first glance. By definition, a process has a beginning and an end—it starts with something and ends up with something else. So process diagrams, unlike structural diagrams, normally have a part that should be looked at first, followed by other parts that are viewed in an orderly progression.

Process diagrams usually give visual clues about where to begin viewing the diagram. There may be *arrows* leading from the start of the process to its end, or the steps of the process may be indicated by *numbered bullets*. Many diagrams make use of both arrows and bulleted items. Color coding is frequently used to help distinguish the parts of a process. Use the arrows, numbered bullets, or both to guide you to the first step in the process, then read each step of the process along with the relevant text. Don't be concerned if you have to look at a process diagram several times to get the main ideas of what it teaches.

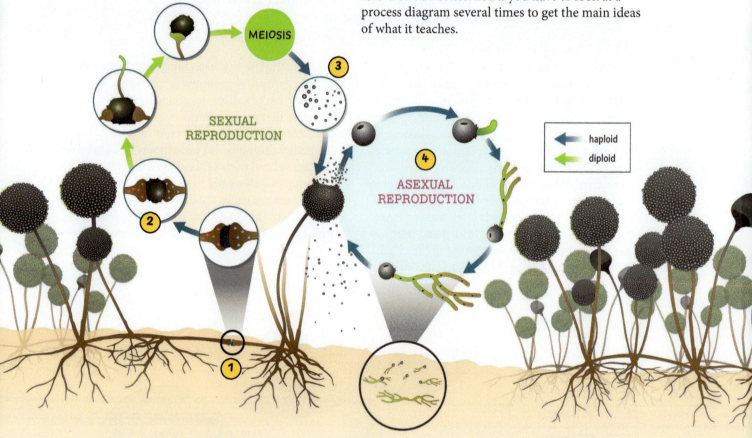

1 Two hyphae from different mating types grow until they touch each other. Nuclei in the tip of each hypha divide several times.

2 The hyphae fuse where they touch to form a zygosporangium, a diploid zygote (2*n*), through sexual reproduction. It forms a thick outer covering and enters dormancy.

3 By means of meiosis, zygosporangia form spores, which germinate to form hyphae through asexual reproduction.

4 As hyphae grow, asexually formed spores called *sporangiospores* form and are released.

APPENDIX E

Math Principles
Metric System

Scientists frequently use very large and very small numbers. They prefer to write these numbers as briefly as possible, especially if they are round numbers. So they developed a system of unit prefixes that indicate which factor of ten should be multiplied by the base unit in the measurement.

Example: 1 kilometer = 1 km = 1 × 1000 m

METRIC PREFIXES

PREFIX	SYMBOL	FACTOR	POWER OF 10	EXAMPLE
giga-	G-	× 1 000 000 000	10^9	gigabyte (GB); the measure of digital data
mega-	M-	× 1 000 000	10^6	megajoule (MJ); the work done by a bulldozer
kilo-	k-	× 1000	10^3	kilometer (km); a distance on Earth's surface
hecto-	h-	× 100	10^2	hectare (ha); the measure of a land area
deka-	da-	× 10	10^1	dekapoise (daP); a unit of viscosity or thickness of a fluid
(base)		× 1	10^0	gram; a standard unit of mass
deci-	d-	× 1/10	10^{-1}	decibel (dB); sound loudness
centi-	c-	× 1/100	10^{-2}	centimeter (cm); the length of a pencil
milli-	m-	× 1/1000	10^{-3}	millivolt (mV); a heart pacemaker signal
micro-	μ-	× 1/1 000 000	10^{-6}	micropascal (μPa); sound wave pressure
nano-	n-	× 1/1 000 000 000	10^{-9}	nanometer (nm); the size of atoms

APPENDIX E

Calculations

Don't tell your math teacher this, but math is for science! Here are some math concepts that you'll need in your science class.

RATIOS

It's really helpful to be able to compare the sizes or numbers of things in science. One way to do this is to find the ratio of one quantity to another.

Let's do an example. Find the ratio of girls to the total number of people in a group, maybe your class or family. To do this, take the number of girls and divide that by the number of everyone in the group. If there are 6 girls and 9 people total, then divide 6 by 9. You get 6/9, or 2/3 after reducing. So the ratio of girls to everyone in the group is 2 to 3, or 2:3. In other words, there are two girls for every three people in the group.

AVERAGING

You and your friends discovered one day that the number of candies in several bags of candies isn't the same. You want to know how many candies you should expect to be in a bag, so you need to find the average number of candies per bag. An average value falls somewhere in the middle of the group of values. It's a value *typical* of the group.

Averaging is straightforward. Add together all the values in the group—the number of candies in each bag of candies, for instance. Then divide the sum by the total number of bags. Suppose you and your four friends counted 38, 42, 40, 37, and 40 candies in the bags in your group. To find the average of these, add the numbers together and divide by 5, the number of bags.

$$38 + 42 + 40 + 37 + 40 = 197$$
$$197 \text{ candies (total)} \div 5 \text{ bags} = 39.4 \text{ candies per bag (average)}$$

So you should expect 39.4 candies per bag. But there aren't fractions of a candy in a bag! What do we do now?

ROUNDING

The solution to our averaging dilemma is called *rounding*. Since there are only whole candies in a bag, it makes sense to round the number to the ones place. Look at the first decimal place to the right, the tenths place. If that number is 5 or larger, then add 1 to the ones place and drop all decimal numbers. If the tenths place value is 0 to 4, just drop the decimal number and leave the ones place unchanged; in this case we round 39.4 candies to 39.

So you should expect about 39 candies per bag of candies. The greater the number of bags you check, the more likely your average will represent a typical value of candies.

GLOSSARY

A

abdomen A body region posterior to the thorax.

abiogenesis The theory that suggests that life can arise from nonlife.

abiotic factor A nonliving element in an ecosystem; not derived from living things.

abscisic acid A plant hormone that acts as an inhibitor of other hormones and causes dormancy in buds and seeds.

accessory organ Of the digestive system, any organ that aids in digestion but is not part of the digestive tract.

acid A substance that can produce hydrogen ions (H^+) in solution.

acidophile Any of a group of organisms in domain Archaea that thrive in very acidic environments, such as acidic hot springs.

actin One of the two types of protein found in muscle fibers.

action potential The rapid change in charge difference across the cell membrane of a neuron that carries a nerve impulse along the axon of the neuron.

active immunity An immunity in which the body makes its own antibodies or has activated T cells for a particular antigen.

active transport The movement of substances across a cell membrane from regions of low concentration to regions of high concentration by any means that requires the use of energy.

adaptation A heritable trait that improves the reproductive success of an organism.

adaptive radiation The change that occurs when a population spreads out into new environments.

adenosine diphosphate (ADP) The molecule produced when ATP is used for energy.

adenosine triphosphate (ATP) A molecule used within all cells for temporary energy storage.

adhesion The attraction of particles in one substance to particles in a different substance.

adrenal gland Either of the endocrine glands located on each kidney that produces several different hormones, including epinephrine and norepinephrine.

aerobic process Any chemical process that requires oxygen.

air sac One of a series of hollow chambers connected to the respiratory system of birds.

albumen The white of an egg.

alga A general term used for autotrophic members of the kingdom Chromista.

algal bloom A local surge in an algae population produced by very favorable environmental conditions.

alimentary canal The tubular passageway from mouth to anus that functions in the intake and digestion of food, the absorption of nutrients, and the elimination of residual waste; also known as the *digestive tract*.

allele One of a pair of genes that have the same position on homologous chromosomes and code for the same trait, though each may code for a different form of the trait.

allele frequency A measurement of how often an allele occurs in a population of organisms.

alternation of generations A plant reproductive strategy that includes alternating spore-forming and gamete-forming stages.

alveolus A small sac within a lung where gas exchange takes place.

amensalism A symbiotic relationship between two organisms that is injurious to one organism without affecting the other organism.

amino acid One of a class of organic compounds that serve as the building blocks of proteins.

amniote An animal whose embryonic development takes place inside a close-fitting membrane (amnion). This group includes reptiles, birds, and mammals.

amniotic fluid The protective fluid that surrounds an embryo within the amniotic sac.

amphibian A chordate animal in the class Amphibia.

amplexus The physical contact of a male and female amphibian that stimulates the female to release eggs into the water.

amylase An enzyme secreted by the salivary glands and pancreas that breaks down starches into simple sugars.

anabolism The phase of metabolism that builds molecules and stores energy.

anaerobic process Any chemical process that does *not* require oxygen.

androgen A male sex hormone produced by the testes.

angiosperm A plant that produces flowers and fruits.

annual ring A visible ring caused by yearly growth in the vascular cambium and seen in a cross-section of a woody stem.

antagonistic pair A pair of muscles that work together to produce a motion and its opposite motion.

antenna An elongated, movable sensory appendage on the head of many invertebrates.

antibody A protein substance produced to eliminate antigens that have entered the body.

anticodon The set of three bases on a molecule of tRNA that corresponds to a complementary codon on an mRNA molecule.

antigen A foreign material in the body that stimulates antibody production or begins cell-mediated immunity.

anus The opening at the end of the alimentary canal through which solid waste is eliminated.

aorta The large artery that carries blood from the left ventricle of the heart out to the body.

appendicular skeleton The bones of the pelvic and pectoral girdles and their appendages.

appendix A short tube attached to the large intestine near the juncture with the small intestine. Its exact function has not been positively determined.

aquifer An underground layer of water-bearing rock or sediment from which groundwater can be extracted.

archaea A member of the domain Archaea, which contains certain kinds of prokaryotic organisms, many of which are extremophiles.

artery Any vessel that carries blood away from the heart.

GLOSSARY

arthropod A member of the phylum Arthropoda.

artificial selection The human practice of breeding organisms that have desirable traits and variations.

atom The building block of matter; the smallest possible particle of an element.

atrium A chamber through which blood enters the heart.

autoimmune disease A disease in which the body's immune system cannot distinguish between body cells and pathogens.

autosome A chromosome other than a sex chromosome.

auxin A plant hormone that promotes cell elongation and lateral root development.

axial skeleton The portion of the skeleton that supports and protects the organs of the head, neck, and trunk.

axon The portion of a neuron that carries impulses away from the cell body.

B

bacteriophage A virus that infects bacteria.

bacterium A member of a large domain of prokaryotic organisms.

basal metabolic rate The amount of calories needed to maintain normal body functions while at rest.

base A substance that can produce hydroxide ions (OH^-) or accept hydrogen ions (H^+) in solution.

base pair A pair of complementary nitrogenous bases that connect the two halves of a DNA molecule.

B cell A type of lymphocyte that functions in humoral immunity.

behavior The way that an animal responds to its environment.

behavioral ecology The study of changes in animal behavior due to ecological pressures.

big bang The event described by the big bang theory, which states that the universe began in a very dense and very hot state and then rapidly expanded.

bilateral symmetry A type of symmetry in animals in which an imaginary plane through the centerline of the organism divides it into mirror-image halves.

bile A greenish fluid produced by the liver that is necessary for the digestion of fats.

binary fission A type of asexual reproduction used by all prokaryotes.

binomial nomenclature A system of naming organisms in which each organism is given a genus and species name.

biodiversity A measurement of the variety of life in a particular ecosystem.

bioethics The study of often-controversial issues that arise as the result of new advances in biology and medicine.

biogeochemical cycle The movement of a specific chemical through the living and nonliving components of Earth's spheres.

biological evolution The idea that organisms can slowly change over time into other kinds of organisms.

biology The study of life.

biome A part of the earth characterized by a particular climate.

bioremediation A technique that uses organisms to remove or neutralize hazardous wastes in the environment, such as oil spills.

biosphere The sum of all the habitable ecosystems on the earth.

biotechnology The use of living systems and organisms to produce new products and technologies, often by manipulating cells or cellular components.

biotic factor A living element in an ecosystem or one derived from living things.

blade The broad part of a plant leaf.

blastocyst A fluid-filled sphere of embryonic cells.

blastula An early stage of embryonic development consisting of a hollow, fluid-filled ball of cells.

blood The liquid carrier of nutrients, hormones, and gases that moves through the circulatory system.

blubber A layer of insulating fat beneath the skin of marine mammals.

body plan The arrangement of physical features in an organism that contributes to its structure and form.

bolus A small mass of chewed food.

bond An electrostatic attraction that forms between atoms when they share or transfer electrons.

bony fish A group of fish whose skeletons and fin rays are made of bone rather than cartilage.

book lung The plate-like respiratory structure found in arachnids.

bottleneck effect The drastic alteration of the allele frequency in a population that is due to a sharp reduction in the size of the popluation.

Bowman's capsule The sac that encloses a glomerulus and collects filtrate from the blood. The filtrate is processed to produce urine.

brain stem The part of the brain that connects to the spinal cord and controls body movements and posture.

bronchiole The passageway through which air flows to an alveolus.

bryophyte A nonvascular plant including mosses, hornworts, and liverworts.

bud An undeveloped shoot located at a node on a plant stem.

budding A type of asexual reproduction, for example in yeast, in which a portion of a parent organism is pinched off and develops into an identical offspring.

C

caecilian A legless amphibian that belongs to the order Gymnophiona.

calcitonin The thyroid hormone that lowers the blood calcium level by causing calcium to be moved to the bones.

Calvin cycle The portion of photosynthesis that uses energy captured during the light-dependent phase to convert carbon dioxide and water into sugars; also known as the *light-independent phase*.

cancer An unrestrained growth of abnormal cells with the potential to spread to other parts of the body.

cap The top of a fungal fruiting body containing the gills and spores.

capillary The smallest vessel of the circulatory system, where diffusion of nutrients and gas exchange occurs.

capillary action The rising of a fluid within a tube due to adhesion between the molecules of the fluid and the tube.

capsid A protein coat that surrounds the nucleic acid of a virus.

capsule A protective coating found outside the cell walls of many bacteria.

carapace The portion of the exoskeleton that covers the cephalothorax in some arthropods.

carbohydrate An organic compound comprised of the elements carbon, hydrogen, and oxygen; includes sugars, starch, and cellulose.

carcinogen A substance that increases the risk of cancer.

cardiac muscle Striated, involuntary muscle tissue found only in the heart.

carpel The female reproductive structure of a flower.

carrier A heterozygous organism that expresses the normal form of a trait but carries the recessive gene for an undesirable form of the trait and can pass it on to its offspring.

carrying capacity The maximum population size that a particular geographic area can support.

cartilage A firm but flexible form of connective tissue.

cartilaginous fish A group of fish whose skeletons are composed of cartilage rather than bone, as in sharks and rays.

catabolism The phase of metabolism that breaks down molecules and releases energy.

catalyst A substance that changes the rate of a chemical reaction but is not itself used up or affected by the reaction.

cell body The part of a neuron containing the nucleus and cytoplasmic organelles.

cell cycle The series of events, including duplication of DNA, leading to the division of a cell into two daughter cells.

cell differentiation The process in multicellular organisms of cells changing in order to perform specialized tasks.

cell membrane A thin layer of phospholipids and proteins that defines the boundary of a cell.

cell theory One of the fundamental models of biology that states that cells are the smallest form of life, come from preexisting cells, are the basic building blocks of all life-forms, and carry out life processes to maintain homeostasis.

cell wall A rigid structure made by cells of plants, fungi, and most bacteria to surround the cell membrane.

cell-mediated immunity An immunity to disease involving activated cells and consisting mainly of T cells.

cellular respiration The process by which cells break down glucose or other nutrients to produce usable energy.

cellulose The primary structural compound in plant cell walls, composed of long chains of glucose molecules.

central nervous system The portion of the nervous system consisting of the brain and spinal cord.

central vacuole An organelle in plant cells that holds materials and helps maintain the cell's shape through turgor pressure.

centromere The attachment point of the two chromatids in a chromosome.

centrosome A cellular organelle that functions in the organization of microtubules.

cephalization The concentration of sensory organs around one end of an organism, usually the head.

cephalothorax A body region in some arthropods consisting of a fused head and thorax.

cerebellum The part of the brain that controls body coordination and that monitors and adjusts body activities involving muscle tone, body posture, and equilibrium.

cerebrospinal fluid The fluid that nourishes and protects the brain and spinal cord and that flows between the two inner meninges.

cerebrum The part of the brain responsible for voluntary muscle activity and conscious activity.

cervix The neck of the uterus where it narrows and joins the vagina.

chaparral A temperate biome characterized by cool, wet winters and warm, dry summers.

chelicera Either of the mouthparts of animals in the subphylum Chelicerata.

cheliped A crustacean's walking leg bearing a small claw at its end.

chemical change A change in a substance that causes it to change its chemical identity.

chitin An organic substance, derived from glucose, found in the cell walls of fungi and the exoskeletons of invertebrates.

chlorophyll The green pigment of plant cells that is necessary for photosynthesis.

chloroplast A cellular organelle found in plant cells that contains the pigment chlorophyll and is the site where photosynthesis takes place.

chordate An organism in the phylum Chordata.

choroid The thin middle layer of the eyeball that contains blood vessels for nourishing the retina.

chromatid One of two identical halves of a chromosome formed prior to cell division.

chromatin The genetic material of eukaryotic cells consisting of DNA and associated proteins.

chromist A plantlike organism in the kingdom Chromista.

chromosome A structure consisting of DNA and supporting proteins, usually found in a cell's nucleus.

chyme The semiliquid mixture of partly digested food and digestive juices in the stomach and small intestine.

cilium A short, hairlike extension of a cell used either for movement or as a sensory organelle.

citric acid cycle The second phase of aerobic cellular respiration in which pyruvate reacts with enzymes to produce acetyl CoA, carbon dioxide, hydrogen ions, ATP, and electrons; also called the *Krebs cycle*.

GLOSSARY

clade A branch on a phylogenetic tree that includes all the descendants of an evolutionary common ancestor.

cladistics The classification of organisms on the basis of shared characteristics thought to be derived from a common ancestor.

class A taxonomic division within a phylum; composed of one or more orders.

cleavage A series of cell divisions in the development of an embryo during which the zygote forms into a multicellular mass.

climate change The change in global temperatures and weather patterns over time.

climax species A stable, long-lived species that marks the end of succession and characterizes a mature ecosystem.

cloaca A single opening for the digestive, urinary, and reproductive tracts in birds, reptiles, amphibians, and some mammals.

clone An organism that is genetically identical to its parent.

closed circulatory system A system of fluid circulation within organisms in which the fluid (typically blood) remains within vessels.

cnidarian A member of the phylum Cnidaria.

coagulation The biochemical process that forms a blood clot.

cochlea The snail-shaped division of the inner ear that functions in sound perception.

codominance A type of inheritance in which different alleles of a gene are expressed simultaneously, producing a phenotype that exhibits contributions from both alleles.

codon A set of three bases in an RNA strand that together code for a specific amino acid.

cohesion The attraction between like particles within polar substances.

cohesion-tension theory The currently accepted theory that explains the flow of water upward in the xylem of plants.

colony (1) A group of individual organisms of the same species living closely together. (2) A group of cells that live and work together. Unlike cells in a multicellular organism, cells in a colony may break away and survive on their own.

commensalism A symbiotic relationship between two organisms that benefits one organism without helping or harming the other organism.

common ancestor An ancestral organism whose offspring diverged into two or more species.

compact bone Dense, hard bone tissue made of tightly packed osteons.

comparative anatomy The study of similarities and differences in the anatomy of different species.

competition The conflict that results when organisms attempt to use the same resources.

complete metamorphosis A common form of insect development that normally includes four stages: egg, larva, pupa, and adult.

compound A pure substance made from two or more elements that are chemically combined.

compound eye An eye composed of many individual lenses.

cone Any one of the receptors in the eye that is sensitive to color.

coniferous forest A biome with cooler temperatures dominated by cone-bearing trees (conifers).

conjugation The transfer of plasmid material from one bacterium to another through a conjugation tube.

connective tissue Any of the tissues of the body that connect, support, cushion, and fill spaces around other tissues or organs.

conservation The preservation and wise use of natural resources, usually with consideration for the current and future needs of humans.

consumer A heterotrophic organism; an organism that cannot produce its own food and must obtain energy by consuming other organisms.

convergence The process of two different species independently evolving the same derived trait.

cork In woody plants, a thick, tough layer of dead epidermal cells that support the weight of the plant.

cork cambium A layer of meristematic cells that replace the epidermis in woody plant stems.

cornea The transparent front portion of the sclera of the eye.

corpus luteum The structure formed in the follicle of the ovary after ovulation; produces several hormones.

cortisol A hormone released by the adrenal glands that helps regulate glucose metabolism during stressful situations.

cotyledon A structure within a plant seed that nourishes the developing embryo and may, upon germination of the seed, become the seedling's first leaves.

courtship The animal behavior that promotes mate selection and breeding.

Creation Mandate The command given by God in Genesis 1:28 to Adam and Eve to fill the earth and have dominion over it.

crocodilian An animal in the order Crocodylia that includes crocodiles, alligators, caimans, and gharials.

crop A thin-walled portion of the digestive tract that temporarily stores food prior to digestion.

cuticle The waxy protective coating secreted by a plant's epidermal cells on the surface of plant stems, leaves, and fruits.

cyst A dormant stage for some bacteria, protists, and a few invertebrates that allows an organism to survive within a protective coating during times of unfavorable environmental conditions.

cytokinesis The process during cell division in which the cytoplasm of a cell is divided between two new daughter cells.

cytokinin A plant hormone that promotes cell division and the development of lateral buds.

cytoplasm A thick fluid inside cells that contains the organelles. Also called *cytosol*.

cytoskeleton The structure within cells that helps define their shape, maintains internal organization, and aids in division and movement.

day-neutral plant A plant that flowers independently of the photoperiod.

deciduous forest A temperate biome characterized by trees that lose their leaves seasonally.

decomposer An organism that breaks down dead organic material, recycling nutrients to be used by other organisms.

dendrite The part of a neuron that receives nerve impulses and transmits them toward the cell body.

deoxyribonucleic acid (DNA) The double-stranded molecule used by most cells for storing genetic information.

derived trait A trait that arises within a clade and is shared by all future members of that clade.

dermal tissue Any of the tissues that form the outer layer of a plant or other organism.

dermis The thick inner layer of the skin.

descent with modification Darwin's idea that all organisms come from common ancestors and change a little with each generation.

desert A biome with dry conditions and little rainfall; may be either hot or cold.

detritus Small particles of dead organic material.

diaphragm The muscle that separates the thoracic and abdominal cavities in mammals and humans.

dicot Any flowering plant whose embryo has two cotyledons.

diffusion The process by which particles in solution are evenly distributed throughout the solvent by Brownian motion.

digestive tract See *alimentary canal*.

dihybrid cross A genetic cross that tests two sets of alleles.

diploid organism An organism having two complete sets of chromosomes.

dissolving The process by which one substance, the solute, is broken up into smaller pieces by and distributed within a second substance, the solvent.

divergence The process of two similar species becoming increasingly different over time.

DNA fingerprinting A technique used to identify an individual that is based on sequences in his or her DNA.

domain The taxonomic division just above the kingdom level that groups organisms on the basis of chemical and genetic analyses.

dominant trait A characteristic that is expressed even in the presence of a recessive allele of the gene that codes for that characteristic.

dopamine A naturally occurring neurotransmitter whose levels can be elevated by some kinds of drug usage.

double fertilization In plants, the fertilization of both the egg cell and the polar bodies, which later form the endosperm.

drug A chemical that causes change in the structure or function of a living tissue.

eardrum See *tympanic membrane*.

echinoderm A member of the phylum Echinodermata.

ecological footprint A measurement of human demand on the earth's ecosystems that is based on the amount of resources used, the amount of waste generated, and the rate at which resources can be renewed.

ecological pyramid A model for showing the flow of energy through the various trophic levels within an ecosystem from producers to top-level consumers.

ecology The study of the interrelationships between living things and their physical environment.

ecosystem The community of organisms in a particular area along with the physical characteristics of their environment.

ectoderm The outer germ layer of an embryo.

ectotherm An animal whose body temperature remains close to the temperature of its environment.

electron transport chain A series of compounds that transfer electrons from electron donor substances to electron acceptors. This chain is an important part of both photosynthesis and cellular respiration.

element A pure substance made of only one kind of atom.

embryo (1) The part of the seed that develops into a plant. (2) The early stage of development in the unborn young of a multicellular organism.

endoderm The inner germ layer of an embryo.

endoplasmic reticulum The cellular organelle that consists of a network of membranes used to transport substances throughout a cell. Rough endoplasmic reticulum is studded with ribosomes; smooth endoplasmic reticulum is not.

endoskeleton An internal skeleton usually made of bone and cartilage; a characteristic of vertebrate animals.

endosperm The stored food inside a plant seed.

endotherm An animal whose body temperature is internally regulated and whose body heat is derived mainly from metabolism.

energy The ability to do work; especially in biology, the driver of the physical and chemical processes necessary for life.

entropy A mathematical measure of the dispersal of energy in a system.

envelope A lipid bilayer that surrounds the capsid in some viruses.

environmentalism A broad philosophical and ethical movement that seeks to improve and protect the quality of the natural environment, sometimes at the expense of considerations for human needs and activities.

enzyme A naturally occurring catalyst, usually a protein.

epidermis (1) In plants, the outer layer of cells that usually lack chlorophyll and serve for protection, including the secretion of the cuticle. (2) In animals and humans, a tissue that covers or lines a structure, particularly the outermost layer of the skin.

GLOSSARY

epididymis A coiled tube on the outer surface of each testis that stores sperm.

epiglottis The flap that protects the trachea during swallowing.

epinephrine The hormone secreted by the adrenal medulla that stimulates the fight-or-flight response needed in an emergency.

epithelial tissue A tissue that covers or lines a body part; functions in absorption, secretion, and protection.

erythrocyte A red blood cell.

esophagus The portion of the alimentary canal connecting the pharynx and stomach.

estivation A state of dormancy that an animal enters into in response to high temperatures and arid conditions.

estrogen A female sex hormone that stimulates the development of secondary sex characteristics.

ethylene A plant hormone that regulates the ripening of fruit, opening of flowers, and shedding of leaves.

eukaryote An organism with cells having a true nucleus as well as membrane-bound organelles.

exoskeleton A system of stiff external plates for support and protection; a characteristic of many invertebrate animals.

exponential growth A rate of population growth in which the population size multiplies at a constant rate at regular intervals.

external auditory canal The passageway from the external ear to the eardrum.

external fertilization The uniting of sperm and egg cells outside of an organism's body.

extracellular digestion The breakdown of substances that occurs in spaces outside of cells, such as within the stomach or intestine.

eyespot A light-sensitive area of pigmentation found in some protozoans.

F

facilitated diffusion A type of passive transport in which substances can move across a cell membrane through special protein channels embedded in the membrane.

fallopian tube A tube in female mammals and humans that leads from an ovary into the uterus.

family A taxonomic division within an order; composed of one or more genera.

fascicle A small bundle of muscle fibers.

feather One of the proteinaceous structures that cover a bird and enable flight.

fermentation The anaerobic breakdown of sugars to pyruvate and then to either lactic acid or carbon dioxide and alcohol.

fetus In humans, an unborn child during the second and third trimesters of pregnancy.

fibrous root system A root system lacking a taproot, consisting of many small roots that come directly from a stem.

flagellum A whiplike cellular organelle similar in structure to a cilium but longer and used primarily for movement in many bacteria.

food web The predator-prey relationships between populations in an ecosystem involving overlapping food chains.

formula unit The smallest part of an ionic compound.

fossil The preserved remains or trace of an organism that lived in the past.

founder effect The loss of genetic variation that results when a new population of organisms is established by a small number of individuals from a larger population.

fragmentation A type of asexual reproduction in fungi in which new organisms grow from pieces of an original.

frameshift mutation A mutation that deletes or adds a nucleotide base to a DNA sequence, causing a shift in the reading of the sequence during transcription.

frog A short-bodied, tailless amphibian in the order Anura.

frond The leaf of a fern plant.

fruit A ripened plant ovary with seeds.

fruiting body A special fungal structure, composed of many hyphae, that is responsible for reproduction.

G

gallbladder A pear-shaped sac on the underside of the liver that concentrates and stores bile.

gamete A haploid reproductive cell that can unite with another gamete to form a zygote.

gametophyte The haploid stage in the alternation of generations life cycle of plants during which gametes are produced.

ganglion A bundle of nerve cells.

gastrulation The phase of embryonic development during which the three germ layers are formed.

gene A segment of DNA that codes for a specific protein, resulting in a particular trait.

gene expression The process by which the information in a gene is used to make a functional gene product, usually a protein.

gene flow The movement of alleles into and out of a population of organisms.

gene pool The sum of all the alleles possessed by all the individuals in a population of organisms.

gene therapy The use of genetic engineering to treat genetic disorders or diseases.

genetic drift A change in allele frequency due to random events, usually within a small population.

genetic engineering The process of manipulating an organism's genes by methods other than natural reproduction.

genetic equilibrium The condition of having a stable allele frequency, usually only in large populations.

genetic load The reduction in fitness of a population with an increasing number of mutations.

genetically modified organism (GMO) An organism whose genetic material has been altered using genetic engineering techniques.

genome The full set of genetic information coded in an organism's DNA.

genotype The genetic makeup of an individual organism, especially regarding its particular combination of alleles for a specific trait.

genus A taxonomic division within a family; composed of one or more species.

geologic timescale A system that relates evolutionary history to rock layers in the geologic column.

germination The beginning of growth by a seed, spore, bud, or other structure after a period of dormancy.

germ layer One of two or three primary layers of cells that form early in embryonic development.

gestation The period of time between conception and birth.

gibberellin One of a group of plant hormones with a role in regulating many aspects of plant growth, mainly that of stem elongation and cell division in shoots and leaves.

gill In some mushrooms, a rib located beneath the cap used to disperse spores.

gizzard A thick-walled digestive organ that grinds food.

glomerulus The filtering unit within a nephron, consisting of a network of capillaries surrounded by the Bowman's capsule.

glucagon A hormone that raises the blood glucose level and stimulates breakdown of fats to form glucose.

glycolysis The first phase of cellular respiration, in which glucose is broken down into pyruvate.

Golgi apparatus The cellular organelle that processes and packages proteins in preparation for secreting them from a cell.

gonad A sex gland, referring to the testes in males and the ovaries in females.

granum A stack of disk-shaped thylakoids within a chloroplast.

grassland A temperate biome dominated by grasses because of insufficient rainfall for trees.

gravitropism The growth movement of a plant in response to gravity.

greenhouse gas A gas that can trap heat in the earth's atmosphere, such as carbon dioxide or water vapor.

ground tissue Any of the tissues in a plant that are neither dermal nor vascular tissue.

growth hormone A hormone produced by the anterior pituitary gland that primarily affects bone growth.

growth plate A plate of cartilage at each end of a long bone where bone growth takes place in children and adolescents.

guard cell A modified epidermal cell that opens and closes stomata in the leaves of plants.

gymnosperm A plant that produces unenclosed seeds, usually in cones.

habitat The physical portion of an ecosystem that an organism prefers.

hair cell A receptor cell in the cochlea that converts vibrations into nerve impulses.

hair follicle The tube from which a hair grows.

halophile Any of a group of organisms in domain Archaea that thrive in extremely salty environments, such as Utah's Great Salt Lake.

haploid organism An organism having only a single set of chromosomes.

Haversian canal The channel within an osteon that contains the capillaries and nerves that service the osteon's osteocytes.

hemoglobin The red pigment molecule that transports oxygen and carbon dioxide in vertebrate red blood cells.

heredity The passing of traits from one generation to another.

hermaphrodite An organism possessing both male and female reproductive organs.

heterozygous organism An organism having different alleles for a gene at the same position on homologous chromosomes.

hibernation A state of inactivity and extremely low metabolism that an animal enters into in order to survive unfavorable environmental conditions.

homeostasis The internal balance within the systems of living organisms that must be kept stable in order to maintain life. Processes that maintain homeostasis include getting nutrition, expelling wastes, and responding to the environment.

homeotic gene A gene that regulates the development of an anatomical structure in an organism.

hominid Any primate in the family Hominidae. Evolutionary theory includes modern humans and the extinct supposed predecessors of humans in this group.

homologous structures Organs that are similar in different organisms and are thought to show evolutionary relationships.

homozygous organism An organism having the same two alleles for a gene at the same position on homologous chromosomes.

hormone A chemical regulator produced in an endocrine gland and carried in blood to its target cell in order to produce a response.

host See *parasitism.*

human anatomy The science that deals with the structure of the human body.

human physiology The science that deals with the various processes and activities that take place in the human body.

humoral immunity An immunity to disease involving antibodies and B cells.

hydroskeleton A support structure in many soft-bodied animals consisting of a fluid-filled cavity surrounded by muscles.

hydrotropism The growth movement of plant roots in response to the presence of water.

hypertonic solution A solution whose solute concentration is greater than the solute concentration of the cytoplasm in a cell.

hypha The slender filament that is the primary structure of a fungal colony.

GLOSSARY

hypothalamus The region of the brain that controls involuntary activities, emotional expressions, and appetite for food and links the nervous and endocrine systems.

hypothesis A simple, testable statement that predicts an answer to a question being investigated using scientific inquiry.

hypotonic solution A solution whose solute concentration is less than the solute concentration of the cytoplasm in a cell.

I

image of God The combination of qualities that God has placed in humans as a reflection of Himself.

implantation The process whereby a fertilized egg attaches to the uterine wall and forms a placenta.

incomplete dominance A type of inheritance in which different alleles of a gene are neither dominant nor recessive and are expressed simultaneously, resulting in a blended, usually intermediate phenotype.

incomplete metamorphosis A type of insect development in which eggs hatch into nymphs that mature into adults.

incubation The process of keeping an egg at a suitable temperature for development and hatching.

incus The second of the three bones in the middle ear that transmit vibrations from the eardrum to the inner ear; also known as the *anvil*.

index fossil A fossil of an accepted age that is used to assign an age to a rock layer in the geologic column.

inferior vena cava The vein that carries blood from the lower body to the right atrium of the heart.

inflammatory response The reaction of tissues to infection or injury; characterized by increased blood flow, redness, pain, and swelling.

inhibitor A substance that binds to an enzyme and reduces its activity, thus slowing a chemical reaction.

instinct An unlearned, often highly complex inborn behavior.

insulin A hormone produced by the pancreas that helps control the glucose level in the blood.

internal fertilization The uniting of sperm and egg cells inside the body of a female organism.

interneuron A neuron located in the central nervous system that transmits an impulse from a sensory neuron to another neuron.

internode The region between nodes on a plant stem.

interphase The period of the cell cycle in between cell divisions; the time of regular growth and activity.

intracellular digestion The breakdown of substances within cells.

invasive species A species that moves into a new habitat, either naturally or by human activity, and then competes with native species for resources.

invertebrate An animal that does not have a backbone or vertebral column.

iris The colored portion of the eye that controls the size of the pupil.

isotonic solution A solution whose solute concentration is equal to the solute concentration of the cytoplasm in a cell.

J

Jacobson's organ An auxiliary sense of smell organ found in many animals, used to detect moisture-borne scent particles and pheromones.

jawless fish A group of fish that includes hagfishes and lampreys.

joint The point where two bones come together.

K

keel The large ridge on a bird's sternum.

keratin A tough, fibrous protein found in the horns, nails, hair, scales, feathers, and other structures of reptiles, birds, mammals, and humans.

kidney The organ in most vertebrates that filters wastes from the blood and excretes them in the form of urine.

kind A group of organisms that God created as distinct from other groups of organisms.

kingdom A taxonomic division within a domain; composed of one or more phyla.

L

large intestine The portion of the alimentary canal following the small intestine, where water and some vitamins are absorbed.

larynx The short passageway between the pharynx and esophagus; functions in breathing and sound production.

law A description (often mathematical) that is based on repeated observations of the relationship between two or more phenomena.

lens The transparent structure within the eye that focuses light rays onto the retina.

lenticel An area of porous tissue in the surface of a woody plant stem that allows for gas exchange with the environment.

leucoplast A nonpigmented cellular organelle found in plant cells whose functions include storing starches, lipids, and proteins.

leukocyte A white blood cell.

lichen A composite organism consisting of an alga and a fungus in a symbiotic relationship.

ligament A band of connective tissue that holds a joint together.

light-dependent phase The portion of photosynthesis that requires light energy.

light-independent phase The portion of photosynthesis that does not require light energy in order to proceed. But the light-independent phase does require the products of the light-dependent phase; therefore, neither phase occurs during darkness. Also known as the *Calvin cycle*.

limiting factor A factor within an ecosystem that prevents a population from growing beyond a certain size.

lipase A pancreatic enzyme that digests fats.

lipid A class of nonpolar organic compounds that are insoluble in water and are used for energy storage and cell membranes in living things.

lipid bilayer The two layers of phospholipids that make up a cell membrane.

liver An organ in the digestive system that secretes bile, purifies blood, metabolizes food molecules, and stores minerals and vitamins.

long-day plant A plant that requires a short period of darkness in order to flower.

loop of Henle The portion of a nephron that connects the proximal convoluted tubule to the distal convoluted tubule.

lung An organ for the exchange of gases between the atmosphere and the blood of an organism.

lung capacity The amount of air that can fill the lungs.

lymph The clear fluid found between body cells that is absorbed by the lymphatic system and returned to the bloodstream.

lymph node A small organ in the lymphatic system through which lymph flows and in which lymphocytes are found.

lymphocyte A type of white blood cell that functions in immunity.

lysogenic cycle A type of viral reproduction in which a virus's genetic material is reproduced by the host cell during normal cell division, resulting in daughter cells that both contain the viral material.

lysosome A cellular organelle that contains digestive enzymes.

lytic cycle The rapid infection and destruction of a host cell by a virus, resulting in more virus particles.

M

macrophage A large, amoeba-like cell found in the lymphatic system and surrounding tissues.

malleus The first of the three bones in the middle ear that transmit vibrations from the eardrum to the inner ear; also known as the *hammer*.

Malpighian tubule One of numerous threadlike tubules in insects that extract wastes from the blood and empty them into the intestine.

mammary gland An organ in female mammals and humans that produces milk to nourish their young.

mandible A chewing mouthpart found in insects and some other arthropods.

mantle The sheath of tissue that covers the body of a mollusk.

matrix Nonliving secretions produced by cells to provide structural and biochemical support.

matter Anything that takes up space and has mass.

medusa The free-swimming, umbrella-shaped stage in the life cycle of cnidarians; reproduces sexually.

meiosis The process by which haploid gametes are produced from diploid cells.

melanin A dark brownish-black pigment.

meninges The protective coverings of the brain and spinal cord.

menstrual cycle The regular changes caused by changes in hormones that take place in the ovaries and lining of the uterus to make pregnancy possible.

menstruation The regular shedding of the uterine lining in the event that pregnancy does not occur during a menstrual cycle.

meristematic tissue Undifferentiated cells found in the fast-growing parts of plants, such as root tips and new leaves.

mesoderm The middle germ layer of an embryo.

messenger RNA (mRNA) The RNA molecule that carries the code for a protein from the DNA to the cytoplasm of a cell.

metabolic waste The substances produced during metabolic activities that cannot be used by the body and must therefore be excreted.

metabolism The sum of all the chemical processes needed within cells to maintain life.

metamorphosis A change in shape or form that an animal undergoes in its development from egg to adult.

methanogen Any of a group of organisms in domain Archaea that live in anaerobic environments and produce methane during metabolism.

microbiome The collective genomes of the microorganisms (microbiota) that live in an environmental niche.

migration The movement of an organism from one location to another, often seasonally and over long distances.

mitochondrion The cellular organelle in which aerobic respiration takes place to produce energy from food.

mitosis The phase of the cell cycle in which a cell's duplicated DNA is divided into two identical sets.

model A simple, workable representation of a usually complex object or concept.

modern synthesis A modern evolutionary theory stating that changes within populations of organisms are due to natural selection and mutations.

molecular clock A technique that uses rates of molecular change to calculate the point in the past in which two species diverged.

molecular evolution A slow, gradual change in the genetic material and proteins of organisms.

molecule The smallest particle of a covalently bonded compound.

mollusk A member of the phylum Mollusca.

molting The periodic shedding of an exoskeleton, scales, feathers, or fur.

monocot Any flowering plant whose embryo has one cotyledon.

monohybrid cross A genetic cross that tests only one set of alleles.

motor neuron A neuron that receives impulses from the central nervous system and stimulates a muscle or gland.

GLOSSARY

multicellular organism An organism consisting of two or more cells. These cells cannot live independently apart from the organism.

multiple alleles A condition in which more than two alleles for a gene are possible.

muscle fiber A muscle cell.

muscle tissue Tissue made of cells that can contract to produce movement.

mutagen A physical or chemical agent that can change the genetic material of an organism.

mutation A random change in the sequence of bases in a DNA molecule.

mutualism A symbiotic relationship between two organisms that benefits both organisms involved.

mycelium A fungal structure made of interwoven hyphae.

mycorrhiza A symbiotic relationship formed between many fungi and plant roots in which the fungi increase the absorptive area of the roots and receive carbohydrates in return.

myofibril One of the functional fibers within a muscle cell that causes contraction by the movement of actin and myosin filaments.

myosin One of the two types of protein found in muscle fibers.

N

nasal cavity A large, air-filled space above and behind the nose.

natural selection The process by which some phenotypes within a species become more common than others because of variations that lead to increased survival rates.

negative feedback A cellular signaling mechanism in which the product of a particular process inhibits the process itself.

nephron One of the microscopic functional units of the kidney.

nerve cord A length of nerve tissue that connects the brain to the rest of a chordate's body.

nerve net A nervous system that lacks a brain or major ganglia.

nervous tissue Tissue made of cells that respond to stimuli and transmit and receive electrical impulses.

neuron The functional unit of the nervous system; a cell that receives and distributes nerve impulses.

neurotransmitter A chemical that a neuron releases into the synaptic cleft for the purpose of stimulating receptor proteins in the membrane of the next neuron.

neutralism A relationship in which two organisms share the same habitat without directly affecting each other.

niche The role of an organism within its habitat.

nitrogen fixation The conversion of nitrogen gas (inorganic nitrogen) by one of several mechanisms into nitrogen compounds (organic nitrogen) that living things can use.

node The point along a plant stem where leaves are produced.

nondisjunction The failure of homologous chromosomes or chromatids to separate properly during meiosis.

nonsteroid hormone A hormone without a large lipid component that affects its target cell with the aid of a receptor in the target cell's membrane.

nonvascular plant Any plant lacking tissues that can transport water.

norepinephrine A hormone secreted by the adrenal medulla that functions with epinephrine during stressful situations.

notochord A tough, flexible rod that provides support for a chordate animal at some time during its life.

nucleic acid An organic molecule that carries genetic information in the form of either DNA or RNA.

nucleoid The non-membrane-bound region in prokaryotic cells that contains most of the genetic material.

nucleolus A spherical structure within a eukaryotic cell's nucleus where ribosomes are manufactured.

nucleotide A member of a class of organic compounds that serve as the building blocks for the information storage molecules DNA and RNA.

nucleus (1) The membrane-bound region of a eukaryotic cell that contains the genetic material. (2) The central portion of an atom where protons and neutrons are located.

nutrient Any of several groups of substances that are necessary for growth and the maintenance of life.

nymph One of the stages in the incomplete metamorphosis of an insect.

O

olfactory sense The sense of smell.

operculum A plate that covers the gills of a fish.

optic nerve The nerve that carries nerve impulses from the photoreceptors to the occipital lobe of the brain.

oral cavity The portion of the mouth behind the teeth and bounded by the tongue and hard and soft palates.

order A taxonomic division within a class; composed of one or more families.

organ A structure made of two or more tissues that work together to perform a particular function.

organ system A group of two or more organs that work together to perform a particular function.

organelle A structure found within the cytoplasm of cells that performs a specific set of functions in a cell that the cell needs to stay alive.

organic compound A covalently bonded compound containing the element carbon.

osmosis The diffusion of water molecules through a semi-permeable membrane.

ossification The laying down of new bone material by osteoblasts.

osteoblast A cell that builds and mineralizes new bone.

osteoclast A type of cell that breaks down bone tissue.

osteocyte A living bone cell resident in an osteon.

osteon The subunit that makes up compact bone.

ovary The primary reproductive organ in females; produces haploid ova.

oviparous species An organism that uses a method of reproduction in which young develop inside eggs that are laid and hatched outside the parent's body.

ovoviviparous species An organism that uses a method of reproduction in which young develop within eggs that hatch inside the parent's body.

ovulation The release of ova from an ovary.

ovule A structure in a plant ovary that contains the egg cell and will mature into a seed.

ovum A female gamete.

oxytocin A hormone released at the posterior lobe of the pituitary gland by the hypothalamus that stimulates smooth muscle to contract.

P

pancreas An organ that secretes enzymes into the small intestine to perform digestion. It also secretes the hormones insulin and glucagon to control the blood glucose level.

parasitism A symbiotic relationship between two organisms in which one organism benefits at the expense of the other organism, known as the *host*.

parasympathetic nervous system The system of neurons that helps the body return to normal processes after a stressful situation.

parathyroid gland The endocrine gland that helps regulate calcium and phosphorus levels in the body.

passive immunity An immunity in which an individual receives antibodies that have been formed by another individual or an animal.

passive transport The movement of substances across a cell membrane without the need for energy.

pathogen An agent that causes a disease.

pedicel The stalk that connects a flower to the rest of the plant.

pedigree A heredity chart that tracks the expression of a trait through several generations.

pedipalp One of the second pair of arachnid appendages used for sensory perception and sperm transfer.

penis The organ that transfers sperm from male to female in humans and many vertebrates.

pepsin An enzyme produced by the stomach to digest proteins.

peptidoglycan A protein found in the cell walls of bacteria but not in archaea.

periosteum A layer of fibrous connective tissue that covers the outer surface of bones, except the joint ends of long bones.

peripheral nervous system The portion of the nervous system consisting of the nerves and ganglia outside the brain and spinal cord.

peristalsis The wavelike, involuntary muscular contractions that move food through the alimentary canal.

petal A flower structure located just inside the sepals and often large, conspicuous, and scented.

petiole The small stalk that connects a leaf to the stem of a plant.

pharyngeal pouch The fold of the skin along the neck region of chordate embryos; in vertebrates, these develop into gills or structures of the lower face and neck.

pharynx The portion of the throat posterior to the mouth and nasal cavity.

phenotype The physical expression of a trait in an organism.

pheromone A chemical released by an animal that influences the behavior of another animal of the same species.

phloem The type of vascular tissue in plants that transports food from the leaves to the rest of the plant.

photoperiodism The response of a plant to changes in duration and intensity of light exposure.

photoreceptor A specialized neuron that is stimulated by light.

photosynthesis The process whereby plants, algae, and some bacteria form simple sugars from carbon dioxide and water in the presence of light and chlorophyll.

phototropism Growth movement of a plant in response to light.

phylogenetic tree A diagram that shows the supposed evolutionary relationships between groups of organisms.

phylogeny The evolutionary history of a species or population.

phylum A taxonomic division within a kingdom, composed of one or more classes.

physical change A change in a substance that does not change the identity of the substance (e.g., a change of form or state).

phytochrome A plant pigment that regulates a plant's response to photoperiod changes.

pineal gland A small endocrine gland in the brain that secretes melatonin.

pioneer species Any of the plants and animals that first colonize an area during primary succession.

pith The spongy ground tissue located in the center of a plant stem.

pituitary gland An endocrine gland attached to the hypothalamus that controls many of the other endocrine glands.

pit viper One of a group of venomous snakes that possess a heat-sensing organ between each eye and nostril.

placenta The structure in most mammals and humans that consists of a portion of the uterine wall and chorion of the embryo that allows nutrient and waste exchange between mother and embryo.

plankton Organisms that float in the water columns of oceans, seas, and large lakes and cannot swim against a current.

plasma The liquid portion of the blood that holds blood cells in suspension.

plasmid A small, circular piece of DNA distinct from chromosomal DNA and usually found in bacteria.

platelet A small, nonnucleated component of the blood that functions in blood clotting.

GLOSSARY

point mutation A mutation that changes a single nucleotide base in a DNA sequence.

polar molecule A molecule having an uneven distribution of electrical charge, resulting in some regions of the molecule having negative charge while others are positive.

polygenic inheritance A type of inheritance in which a trait is determined by more than one gene.

polymerase chain reaction A laboratory process used to quickly generate many copies of a piece of DNA for medical or research purposes.

polyp A sessile, cup-shaped cnidarian with a mouth and tentacles at one end and a basal disc at the other; reproduces asexually.

polyploidy The condition of having three or more complete sets of chromosomes.

population A group of organisms of the same species living and interacting within the same geographic area.

population density The number of organisms in a defined unit of area.

positive feedback A cellular signaling mechanism in which the product of a particular process enhances the process itself.

post-anal tail A structure possessed by all chordate embryos that extends beyond the anus.

prehensile tail A tail that can be flexed and is used for gripping or holding things.

pressure-flow hypothesis The best-supported theory for explaining the flow of sap in the phloem of a plant.

primary succession The stage of succession characterized by organisms first colonizing a previously barren ecosystem, such as a lava field or ground exposed by a retreating glacier.

prion Short for the term "proteinaceous infectious particle"; an infectious agent made entirely of protein.

producer An autotrophic organism, that is, one that can produce its own food from light or chemical energy.

product A substance formed during a chemical reaction, usually indicated on the right-hand side of a chemical equation.

progesterone A female sex hormone involved in the menstrual cycle and in pregnancy.

prokaryote An organism with cells lacking a true nucleus and membrane-bound organelles.

promoter A short sequence of DNA that determines where an RNA polymerase molecule is to begin the transcription of a strand of messenger RNA.

prostate gland A structure in males that produces a portion of the semen.

protein A member of the class of large organic compounds that are made of amino acids linked by peptide bonds; used for structure and as organic catalysts in living things.

protein synthesis The process by which ribosomes build a sequence of linked amino acids, the basic, or primary, structure of a protein.

protist A single-celled or multicellular eukaryotic microorganism.

protozoan An animal-like organism in the kingdom Protozoa.

pseudopodium A temporary extension of a cell membrane used by some protozoans either to move about or to engulf food particles.

puberty The period of hormone-induced change during which secondary sex characteristics develop.

pulmonary artery The artery that carries blood from the right ventricle of the heart to the lungs.

pulmonary circulation The flow of blood from the heart to the lungs and back.

pulmonary vein The vein that carries blood from the lungs to the left atrium of the heart.

Punnett square A diagram used to visualize genetic crosses.

pupa The resting or inactive stage of development during complete metamorphosis in an insect.

pupil The circular opening in the iris of the eye.

R

radial symmetry A type of symmetry in animals that do not have left and right halves. Mirror image halves can be produced by several imaginary planes through the center of the body.

radiometric dating A process that analyzes radioactive elements in rocks and uses the rate at which these elements decay to calculate a sample's age.

radula A platelike structure near the mouth of many mollusks; bears many tiny teeth used for scraping up food into the animal's mouth.

reactant A substance that is present before a chemical reaction and takes part in it. It is usually shown on the left-hand side of a chemical equation.

receptacle The thickened top portion of a pedicel that supports the actual parts of a flower.

recessive trait A characteristic that is expressed only in the absence of a dominant allele of the gene that codes for that characteristic.

recombinant DNA DNA molecules produced in a laboratory by combining sequences of DNA from different sources, yielding a new sequence not normally found in any genome.

rectum The final section of the large intestine.

relative dating The process of estimating the age of rock strata above and below a given stratum.

REM sleep The period of the sleep cycle during which dreaming takes place.

replication The process by which a strand of DNA is copied to produce an identical strand.

reproductive barrier Anything that prevents two species or two populations of the same species from interbreeding.

resting potential The relative charge difference between the inside and outside of a neuron.

restriction enzyme An enzyme used to cut DNA into pieces at specific places in the DNA sequence.

retina The innermost layer of the eyeball that is composed of light-sensitive photoreceptors and their fibers.

retrovirus A virus that can force a host cell to transcribe the viral RNA into the host's DNA.

ribonucleic acid (RNA) The single-stranded genetic information molecule, copied from DNA, whose main function is, in conjunction with ribosomes, to build protein molecules. See also *messenger RNA*, *ribosomal RNA*, and *transfer RNA*.

ribosomal RNA (rRNA) The RNA molecule that combines with proteins to form a ribosome.

ribosome The cellular organelle that directs the protein-building process.

rod One of the sensory receptors in the eye that is not sensitive to color but can rapidly discern movement and is important in low-light vision.

rough endoplasmic reticulum See *endoplasmic reticulum*.

rumen The first chamber in a ruminant stomach.

S

salamander A lizard-like amphibian in the order Urodela.

salivary gland A gland that secretes saliva and amylase into the mouth to lubricate food and begin the breakdown of starches.

salt gland A gland located near the eye of a reptile that excretes excess salt from the animal.

sap The fluid transported by xylem and phloem in a plant.

sarcomere One of the functional segments of a muscle.

savanna A biome consisting of grassland mixed with shrubs and widely spaced trees.

scale A small body-covering plate on fish and reptiles.

scientific inquiry An orderly way of investigating phenomena by using measurable and repeatable observations to test a hypothesis.

scientific name A two-part name, consisting of a genus name and species epithet, that is unique to a particular species.

sclera The outer layer of the eye; the "white of the eye."

scrotum The skin pouch in males within which the testes are located.

scute A bony external plate covered with keratin.

sebaceous gland A gland of the skin that releases a mixture of oils, waxes, and metabolic wastes.

secondary succession Succession that takes place in a disturbed area with soil and perhaps a few plants.

semen The fluid that contains sperm.

semicircular canal The structure in the inner ear that maintains dynamic equilibrium.

seminiferous tubule One of the tiny tubes that produces sperm within the testes.

sensory neuron A neuron that carries impulses toward the spinal cord or brain.

sensory receptor A specialized dendrite found in sensory neurons.

sepal The leaflike structure that protects a flower bud during formation and supports the flower after blooming.

septum A biological wall that divides a cavity or structure into smaller chambers.

sequencing The process of determining the order of nucleotides in an organism's DNA.

sessile Describes an organism that lives and grows while attached to something else; nonmotile.

sex chromosome One of the chromosomes that determine whether an organism will be male or female.

sex-linked trait A trait coded for by a gene located on a sex chromosome.

short-day plant A plant that requires a long period of darkness in order to flower.

siphon One of a pair of tubes in many aquatic mollusks used to draw in or expel water.

skeletal muscle Striated, voluntary muscle tissue that is attached to and moves the skeleton.

small intestine The portion of the alimentary canal immediately after the stomach where most of the digestion and absorption of nutrients takes place.

smooth endoplasmic reticulum See *endoplasmic reticulum*.

smooth muscle Nonstriated, involuntary muscle tissue that is not composed of sarcomeres.

sorus A structure on the underside of a fern frond that houses and protects the fern's sporangia.

spawning The process by which aquatic animals release eggs and sperm.

speciation The process of forming new and distinct species.

species A taxonomic division within a genus that consists of a single type of organism.

spike A protein that attaches to specific cell surfaces.

spinneret An organ in spiders and some insects that produces a filament made from secretions produced in silk glands.

spiracle One of the small pores in an insect's body that opens into the tracheae.

spongy bone The type of bone that contains many small spaces, usually located in the ends of long bones.

spore A type of reproductive cell protected by a hard covering.

sporophyte The diploid stage in the alternation of generations life cycle of plants during which spores are produced.

squamate An animal in the order Squamata, including snakes and lizards.

stamen The male reproductive structure of a flower; produces pollen.

stapes The third of the three bones in the middle ear that transmit vibrations from the eardrum to the inner ear; also known as the *stirrup*.

stem cell A generalized cell that has the potential to differentiate into a cell that performs a specific function.

steroid hormone A hormone composed largely of lipids that diffuses through its target cell's cell membrane.

stipe The stalk of a fungal fruiting body supporting the cap.

stoma A small opening in the undersides of leaves that regulates gas exchange.

GLOSSARY

stomach The muscular pouch of the digestive system that connects to the esophagus.

stroma The material within a chloroplast that surrounds the thylakoids.

subcutaneous layer The layer of fat and connective tissue below the dermis of the skin.

sugar One of a class of simple organic compounds important in living things as a source of both energy and structure.

superior vena cava The vein that carries blood from the upper body to the right atrium of the heart.

survey A subset of observations gathered from within a larger possible set of data.

sustainability The ability of the biosphere to maintain its balance indefinitely.

sweat gland One of the glands in the skin that releases perspiration to cool the body and release wastes.

swim bladder A gas-filled structure in many fish that enables them to maintain neutral buoyancy.

swimmeret One of the small appendages on some crustaceans that is used for swimming and reproduction.

symbiosis The interaction between two different organisms within an ecosystem.

sympathetic nervous system The system of neurons that helps the body adjust to stressful situations.

synapse The space between an axon and a dendrite or between the end of an axon and the body structure that it affects.

syrinx The voice organ in birds.

systematics The study of evolutionary relationships among living things through time.

systemic circulation The flow of blood from the heart to all parts of the body, except the lungs, and back to the heart.

T

taproot system A root system consisting of a large, main root (the taproot), from which smaller roots branch.

target cell The cell that is activated by a hormone.

TATA box A common promoter sequence.

taxon A grouping within a classification scheme.

taxonomy The science of classifying organisms.

T cell Any of several types of lymphocytes involved in cell-mediated immunity.

telomere The repetitive nucleotide sequence that serves as a protective cap at the end of a chromosome.

temperature A measurement of the average kinetic energy of the particles within a substance.

tendon A type of connective tissue that connects muscle to bone.

testis The primary male reproductive organ; produces sperm and male sex hormones.

testosterone The hormone that promotes the development of secondary sex characteristics in males.

tetrapod An animal that has four limbs.

thalamus The brain region that receives general sensations and relays impulses to the cerebrum; regulates consciousness and sleep.

theory A model that attempts to explain a set of observations.

thermophile Any of a group of organisms in domain Archaea that thrive in hot environments, such as around hydrothermal vents.

thigmotropism Growth movement of a plant in response to touch, primarily found in climbing plants such as vines.

thorax The body region between the head and abdomen.

threshold The strength that a stimulus must reach before it causes an action potential.

thylakoid A flattened, membrane-bound sac inside a chloroplast where the light-dependent phase of photosynthesis takes place.

thymus A mass of lymphatic tissue thought to produce the hormones called thymosins; degenerates prior to puberty.

thyroid The endocrine gland that controls the body's overall rate of metabolism and affects bone growth.

thyroxine The thyroid hormone that regulates metabolic rate.

tissue A group of cells that are specialized to perform a particular function.

trachea (1) The passageway in many animals that connects the larynx to the lungs; (2) One of the chitinous tubules in insects that transfer air directly to its tissues.

transcription The process by which a particular segment of DNA is copied into RNA.

transduction The transfer of DNA from one bacterium to another by a bacteriophage.

transfer RNA (tRNA) The RNA molecule that carries a specific amino acid to a ribosome during protein synthesis.

transformation The process of a bacterium taking in free-floating DNA from its environment, after which the bacterium can express traits coded for by the new DNA.

transgenic organism An organism containing genes that have been introduced from a different kind of organism.

transitional form One of the hypothetical organisms whose structures are intermediate between their ancestors and their descendants.

translation The process in which a ribosome builds a protein that is based on the sequence of codons in an mRNA molecule.

transpiration The release of water through the leaves of a plant.

tropical rainforest A biome characterized by a hot, wet climate and high levels of biodiversity.

tropism A directionally dependent growth response of plants to external stimuli such as light, gravity, touch, and the presence of water.

tuatara A lizard-like reptile native to New Zealand.

tube foot A small, soft, tubular structure in echinoderms used for locomotion and food capture; part of the water-vascular system.

tumor An abnormal growth of cells.

tundra A biome with low temperatures for most of the year. Long, cold winters and a short growing season prevent the growth of most plants, especially trees.

turgor pressure The pressure exerted by water in a full central vacuole upon the cytoplasm and cell membrane of a plant cell; maintains plant rigidity.

turtle A shell-covered reptile in the order Testudines.

tympanic membrane The circular membranous structure that transmits sound vibrations to the middle ear cavity; also known as the *eardrum*.

U

umbilical cord The flexible structure containing blood vessels that conducts the blood of the fetus to the placenta for the exchange of food, wastes, and gases.

ungulate A hoofed mammal.

unicellular organism An organism consisting of only one cell.

uniformitarianism The theory that suggests that the currently observable processes have occurred at the same rate in the past as they do in the present.

ureter One of the tubes that carries urine from the kidney to the bladder.

urethra The passageway by which urine is conducted out of the body from the bladder.

urinary bladder The organ that serves as a storage reservoir for urine.

urine A liquid excreted from the body containing metabolic wastes from the blood.

uterus The organ in most mammals and in humans in which an embryo develops.

V

vaccination A method of exposing a person to a controlled amount of a disease-causing factor for the purpose of developing an immunity.

vagina The elastic canal in females that leads from outside the body to the cervix of the uterus.

valve A flap of tissue within a vein or in the heart that ensures the one-way flow of blood.

variation The range of genotypic differences possible between individuals in the same gene pool.

vascular bundle In plants, a bundle composed of both xylem and phloem; also called a *vein*.

vascular cambium A layer of meristematic cells located between the xylem and phloem in the woody stems of plants.

vascular plant A plant having tissues capable of transporting water.

vascular tissue In plants, the tissue that transports water and nutrients throughout the plant.

vas deferens The tube that carries sperm from the testes to the urethra.

vegetative propagation Any form of asexual reproduction in a plant.

vein (1) In plant leaves, a bundle composed of both xylem and phloem. See *vascular bundle*. (2) In animals, any blood vessel that carries blood toward the heart.

ventricle A chamber of the heart from which blood is pumped into arteries.

vertebrate An animal that has a backbone or vertebral column.

vertical zonation Changes in the plant and animal communities within a biome due to changes in elevation that produce differences in light, precipitation, and temperature.

vestigial structure A structure that seems to have no function in an organism and is thought to be left over from the evolutionary process.

virus A small infectious agent that can reproduce only inside the living cells of another organism.

visceral mass The portion of a mollusk's body that contains its internal organs.

viviparous species An organism that uses a method of reproduction in which young are born alive after developing inside the female's reproductive tract.

W

walking leg An arthropod appendage used for locomotion.

water-vascular system A system of canals and tubes within echinoderms that functions in movement, circulating nutrients, and capturing food.

withdrawal The symptoms that occur when the use of a drug is discontinued.

worldview A way of seeing and interpreting all aspects of life that arises out of an overarching narrative and that shapes how a person thinks and acts.

X

xylem In plants, the type of vascular tissue that transports water and dissolved minerals from the roots to the rest of the plant.

Z

zygote A diploid cell formed by the union of two gametes.

INDEX

A **boldface** page number denotes the page on which the term is defined.

INDEX

INDEX

INDEX

INDEX

INDEX

INDEX

tubers, 335

tumors, **184**

tumor suppressor gene, 185

tundra, **58**

turgor pressure, **101**, 327

turtles, **415**, 417, 420

tympanic membrane, **539**, 542

U

umbilical cord, **563**, 565

ungulates, **439**

unicellular organisms, **99**

uniformitarianism, **202**, 217, 228

unsaturated fats, 41

uplift, geological, 201, 205

uracil, 123

urea, 404, 520–22

ureter, **520**, 521

urethra, **520**, 556

uric acid, 418, 433, 434, 444, 522

urinary bladder, **520**, 536, 556

urinary system, human, 457, 463, 520–23

urine, **443**, 520, 547

uterus, **437**, 557–60, 563, 565

utricle, 539, 542

V

vaccinations, **468**

vacuole

central, 101, 327

contractile, 278

food, 278

vagina, **557**, 565

valence electrons, 27, 30–31

valves, **500**, 502

Van Leeuwenhoek, Antonie, 97, 275

vaping, 506

variation, genetic, **177**–78

varicella-zoster, 265

vascular bundles, 305

vascular cambium, **306**

vascular plants, **298**–99

vascular tissue, **301**

vas deferens, **556**

vectors, 190, 261

vegetative propagation, 302, 307, 311, **335**, 337–38

veins, **500**

pulmonary, **503**

renal, 521

vena cava, superior and inferior, **502**–3, 505

ventricles, **502**–3, 505

vertebrates, **350**, 400

vessel cells, 301

vestibular nerve, 542

vestigial structure, **203**, 204

villi, 512, 518

Virchow, Rudolf, 98

viruses, **263**

function of, in the environment, 266, 268

reproduction in, 264–66

structure of, 263

visceral mass, **365**

vitamins, as nutrients, 515, 518

vitreous humor, 540, 543

viviparous species, **352**, 405, 445

vocal folds, 495, 566

voluntary muscle tissue, 484

W

wading birds, 428

walking legs, **386**

water

as a nutrient, 514, 518, 522

importance of, 38–39

water cycle, 74

waterfowl, 429

water molds, 281–82

water-vascular system, **370**

Watson, James, 14–15

white matter, 530

wings

bird, 432

insect, 388–89

withdrawal, drug, **570**

wolf reintroduction at Yellowstone, 446

wood louse, 385

worldview, **5**

classification and, 248–50

importance of, 5–6, 18

perspectives stemming from, 82, 84, 88, 215, 561

worms, 351, 361–64

X

Xenarthra (order), 441

xylem, **301**, 304–5, 306, 325–28

Y

Yellowstone wolf reintroduction, 446

yolk sacs, 414, 445

Z

Z lines, 486

zooplankton, 284

Zygomycota (phylum), 290

zygotes, **155**, 169, 557, 559

PHOTO CREDITS

Key: (t) top; (c) center; (b) bottom; (l) left; (r) right; (i) inset; (bg) background; (fg) foreground

COVER

Tim Platt/Stone via Getty Images

FRONT MATTER

i, xcr Eric Isselee/Shutterstock.com; **iit, iib, iv** BOONCHUAY PROMJIAM/Shutterstock.com; **vi–vii** Gfed/iStock/Getty Images Plus via Getty Images; **vii (insert)** TIM SLOAN/AFP via Getty Images; **ix** sinology/Moment via Getty Images; **ixi** binik/Shutterstock.com; **xtl** luoman/iStock/Getty Images Plus via Getty Images; **xtr** Westend61 via Getty Images; **xcl** © Leo Malsam|Dreamstime.com; **xbl** Insectpedia/Shutterstock.com; **xbr** eye-blink/Shutterstock.com

CHAPTER 1

2–3 IstvanKadarPhotography/Moment via Getty Images; **4** SciePro/Shutterstock.com; **6** ChristineKohler/iStock /Getty Images Plus via Getty Images; **8t, 23t** jack0m /DigitalVisionVectors via Getty Images; **8c** AlexanderYershov /iStock/Getty Images Plus via Getty Images; **8b, 19** FatCamera /E+ via Getty Images; **9t** CSA Images/Vetta via Getty Images; **9c** Siri Stafford/The Image Bank via Getty Images; **9b** CSA-Archive/DigitalVision Vectors via Getty Images; **10t** leonello /iStock/Getty Images Plus via Getty Images; **10b** MorphartCreation/Shutterstock.com; **11t** DrAfter123 /DigitalVision Vectors via Getty Images; **11b** Damir Khabirov /iStock/Getty Images Plus via Getty Images; **12t, bl, br** ©Jeffery R. Werner/Incredible Features ALL RIGHTS RESERVED.; **13t** Diamond Dogs/iStock/Getty Images Plus via Getty Images; **13bl** Rattiya Thongdumhyu/Shutterstock.com; **13br** fotograzia /Moment via Getty Images; **14–15bg** LHFGraphics/Shutterstock .com; **14l, c** Science History Images/Alamy Stock Photo; **14tr, br, 15t** Ava Helen and Linus Pauling Papers (MSS Pauling), Oregon State University Special Collections and Archives Research Center, Corvallis, Oregon.; **15b (photo)** Heritage Auctions, HA .com; **15b (figures)** Figures reprinted by permission from Springer Nature Customer Service Centre GmbH: Springer Nature. *Nature*, "Genetical Implications of the Structure of Deoxyribonucleic Acid", WATSON, J., CRICK, F., Copyright © 1953, Nature Publishing Group; **16** Antonio Guillem/Shutterstock.com; **17, 23b** tommy/DigitalVision Vectors via Getty Images; **18t** Nastasic/iStock/Getty Images Plus via Getty Images; **18c** ninjaMonkeyStudio/E+ via Getty Images; **18bl** stocksnapper /iStock/Getty Images Plus via Getty Images; **18br** jsolie/E+ via Getty Images; **21** Monkey Business Images/Shutterstock.com; **22t** Ugis Riba/Shutterstock.com; **22b** monkeybusinessimages /iStock/Getty Images Plus via Getty Images

CHAPTER 2

26 mikroman6/Moment via Getty Images; **28t** Çağla Köhserli /iStock/Getty Images Plus via Getty Images; **28bl** Historic Illustrations/Alamy Stock Photo; **28br** bauhaus1000 /DigitalVision Vectors via Getty Images; **29l** MARK GARLICK /SCIENCE PHOTO LIBRARY via Getty Images; **29r** Roman Novitskii/iStock/Getty Images Plus via Getty Images; **30t** MiroNovak/iStock/Getty Images Plus via Getty Images; **30c** fcafotodigital/E+ via Getty Images; **30b** maystra/iStock/Getty Images Plus via Getty Images; **31** Maisei Raman/Shutterstock .com; **33t** mel-nik/iStock/Getty Images Plus via Getty Images; **33b** tonaquatic/iStock/Getty Images Plus via Getty Images; **34** Kara Capaldo/iStock Editorial/Getty Images Plus via Getty Images; **36t–37** Ghislain & Marie David de Lossy/The Image Bank via Getty Images; **36b** Laguna Design/Science Source; **38** David Sacks/DigitalVision via Getty Images; **39** Universal Images Group North America LLC/Alamy Stock Photo; **41** PhotoAlto/Laurence Mouton/PhotoAlto Agency RF Collections via Getty Images; **43** LHFGraphics/Shutterstock.com; **48** SEBASTIAN KAULITZKI/SCIENCE PHOTO LIBRARY via Getty Images; **49** Melpomenem/iStock/Getty Images Plus via Getty Images

CHAPTER 3

50 Eye of Science/Science Source; **52t–53tl** milehightraveler/E+ via Getty Images; **52c** Ed Reschke/Stone via Getty Images; **52b** treasurephoto/iStock/Getty Images Plus via Getty Images; **53tr, 57cr, 71** titoOnz/iStock/Getty Images Plus via Getty Images; **53ct** Kitnha/Shutterstock.com; **53cb, 54tl** Marek Stefunko /EyeEm via Getty Images; **53bl, 54tr** © RAZVAN CIUCA /Moment via Getty Images; **53br** Wasan Amornsang/iStock /Getty Images Plus via Getty Images; **54b** Georgette Douwma /Stone via Getty Images; **55** Floortje/iStock/Getty Images Plus via Getty Images; **56** clu/DigitalVision Vectors via Getty Images; **57t** FrankvandenBergh/E+ via Getty Images; **57cl, 58t, 61t** Map Resources; **57b** "Köppen-Geiger Climate Classification Map" by Beck, H.E., Zimmermann, N. E., McVicar, T. R., Vergopolan, N., Berg, A., & Wood, E. F./"Present and future Köppen-Geiger climate classification maps at 1-km resolution". *Nature Scientific Data.*/Wikimedia Commons/modified/CC By 4.0/Map Resources; **58b** Bruno Guerreiro/Moment via Getty Images; **58fg** BearFotos/Shutterstock.com; **59t, 63 (trees)** Jose F. Donneys/Shutterstock.com; **59fg** GlobalP/iStock/Getty Images Plus via Getty Images; **59bl** johnaudrey/iStock/Getty Images Plus via Getty Images; **59br, 63 (cactus)** Thomas Roche /Moment via Getty Images; **60t** Ayzenstayn/Moment via Getty Images; **60c, 63 (grassland)** Claudia Totir/Moment via Getty Images; **60fg** DieterMeyrl/E+ via Getty Images; **60b** KJELL LINDER/Moment via Getty Images; **61c** CampPhoto/iStock /Getty Images Plus via Getty Images; **61b, 68** ekkawit998 /iStock/Getty Images Plus via Getty Images; **62** Gary Hincks /Science Source; **64** ugurhan/E+ via Getty Images; **65 (animals), 69t** mamita/Shutterstock.com; **65 (shrubs), 69b** Kancerina /Shutterstock.com; **65l** weisschr/iStock/Getty Images Plus via Getty Images; **65r** Vicki Jauron, Babylon and Beyond Photography/Moment via Getty Images; **66t** Nick Dale/EyeEm via Getty Images; **66c** slowmotiongli/iStock/Getty Images Plus via Getty Images; **66b** Cormac Price/Shutterstock.com, **67** Claude Huot/Shutterstock.com

CHAPTER 4

72 ARoxo/Moment via Getty Images; **73, 85b** CSA Images /Vetta via Getty Images; **78–79t** BriBar/iStock/Getty Images Plus via Getty Images; **79b** Gannie/Shutterstock.com; **80** Kerry Hargrove/iStock/Getty Images Plus via Getty Images; **81t** Steve Satushek/The Image Bank via Getty Images; **81b** John Callery/500px Plus via Getty Images; **82** Nigel Cattlin/FLPA /Minden Pictures; **84** FilippoBacci/E+ via Getty Images; **85t** milehightraveler/iStock Unreleased via Getty Images; **85c, 91** seamartini/iStock/Getty Images Plus via Getty Images; **86t, 87tl** Kryssia Campos/Moment via Getty Images; **86cl** THEPALMER/DigitalVision Vectors via Getty Images; **86cr** Gianfranco Vivi/Moment via Getty Images; **86b** David H. Wells /The Image Bank Unreleased via Getty Images; **87tr** titoOnz /iStock/Getty Images Plus via Getty Images; **87b** Danny Lehman/The Image Bank via Getty Images; **88t** Suzi Eszterhas /Minden Pictures; **88b** arquiplay77/iStock/Getty Images Plus via Getty Images; **89t** Stephan Kogelman/iStock Editorial/Getty Images Plus via Getty Images; **89b** Michel Gunther/Biosphoto /Minden Pictures; **90** Westend61 via Getty Images

PHOTO CREDITS

CHAPTER 11

236–37 ArtSvetlana/Shutterstock.com; **238** Nastasic/DigitalVision Vectors via Getty Images; **239l** "Aegiphila pernambucensis" by Tarciso Leão/Flickr/modified/CC By 2.0; **239r** Ian Peter Morton/Shutterstock.com; **240tl** Tamara Kulikova/Shutterstock.com; **240tr** Gerard Soury/The Image Bank via Getty Images; **240bl** Eye of Science/Science Source; **240bc, br** Science Photo Library - STEVE GSCHMEISSNER/Brand X Pictures via Getty Images; **241t** AP Photo; **241bl** NNehring/iStock/Getty Images Plus via Getty Images; **241bcl** oksana2010/Shutterstock.com; **241bcr** Food Impressions/Shutterstock.com; **241br** tiero/iStock/Getty Images Plus via Getty Images; **242t** Peter Leahy/Shutterstock.com; **242c** indigojt/iStock/Getty Images Plus via Getty Images; **242bl** photomaster/Shutterstock.com; **242bc** YK/Shutterstock.com; **242br** Panu Ruangjan/Shutterstock.com; **243t** Suzanne Renfrow/Shutterstock.com; **243b** GL Archive/Alamy Stock Photo; **244l, 252tl** Dora Zett/Shutterstock.com; **244cl, 252tc** Ana Gram/Shutterstock.com; **244cr, 252tr** ArCaLu/Shutterstock.com; **244r–45, 252b** Dudley Simpson/Shutterstock.com; **246t** JohanSjolander/E+ via Getty Images; **246bl** FineArt/Alamy Stock Photo; **246br** ART Collection/Alamy Stock Photo; **247** Universal Images Group North America LLC/Alamy Stock Photo; **248t** vusta/E+ via Getty Images; **248ct, cb** Eric Isselee/Shutterstock.com; **248b** cynoclub/Shutterstock.com; **250t** © Colette6|Dreamstime.com; **250b** Robert Wyatt/Alamy Stock Photo

CHAPTER 12

254 National Institutes of Health/Stocktrek Images via Getty Images; **256tl** Daniela Duncan/Moment via Getty Images; **256tli, tri, bri, 261** (*Borrelia*), **269b, 270tr** Eye of Science/Science Source; **256tr, 270tl** MARUM – Center for Marine Environmental Sciences, University of Bremen/modified/CC-By 4.0; **256bl** vvvita/Shutterstock.com; **256bli, 257tl** Dennis Kunkel Microscopy/Science Source; **256br** Daphne Zheng/iStock/Getty Images Plus via Getty Images; **257tr, 271** Mediscan/Alamy Stock Photo; **257ct** STEVE GSCHMEISSNER/SCIENCE PHOTO LIBRARY via Getty Images; **257cb** Science Photo Library - HEATHER DAVIES/Brand X Pictures via Getty Images; **257b** © Grazia Natalotto|Dreamstime.com; **260–61t** coldsnowstorm/E+ via Getty Images; **261** (*Strep. pyogenes*), (*Rickettsia*), **268l** Science History Images/Alamy Stock Photo; **261** (*Strep. pneumoniae*), **263b, 268r** BSIP SA/Alamy Stock Photo; **261** (*Staph.*) Callista Images/Image Source via Getty Images; **261** (*Listeria*), **262b** SCIMAT/Science Source; **261** (*Mycoplasma*) Kevin Mackenzie/UNIVERSITY OF ABERDEEN/Science Source; **262t** © Tatiana Blinova|Dreamstime.com; **263t, 270b** Andrea Danti/Shutterstock.com; **264** SimpleImages/Moment via Getty Images; **265t** Dragana Gordic/Shutterstock.com; **265b** Veni vidi...shoot/iStock/Getty Images Plus via Getty Images; **266t** Dejan Dundjerski/Shutterstock.com; **266b** THOM LEACH/SCIENCE PHOTO LIBRARY via Getty Images; **269t** JazzIRT/E+ via Getty Images

CHAPTER 13

274 mikroman6/Moment via Getty Images; **276t** E.R. Degginger/Alamy Stock Photo; **276c** Jarun Ontakrai/Shutterstock.com; **276b** Scott Camazine/Alamy Stock Photo; **277tl, 278t** blickwinkel/Alamy Stock Photo; **277tr** Tom Stack/Alamy Stock Photo; **277ct, 293t** Alf Jacob Nilsen/Alamy Stock Photo; **277cb** Adobe Stock/ll911; **277b** Lebendkulturen.de/Shutterstock.com; **278b** Designua/Shutterstock.com; **279** "Entamoeba histolytica" by Stefan Walkowski/Wikimedia Commons/modified/CC By-SA 3.0; **280** Biophoto Associates/Science Source;

282t–83 divedog/Shutterstock.com; **282ctl** © Mykola Ohorodnyk|Dreamstime.com; **282ctr, 293c** Science History Images/Alamy Stock Photo; **282cbl** Natural Visions/Alamy Stock Photo; **282cbr** Scenics & Science/Alamy Stock Photo; **282bl** © Daniel Poloha|Dreamstime.com; **282br** Universal Images Group North America LLC/DeAgostini/Alamy Stock Photo; **284** © Michael Williams|Dreamstime.com; **286–87l** ON-Photography Germany/Shutterstock.com; **287r** Kichigin/Shutterstock.com; **288t** adamikarl/Shutterstock.com; **288cl** Simlinger/Shutterstock.com; **288cr** godi photo/Shutterstock.com; **288bl** Adobe Stock/vincentpremel; **288bc, 293b** Shutter_arlulu/Shutterstock.com; **288br** Gertjan Hooijer/Shutterstock.com; **289t** THEPALMER/DigitalVision Vectors via Getty Images; **289b** LDarin/Shutterstock.com; **290** M.Baturitskii/Shutterstock.com; **291t** knelson20/Shutterstock.com; **291b** inga spence/Alamy Stock Photo; **292** onkachura/Shutterstock.com; **295t** Ryan McGill/Shutterstock.com; **295bl** Ray Wilson/Alamy Stock Photo; **295br** ggw/Shutterstock.com

CHAPTER 14

296 bauhaus1000/DigitalVision Vectors via Getty Images; **297, 298–99bg** aperturesound/Shutterstock.com; **297i** Dr. Norbert Lange/Shutterstock.com; **298t, 315** (dicot, monocot) Fancy Tapis/Shutterstock.com; **298b** dpa picture alliance/Alamy Stock Photo; **299tl** Moore-Moore/Shutterstock.com; **299tr** iiokua/iStock/Getty Images Plus via Getty Images; **299bl** KenCanning/iStock/Getty Images Plus via Getty Images; **299br** ALEXANDER V EVSTAFYEV/Shutterstock.com; **300t** Ted Kinsman/Science Source; **300b** Jaroslav Nesvadba/500px via Getty Images; **301tl, 320** showcake/Shutterstock.com; **301tr** BlueRingMedia/Shutterstock.com; **301b** Steve Gschmeissner/Science Source; **302tl** Valery121283/Shutterstock.com; **302tr, cr** VectorMine/Shutterstock.com; **302cl** White ground/Shutterstock.com; **302bl** Westend61 via Getty Images; **302br** Ed Reschke/Stone via Getty Images; **303tl** LunaKate/Shutterstock.com; **303tc, 322t** jopelka/Shutterstock.com; **303tr** allstars/Shutterstock.com; **303bl** Brzostowska/Shutterstock.com; **303bc** sevenke/Shutterstock.com; **303br** Africa Studio/Shutterstock.com; **304l** Kondor83/Shutterstock.com; **304c** fotofermer/iStock/Getty Images Plus via Getty Images; **304r, bg, 305** Sodel Vladyslav/Shutterstock.com; **306t** Mehmet Gokhan Bayhan/Shutterstock.com; **306b** Science Stock Photography/Science Source; **307** imageBROKER/Mara Brandl via Getty Images; **308** bergamont/Shutterstock.com; **309** Rex Jones/Shutterstock.com; **313** Le Do/Shutterstock.com; **316t** © iStock.com/BogWan; **316b** grey_and/Shutterstock.com; **317tl** WIPHARAT CHAINUPAPHA/Shutterstock.com; **317tr** Anna Kucherova/Shutterstock.com; **317b** Tim UR/Shutterstock.com; **318t–19l** mtreasure/iStock/Getty Images Plus via Getty Images; **318b** Daniel Prudek/Shutterstock.com; **319r** gustavo ramirez/Moment via Getty Images; **321, 322b** Morphart Creation/Shutterstock.com; **323** NIKCOA/Shutterstock.com

CHAPTER 15

324 Kristine Dzalbe/Shutterstock.com; **325** Irina Zharkova31/Shutterstock.com; **326tl** showcake/Shutterstock.com; **326tr–27tl** Andreas C. Fischer/Shutterstock.com; **326ti** Adobe Stock/Aldona; **326b, 340t** Menno van der Haven/Shutterstock.com; **327tc** Aldona/iStock/Getty Images Plus via Getty Images; **327r, 328l** t_kimura/iStock/Getty Images Plus via Getty Images; **327bl, bc** David Cook/blueshiftstudios/Alamy Stock Photo; **327** (turgor pressure) © Lukaves|Dreamstime.com; **328c** Biology Education/Shutterstock.com; **328r–29** segawa7/Shutterstock.com; **330t** huePhotography/E+ via Getty Images; **330ctl** Ed Reschke/Stone via Getty Images; **330ctr** Kallayanee Naloka

CHAPTER 16

CHAPTER 17

CHAPTER 18

PHOTO CREDITS

via Getty Images; **499** STEVE GSCHMEISSNER/Science Source; **500**tl, **507**b Sashkin/Shutterstock.com; **500**bl mmutlu /Shutterstock.com; **501** DiBtv/Shutterstock.com; **501**i Micro Discovery/Corbis Documentary via Getty Images; **502**tl Macrovector/Shutterstock.com; **502**tr, br, **503**b, **508**r ilusmedical /Shutterstock.com; **502**bl SciePro/Shutterstock.com; **503**t Olga Bolbot/Shutterstock.com; **505**l VectorMine/Shutterstock.com; **505**c THEPALMER/DigitalVision Vectors via Getty Images; **505**r hadynyah/E+ via Getty Images; **506** kitiara65/iStock/Getty Images Plus via Getty Images; **508**l peakanucha/Shutterstock .com; **509** Alfa MD/Shutterstock.com

CHAPTER 23

510 inter reality/Shutterstock.com; **511** CSA Images/Vetta via Getty Images; **512**tl Andrea Danti/Shutterstock.com; **512**r leonello/iStock/Getty Images Plus via Getty Images; **512**cl Aldona Griskeviciene/Shutterstock.com; **512**bl OLEKSANDRA TROIAN/Moment via Getty Images; **513**t, **524**l Olga Bolbot /Shutterstock.com; **513**c © Axel Kock|Dreamstime.com; **513**b Crevis/Shutterstock.com; **514–15**bg Konstantin Kopachinsky /Shutterstock.com; **514** Carlyn Iverson/Science Source; **515** Designua/Shutterstock.com; **518**t lechatnoir/E+ via Getty Images; **518**b THEPALMER/DigitalVision Vectors via Getty Images; **520**bg, **524**r rob3000/Alamy Stock Vector; **520, 524**ri MatoomMi /Shutterstock.com; **521**t BlueRingMedia/Shutterstock.com; **521**b, **522**tl Axel_Kock/Shutterstock.com; **522**tr**–23** kazoka /Shutterstock.com; **522**b CSA-Printstock/DigitalVision Vectors via Getty Images; **526** Adobe Stock/Rungruedee; **527** gorodenkoff/iStock/Getty Images Plus via Getty Images

CHAPTER 24

528 Ed Reschke/Stone via Getty Images; **529** THEPALMER /DigitalVision Vectors via Getty Images; **530**tl Jose Luis Calvo /Shutterstock.com; **530**cl, **546**li, ri VectorMine/Shutterstock .com; **530**bl, **531**tr S K Chavan/Shutterstock.com; **530**r, **531**l, **551**l Adobe Stock/Gatria; **531**cr, **535**l, **547**bi Aldona Griskeviciene/Shutterstock.com; **531**br Rattiya Thongdumhyu /Shutterstock.com; **532** Springer Medizin/Science Source; **533** SciePro/Shutterstock.com; **533**tli, tri Vlada Young/Shutterstock .com; **533**bli, bri Net Vector/Shutterstock.com; **534**t PIXOLOGICSTUDIO/SCIENCE PHOTO LIBRARY via Getty Images; **534**b BSIP SA/Alamy Stock Photo; **535**r first vector trend/Shutterstock.com; **536** VikiVector/Shutterstock .com; **538** Lugaaa/iStock/Getty Images Plus via Getty Images; **539** Universal Images Group North America LLC/Alamy Stock Photo; **540**t CLAUS LUNAU/Science Source; **540**bl Alexander_P/Shutterstock.com; **540**br Marochkina Anastasiia /Shutterstock.com; **541**t, **546**r, **547**l, **551**r Macrovector /Shutterstock.com; **541**c, **545, 552** Pikovit/Shutterstock.com; **541**bl, bc, br PushprajP/Shutterstock.com; **542**tl, b Stocktrek Images via Getty Images; **542**tc STEVE GSCHMEISSNER /Science Photo Library via Getty Images; **542**tr**–43** TheCrimsonMonkey/iStock/Getty Images Plus via Getty Images; **543**i Designua/Shutterstock.com; **544**t Ground Picture /Shutterstock.com; **544**b Nastasic/DigitalVision Vectors via Getty Images; **546**l TimeLineArtist/Shutterstock.com; **547**c, r Science History Images/Alamy Stock Photo; **548**t Andrea Danti/Shutterstock.com; **548**b BSIP/Collection Mix: Subjects via Getty Images; **549**t Rawpixel.com/Shutterstock.com; **549**b Magicleaf/Shutterstock.com

CHAPTER 25

554 Steve Allen/Shutterstock.com; **556**tl BlueRingMedia /Shutterstock.com; **556**r, **557**t, **560**l, **563**ctl, **570**b, **574** VectorMine/Shutterstock.com; **556**bl, **563**cb, bl, bc, br, **564**tl, c, bl, br SciePro/Shutterstock.com; **557**bl Medical Art Inc /Shutterstock.com; **557**br Lars Neumann/iStock/Getty Images Plus via Getty Images; **558**t, c, **559**bc Karina Glukhanyk /Shutterstock.com; **558**b, **560**r, **570**t Designua/Shutterstock.com; **559**t (all), br, **565**t, cl, cc, cr Alila Medical Media/Shutterstock .com; **559**bl Ana Krasavina/Shutterstock.com; **560** (brain) Magicleaf/Shutterstock.com; **561** monkeybusinessimages /iStock/Getty Images Plus via Getty Images; **562** StockPlanets /E+ via Getty Images; **563**tl fixer00/Shutterstock.com; **563**tr Design_Cells/Shutterstock.com; **563**ctr Sakurra/Shutterstock .com; **564**tr UrsaHoogle/E+ via Getty Images; **564**bc Dr. Najeeb Layyous/Science Source; **565**b SDI Productions/E+ via Getty Images; **566**t Jose Luis Pelaez Inc/DigitalVision via Getty Images; **566**ct jaroon/E+ via Getty Images; **566**cc Sean Murphy /Photodisc via Getty Images; **566**cb Yuri_Arcurs/iStock/Getty Images Plus via Getty Images; **566**b A Lot Of People/Shutterstock .com; **567** skynesher/E+ via Getty Images; **568** melitas /Shutterstock.com; **569** Alexander Spatari/Moment via Getty Images; **571**t SolStock/E+ via Getty Images; **571**c Boy_Anupong /Moment via Getty Images; **571**b CORDELIA MOLLOY/Science Source; **572** Med_Ved/Shutterstock.com; **572** (icons) Real Vector /Shutterstock.com; **577** AP Photo/Rich Pedroncelli, File

BACK MATTER

578 kali9/E+ via Getty Images; **579**t Malachi Jacobs/Shutterstock .com; **579**b "Ptolemaic system 2 (PSF)" by Pearson Scott Foresman/Wikimedia Commons/Public Domain; **580** Anastasiia Stiahailo/iStock/Getty Images Plus via Getty Images; **582** Photo Disc

PERIODIC TABLE OF THE ELEMENTS

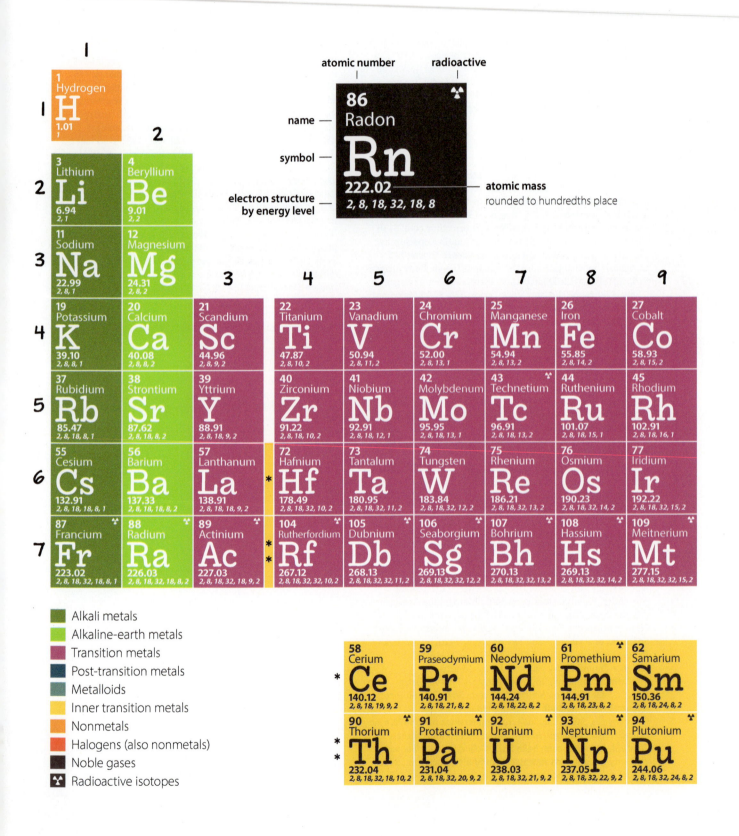

1

| 1
Hydrogen
H
1.01
1 |

2

| 3
Lithium
Li
6.94
2, 1 | 4
Beryllium
Be
9.01
2, 2 |

atomic number — **86** — radioactive
name — Radon
symbol — **Rn**
222.02 — atomic mass (rounded to hundredths place)
electron structure by energy level — 2, 8, 18, 32, 18, 8

| 11
Sodium
Na
22.99
2, 8, 1 | 12
Magnesium
Mg
24.31
2, 8, 2 |

| **3** | **4** | **5** | **6** | **7** | **8** | **9** |

| 19
Potassium
K
39.10
2, 8, 8, 1 | 20
Calcium
Ca
40.08
2, 8, 8, 2 | 21
Scandium
Sc
44.96
2, 8, 9, 2 | 22
Titanium
Ti
47.87
2, 8, 10, 2 | 23
Vanadium
V
50.94
2, 8, 11, 2 | 24
Chromium
Cr
52.00
2, 8, 13, 1 | 25
Manganese
Mn
54.94
2, 8, 13, 2 | 26
Iron
Fe
55.85
2, 8, 14, 2 | 27
Cobalt
Co
58.93
2, 8, 15, 2 |

| 37
Rubidium
Rb
85.47
2, 8, 18, 8, 1 | 38
Strontium
Sr
87.62
2, 8, 18, 8, 2 | 39
Yttrium
Y
88.91
2, 8, 18, 9, 2 | 40
Zirconium
Zr
91.22
2, 8, 18, 10, 2 | 41
Niobium
Nb
92.91
2, 8, 18, 12, 1 | 42
Molybdenum
Mo
95.95
2, 8, 18, 13, 1 | 43
Technetium
Tc
96.91
2, 8, 18, 13, 2 | 44
Ruthenium
Ru
101.07
2, 8, 18, 15, 1 | 45
Rhodium
Rh
102.91
2, 8, 18, 16, 1 |

| 55
Cesium
Cs
132.91
2, 8, 18, 18, 8, 1 | 56
Barium
Ba
137.33
2, 8, 18, 18, 8, 2 | 57
Lanthanum
La
138.91
2, 8, 18, 18, 9, 2 | 72
Hafnium
Hf
178.49
2, 8, 18, 32, 10, 2 | 73
Tantalum
Ta
180.95
2, 8, 18, 32, 11, 2 | 74
Tungsten
W
183.84
2, 8, 18, 32, 12, 2 | 75
Rhenium
Re
186.21
2, 8, 18, 32, 13, 2 | 76
Osmium
Os
190.23
2, 8, 18, 32, 14, 2 | 77
Iridium
Ir
192.22
2, 8, 18, 32, 15, 2 |

| 87
Francium
Fr
223.02
2, 8, 18, 32, 18, 8, 1 | 88
Radium
Ra
226.03
2, 8, 18, 32, 18, 8, 2 | 89
Actinium
Ac
227.03
2, 8, 18, 32, 18, 9, 2 | 104
Rutherfordium
Rf
267.12
2, 8, 18, 32, 32, 10, 2 | 105
Dubnium
Db
268.13
2, 8, 18, 32, 32, 11, 2 | 106
Seaborgium
Sg
269.13
2, 8, 18, 32, 32, 12, 2 | 107
Bohrium
Bh
270.13
2, 8, 18, 32, 32, 13, 2 | 108
Hassium
Hs
269.13
2, 8, 18, 32, 32, 14, 2 | 109
Meitnerium
Mt
277.15
2, 8, 18, 32, 32, 15, 2 |

Legend:
- Alkali metals
- Alkaline-earth metals
- Transition metals
- Post-transition metals
- Metalloids
- Inner transition metals
- Nonmetals
- Halogens (also nonmetals)
- Noble gases
- Radioactive isotopes

*
| 58
Cerium
Ce
140.12
2, 8, 18, 19, 9, 2 | 59
Praseodymium
Pr
140.91
2, 8, 18, 21, 8, 2 | 60
Neodymium
Nd
144.24
2, 8, 18, 22, 8, 2 | 61
Promethium
Pm
144.91
2, 8, 18, 23, 8, 2 | 62
Samarium
Sm
150.36
2, 8, 18, 24, 8, 2 |

**
| 90
Thorium
Th
232.04
2, 8, 18, 32, 18, 10, 2 | 91
Protactinium
Pa
231.04
2, 8, 18, 32, 20, 9, 2 | 92
Uranium
U
238.03
2, 8, 18, 32, 21, 9, 2 | 93
Neptunium
Np
237.05
2, 8, 18, 32, 22, 9, 2 | 94
Plutonium
Pu
244.06
2, 8, 18, 32, 24, 8, 2 |

18

						18
						2 Helium **He** 4.00 2

13   **14**   **15**   **16**   **17**

13	14	15	16	17	18
5 Boron **B** 10.81 2, 3	**6** Carbon **C** 12.01 2, 4	**7** Nitrogen **N** 14.01 2, 5	**8** Oxygen **O** 16.00 2, 6	**9** Fluorine **F** 19.00 2, 7	**10** Neon **Ne** 20.18 2, 8
13 Aluminum **Al** 26.98 2, 8, 3	**14** Silicon **Si** 28.09 2, 8, 4	**15** Phosphorus **P** 30.97 2, 8, 5	**16** Sulfur **S** 32.06 2, 8, 6	**17** Chlorine **Cl** 35.45 2, 8, 7	**18** Argon **Ar** 39.95 2, 8, 8

10   **11**   **12**

10	11	12	13	14	15	16	17	18
28 Nickel **Ni** 58.69 2, 8, 16, 2	**29** Copper **Cu** 63.55 2, 8, 18, 1	**30** Zinc **Zn** 65.38 2, 8, 18, 2	**31** Gallium **Ga** 69.72 2, 8, 18, 3	**32** Germanium **Ge** 72.63 2, 8, 18, 4	**33** Arsenic **As** 74.92 2, 8, 18, 5	**34** Selenium **Se** 78.97 2, 8, 18, 6	**35** Bromine **Br** 79.90 2, 8, 18, 7	**36** Krypton **Kr** 83.80 2, 8, 18, 8
46 Palladium **Pd** 106.42 2, 8, 18, 18	**47** Silver **Ag** 107.87 2, 8, 18, 18, 1	**48** Cadmium **Cd** 112.41 2, 8, 18, 18, 2	**49** Indium **In** 114.82 2, 8, 18, 18, 3	**50** Tin **Sn** 118.71 2, 8, 18, 18, 4	**51** Antimony **Sb** 121.76 2, 8, 18, 18, 5	**52** Tellurium **Te** 127.60 2, 8, 18, 18, 6	**53** Iodine **I** 126.90 2, 8, 18, 18, 7	**54** Xenon **Xe** 131.29 2, 8, 18, 18, 8
78 Platinum **Pt** 195.08 2, 8, 18, 32, 17, 1	**79** Gold **Au** 196.97 2, 8, 18, 32, 18, 1	**80** Mercury **Hg** 200.59 2, 8, 18, 32, 18, 2	**81** Thallium **Tl** 204.38 2, 8, 18, 32, 18, 3	**82** Lead **Pb** 207.24 2, 8, 18, 32, 18, 4	**83** Bismuth **Bi** 208.98 2, 8, 18, 32, 18, 5	**84** Polonium **Po** 208.98 2, 8, 18, 32, 18, 6	**85** Astatine **At** 209.99 2, 8, 18, 32, 18, 7	**86** Radon **Rn** 222.02 2, 8, 18, 32, 18, 8
110 Darmstadtium **Ds** 282.17 2, 8, 18, 32, 32, 16, 2	**111** Roentgenium **Rg** 282.17 2, 8, 18, 32, 32, 17, 2	**112** Copernicium **Cn** 286.18 2, 8, 18, 32, 32, 18, 2	**113** Nihonium **Nh** 286.18 2, 8, 18, 32, 32, 18, 3	**114** Flerovium **Fl** 290.19 2, 8, 18, 32, 32, 18, 4	**115** Moscovium **Mc** 290.20 2, 8, 18, 32, 32, 18, 5	**116** Livermorium **Lv** 293.21 2, 8, 18, 32, 32, 18, 6	**117** Tennessine **Ts** 294.21 2, 8, 18, 32, 32, 18, 7	**118** Oganesson **Og** 295.22 2, 8, 18, 32, 32, 18, 8

63	64	65	66	67	68	69	70	71
63 Europium **Eu** 151.96 2, 8, 18, 25, 8, 2	**64** Gadolinium **Gd** 157.25 2, 8, 18, 25, 9, 2	**65** Terbium **Tb** 158.93 2, 8, 18, 27, 8, 2	**66** Dysprosium **Dy** 162.50 2, 8, 18, 28, 8, 2	**67** Holmium **Ho** 164.93 2, 8, 18, 29, 8, 2	**68** Erbium **Er** 167.26 2, 8, 18, 30, 8, 2	**69** Thulium **Tm** 168.93 2, 8, 18, 31, 8, 2	**70** Ytterbium **Yb** 173.05 2, 8, 18, 32, 8, 2	**71** Lutetium **Lu** 174.97 2, 8, 18, 32, 9, 2
95 Americium **Am** 243.06 2, 8, 18, 32, 25, 8, 2	**96** Curium **Cm** 247.07 2, 8, 18, 32, 25, 9, 2	**97** Berkelium **Bk** 247.07 2, 8, 18, 32, 27, 8, 2	**98** Californium **Cf** 251.08 2, 8, 18, 32, 28, 8, 2	**99** Einsteinium **Es** 252.08 2, 8, 18, 32, 29, 8, 2	**100** Fermium **Fm** 257.10 2, 8, 18, 32, 30, 8, 2	**101** Mendelevium **Md** 258.10 2, 8, 18, 32, 31, 8, 2	**102** Nobelium **No** 259.10 2, 8, 18, 32, 32, 8, 2	**103** Lawrencium **Lr** 262.11 2, 8, 18, 32, 32, 8, 3